Michel W. Barsoum

MAX Phases

Related Titles

Riedel, R., Chen, I-W. (eds.)

Ceramics Science and Technology

Volume 3: Synthesis and Processing

2012

ISBN: 978-3-527-31157-6

Riedel, R., Chen, I-W. (eds.)

Ceramics Science and Technology

Volume 2: Materials and Properties

2010

ISBN: 978-3-527-31156-9

Heimann, R. B.

Classic and Advanced Ceramics

From Fundamentals to Applications

2010

ISBN: 978-3-527-32517-7

Aldinger, F., Weberruss, V. A.

Advanced Ceramics and Future Materials

An Introduction to Structures, Properties, Technologies, Methods

2010

ISBN: 978-3-527-32157-5

Michel W. Barsoum

MAX Phases

Properties of Machinable Ternary Carbides and Nitrides

Verlag GmbH & Co. KGaA

The Author

Prof. Michel W. Barsoum
Drexel University
Department of Material Science &
Engineering
3141, Chestnut Street
Philadelphia, PA 19104
USA

Cover
This colorized scanning electron micrograph is of a fractured surface of a Mg/MAX composite material. The micrograph highlights the propensity of the MAX phases to kink upon deformation. The micrograph was taken and colorized by Mr. Babak Anasori of Drexel University.

Library of Congress Card No.: applied for

British Library Cataloguing-in-Publication Data
A catalogue record for this book is available from the British Library.

Bibliographic information published by the Deutsche Nationalbibliothek
The Deutsche Nationalbibliothek lists this publication in the Deutsche Nationalbibliografie; detailed bibliographic data are available on the Internet at <http://dnb.d-nb.de>.

Print ISBN: 978-3-527-33011-9
ePDF ISBN: 978-3-527-65461-1
ePub ISBN: 978-3-527-65460-4
mobi ISBN: 978-3-527-65459-8
oBook ISBN: 978-3-527-65458-1

Cover Design Grafik-Design Schulz, Fußgönheim
Typesetting Laserwords Private Limited, Chennai, India
Printing and Binding Markono Print Media Pte Ltd, Singapore

Printed in Singapore
Printed on acid-free paper

Dedicated to my wonderful wife, close confidant, and best friend, Patricia.

Contents

Preface

The MAX phases are a fascinating class of layered solids that are relatively young. Interest in these 50+ phases has increased recently because they combine an unusual and very often unique combination of properties. For example, some are stiff and light and yet are readily machinable. Some are oxidation and creep resistant while also being metallic conductors and exceptionally thermal shock resistant. At this time, there are a number of good review articles on the MAX phases. However, the articles either focus on a few MAX phases, most notably Ti_3SiC_2, Ti_3AlC_2, Ti_2AlC, and Cr_2AlC, or try to tackle the entire subject in which much per force has to be glossed over. Said otherwise, there is no comprehensive compact monograph that renders these phases justice.

In this book, I attempt to summarize and explain, from both an experimental and a theoretical viewpoint, all the features that are necessary to understand the properties of these new materials. The book covers elastic, electrical, thermal, chemical, and mechanical properties in different temperature regimes. As much as possible, I tried to emphasize the physics.

One of the joys of working with the MAX phases is the ease by which one can change chemistry, while keeping the structure the same. As I anticipated many years ago, this has proven to be a real boon; I have a hunch, with no data to back me up, that the progress the MAX phase community has made in understanding their properties, in the past decade or so, can be traced directly to this feature. The range of experimental and theoretical techniques currently available has also indubitably made a big difference. In today's world, like much else, we have Science on steroids. We are quickly reaching the point – if we have not it already – at which the rate of data generation far exceeds our capability to make sense of them. In this book, I tried to buck the tide and make sense of what we currently know. The reader of this book will quickly realize from the sheer volume of data tabulated and plotted that this was not a trivial task. I do believe, however, that to truly understand properties and what influences them, one needs, every now and then, to step backward and make out the forest from the trees.

As shown in this book, this systemic approach, while tedious, is quite gratifying and edifying. For example, one of the leitmotivs of this book is the idea that above a certain concentration of valence electrons per unit volume, n_{val}, the MAX phases are somehow destabilized. While plotting one set of properties versus n_{val} does not necessarily make a compelling case, but when this destabilization is repeated and recognized in several different properties, the idea becomes harder to dismiss. Another important idea of this book is that we can roughly subdivide the MAX phases into four categories: (i) those with exceptionally low c-parameters, such as Ti_2SC; (ii) those with large atoms, such as Sn, Hf, Zr; (iii) those in between but with low n_{val} values; and (iv) those in between, but with high n_{val} values that are relatively unstable, such as some of the Cr-containing MAX phases. Hopefully, this idea comes across.

The other joy of working with the MAX phases is their two-dimensional nature, especially when it comes to mechanical properties. The fact that dislocations are, for the most part, confined to 2D and that the orientation of the basal planes on which these dislocations glide are in many cases readily determined from optical microscope micrographs has rendered understanding their mechanical response rather straightforward. In solid-state physics, the pedagogy is well established; first you solve the one-dimensional problem, move on to the 2D, and then, and only then, generalize to the most complicated 3D situation. In dealing with the deformation of solids, however, the hapless metallurgy or materials science undergraduate is immediately asked to deal with more than five independent slip systems, a daunting task that certainly biased me toward ceramics, where I thought I would be safe. That I can now talk somewhat intelligently about dislocations is, in my case, not a mark of any intellectual prowess, but rather a reflection of the simplicity of the problem at hand. Basically, dislocations in the MAX phases, and in the much larger class of solids that we identified as kinking nonlinear elastic (KNE), appear either in dislocation pileups (DPs) and/or dislocation walls normal to the pileups or arrays. Confining the dislocations to 2D also helped us identify a new micromechanism in solids, namely, incipient kink bands (IKBs). As discussed in Chapters 8 and 9, IKBs are the yin to the yang of DPs. IKBs absorb significant amounts of energy at low strains; DPs result in large strains, but little stored energy. It follows that Nature's first line of defense in the case of KNE solids is to nucleate IKBs.

By bringing together, in a unified, self-contained manner, all the information on MAX phases hitherto only found scattered in the journal literature, I hope to help move the field along to the next stage. I have also tried to critically assess the now voluminous literature. The number of papers in the field has increased recently and the task of anybody attempting to review this body of work is becoming daunting. In 2000, when I wrote an early review article on the subject, the situation was significantly easier.

In addition to outlining the contents of this book, it is important to stress what it is not about and what it does not cover. This book is geared to understand the physics of the MAX and hence the synthesis of these phases is not discussed. Thin films are for the most part not covered. A recent review has done this topic justice. When thin films are discussed, it is only to make an important point for which the information is lacking in bulk solids. Composites of MAX phases with other compounds and second phases are also mostly not discussed, except in instances where comparing the properties of the composites with the pure bulk materials sheds light on the properties of the latter, which is the main focus of this book.

A perusal of the figures in this book will quickly establish that most of the figures originate from papers we wrote. This does not imply that other work is less important. It simply reflects the fact that the information was more easily accessible. In many cases, results and data have been grouped/replotted and in that case having access to the raw data is invaluable and time saving. I have assiduously tried to assign credit where credit is due. It follows that to the best of my abilities, I carefully combed the literature to make sure that when new information on the behavior of the MAX phases was reported, the original paper was cited. The record is out there and I tried my best. If at any time, such attribution is incorrect or lacking, I sincerely apologize and please contact me and I will try to set the record straight in any future editions of this book or any papers I write.

This book is divided into 11 chapters. The first chapter is an introductory chapter where the history of the MAX phases is outlined. Chapter 2 reviews the atomic structures and bonding commonalties and trends in these phases. This chapter also summarizes ab initio or density functional theory (DFT) calculations that, for the most part, capture the essence of the bonding in these solids. Chapter 3 deals with their elastic properties, both experimental and those calculated from DFT. Chapter 4 summarizes the thermal properties, including thermal expansion, conductivity heat capacities, atomic displacement parameters, and stability. Chapter 5 deals with the electrical transport, including conductivity, and Hall and Seebeck coefficient measurements. Their optical and magnetic properties are also touched upon.

Chapter 6 deals with the reactivity of the MAX phases with oxygen and other gases. The reactivities of the MAX phases with solids and liquids, including molten metals and common acids and bases, are reviewed in Chapter 7.

Chapters 8–10 deal with the mechanical properties. Chapter 8 deals with kinking nonlinear elasticity and damping. How the MAX phases respond to stresses – compressive, shear, tensile, and so on – at ambient temperature are discussed in Chapter 9. Chapter 10 deals with their response to stresses at elevated temperatures, including creep. Chapter 11 summarizes some of the outstanding scientific issues and outlines some of the potential applications and what needs to be done, research-wise, for these solids to be more widely used.

The quality and quantity of the papers one publishes in academia depend critically on the quality, resourcefulness, imagination, and hardwork of one's students. I would thus like to sincerely thank all my students who have worked with me on the MAX phases over the past 15 or so years. In rough chronological order, they are: T. El-Raghy, D. Brodkin, M. Radovic, S. Chakraborty, A. Procopio, J. Travaglini, L. H. Ho-Duc, I. Salama, P. Finkel, A. Murugaiah, T. Zhen, A. Ganguly, E. Hoffman, S. Gupta, S. Basu, A. Zhou, S. Amini, T. Scabarosi, J. Lloyd, I. Albaryak, C. J. Spencer, M. Shamma, N. Lane, D. Tallman, B. Anasori, M. Naguib, G. Bentzel, and J. Halim. It was a distinct pleasure to work with each and every one of them. Their productivity and contributions to the field cannot be overemphasized.

The number of postdocs that worked with me over the years is not as numerous as my students, but their input and insights were as important and appreciated. In chronological order, I would like to thank L. Farber, N. Tzenov, D. Filimonov, J. Córdoba, and V. Presser. I also had the distinct pleasure of working with a few visiting scholars who spent some time with me at Drexel. I would thus like to thank Drs. Z.-M. Sun, O. Yeheskel, V. Jovic, T. Cabioch, and E. Caspi.

I like to collaborate and I have sought out collaborators in many countries and on many continents. In that vein, I would like to profusely thank the following colleagues and friends with whom I have worked with over the years on the MAX phases and from whom I learned quite a bit. I am greatly indebted to G. Hug, M. Jaouen, L. Thilly, S. Dubois, M. Le Flem, J.-L. Béchade, and J. Fontaine in France; J. Hettinger and S. Lofland at the Rowan University; L. Hultman, M. Magnuson, P. Eklund, J. Rosen, J. Lu, and R. Ahuja in Sweden; and J. Schneider in Germany.

Much of this work would not have been possible without funding. The ceramics program of the Division of Materials Research of the National Science Foundation funded much of the early MAX phase work. I would like to especially thank Drs. L. Madsen and L. Schioler for their support. The Army Research Office has also funded our MAX phase work over the years. Here I am indebted to Drs. D. Stepp and S. Mathaudhu who have supported, and are still supporting, the work we are doing.

I would also like to acknowledge the support of the Swedish Foundation for Strategic Research (SSF) and the Linkoping University for funding my numerous visits to Linkoping since 2008. Prof. Lars Hultman must get the lion's share of the credit for arranging this very fruitful collaboration that is still ongoing. I would also like to thank the University of Poitiers, Poitiers, France, for hosting me for a few extended visits over the years. I would especially like to thank Profs. M. Jaouen and T. Cabioch for arranging the visits and their wonderful hospitality.

I would also be remiss if I did not acknowledge the many very fruitful discussions I have had over the years with my colleagues in the Department of Materials Science and Engineering at the Drexel University. Special thanks are due to R. Doherty, Y. Gogotsi. S. Tyagi, A. Zavaliangos, J. Spanier, G. Friedman, A. Kontsos, A. Zavaliangos, and S. Kalidindi.

I have coauthored papers with a large number of colleagues in many corners of the world. This list (again somewhat chronologically) includes Drs. M. Amer, M. Gamarnik, E. H. Kisi, J. A. Crossley, S. Myhra, L. Ogbuji, S. Wiederhorn, R. O. Ritchie, H.-I. Yoo, H. Seifert, F. Aldinger, J. Th. M. De Hosson, H. Drulis, M. Drulis, B. Manoun, J. Fontaine, J. Schuster, S. K. Saxena, D. Jurgens, M. Uhrmacher, P. Schaaf, B. Yang, D. Brown, S. Vogel, B. Clausen, X. He, and Y. Bai. I am indebted to all of them for the excellent papers we published together.

1
Introduction

1.1
Introduction

The $M_{n+1}AX_n$, or MAX, phases are layered, hexagonal, early transition-metal carbides and nitrides, where $n = 1$, 2, or 3 "M" is an early transition metal, "A" is an A-group (mostly groups 13 and 14) element, and "X" is C and/or N. In every case, near-close-packed M layers are interleaved with layers of pure group-A element with the X atoms filling the octahedral sites between the former (Figure 1.1a–c). The M_6X octahedra are edge-sharing and are identical to those found in the rock salt structure. The A-group elements are located at the center of trigonal prisms that are larger than the octahedral sites and thus better able to accommodate the larger A atoms. The main difference between the structures with various n values (Figure 1.1a–c) is in the number of M layers separating the A layers: in the M_2AX, or 211, phases, there are two; in the M_3AX_2, or 312, phases there are three; and in the M_4AX_3, or 413, phases, there are four. As discussed in more detail in later chapters, this layering is crucial and fundamental to understanding MAX-phase properties in general, and their mechanical properties in particular. Currently, the MAX phases number over 60 (Figure 1.2) with new ones, especially 413s and solid solutions, still being discovered.

Most of the MAX phases are 211 phases, some are 312s, and the rest are 413s. The M group elements include Ti, V, Cr, Zr, Nb, Mo, Hf, and Ta. The A elements include Al, Si, P, S, Ga, Ge, As, Cd, In, Sn, Tl, and Pb. The X elements are either C and/or N.

Thermally, elastically, and electrically, the MAX phases share many of the advantageous attributes of their respective binary metal carbides or nitrides: they are elastically stiff, and electrically and thermally conductive. Mechanically, however, they cannot be more different: they are readily machinable – remarkably a simple hack-saw will do (Figure 1.3) – relatively soft, resistant to thermal shock, and unusually damage-tolerant. They are the only polycrystalline solids that deform by a combination of kink and shear band formation, together with the delaminations of individual grains. Dislocations multiply and are mobile at room temperature, glide exclusively on the basal planes, and are overwhelmingly arranged either

MAX Phases: Properties of Machinable Ternary Carbides and Nitrides, First Edition. Michel W. Barsoum.

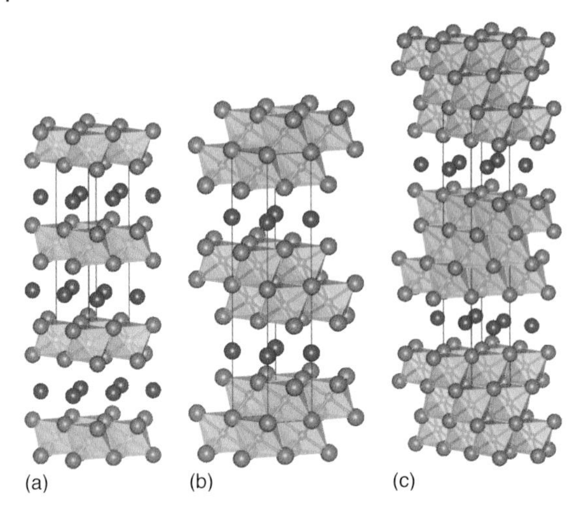

Figure 1.1 Atomic structures of (a) 211, (b) 312, and (c) 413 phases, with emphasis on the edge-sharing nature of the MX_6 octahedra.

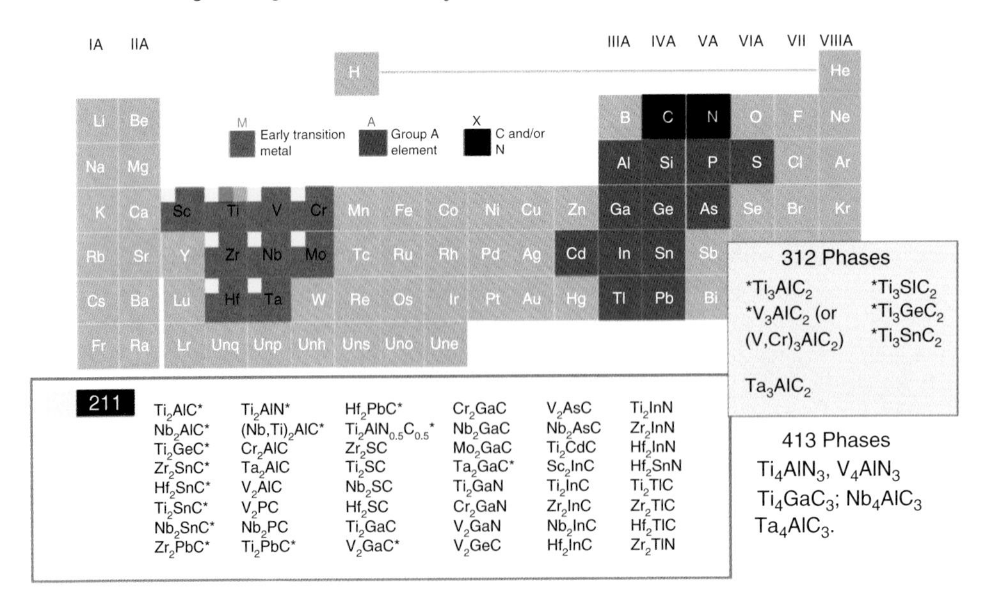

Figure 1.2 List of known MAX phases and elements of the periodic table that react to form them.

in arrays or kink boundaries. They combine ease of machinability with excellent mechanical properties, especially at temperatures $>1000\,^{\circ}C$. Some, such as Ti_3SiC_2 and Ti_4AlN_3, combine mechanical anisotropy with thermal properties that are surprisingly isotropic.

As discussed in this book, this unusual combination of properties is traceable to their layered structure, the mostly metallic – with covalent and ionic

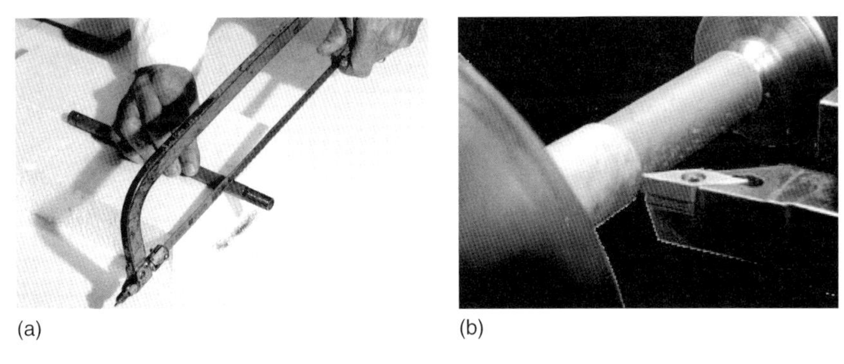

(a) (b)

Figure 1.3 One of the hallmarks of the MAX phases is the ease with which they can be machined with (a) a manual hack-saw and (b) lathe.

contributions – nature of the MX bonds that are exceptionally strong, together with M–A bonds that are relatively weak, especially in shear. The best characterized ternaries to date are Ti_3SiC_2, Ti_3AlC_2, and Ti_2AlC. We currently know their compressive and flexural strengths and their temperature dependencies, in addition to their hardness, oxidation resistance, fracture toughness and R-curve behavior, and tribological properties. Additionally, their electrical conductivities, Hall and Seebeck coefficients, heat capacities (both at low and high temperatures), elastic properties and their temperature dependencies, thermal expansions, and thermal conductivities have been quantified.

1.2 History of the MAX Phases

The MAX phases have two histories. The first spans the time they were discovered in the early and mid-1960s to roughly the mid-1990s, when, for the most part, they were ignored. The second is that of the last 15 years or so, when interest in these phases has exploded. Table 2.2 lists the first paper in which each of the MAX phases was first reported.

1.2.1 History before 1995

Roughly 40 years ago, Nowotny published a review article (Nowotny, 1970) summarizing some of the work that he and his coworkers had carried out during the 1960s on the syntheses of a large number of carbides and nitrides. It was an impressive accomplishment; during that decade, his group discovered over 100 new carbides and nitrides. Amongst them were more than 30 Hägg phases which are of specific interest to this work. These phases, also called H-phases – but apparently not after Hägg (Eklund *et al.*, 2010) – had the M_2AX chemistry (Figures 1.1a and 1.2).

The history of the H-phases, henceforth referred to as 211s, before 1997 is short. Surprisingly, from the time of their discovery until our first report (Barsoum, Brodkin, and El-Raghy, 1997) – and apart from four Russian papers in the mid-1970s (Ivchenko and Kosolapova, 1975, 1976; Ivchenko *et al.*, 1976a,b) in which it was claimed that 90–92% dense compacts of Ti_2AlC and Ti_2AlN were synthesized – they were totally ignored. The early Russian results have to be interpreted with caution as their reported microhardness values of $\approx$21–24 GPa are difficult to reconcile with the actual values, which range from 3 to 4 GPa. Some magnetic permeability measurements on Ti_2AlC and Cr_2AlC were also reported in 1966 (Reiffenstein, Nowotny, and Benesovsky, 1966).

In 1967, Nowotny's group discovered the first two 312 phases, Ti_3SiC_2 (Jeitschko and Nowotny, 1967) and Ti_3GeC_2 (Wolfsgruber, Nowotny, and Benesovsky, 1967), both of which are structurally related to the H-phases in that M_3X_2 layers now separate the A layers (Figure 1.1b). It was not until the early 1990s, however, that Pietzka and Schuster, the latter a student of Nowotny's, added Ti_3AlC_2 to the list (Pietzka and Schuster, 1994, 1996).

The history of Ti_3SiC_2 is slightly more involved. The first hint that Ti_3SiC_2 was atypical came as early as 1972, when Nickl, Schweitzer, and Luxenberg (1972) working on single crystals grown by chemical vapor deposition (CVD) showed that Ti_3SiC_2 was anomalously soft for a transition-metal carbide. The hardness was also anisotropic, with the hardness normal to the basal planes being roughly three times that parallel to them. When the authors used a solid-state reaction route, the resulting material was no longer "soft." In 1987, Goto and Hirai (1987) confirmed the results of Nickl *et al.* A number of other studies on CVD-grown films also exist (Fakih *et al.*, 2006; Pickering, Lackey, and Crain, 2000; Racault, Langlais, and Naslain, 1994; Racault *et al.*, 1994).

The fabrication of single-phase, bulk, dense samples of Ti_3SiC_2 proved to be more elusive, however. Attempts to synthesize them in bulk always resulted in samples containing, in most cases, TiC, and sometimes SiC, as ancillary, unwanted phases (Lis *et al.*, 1993; Morgiel, Lis, and Pampuch, 1996; Pampuch and Lis, 1995; Pampuch *et al.*, 1989). Consequently, before our breakthrough in synthesis (Barsoum and El-Raghy, 1996), little was known about Ti_3SiC_2, and much of what was known has since been shown to be incorrect. For example, despite a sentence buried in one of Nowotny's papers claiming Ti_3SiC_2 does not melt but dissociates at 1700 °C into TiC and a liquid (Nowotny and Windisch, 1973), the erroneous information that it has a melting point of over 3000 °C is still being disseminated by some.

Furthermore, and before our work, many of the Ti_3SiC_2 samples fabricated in bulk form were unstable above $\approx$1450 °C (Pampuch *et al.*, 1989; Racault, Langlais, and Naslain, 1994). It is now established that Ti_3SiC_2, if pure, is thermally stable to at least 1700 °C in inert atmospheres (Chapter 4). Another important misconception that tempered the enthusiasm for Ti_3SiC_2 was, again, the erroneous belief that its oxidation resistance above 1200 °C was poor (Okano, Yano, and Iseki, 1993; Racault, Langlais, and Naslain, 1994; Tong *et al.*, 1995).

Despite the aforementioned pitfalls, Pampuch, Lis, and coworkers (Lis *et al.*, 1993; Morgiel, Lis, and Pampuch, 1996; Pampuch, 1999; Pampuch and Lis, 1995; Pampuch *et al.*, 1989) came closest to fabricating pure bulk samples; their best samples were ≈80–90 vol% pure (balance TiC). Nevertheless, using these samples they were the first to show that Ti_3SiC_2 was elastically quite stiff, with Young's and shear moduli of 326 and 135 GPa, respectively, and yet machinable (Lis *et al.*, 1993). They also confirmed its relative softness (Vickers hardness of 6 GPa) and noted that the high stiffness-to-hardness ratio was more in line with ductile metals than ceramics, and labeled it a "ductile" ceramic. Apart from these properties, and a report that the thermal expansion of Ti_3SiC_2 was 9.2×10^{-6} K^{-1} (Iseki, Yano, and Chung, 1990), no other properties were known.

For the sake of completeness, it should be noted that reference to Ti_3SiC_2 in the literature occurs in another context. This phase was sometimes found at Ti/SiC interfaces annealed at high temperatures (Iseki, Yano, and Chung, 1990; Morozumi *et al.*, 1985; Wakelkamp, Loo, and Metselaar, 1991). Ti_3SiC_2 was also encountered when Ti was used as a braze material to bond SiC to itself, in SiC-Ti-reinforced metal matrix composites, or as potential electrodes in SiC-based semiconductor devices (Goesmann, Wenzel, and Schmid-Fetzer, 1998).

The history of the 413 compounds before 1999 is the shortest: they had not been discovered!

1.2.2
History since 1995

In 1996, we made use of a reactive hot pressing (HPing), process termed *transient plastic phase processing* (Barsoum and Houng, 1993) to fabricate, in one step, starting with TiH_2, SiC, and graphite, fully dense predominantly single-phase samples of Ti_3SiC_2. Significantly, the processing temperature (1600 °C) was, at that time, considered to be above the decomposition temperature of this phase. Armed with these samples, we started characterizing the phase and were truly surprised by the combination of properties observed, so much so, that our enthusiasm carried over into the title of our paper: "Synthesis and Characterization of a Remarkable Ceramic: Ti_3SiC_2."

The compound we made was a better electrical and thermal conductor than either Ti or TiC. It was relatively soft and most readily machinable, despite having a Young's modulus of 320 GPa and a density comparable to that of Ti. It was oxidation resistant at least up to 1400 °C in air. It was so resistant to thermal shocks that quenching in water from 1400 °C actually slightly increased its flexural strength compared to unquenched samples. That, by itself, was remarkable enough. Lastly, scanning electron microscopy (SEM) and optical microscopy (OM) images left little doubt as to its layered nature. When in 1997 (El-Raghy *et al.*, 1997) we further showed that its fracture toughness was >6 MPa $m^{1/2}$ and that it was exceedingly damage tolerant, the unique combination of properties of this compound became apparent. As we have argued over the years, what renders the MAX phases remarkable is not any one property *per se*, but a combination of properties.

Once this unique combination of properties was recognized, and it was realized that Ti_3SiC_2 was first cousin to a large family of ternaries, viz. the 211 or H-phases, the next logical step was to fabricate and characterize some of the latter. In 1997 (Barsoum, Brodkin, and El-Raghy, 1997), we published a paper in which we showed that Ti_2AlC and Ti_2AlN, Ti_2GeC and Ti_3GeC_2 – not surprisingly – had properties that were quite similar to those of Ti_3SiC_2. In that paper, we also reported the fabrication of V_2AlC, Nb_2AlC, and Ta_2AlC and showed them to be machinable as well. In the same year (Barsoum, Yaroschuck, and Tyagi, 1997), we published a paper on the fabrication and characterization of Ti_2SnC, Zr_2SnC, Nb_2SnC, and Hf_2SnC and showed them to also possess property combinations that were similar to those of other MAX phases.

From this body of work, we established beyond any doubt that we were dealing with a large family of layered compounds that all had some features in common: machinability – the true hallmark of these compounds and what, to date, distinguishes them from all other structural ceramics and other ternary and quaternary ceramics – damage tolerance, relatively low thermal expansion coefficients, and good thermal and electrical conductivities. We also showed that these compounds did not melt congruently, but peritectically decomposed into an A-rich liquid and the transition-metal carbide or nitride (Barsoum, Yaroschuck, and Tyagi, 1997).

In 1997, we published the first detailed study of the oxidation of a MAX phase, namely, Ti_3SiC_2 and showed that the oxidation kinetics were parabolic at least up to 10 h and that the oxidation occurred by the inward diffusion of oxygen and the simultaneous outward diffusion of Ti.

In 1998 and 1999, we showed by transmission electron microscopy (TEM) that dislocations multiplied and were mobile at room temperature (Barsoum *et al.*, 1999a; Farber *et al.*, 1998; Farber, Levin, and Barsoum, 1999). In 1999, (Barsoum and El-Raghy, 1999) we showed that, when the grains were very large and oriented, Ti_3SiC_2 cubes were ductile when compressed at room temperature and that the deformation occurred by the formation of kink bands in individual grains as well as shear bands across entire samples.

In 1999, we became interested in the MAX phases in the Ti–Al–N system. During our literature search, we came across a paper that claimed the existence of Ti_3AlN_2. When we compared the *c* lattice parameters reported for that phase – 23 Å – to those of its ostensibly isostructural first cousin Ti_3AlC_2 – whose *c* lattice parameter is closer to 18 Å – we wrote a comment (Barsoum and Schuster, 1998) making the case that the reported structure could simply not be a 312 structure. A year later, we showed by high-resolution transmission electron microscopy (HRTEM) that the phase reported as Ti_3AlN_2 was in fact Ti_4AlN_3 (Barsoum *et al.*, 1999b). In the latter phase, four layers of Ti are N are separated by layers of Al (Figure 1.3c).

This was an important discovery, not because it was necessarily a new phase but because it opened the door to a mini gold rush. In the past five years, the following 413 phases have been discovered: Ta_4AlC_3 (Manoun *et al.*, 2006), V_4AlC_3 (Hu *et al.*, 2008), and Ti_4GaC_3 (Etzkorn *et al.*, 2009). At this time, there is no reason to believe that their number will not increase, especially in terms of solid solutions, an area that is quite ripe for new discoveries.

With the discovery of the 413 phases, it became apparent that we were dealing with a family of ternary layered compounds with the general formula $M_{n+1}AX_n$. During the past 15 years, we and many others have shown that these phases represent a new class of solids that can be best described as *thermodynamically stable nanolaminates.* It has long been predicted, and our results fully confirm, that nanoscale solids, especially laminates, should exhibit unusual and exceptional mechanical properties. Full-scale exploitation of this idea, however, had been hindered by two fundamental problems. The first had to do with the cost of manufacturing bulk samples; making large parts by molecular beam epitaxy, for instance, is not commercially viable. The second problem is more fundamental; even if fabricated, such fine-scale assemblages would not be thermodynamically stable and as such would be of limited use at elevated temperatures. It follows that the exceptional thermal stability of the dissimilar atomic layers in the MAX phases is truly extraordinary. And it is thus only logical that many of the remarkable properties of the MAX phases, such as their mechanical properties at ambient and elevated temperatures and the ease with which they can be machined, can be directly traceable to their layered nature.

The second powerful idea to emerge in the last couple of decades in the materials science community is that of biomimetics, wherein Nature's splendid designs that had evolved over millions of years would be imitated. For example, abalone shell (Figure 1.4b), mainly comprising a brittle calcium carbonate, is quite tough. This toughness arises from a submicrometer polymer film that lies between the calcium carbonate layers. The microstructural similarities between the fractured surfaces of abalone shell, for example, and those of the $M_{n+1}AX_n$ phases are noteworthy (Figure 1.4). The layering in abalone, however, is on a much coarser scale. Another fundamental distinction is that Nature optimized the properties of abalone for room-temperature use. Heating an abalone shell to a couple of hundred degrees destroys the polymer and thereby its toughness. Wood is another example, where, again, there is a marked resemblance to the MAX phases (see e.g. Figure 9.24b).

Since 1996, when our first paper on Ti_3SiC_2 was published, the MAX-phase community has embarked on an ambitious program of synthesizing and characterizing as many of the MAX phases as possible. To date, most of the MAX phases have been fabricated, and at least preliminarily characterized. There are numerous research groups exploring various facets and applications of the MAX phases in Europe, Japan, China, India, South Korea, other South Asian countries, Australia, and South America. This book attempts to summarize this global effort.

Probably, the best indicator as to how dynamic and fecund MAX phase research is globally, is to refer to Figure 1.5a,b, in which the number of papers published and the number of citations garnered when Ti_3SiC_2 is entered as the keyword in Thomas Reuters (formerly ISI) Web of Knowledge data base is plotted as a function of time, starting from 1993. Similar trends are found when the keywords chosen

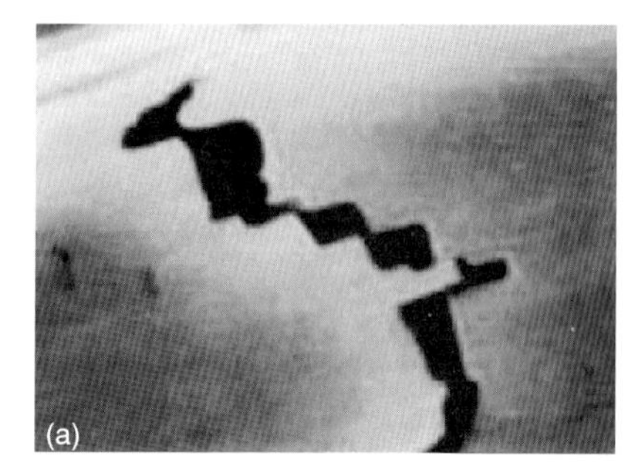

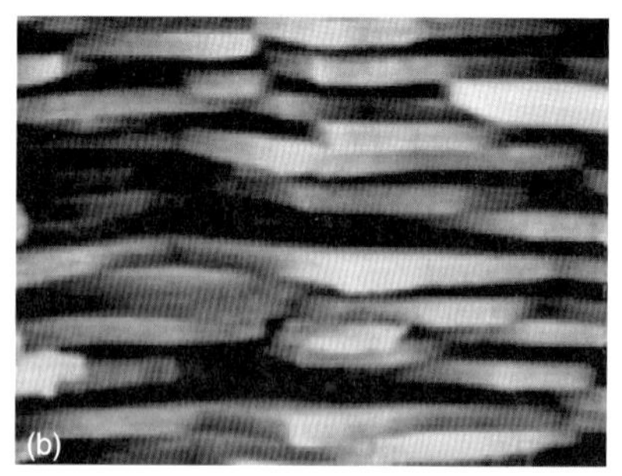

Figure 1.4 Typical SEM images of tortuous paths a crack takes in (a) a Ti_3SiC_2 grain and (b) an abalone shell (Barsoum and El-Raghy, 2001). Note that the two images are of quite different magnifications. (Source: Abalone micrograph courtesy of D. E. Morse, University of California, Santa Barbara.)

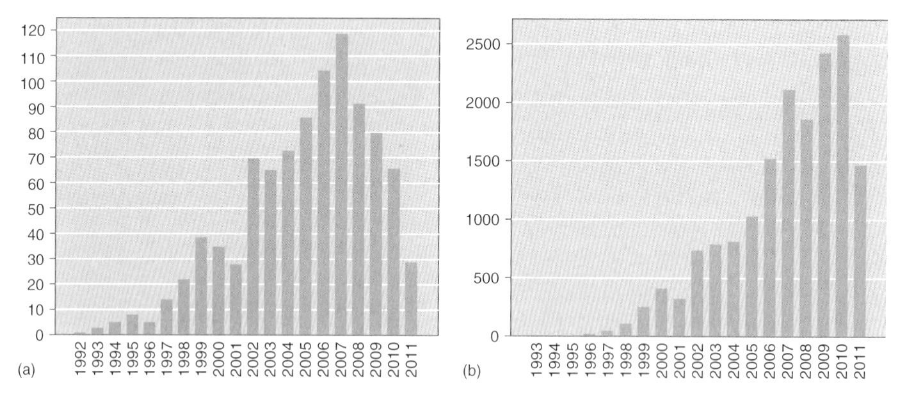

Figure 1.5 (a) Number of papers published and (b) number of citations each year as a function of time starting in 1992, when Ti_3SiC_2 is entered as the keyword, according to the Web of Science. (Source: Thomson Reuters.)

are Ti_3AlC_2 or Ti_2AlC. As noted above, today the research and interest in the MAX phases have exploded.

In brief, and despite the relatively short time since these compounds have been identified as having unusual and sometime unique properties, we have come a long way in understanding their physical, mechanical, and chemical properties. In this book, I try to summarize this understanding. This book is divided into 11 chapters. Chapter 2 reviews the atomic structures and bonding commonalties and trends in these phases. This chapter also summarizes *ab initio* calculations that, for the most part, capture the essence of the bonding in these solids. Chapter 3 deals with their elastic properties, both experimental and those calculated from density functional theory (DFT). Chapter 4 summarizes the thermal properties, including thermal expansion, conductivity, heat capacity and stability. Chapter 5 deals with the electrical transport, including conductivity, as well as Hall and Seebeck coefficient measurements. Their optical and magnetic properties are also touched upon. Chapter 6 deals with the reactivity of the MAX phases with oxygen and other gases. The reactivities of the MAX phases with solids and liquids, including molten metals and common acids and bases, are reviewed in Chapter 7.

Chapters 8–10 deal with the mechanical properties. Chapter 8 deals with kinking nonlinear elasticity and damping. How the MAX phases respond to stresses – compressive, shear, tensile, and so on – at ambient temperature is discussed in Chapter 9. Chapter 10 deals with their response to stresses at elevated temperatures, including creep. The final chapter outlines some of the outstanding scientific issues and potential applications, in addition to what needs to be done researchwise for these solids to be more widely used.

A number of review articles have been published on various aspects of the MAX phases over the years. In 2000, I published the first review on the MAX phases (Barsoum, 2000). In 2004, the first entry on the MAX phases in the *Encyclopedia of Materials Science and Technology* appeared (Barsoum and Radovic, 2004), followed 2 years later by a review, in the same publication, of their physical properties (Barsoum, 2006). A review on Cr_2AlC appeared in 2007 (Lin, Zhou, and Li, 2007). In 2009, Zhang *et al.* reviewed Ti_3SiC_2 (Zhang, Bao, and Zhou, 2009); in 2010, Eklund *et al.* published a review on the processing and properties of MAX-phase thin films (Eklund *et al.*, 2010). In the same year, Wang and Zhou reviewed Ti_2AlC and Ti_3AlC_2 (Wang and Zhou, 2010). In 2010, I published a general review (Barsoum, 2010). A third entry in the *Encyclopedia of Materials Science and Engineering* on kinking nonlinear elasticity – first discovered in the MAX phases (Chapter 8) – was published (Barsoum and Basu, 2010). More recently, Barsoum and Radovic reviewed their mechanical and elastic properties (Barsoum and Radovic, 2011). Sun published a general review in the same year (Sun, 2011).

1.2.3
Discovery of the MAX Phases

Lastly, a book on the MAX phases would surely be incomplete without the story of how they were discovered, and how they acquired their name. The details of how we stumbled on the MAX phases can be found elsewhere (Barsoum and El-Raghy, 2001). In short, like many such discoveries, it was by pure luck. However, as they say, providence every now and then rewards the well prepared. In retrospect, it was our processing breakthrough (Barsoum and Houng, 1993) – a few years before our 1996 paper – that ultimately paved the way to the discovery.

As for how they got their name – that was also due to the twists and turns of fate. Nowotny and coworkers discovered these phases and referred to them as either the H-phases or the M_2BX phases (Nowotny, 1970). Being new to the field, we followed the pioneer's lead and also referred to these phases as M_2BX. The title of our first progress report on these phases was "A Progress Report on Ti_3SiC_2, Ti_3GeC_2, and the H-Phases, M_2BX" (Barsoum and El-Raghy, 1997). This paper was written before the 413 phases were discovered so the full richness of the family was still not apparent.

A few years later, however, we realized that in the 1980s the International Union of Pure and Applied Chemistry had scrambled the periodic table. In the new periodic table, the A elements became B, and the B elements became A. And that is how MAX was born. Since then, the periodic table has been modified yet again. In the latest version, chemists eschew letters altogether and simply number the periodic table columns from 1 to 18. In this latest, and probably final, tinkering with this foundation of chemistry and materials science, the A-group elements of the MAX phases now belong to groups 13–16.

Once the change was made from M_2BX to M_2AX, I knew immediately that the latter would, most likely, be perceived as a "Madison Avenue" type branding effort that in turn would cast disparaging shadows on the discovery. In my first major review article, I tried to buck the tide and made sure that the title referred to the $M_{n+1}AX_n$ phases and not MAX (Barsoum, 2000). But when, a year later, *American Scientist*, a mass circulation magazine for people interested in science, came calling (Barsoum and El-Raghy, 2001) – and despite my objections – there was no talking the editor into using the clumsy $M_{n+1}AX_n$ phases instead of MAX. That was the point of no return. Over the years, some have tried to change the name; one proposal was to change the "M" to a "T". Needless to add, that proposal did not catch on especially in America.

In retrospect, I believe this happy accident is most appropriate for a discovery that was made by chance, and that, in some respects, the MAX phases do indeed live up to their name.

The MAX phases are truly a fascinating family of compounds. The fact that one can readily change chemistry while keeping the structure fixed has allowed us to quickly understand what affects many of the more important properties. From a scientific point of view, the fact that dislocations are confined to two dimensions has also shed important light on the motion of dislocations and their interactions with dislocation arrays.

It is my sincere hope that this book will inspire the next generation of MAX-phase researchers to take these phases to the next level and to develop as many applications as possible. The future looks bright indeed.

References

Barsoum, M.W. (2000) The $M_{n+1}AX_n$ phases: a new class of solids; thermodynamically stable nanolaminates. *Prog. Solid State Chem.*, **28**, 201–281.

Barsoum, M.W. (2006) Physical properties of the MAX phases, in *Encyclopedia of Materials Science and Technology* (eds K.H.J. Buschow, R.W. Cahn, M.C. Flemings, E.J. Kramer, S. Mahajan, and P. Veyssiere), Elsevier, Amsterdam.

Barsoum, M.W. (2010) The MAX phases and their properties, in *Ceramics Science and Technology*, Properties, Vol. **2** (eds R.R. Riedel and I.-W. Chen), Wiley-VCH Verlag GmbH.

Barsoum, M.W. and Basu, S. (2010) in *Encyclopedia of Materials Science and Technology* (eds K.H.J. Buschow, R.W. Cahn, M.C. Flemings, B. Ilschner, E.J. Kramer, S. Mahajan, and P. Veyssiere), Elsevier, Amsterdam.

Barsoum, M.W., Brodkin, D., and El-Raghy, T. (1997) Layered machinable ceramics for high temperature applications. *Scr. Met. Mater.*, **36**, 535–541.

Barsoum, M.W., Yaroschuck, B.G., and Tyagi, S. (1997) Fabrication and characterization of M_2SnC ($M = Ti, Zr, Hf$ and Nb). *Scr. Mater.*, **37**, 1583–1591.

Barsoum, M.W. and El-Raghy, T. (1996) Synthesis and characterization of a remarkable ceramic: Ti_3SiC_2. *J. Am. Ceram. Soc.*, **79**, 1953–1956.

Barsoum, M.W. and El-Raghy, T. (1997) A progress report on Ti_3SiC_2, Ti_3GeC_2 and the H-phases, M_2BX. *J. Mater. Synth. Process.*, **5**, 197–216.

Barsoum, M.W. and El-Raghy, T. (1999) Room temperature ductile carbides. *Metall. Mater. Trans.*, **30A**, 363–369.

Barsoum, M.W. and El-Raghy, T. (2001) The MAX phases: unique new carbide and nitride materials. *Am. Sci.*, **89**, 336–345.

Barsoum, M.W., Farber, L., El-Raghy, T., and Levin, I. (1999a) Dislocations, kink bands and room temperature plasticity of Ti_3SiC_2. *Metall. Mater. Trans.*, **30A**, 1727–1738.

Barsoum, M.W., Farber, L., Levin, I., Procopio, A., El-Raghy, T., and Berner, A. (1999b) High-resolution transmission electron microscopy of Ti_4AlN_3, or $Ti_3Al_2N_2$ revisited. *J. Am. Ceram. Soc.*, **82**, 2545–2547.

Barsoum, M.W. and Houng, B. (1993) Transient plastic phase processing of Ti-B-C composites. *J. Am. Ceram. Soc.*, **76**, 1445–1446.

Barsoum, M.W. and Radovic, M. (2004) Mechanical properties of the MAX phases, in *Encyclopedia of Materials Science and Technology* (eds K.H.J. Buschow, R.W. Cahn, M.C. Flemings, E.J. Kramer, S. Mahajan, and P. Veyssiere), Elsevier, Amsterdam.

Barsoum, M.W. and Radovic, M. (2011) The elastic and mechanical properties of the MAX phases. *Annu. Rev. Mater. Res.*, **41**, 9.1–9.33.

Barsoum, M.W. and Schuster, J.C. (1998) Comment on "new ternary nitride in the Ti-Al-N system". *J. Am. Ceram. Soc.*, **81**, 785–786.

Eklund, P., Beckers, M., Jansson, U., Högberg, H., and Hultman, L. (2010) The $M_{n+1}AX_n$ phases: materials science and thin-film processing. *Thin Solid Films*, **518**, 1851–1878.

El-Raghy, T., Zavaliangos, A., Barsoum, M.W., and Kalidindi, S.R. (1997) Damage mechanisms around hardness indentations in Ti_3SiC_2. *J. Am. Ceram. Soc.*, **80**, 513–516.

Etzkorn, J., Ade, M., Kotzott, D., Kleczek, M., and Hillebrecht, H. (2009) Ti_2GaC, Ti_4GaC_3 and Cr_2GaC – synthesis, crystal growth and structure analysis of Ga-containing MAX-phases $M_{n+1}GaC_n$ with M = Ti, Cr and n = 1–3. *J. Solid State Chem.*, **182**, 995.

Fakih, H., Jacques, S., Berthet, M.-P., Bosselet, F., Dezellus, O., and Viala, J.-C. (2006) The growth of Ti_3SiC_2 coatings onto SiC by reactive chemical vapor deposition using H_2 and $TiCl_4$. *Surf. Sci. Tech.*, **201**, 3748–3755.

Farber, L., Barsoum, M.W., Zavaliangos, A., El-Raghy, T., and Levin, I. (1998) Dislocations and stacking faults in Ti_3SiC_2. *J. Am. Ceram. Soc.*, **81**, 1677–1681.

Farber, L., Levin, I., and Barsoum, M.W. (1999) HRTEM study of a low-angle boundary in plastically deformed Ti_3SiC_2. *Philos. Mag. Lett.*, **79**, 163.

Goesmann, F., Wenzel, R., and Schmid-Fetzer, R. (1998) Diffusion barriers in gold-metallized titanium-based contacts on SiC. *J. Mater. Sci. - Mater. Electron.*, **9**, 109.

Goto, T. and Hirai, T. (1987) Chemically vapor deposited Ti_3SiC_2. *Mater. Res. Bull.*, **22**, 1195–1202.

Hu, C., Zhang, J., Wang, J., Li, F., Wang, J., and Zhou, Y. (2008) Crystal structure of V_4AlC_3: a new layered ternary carbide. *J. Am. Ceram. Soc.*, **91**, 636–639.

Iseki, T., Yano, T., and Chung, Y.-S. (1990) Wetting and properties of reaction products in active metal brazing of SiC. *J. Ceram. Soc. Jpn. (Int. Ed.)*, **97**, 47.

Ivchenko, V.I. and Kosolapova, T.Y. (1975) Conditions of preparation of ternary Ti-C-Al alloy powders. *Porosh. Metall.*, **150**, 1.

Ivchenko, V.I. and Kosolapova, T.Y. (1976) Abrasive properties of the ternary compounds in the system Ti-Al-C and Ti-Al-N. *Porosh. Metall.*, **164**, 56.

Ivchenko, V.I., Lesnaya, M.I., Nemchenko, V.F., and Kosolapova, T.Y. (1976a) Preparation and some properties of the ternary compound Ti_2A1N. *Porosh. Metall.*, **160**, 60.

Ivchenko, V.I., Lesnaya, M.I., Nemchenko, V.F., and Kosolapova, T.Y. (1976b) Some physical properties of ternary compounds

in the system Ti-Al-C. *Porosh. Metall.*, **161**, 45.
Jeitschko, W. and Nowotny, H. (1967) Die Kristallstructur von Ti_3SiC_2 - Ein Neuer Komplxcarbid-Typ. *Monatsh. Chem.*, **98**, 329–337.
Lin, Z., Zhou, Y., and Li, M. (2007) Synthesis, microstructure and property of Cr_2AlC. *J. Mater. Sci. Technol.*, **23**, 721.
Lis, J., Pampuch, R., Piekarczyk, J., and Stobierski, L. (1993) New ceramics based on Ti_3SiC_2. *Ceram. Int.*, **19**, 91–96.
Manoun, B., Saxena, S.K., El-Raghy, T., and Barsoum, M.W. (2006) High-pressure x-ray study of Ta_4AlC_3. *Appl. Phys. Lett.*, **88**, 201902.
Morgiel, J., Lis, J., and Pampuch, R. (1996) Microstructure of Ti_3SiC_2,-based ceramics. *Mater. Lett.*, **27**, 85–89.
Morozumi, S., Endo, M., Kikuchi, M., and Hamajima, K. (1985) Bonding mechanism between SiC and thin foils of reactive metals. *J. Mater. Sci.*, **20**, 3976.
Nickl, J.J., Schweitzer, K.K., and Luxenberg, P. (1972) Gasphasenabscheidung im Systeme Ti-C-Si. *J. Less-Common Metals*, **26**, 335–353.
Nowotny, H. (1970) Strukturchemie Einiger Verbindungen der Ubergangsmetalle mit den elementen C, Si, Ge, Sn. *Prog. Solid State Chem.*, **2**, 27–70.
Nowotny, H. and Windisch, S. (1973) High temperature compounds. *Annu. Rev. Mater. Sci.*, **3**, 171.
Okano, T., Yano, T., and Iseki, T. (1993) Synthesis and mechanical properties of Ti_3SiC_2. *Trans. Met. Soc. Jpn.*, **14A**, 597.
Pampuch, R. (1999) Advanced HT ceramic materials via solid combustion. *J. Eur. Ceram. Soc.*, **19**, 2395–2404.
Pampuch, R. and Lis, J. (1995) *Ti_3SiC_2 – A Plastic Ceramic Material*, in Advances in Science and Technology, Vol. **3B**, P. Vincenzini, ed. (Faenza, Techna Srl), pp. 725–732.
Pampuch, R., Lis, J., Stobierski, L., and Tymkiewicz, M. (1989) Solid combustion synthesis of Ti_3SiC_2. *J. Eur. Ceram. Soc.*, **5**, 283.
Pickering, E., Lackey, W.J., and Crain, S. (2000) Microstructure of Ti_3SiC_2 Coatings Synthesized by CVD. *Chem. Vap. Deposition*, **6**, 289–295.
Pietzka, M.A. and Schuster, J. (1994) Summary of constitution data of the system Al-C-Ti. *J. Phase Equilib.*, **15**, 392.
Pietzka, M.A. and Schuster, J.C. (1996) Phase equilibria in the quaternary system Ti-Al-C-N. *J. Am. Ceram. Soc.*, **79**, 2321.
Racault, C., Langlais, F., and Naslain, R. (1994) Solid-state synthesis and characterization of the ternary phase Ti_3SiC_2. *J. Mater. Sci.*, **29**, 3384–3392.
Racault, C., Langlais, F., Naslain, R., and Kihn, Y. (1994) On the chemical vapour deposition of Ti_3SiC_2 from $TiCl_4$-$SiCl_4$-CH_4-H_2 gas mixtures Part II An experimental approach. *J. Mater. Sci.*, **29**, 3941–3948.
Reiffenstein, R., Nowotny, H., and Benesovsky, F. (1966) Strukturchemische und magnetochemische Untersuchungen an Komplexcarbiden. *Monatsh. Chem.*, **97**, 1428.
Sun, Z.-M. (2011) Progress in research and development on MAX phases -a family of metallic ceramics. *Int. Mater. Rev.* **56**, 143–166.
Tong, X., Okano, T., Iseki, T., and Yano, T. (1995) Synthesis and high temperature mechanical properties of Ti_3SiC_2/SiC composites. *J. Mater. Sci.*, **30**, 3087.
Wakelkamp, W.J.J., van Loo, F.J., and Metselaar, R. (1991) Phase relations in the titanium-silicon-carbon system. *J. Eur. Ceram. Soc.*, **8**, 135.
Wang, X.H. and Zhou, Y.C. (2010) Layered machinable and electrically conductive Ti_2AlC and Ti_3AlC_2 ceramics: a review. *J. Mater. Sci. Technol.*, **26**, 385–416.
Wolfsgruber, H., Nowotny, H., and Benesovsky, F. (1967) Die Kristallstuktur von Ti_3GeC_2. *Monatsh. Chem.*, **98**, 2401.
Zhang, H.B., Bao, Y.W., and Zhou, Y.C. (2009) Current status in layered ternary carbide Ti_3SiC_2. *J. Mater. Sci. Technol.*, **25**, 1–38.

2
Structure, Bonding, and Defects

2.1
Introduction

The $M_{n+1}AX_n$ phases are layered, hexagonal (space group D_{6h}^4-$P6_3/mmc$), with two formula units per unit cell (Figure 2.1). The X atoms reside in M octahedral cages that are linked together by edges. The A atoms reside in slightly larger right prisms. High-resolution transmission electron microscopy (HRTEM) images of Ti_2AlN, Ti_3SiC_2, and Ti_4AlN_3 are shown in Figure 2.2a–c, respectively. In all cases, not only is the layered nature of these compounds clearly visible but, as important, so is the characteristic herringbone or zig-zag pattern of the M–X blocks.

Table 2.1 lists all the bulk MAX phases known to date, together with their lattice parameters and theoretical densities. The A-group elements are mostly from groups 13 and 14 (i.e., formerly groups IIIA and IVA). The vast majority of compounds are 211s. The most versatile A-group element with thirteen compounds, including nitrides, 312s, and a few 413s phases, is Al. With nine, Ga forms the most 211 phases. As noted in Chapter 1, the vast majority of these phases were discovered in the 1960s by Nowotny and coworkers (Nowotny, 1970).

Currently, there are roughly 50 M_2AX phases (Nowotny, 1970) and five M_3AX_2 phases: Ti_3SiC_2 (Jeitschko and Nowotny, 1967), Ti_3GeC_2 (Wolfsgruber, Nowotny, and Benesovsky, 1967), Ti_3AlC_2 (Pietzka and Schuster, 1994) Ti_3SnC_2 (Dubois *et al.*, 2007), and Ta_3AlC_2 (Etzkorn, Ade, and Hillebrecht, 2007a). The number of M_4AX_3 phases is also growing since that structure was first established in Ti_3AlN_4 (Barsoum *et al.*, 1999c; Rawn *et al.*, 2000). This phase was initially believed to be $Ti_3Al_2N_2$. However, using HRTEM and Rietveld analysis of neutron diffraction (ND) data, and by careful mapping of the ternary phase diagram, we showed that this phase has the unit cell shown in Figure 2.1c (Barsoum *et al.*, 1999c; Rawn *et al.*, 2000).

Since the discovery of Ti_4AlN_3, the following 413 phases have been discovered: Ta_4AlC_3 (Manoun *et al.*, 2006), Nb_4AlC_3 (Hu *et al.*, 2007a), and V_4AlC_{3-x} (Etzkorn, Ade, and Hillebrecht, 2007b). However, since Ti_3SnC_2, Ta_3AlC_2, or V_4AlC_{3-x} have not been synthesized in a predominantly pure form, their characterization awaits a processing breakthrough, and henceforth, for the most part, will not be discussed further in this book.

MAX Phases: Properties of Machinable Ternary Carbides and Nitrides, First Edition. Michel W. Barsoum.

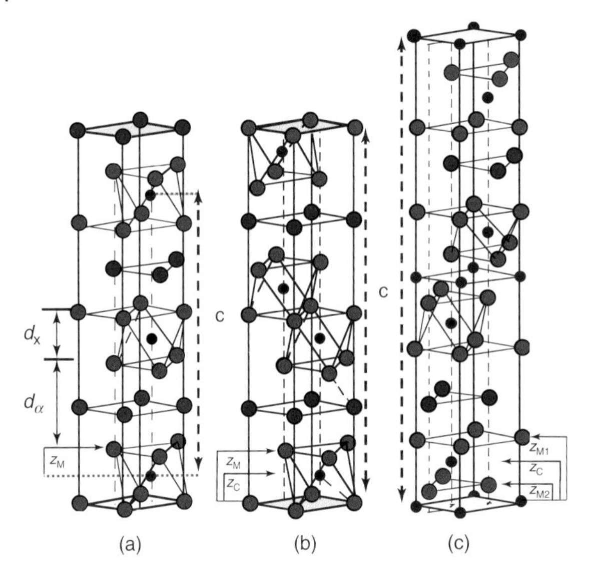

Figure 2.1 Unit cells of (a) 211, (b) 312, and (c) 413 phases. The c parameters are depicted by vertical dashed lines. d_X denotes the thickness – from atom center to atom center – of the $M_{n+1}X_n$ layers; d_α that of the A layers. It follows that for the 211 phases $c = 2d_\alpha + 2d_X$. Also shown are the various z values (see Table 2.2).

2.2
Atom Coordinates, Stacking Sequences, and Polymorphic Transformations

For the 211 structure, there are three inequivalent atoms; in the 312, there are four, and in the 413, there are five. The coordinates of all the atoms are listed in Table 2.2 for the various polymorphs. For the 211 phases, there is only one polymorph (Figure 2.3a). In the 312 case there are two, α and β, shown in Figures 2.3b and c, respectively. For the 413 phases, there are three polymorphs, viz. α, β, and γ, shown in Figures 2.3d, e and f, respectively.

When describing the atomic stacking sequences, it is important to label the various atoms. For the 312 and 413 phases, the existence of two locations for the M and X atoms can lead to confusion in their labeling. For example, Kisi *et al.* (1998) labeled the Ti in the center of the unit cell of Ti_3SiC_2 as Ti_I, while others labeled the same atom as Ti_{II}. In this book, we follow the crystal chemists' (Etzkorn *et al.*, 2009) lead and label the M atoms bonded to the A atoms as M_I, and those that are only bonded to the X atoms as M_{II}. The same holds for the X atoms; those bonded to an M_I atom are labeled X_I, and those bonded to only M_{II} as X_{II}. If the community of MAX-phase researchers can adhere to this scheme, it would greatly reduce the propensity for confusion.

On the basis of Table 2.2 and the HRTEM images, the stacking sequences for the most common, that is, α-polymorphs, of the 211, 312, and 413 phases are, respectively,

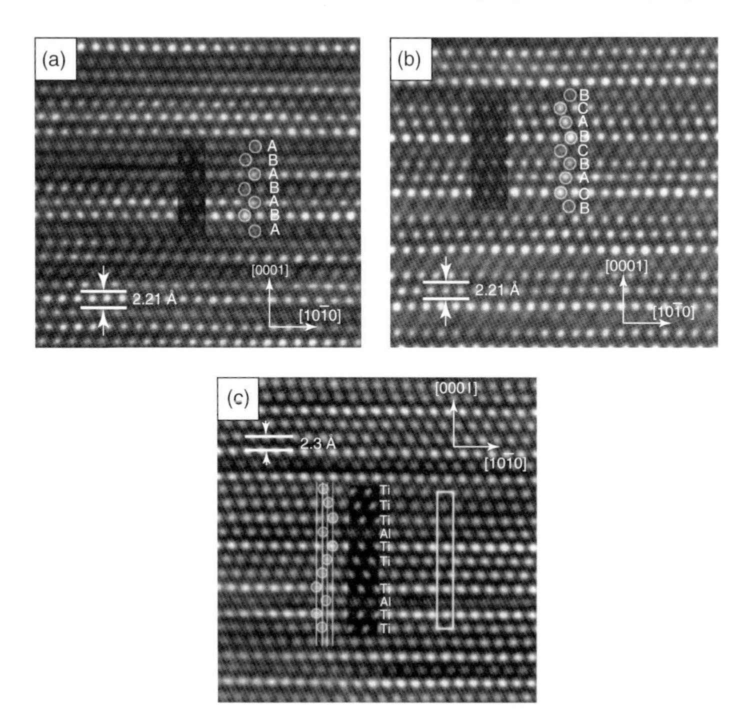

Figure 2.2 High-resolution TEM micrographs along the $(11\bar{2}0)$ zone for (a) Ti_2AlN (Farber *et al.*, 1999), (b) β-Ti_3SiC_2 (Farber *et al.*, 1999), and (c) α-Ti_4AlN_3. (Barsoum *et al.*, 1999c). In each figure, a simulated contrast is shown as an inset superimposed on the experimental image.

$$\text{B}\gamma\text{AbA}\gamma\text{BaB}\gamma\text{AbA}\gamma\text{B} \ldots \text{M}_2\text{AX} \quad (2.1)$$

$$\text{A}\gamma\text{BaB}\gamma\text{A}\beta\text{Ca}C\beta\text{A}\gamma\text{BaB}\gamma\text{A} \ldots \alpha\text{-M}_3\text{AX}_2 \quad (2.2)$$

$$\alpha\text{BcB}\alpha C\beta\text{A}\gamma\text{BaB}\gamma\text{A}\beta\text{C}\alpha\text{BcB}\alpha\text{C} \ldots \alpha\text{-M}_4\text{AX}_3 \quad (2.3)$$

In these arrangements, the capital and lower-case letters correspond to the M and A layers, respectively. The Greek letters refer to the X positions corresponding to their Roman letter counterpart: that is, α is an A site, β is a B site, and so on. These sequences are best appreciated when the $(11\bar{2}0)$ planes are sketched, respectively, in Figure 2.3a,b,d. Note that in all cases, the A-element planes are mirror planes and the MX blocks are zig-zagged, in agreement with the HRTEM micrographs (Figure 2.2).

The stacking sequences for the corresponding β-phases are

$$\text{A}\gamma\text{BcB}\gamma\text{A}\beta\text{CbC}\beta\text{A}\gamma\text{BcB}\gamma\text{A} \ldots \beta\text{-M}_3\text{AX}_2 \quad (2.4)$$

$$\gamma\text{BcB}\gamma\text{A}\gamma\text{B}\gamma\text{AbA}\gamma\text{B}\gamma\text{A}\gamma\text{BcB}\gamma\text{A} \ldots \beta\text{-M}_4\text{AX}_3 \quad (2.5)$$

The $(11\bar{2}0)$ planes of the β-M_3AX_2 and β-M_4AX_3 phases are sketched, respectively, in Figure 2.3c,e. To transform α-M_3AX_2 to the β form, one needs to only shear the A-group layers (compare Figure 2.3b,c). Since the transformation does not involve

Table 2.1 Summary of $M_{n+1}AX_n$ phases known to date. The theoretical density ($Mg\ m^{-3}$) is in bold. The *a* and *c* lattice parameters in angstrom are in parentheses. Most of this list appeared in the review paper by Nowotny (1970). The 211 phases are indicated in bold and the 312s are indicated in bold italics. The 413 are in normal font.

12	13	14	15	16
	Al **Ti_2AlC, 4.11** (3.04, 13.60) **V_2AlC, 4.87** (2.91, 13.10) **Cr_2AlC, 5.24** (2.86, 12.8) **Nb_2AlC, 6.50** (3.10, 13.8) **Ta_2AlC, 11.82** (3.07, 13.8) **Ti_2AlN, 4.31** (2.99, 13.61) ***Ti_3AlC_2***, **4.50** (3.07, 18.58) ***Ta_3AlC_2***, **12.20** (3.09, 19.16) Ti_4AlN_3, 4.76 (2.99, 23.37) α-Ta_4AlC_3, 12.92 (3.11, 24.10) β-Ta_4AlC_3, 13.36 (3.09, 23.71) Nb_4AlC_3, 7.06 (3.13, 24.12) $V_4AlC_{2.7}$, **5.16** (2.93, 22.74)	Si ***Ti_3SiC_2*** **4.52** (3.07, 17.67)	P **V_2PC** **5.38** (3.08, 10.91) **Nb_2PC** **7.09** (3.28, 11.5)	S **Ti_2SC, 4.62** (3.21, 11.22) **Zr_2SC, 6.20** (3.40, 12.13) **$Nb_2SC_{0.4}$**, (3.27, 11.4) **Hf_2SC**, (3.36, 11.99)
Zn	Ga **Ti_2GaC, 5.53** (3.07, 13.52) **V_2GaC, 6.39** (2.93, 12.84) **Cr_2GaC, 6.81** (2.88, 12.61) **Nb_2GaC, 7.73** (3.13, 13.56) Mo_2GaC, 8.97 (3.01, 13.18) **Ta_2GaC, 13.05** (3.10, 13.57) **Ti_2GaN, 5.75** (3.00, 13.3) **Cr_2GaN, 6.82** (2.87, 12.70) **V_2GaN, 5.94** (3.00, 13.3) Ta_4GaC_3, 13.99 (3.07, 23.44)	Ge **Ti_2GeC, 5.68** (3.07, 12.93) **V_2GeC, 6.49** (3.00, 12.25) **Cr_2GeC, 6.88** (2.95, 12.08) ***Ti_3GeC_2***, **5.55** (3.07, 17.76)	As **V_2AsC** **6.63** (3.11, 11.3) **Nb_2AsC** **8.025** (3.31, 11.9)	 Se
Cd **Ti_2CdC** **9.71** (3.1, 14.41)	In **Sc_2InC** **Ti_2InC, 6.2** (3.13, 14.06) **Zr_2InC, 7.1** (3.34, 14.91) **Nb_2InC, 8.3** (3.17, 14.37) **Hf_2InC, 11.57** (3.30, 14.73) **Ti_2InN, 6.54** (3.07, 13.97) **Zr_2InN, 7.53** (3.27, 14.83)	Sn **Ti_2SnC, 6.36** (3.163, 13.679) **Zr_2SnC, 7.16** (3.3576, 14.57) **Nb_2SnC, 8.4** (3.24, 13.80) **Hf_2SnC, 11.8** (3.32, 14.39) **Hf_2SnN, 7.72** (3.31, 14.3) ***Ti_3SnC_2***, **5.95** (3.14, 18.65)	Sb	Te
	Tl **Ti_2TlC, 8.63** (3.15, 13.98)	Pb **Ti_2PbC, 8.55**	Bi	

Table 2.1 (*Continued*).

12	13	14	15	16
	Zr_2TlC, **9.17** (3.36, 14.78) Hf_2TlC **13.65** (3.32, 14.62) Zr_2TlN, **9.60** (3.3, 14.71)	(3.20, 13.81) Zr_2PbC, **9.2** 3.38, 14.66 Hf_2PbC, **12.13** (3.55, 14.46)		

breaking the strong M–X bonds comprising the octahedra, this transformation is a relatively low activation energy process (He *et al.*). Not surprisingly, and as shown in Figure 2.4a, this transformation has an effect on the X-ray diffraction (XRD) diffractograms from the two polymorphs. This transformation can be induced – and was first discovered – when thinning samples for transmission electron microscophic (TEM) observation (Farber *et al.*, 1999). The same transformation was also observed when α-Ti_3GeC_2 was compressed in an anvil cell in which the shear stresses were not insignificant (Wang, Zha, and Barsoum, 2004).

The differences between the α and β polymorphs of M_4AX_3, however, are more profound because they involve a restructuring of the MX blocks (compare Figure 2.3d,e), which is why the activation energy for this transformation is quite high (He *et al.*). The differences in their XRD patterns (Figure 2.4b) allow these two polymorphs to be distinguished. One key differentiating peak is the one that appears at $2\theta = 42.07^\circ$ in the α-polymorph, but does not appear in the β-polymorph.

Lastly, the sequence for the γ-M_4AX_3 structure is

$$\alpha BaB\alpha C\beta A\gamma BcB\gamma A\beta C\alpha BaB\alpha C \ldots \gamma\text{-}M_4AX_3 \quad (2.6)$$

The γ-M_4AX_3 $(11\bar{2}0)$ planes are sketched in Figure 2.3f. The α- to γ-M_4AX_3 transformation only involves the shearing of A layers (compare Figure 2.3d,f) and is thus a low activation energy transformation (He *et al.*, 2011). Note that in *all* polymorphs, the A atoms are always on mirror planes.

There have been recent reports of MAX phases with $n = 4$ and 5. However, what is unclear at this time is whether these compositions can be synthesized as single phases in bulk form or whether they are simply regions with high densities of stacking faults (SFs).

Hybrid structures also exist. The first ones were reported in the Ti–Si–C system by Palmquist *et al.* (2004). They showed that depending on the concentration of the Si in their thin films, they could grow, in addition to the 312 and 413 phases, $Ti_5Si_2C_3$ and $Ti_7Si_2C_5$ regions. The latter two can be considered to be combinations of half unit cells of the more common MAX phases. For example, the 523 phase can be considered to be composed of two half-cells of 312 and 211 unit cells (Figure 2.5). Similarly, the 725 phase has a structure that can be described as two half-cells of the 312 and 413 phases, that is, alternating three and four M layers between the A layers.

Until very recently, these structures were observed as minority phases in thin films and were thus hypothesized to be metastable (Palmquist *et al.*, 2004). Recently,

Table 2.2 Sites and idealized coordinates of the $M_{n+1}AX_n$ phases for $n = 1$–3. Also listed are the currently known polymorphs. The fifth column lists the canonical positions.

Atoms	Chemistry/archetypical phase				z_M Range	Notes and references
	M_2AX/Ti_2SC				—	Kudielka and Rohde (1960)
	Wyckoff	x	y	z_i (canonical)	—	—
A	2d	1/3	2/3	3/4	—	
M	4f	2/3	1/3	1/12 (0.083)	0.07–0.1	z_M in Figure 2.1a
X	2a	0	0	0	—	
α-M_3AX_2/Ti_3SiC_2						Jeitschko and Nowotny (1967)
A	2b	0	0	4/16	—	—
M_I	4f	1/3	2/3	2/16 (0.125)	0.131–0.138	z_{M1} in Figure 2.1b
M_{II}	2a	0	0	0	—	—
X_I	4f	2/3	1/3	1/16 (0.0625)	0.0722	z_C in Figure 2.1b
β-M_3AX_2/Ti_3SiC_2					—	Farber *et al.* (1999)
A	2d	1/3	2/3	4/16	—	Should be quite similar in properties to α-M_3AX_2
M_I	4f	1/3	2.3	2/16(0.125)	0.1355	
M_{II}	2a	0	0	0	—	
X_I	4f	2/3	1/3	1/16 (0.0625)	0.072	
α-M_4AX_3/Ti_4AlN_3					—	Barsoum *et al.* (1999c) and Rawn *et al.* (2000)
A	2c	1/3	2/3	5/20	—	—
M_I	4e	0	0	3/20 (0.15)	0.155–0.158	z_{M1} in Figure 2.1c
M_{II}	4f	1/3	2/3	1/20 (0.05)	0.052–0.055	z_{M2} in Figure 2.1c
X_I	2a	0	0	0	—	—
X_{II}	4f	2/3	1/3	2/20	0.103–0.109	z_C in Figure 2.1c
β-M_4AX_3/Ta_4AlC_3					—	Eklund *et al.* (2007)
A	2c	1/3	2/3	5/20	—	—
M_I	4e	1/3	2/3	12/20 (0.6)	0.658	—
M_{II}	4f	1/3	2/3	1/20	0.055	Eklund *et al.* (2007)
X_I	2a	0	0	0	—	—
X_{II}	4e	2/3	1/3	2/20	0.103	—
γ-M_4AX_3/Ta_4GaC_3					—	Etzkorn *et al.* (2009)
A	2c	1/3	2/3	5/20	—	—
M_I	4e	0	0	3/20 (0.15)	0.156	—
M_{II}	4f	1/3	2/3	1/20	0.056	Etzkorn *et al.* (2009) and He *et al.* (2011)
X_I	2a	0	0	0	—	—
X_{II}	4f	2/3	1/3	2/20	0.1065	—

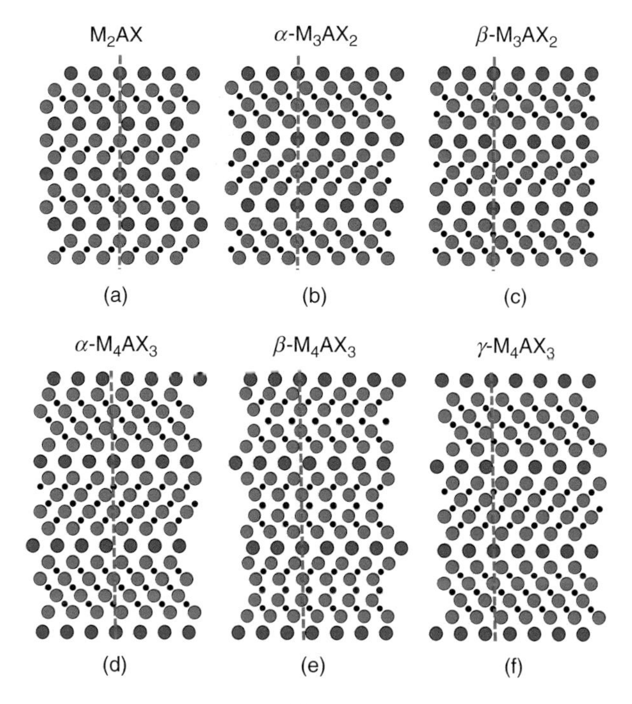

Figure 2.3 Schematics of $(11\bar{2}0)$ planes in (a) M_2AX, (b) α-M_3AX_2, (c) β-M_3AX_2, (d) α-M_4AX_3, (e) β-M_4AX_3, and (f) γ-M_4AX_3. The dashed vertical lines are guides to the eyes. Note that it is only in the α-M_3AX_2 structure that the A atoms lie on top of each other.

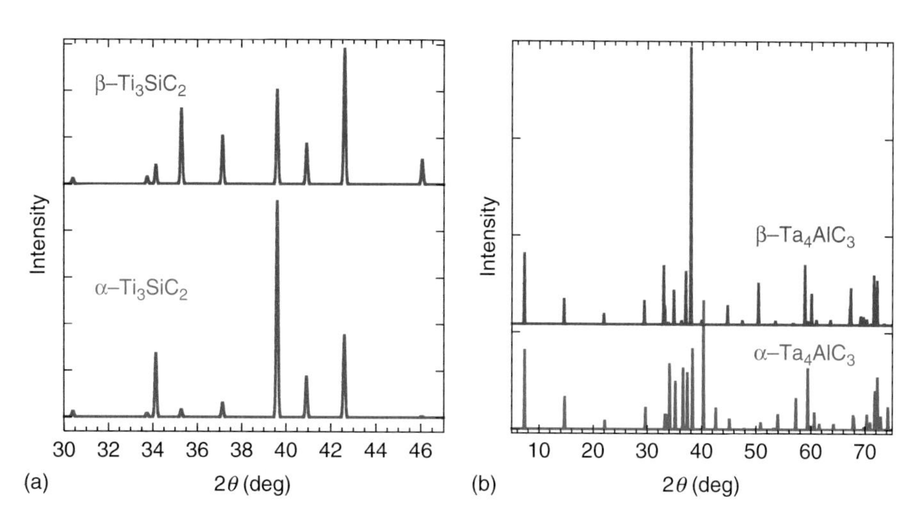

Figure 2.4 Simulated XRD patterns for Cu Kα radiation of (a) α- and β-Ti_3SiC_2 (Farber *et al.*, 1999) and (b) α-Ta_4AlC_3 and β-Ta_4AlC_3.

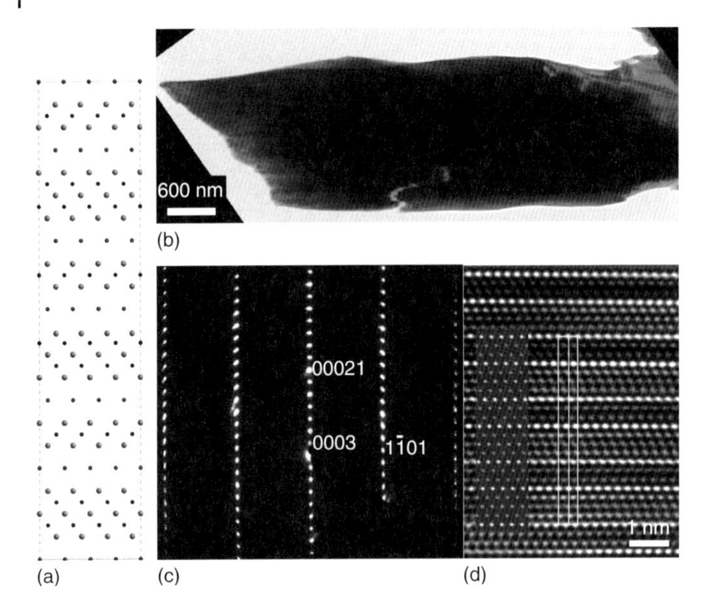

Figure 2.5 $Ti_5Al_2C_3$ (a) unit cell, (b) particle imaged in TEM edge-on, (c) selected area diffraction, and (d) HRTEM image and simulation of structure (Lane *et al.*, 2012b).

we were able to partially convert Ti_2AlC powders into a predominantly 523 phase by annealing the former in Ar at 1500 °C for 8 h (Lane *et al.*, 2012b). The 532 unit cell consists of three formula units (Figure 2.5a), and has a trigonal structure, with *a* and *c* cell parameters of 3.067 and 48.16 Å, respectively. Note that the herringbone pattern of the MX layers is maintained. In a paper submitted about two weeks after ours, Wang *et al.* (2012) also reported on this structure. Their structure, however, is wrong and wholly unsupported by the evidence presented in their paper (Lane *et al.*, 2012a).

2.2.1
Polymorphic Transformations

There are several polymorphic transformations that occur in the MAX phases, amongst which are the following:

1) **Thinning-induced**: When samples of Ti_3SiC_2, $Ti_4AlN_{2.9}$, and Ti_3AlC_2 were thinned for TEM, some of the A layers sheared relative to each other, as shown schematically in Figure 2.3c (Barsoum *et al.*, 1999c; Farber *et al.*, 1999). In other words, thinning the samples resulted in a dimensionally induced polymorphic phase transformation, which involved the shearing of the Si or Al planes in opposite directions.

   These transformations resemble the dimensionally induced hexagonal close-packed (hcp) to face-centered cubic (fcc) transformations observed in Ti-based thin multilayers including, notably, Ti–Al multilayers (Banerjee, Ahuja, and Fraser, 1996; Dregia, Banerjee, and Fraser, 1998; Shechtman, van Heerden, and Josell, 1994; Tepper *et al.*, 1997; Van Heerden, Josell, and Shechtman, 1996). More work, however, is required to establish the commonality of these phenomena and to understand the nature of the driving forces and transformation mechanisms, neither of which is totally understood even for the Ti multilayers.

2) **Thermally induced**: At this time there are no credible accounts of a thermally induced phase transition that is not accompanied by changes in chemistry. Using *ab initio* calculations, Wang *et al.* (2008) predicted that α-Ta_4AlC_3 should transform to β-Ta_4AlC_3 at temperatures higher than 1600 °C. We repeatedly tried, but failed, to induce that transformation (Lane *et al.*, 2011). One difficulty with the experiment is that heating the MAX phases at such temperatures typically results in the loss of the A element. Interestingly, the first Ta_4AlC_3 sample reported – using a powder made by Kanthal – was of the β-type (Manoun *et al.*, 2006). More recently, we tried synthesizing the structure starting with elemental powders, following the procedure outlined by Hu *et al.* (2007b), but the only phase obtained was α. How, and under what conditions, the β-phase forms remains a mystery at this time.
3) **Shear-induced**: The only shear-induced phase transition reported to date occurred in Ti_3GeC_2 when it was compressed in a diamond anvil cell – in which the shear stresses were significant – to a stress of 27 GPa (Wang, Zha, and Barsoum, 2004). In this α–β transformation, the atomic arrangement changes from the one shown in Figure 2.3b to that shown in Figure 2.3c. Note that in that experiment the stress was *not* truly hydrostatic. Not surprisingly, hydrostatic pressure alone does not result in a transformation (Manoun *et al.*, 2007b).

2.3 Lattice Parameters, Bond Lengths, and Interlayer Thicknesses

2.3.1 Lattice Parameters

Given that one of the fundamental building blocks of the MAX phases are the edge-sharing MX_6 octahedra (Figure 1.1), it is unsurprising that an excellent, almost one-to-one, correlation exists (Figure 2.6a) between the *a* lattice parameters of the former, and the M–M distances in the corresponding MX binaries, or d_{M-M} (Barsoum, 2000). The inclined line in Figure 2.6a has a slope of unity. Least squares fit yields an $R^2 > 0.75$, implying that, for the most part, the correlation is good. Note that in contradistinction of all other MX phases, CrC does not crystallize in the rock salt structure. The M–M distances for Cr plotted in Figure 2.6a are those for CrN, which does crystallize in the rock salt structure (Haglund *et al.*, 1991).

When the dependencies of the *c* lattice parameters on d_{M-M} and the diameter of the A-group element, d_A, are plotted (in Figure 2.6b,c, respectively), the following can be concluded:

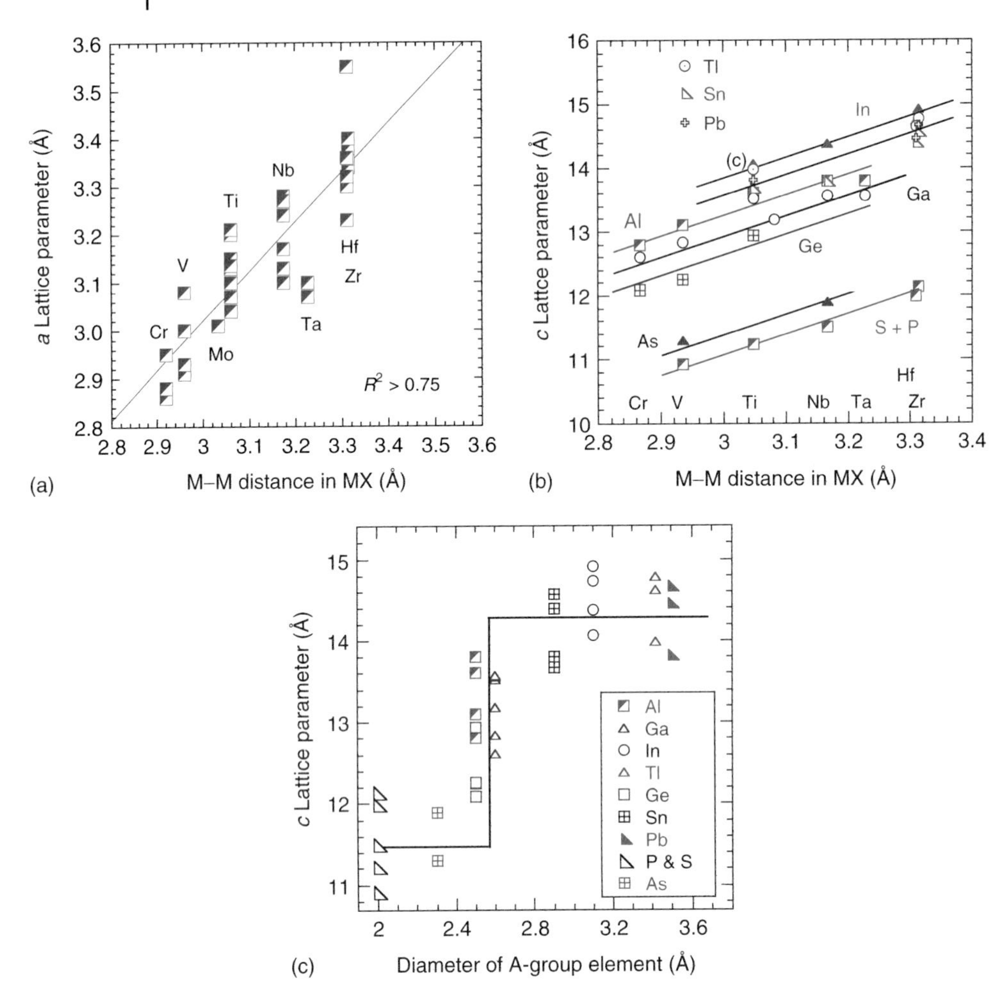

Figure 2.6 Functional dependence of (a) a lattice parameters and (b) c lattice parameters in the 211 phases on the M–M distance in the corresponding stoichiometric MX compounds (Barsoum, 2000). With the exception of the Cr values, the latter are taken from Cottrell (1995). The M–M distances for Cr are those for CrN, which crystallizes in the rock salt structure. (c) c lattice parameters on diameter d_A of the A group element. In (b) and (c), the lines are guides to the eye.

- For each A-group element, the c parameter increases linearly with increasing d_{M-M} (Figure 2.6b). The same was found for the M_3AlC_2 compounds (He *et al.*, 2010).
- The P-, S-, and As-containing ternaries have significantly shorter c lattice parameters than those of Ge, which in turn are lower than those of the remaining compounds (Figure 2.6c). Not surprisingly, the Sn-, Pb-, and Tl-containing phases have the highest values of c.

- The relationship between the c parameter and d_A is complicated, highly nonlinear, and separable into three regimes (Figure 2.6c). For $d_A < 2.5$ Å, the c lattice parameters are relatively small, and increase slightly with increasing d_A. Around 2.5 Å, there is strong, almost step-like, dependence of the c parameter on d_A. For $d_A > 2.6$ Å, the c lattice parameters are again more or less independent of d_A. The most likely explanation for these observations is that for $d_A < 2.5$ Å, the A atoms are small enough to allow the formation of stronger d–d bonds between the M layers on either side of them (see subsequent text). Presumably, for $d_A > 2.5$ Å, these bonds do not form, or, if they form, are weaker, and the c-axis increases accordingly. Why the c lattice parameter saturates at around 14.4 Å is unclear at this time, but must be related to the bonding intricacies and quite possibly the polarizabilities of the larger A-atoms.

The c/a ratio for the 211s range from 3.5 to 4.6, with the lowest values belonging to the P-, S-, and As-containing ternaries, followed by those for Ge, and so on. The simplest explanation of this fact is to assume the ternaries to be interstitial compounds in which the A and X atoms fill the interstitial sites between the M atoms. In such a scheme, the c parameter, comprising four M layers per unit cell, should be ≈4 times the a parameter, which in turn is $\approx d_{M-M}$. In other words, the 211s should have a c/a ratio of ≈4, as observed. Similar arguments for the 312s, with six layers of Ti atoms per unit cell, predict a ratio of ≈6; for the 413s, a ratio of ≈8. The actual c/a ratios for the latter two are, respectively, ≈5.8–6 and 7.7–7.8, supporting this simple interstitial compound notion.

2.3.2 Relationship between the MAX and MX Phases

In several important properties, such as elastic, electric and thermal, the MAX and their corresponding MX phases share much in common. To convert a typical MAX phase (Figure 2.7a) to MX (Figure 2.7c), the first step is to replace the A atoms by X atoms (Figure 2.7b) (Barsoum *et al.*, 1999a). This results in a highly twinned rock salt structure, with the A mirror planes converting to X mirror planes (Figure 2.7b). The second step is to de-twin every other M_3X_2 block of layers (by rotation around an axis normal to the c-axis). The resulting plane is the (110) plane of the rock salt structure (Figure 2.7c).

2.3.3 Bond Lengths and Interlayer Distances

Table 2.2 lists the atoms' idealized coordinates as a function of n for the various polymorphs. In these structures, the z_i values – one for the M atoms and the other for the X atoms – are the only variables. The various z_Ms and z_Xs values associated with the MAX phases are shown in Figure 2.1.

Once the z_i values and lattice parameters are known, all the bond lengths and bond angles in the structures can be determined. For example, for the 211 phase,

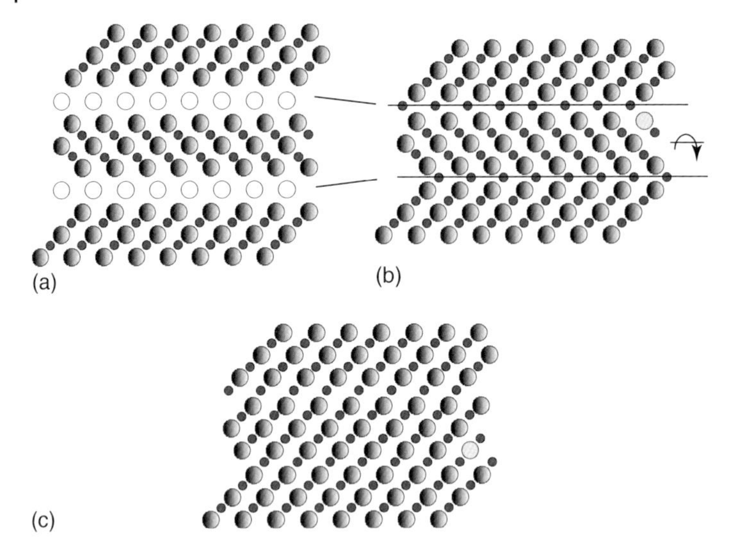

Figure 2.7 Structural relationship between the MAX and MX phases. (a) MAX phase $(11\bar{2}0)$ plane. (b) Substitution of A atoms by X. (c) De-twinning of every other mirror plane. The end result is the (110) plane in the rock salt structure (Barsoum *et al.*, 1999a).

the following applies:

$$R_{\mathrm{M-X}} = \sqrt{\frac{a^2}{3} + c^2 z_{\mathrm{M}}^2} \tag{2.7}$$

$$R_{\mathrm{M-A}} = \sqrt{\frac{a^2}{3} + c^2 \left(\frac{1}{4} - z_{\mathrm{M}}\right)^2} \tag{2.8}$$

$$R_{\mathrm{M-M}} = a \quad \text{(in plane M atoms)} \tag{2.9}$$

For the out-of-plane atoms,

$$R_{\mathrm{M1-M2}} = \sqrt{\frac{a^2}{3} + 4c^2 z_{\mathrm{M}}^2} \tag{2.10}$$

Appendix A1 lists the equations for the 312 and 413 phases. Table 2.3 compares the various bond lengths in select Ti-containing MAX phases with those in TiC and TiN. In general, the bond lengths in the binaries are similar to those in the ternaries. There are important differences and trends, however. According to Table 2.3, and with no exceptions, the M_I–X bonds in the 211 and 312 phases are all shorter than those in the binaries.

The situation for the 413 phases is slightly more complicated than for the 211 and 312 phases in that a M layer that is bonded solely to X atoms exists. At ≈2.09 Å, the Ti–N bonds in the octahedra nearest the A layers, namely, the Ti_I–X_I bonds, are significantly shorter than the ones in the center of the unit cell – namely, Ti_{II}–N_{II} – at 2.141 Å.

To further illustrate the generality of this conclusion, it is instructive to compare the various interlayer distances in the 211, 312, and 413 phases comprised of the

Table 2.3 Summary of experimental bond distances (Å), deduced from the lattice parameters and the z_i parameters for select Ti-containing MAX and MX phases.

	M_I-A	M_I-X_I	M_I-M_{II}	M_I-M_I	$M_{II}-X_I$	$M_{II}-X_{II}$	References
Ti_2AlC	2.846	2.116	2.905	3.064	—	—	@10 K (Hug, Jaouen, and Barsoum, 2005)
	2.855	2.119	2.942	3.058	—	—	400 K unpub.
Ti_2AlN	2.823	2.087	2.910	2.994	—	—	Jeitschko, Nowotny, and Benesovsky (1963d)
	2.837	2.086	2.905	2.995	—	—	@10 K (Hug, Jaouen, and Barsoum, 2005)
Ti_3SiC_2	2.696	2.135	2.971	3.068	2.135	—	Jeitschko and Nowotny (1967)
	2.693	2.085	2.963	3.066	2.181	—	Barsoum *et al.* (1999b)
	2.681	2.088	2.963	3.058	2.176	—	Kisi *et al.* (1998)
$Ti_4AlN_{2.9}$	2.818	2.087	2.913	2.988	2.093	2.141	Barsoum *et al.* (2000)
TiC	—	2.165	—	3.062	—	—	Lye and Logothetis (1966)
TiN	—	2.120	—	2.997	—	—	—

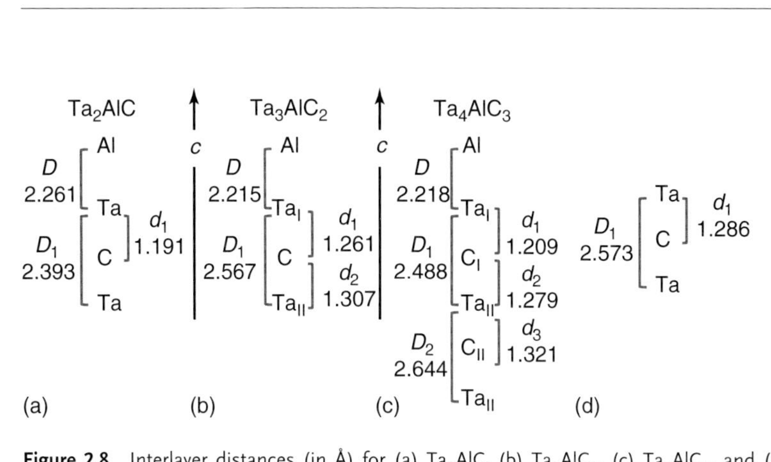

Figure 2.8 Interlayer distances (in Å) for (a) Ta_2AlC, (b) Ta_3AlC_2, (c) Ta_4AlC_3, and (d) TaC. The values for Ta_2AlC were calculated assuming $z_{Ta} = 0.0865$; the other values were taken from Etzkorn, Ade, and Hillebrecht, 2007a. The values for TaC were taken from Cottrell (1995).

same elements. The only system in which this is true is the Ta–Al–C system. Figure 2.8 compares the interlayer thicknesses in Ta_2AlC, Ta_3AlC_2, Ta_4AlC_3, and TaC (Etzkorn, Ade, and Hillebrecht, 2007a). On the basis of this figure, the following conclusions – valid for many α-MAX phases – can be reached:

1) The compression along [0001] of the octahedra increases as the latter are closer to an A layer. This is manifested by the fact that in all cases, $d_1 < d_2 < d_3$ (Figure 2.8c).
2) The M–A interlayer distance – D in Figure 2.8 – is a weak function of n.

3) The M_I–X_I interlayer distance – D_1 in Figure 2.8 – is less than the corresponding distance in the binary (Figure 2.8d). The opposite is true of D_2. Note that the Ta-C distance in the binary is roughly the average of the Ta-C distances in the ternaries. This is not only true in this system, but in the Ti-Al-C system as well.

2.4 Theoretical Considerations

This section summarizes the extensive body of work that now exists on theoretical modeling of the MAX phases. Before discussing atomic bonding in the MX and MAX phases as determined by density functional theory (DFT), and the complications therein, it is important to check the validity of some simple models put forth, roughly a century ago, by Drude and Sommerfeld to understand the properties of metallic conductors. Despite being crude approximations in some important respects, these early models remain, when properly used, of considerable value even today.

2.4.1 Drude and Sommerfeld Models

Drude applied the kinetic theory of gases to metals. He assumed that the electrons in a metal, similar to gas atoms in a room, were free to wander from one end to another, while, occasionally, being scattered by the atomic cores and other electrons. In other words, the electrons formed an *electron gas* that moved against a background of heavy immobile ions.

It follows that in the Drude model, and the more sophisticated Sommerfeld model (see subsequent text), one of the most important parameters is the *number of valence electrons per unit volume,* n_{val}. Once n_{val} is known, then a number of properties can be estimated (Ashcroft and Mermin, 1976). To calculate n_{val} for the MAX phases, the following expression is used:

$$n_{val} = \frac{4(n+1)\,z_{av}}{V_{UC}} \tag{2.11}$$

where V_{UC} is the unit cell volume and z_{av} is the average number of *formal* valence electrons in a unit cell. The latter for Ti, Zr, Hf, Si, C, Ge, and Sn is 4; for V, Nb, Ta, N, P, and As it is 5; for Al, Ga, In, and Tl, it is 3; for S, it is 6, and so on. It follows that z_{av} for Ti_2AlC is 3.75, for V_4GeN_3 it is 4.875, and so on. Table 2.4 lists the values of n_{val} for all known MAX phases. It is crucial to note that in this book z_{av} is the *formal* valence state of the elements. In Drude's model, there was no rational basis for the choice of z_{av}, especially for the transition metals (Ashcroft and Mermin, 1976). For example, in the Drude model, z_{av} for Nb was assumed to be 1; here it is assumed to be 5. The rationale and evidence for the validity of this choice of z_{av} is presented in Chapters 3–5.

Table 2.4 Summary of electronic parameters derived from the Drude and Sommerfeld models. The last column lists the first report of the MAX phase listed. At the end of the table, values for select MX phases and M elements are also listed for comparison.

	n_{val} 10^{29} (m^{-3})	r_s/r_B	E_F (eV)	k_F ($Å^{-1}$)	v_F ($Mm\ s^{-1}$)	B (GPa)	$\hbar\omega_{p,val}$ (eV)	References
				413 phases				
Ti_4AlN_3	3.76	1.62	18.98	2.07	2.59	76	22.7	Barsoum *et al.* (1999c)
Ta_4AlC_3	3.47	1.67	17.99	2.01	2.52	67	21.8	Manoun *et al.* (2006)
Ta_4GaC_3	3.66	1.64	18.64	2.05	2.56	73	22.4	Etzkorn *et al.* (2009)
$V_4AlC_{2.7}$	4.00	1.59	19.78	2.11	2.64	85	23.4	Etzkorn, Ade, and Hillebrecht (2007b)
Nb_4AlC_3	3.42	1.67	17.82	2.00	2.50	65	21.6	Hu *et al.* (2007a)
				312 phases				
Ti_3SiC_2	3.34	1.69	17.54	1.99	2.48	63	21.4	Jeitschko and Nowotny (1967)
Ti_3GeC_2	3.31	1.70	17.43	1.98	2.48	62	21.3	Wolfsgruber, Nowotny, and Benesovsky (1967)
Ti_3AlC_2	3.02	1.75	16.40	1.92	2.40	53	20.3	Pietzka and Schuster (1994)
Ti_3SnC_2	3.01	1.75	16.36	1.92	2.40	53	20.3	Dubois *et al.* (2007)
Ta_3AlC_2	3.28	1.70	17.397	1.98	2.47	61.5	21.29	Etzkorn, Ade, and Hillebrecht (2007a).
				211 phases				
Ti_2AlN	3.04	1.74	16.47	1.93	2.41	54	20.4	Jeitschko Nowotny, and Benesovsky (1963d)
Ti_2AlC	2.76	1.80	15.44	1.87	2.33	46	19.4	Jeitschko Nowotny, and Benesovsky (1963b)
Ti_2SnC	2.75	1.80	15.41	1.86	2.33	45	19.4	Jeitschko, Nowotny, and Benesovsky (1963b)
Ti_2GeC	3.03	1.75	16.44	1.92	2.41	53	20.4	Jeitschko, Nowotny, and Benesovsky (1963a)
Ti_2GaC	2.72	1.81	15.29	1.86	2.32	45	19.3	Jeitschko, Nowotny, and Benesovsky (1964c)

(continued overleaf)

Table 2.4 *(Continued)*.

	n_{val} 10^{29} (m^{-3})	r_s/r_B	E_F (eV)	k_F ($Å^{-1}$)	v_F ($Mm\ s^{-1}$)	B (GPa)	$\hbar\omega_{p,val}$ (eV)	References
Ti_2GaN	3.09	1.73	16.65	1.94	2.42	55	20.6	Jeitschko, Nowotny, and Benesovsky (1964c)
Ti_2SC	3.59	1.65	18.40	2.04	2.55	71	22.2	Kudielka and Rohde (1960)
Ti_2InC	2.51	1.86	14.50	1.81	2.26	39	18.5	Jeitschko, Nowotny, and Benesovsky (1963a)
Ti_2InN	2.81	1.79	15.63	1.88	2.35	47	19.6	Jeitschko, Nowotny, and Benesovsky (1964c)
Ti_2PbC	2.61	1.83	14.88	1.83	2.29	42	18.9	Jeitschko, Nowotny, and Benesovsky (1964b)
Ti_2TlC	2.50	1.86	14.46	1.80	2.26	39	18.5	Jeitschko, Nowotny, and Benesovsky (1964b)
V_2AlC	3.54	1.66	18.23	2.03	2.53	69	22.0	Nowotny (1970)
V_2GeC	3.77	1.62	19.01	2.07	2.59	77	22.7	Jeitschko, Nowotny, and Benesovsky (1963c)
V_2GaC	3.56	1.65	18.30	2.03	2.54	70	22.1	Nowotny (1970)
V_2GaN	3.47	1.67	17.99	2.01	2.52	67	21.8	Nowotny (1970)
V_2AsC	4.01	1.59	19.81	2.11	2.64	85	23.4	Boller and Nowotny (1966)
V_2PC	4.24	1.56	20.56	2.15	2.69	93	24.1	Boller and Nowotny (1968)
Cr_2AlC	4.19	1.57	20.40	2.14	2.68	92	24.0	Jeitschko, Nowotny, and Benesovsky (1963b)
Cr_2GeC	4.39	1.54	21.04	2.18	2.72	99	24.5	Jeitschko, Nowotny, and Benesovsky (1963c)
Cr_2GaC	4.20	1.57	20.43	2.15	2.68	92	24.0	Jeitschko, Nowotny, and Benesovsky (1963c)
Cr_2GaN	4.41	1.54	21.11	2.18	2.73	100	24.6	Beckmann *et al.* (1969)
Mo_2GaC	3.76	1.62	18.98	2.07	2.59	76	22.7	Toth (1967)
Nb_2AlC	2.96	1.76	16.18	1.91	2.39	51	20.1	Schuster and Nowotny (1980)
Nb_2GaC	2.96	1.76	16.18	1.91	2.39	51	20.1	Jeitschko, Nowotny, and Benesovsky (1964c)
Nb_2SnC	2.87	1.78	15.85	1.89	2.36	49	19.8	Jeitschko, Nowotny, and Benesovsky (1964b)
Nb_2InC	2.72	1.81	15.29	1.86	2.32	45	19.3	Jeitschko, Nowotny, and Benesovsky (1964b)
Nb_2AsC	3.37	1.69	17.64	1.99	2.49	64	21.5	Beckmann, Boller, and Nowotny (1968)
Nb_2PC	3.55	1.66	18.27	2.03	2.54	69	22.0	Beckmann, Boller, and Nowotny (1968)
Nb_2SC	3.76	1.62	18.98	2.07	2.59	76	22.7	Beckmann, Boller, and Nowotny (1968)
Hf_2SC	3.07	1.74	16.58	1.93	2.42	55	20.5	Kudielka and Rohde (1960)
Hf_2SnC	2.33	1.91	13.79	1.76	2.20	34	17.9	Jeitschko, Nowotny, and Benesovsky (1963b)
Hf_2SnN	2.51	1.86	14.50	1.81	2.26	39	18.5	Nowotny (1970)
Hf_2InC	2.16	1.95	13.12	1.72	2.15	30	17.2	Jeitschko, Nowotny, and Benesovsky (1963a)
Hf_2PbC	2.03	2.00	12.58	1.68	2.10	27	16.7	Jeitschko, Nowotny, and Benesovsky (1964a)

Hf_2TlC	2.27	1.92	13.56	1.75	2.18	33	17.6	Jeitschko, Nowotny, and Benesovsky (1964a)
Ta_2AlC	3.02	1.75	16.40	1.92	2.40	53	20.3	Schuster and Nowotny (1980)
Ta_2GaC	3.01	1.75	16.36	1.92	2.40	53	20.3	Jeitschko, Nowotny, and Benesovsky (1964b)
Ta_2GaN	3.47	1.67	17.99	2.01	2.52	67	21.8	Jeitschko, Nowotny, and Benesovsky (1964b)
Zr_2SnC	2.25	1.93	13.48	1.74	2.18	32	17.6	Nowotny (1970)
Zr_2SC	2.96	1.76	16.18	1.91	2.39	51	20.1	Kudielka and Rohde (1960)
Zr_2InC	2.08	1.98	12.79	1.70	2.12	28	16.9	Nowotny (1970)
Zr_2InN	2.33	1.91	13.79	1.76	2.20	34	17.9	Jeitschko, Nowotny, and Benesovsky (1964c)
Zr_2PbC	2.21	1.94	13.32	1.73	2.17	32	17.4	Jeitschko, Nowotny, and Benesovsky (1964a)
Zr_2TlC	2.07	1.98	12.80	1.98	2.47	61	16.9	Jeitschko, Nowotny, and Benesovsky (1964a)
M elements and MX compounds								
TiC	4.00	1.59	19.78	2.11	2.64	85	23.4	—
TiN	4.70	1.51	22.02	2.23	2.78	112	25.4	—
$TiN_{0.8}$	4.17	1.57	20.33	2.14	2.68	91	23.9	—
Ti	17.6	1.93	13.52	1.75	2.18	33	17.6	Some of these values may be different from those assumed by Drude because here the formal valence state was assumed in all cases.
V	22.2	1.65	18.44	2.04	2.55	71	22.2	
Cr	25.8	1.49	22.55	2.25	2.82	120	25.8	
Nb	19.5	1.80	15.52	1.87	2.34	46	19.5	
Hf	15.7	2.08	11.57	1.61	2.02	22	15.7	

Knowing n_{val}, the radius of the sphere r_s whose volume is equal to the volume per conduction electron given by

$$r_s = \left(\frac{3}{4\pi n_{\text{val}}}\right)^{\frac{1}{3}} \tag{2.12}$$

can be determined. According to the free electron model, r_s/r_B, where r_B is the Bohr radius (0.529×10^{-10} m), can then be used to estimate several properties of metallic conductors. The r_s/r_B values obtained herein (see column 3 in Table 2.4) are in good agreement with those for other metallic conductors. For example, the values of r_s/r_B, for the transition metals range from 1.5 to $\approx$2 (see the bottom of Table 2.4).

Once r_s/r_B is known, the Fermi level energy E_F, in electron volts, is given by (Ashcroft and Mermin, 1976)

$$E_F = \frac{50.1}{(r_s/r_B)^2} \tag{2.13}$$

The Fermi wave vector, in Å^{-1}, is given by

$$k_F = \frac{3.36}{r_s/r_B} \tag{2.14}$$

The Fermi velocity v_F, in meters per second, is given as

$$v_F = \frac{4.2 \times 10^6}{r_s/r_B} \tag{2.15}$$

The bulk modulus B, in gigapascals, is given as

$$B = 10\left(\frac{6.13}{r_s/r_B}\right)^5 \tag{2.16}$$

And lastly, the plasma frequency $\omega_{\text{p,val}}$ is given by

$$\hbar\omega_{\text{p,val}} = \hbar\sqrt{\frac{n_{\text{val}}}{\varepsilon_o m_e}} \tag{2.17}$$

The values of E_F, k_F, v_F, B, and $\hbar\omega_{\text{p,val}}$ in electron volts calculated for the MAX and MX phases, as well as select M elements are listed in Table 2.4.

This exercise begs the question: how valid or useful are the results listed in Table 2.4? The most compelling evidence for the validity of using n_{val} is presented in Chapter 5, where the optical properties of the MAX phases are discussed. In brief, Figure 5.13a shows that the plasma energies of the MAX phases derived from Eq. (2.17) are in remarkably good agreement with those measured by electron energy loss spectroscopy (EELS). In this chapter, the discussion will be restricted to other properties.

The average for v_F (column 6 in Table 2.4) for the MAX phases is $\approx 2.4 \pm 0.2$ Mm s^{-1}. Not surprisingly – since the same simple Drude model applies – these values are typical for the transition metals that comprise the MAX phases as evidenced by the last five entries in Table 2.4. More importantly, in Chapter 3, v_F is used in the Bohm–Staver relationship to yield very reasonable values for the velocities of sound in the MAX phases.

As another check, it can be shown (Ashcroft and Mermin, 1976) that the ratio of the Debye temperature θ_D to the Fermi temperature T_F – which is equal to E_F/k_B – should be roughly equal to the ratio of the velocity of sound v_L in the solid to v_F. For Ti_3SiC_2, θ_D, T_F, v_L, and v_F are, respectively, 700 K, 2×10^5 K, 9000 m s^{-1}, and 2.5×10^6 m s^{-1}. It follows that, indeed, $\theta_D/T_F \approx v_L/v_F \approx 3.5 \times 10^{-3}$.

The bulk moduli B listed in Table 2.4 are roughly half the measured values (Table 3.5). Nevertheless, the agreement between the predictions of this simple model and experiment is as good as, or better, than the agreement obtained for the early transition metals (Table 2.4).

As noted above, it is worth reemphasizing that some of Drude's assignment of the z_{av} values for some elements was done with no justification. The results shown in the preceding text show that the formal valences of, at least, the early transition metals should, and can, be used in the Drude and/or Sommerfeld models. It follows that, when predicting E_F, k_F, v_F, T_F, B, and $\hbar\omega_{p,val}$, it is appropriate to include all the valence electrons. It is for this reason that when, say, the elastic moduli of the MAX phases are plotted as a function of n_{val} – or its near equivalent, the average number of valence electrons per formula unit, z_{av} – decent correlations are obtained, for the most part (e.g., see Figure 3.5a). Another example is the good correlation between the experimental density of states (DOS) at E_F, namely, $N(E_F)$, and z_{av} (Figure 4.17b). When dealing with transport properties (Chapter 5), however, the appropriate number of electrons to use is about two orders of magnitude lower.

2.4.2 The MX Compounds

Before delving into the details of the *ab initio* calculations for the MAX phases, it is instructive to review the electronic structures of the stoichiometric MX phases (Calais, 1977; Cottrell, 1995; Fernandez-Guillermet, Haglund, and Grimvall, 1992; Haglund *et al.*, 1991; Neckel, 1990; Schwarz, 1987), in which the bonding, as in the MAX phases, is a mixture of metallic, covalent, and ionic. Each is discussed separately below.

2.4.3 Covalent Bonds

Three main types of covalent bonds have been identified in the MX compounds. The lobes of the M t_{eg} orbitals extend toward the neighboring nonmetal atoms and can form pd_σ bonds with the 2p orbitals of the latter (Figure 2.9a). The lobes of the M t_{2g} orbitals are also capable of forming pd_π bonds with the 2p orbitals of the adjacent nonmetal atoms (Figure 2.9b). These interactions are mostly bonding. Lastly, the same lobes can form metal–metal dd_σ bonds with t_{2g} orbitals of adjacent M atoms (Figure 2.9c). The latter type of bonding occurs primarily at energies above E_F for TiC, and near E_F for TiN and the MAX phases.

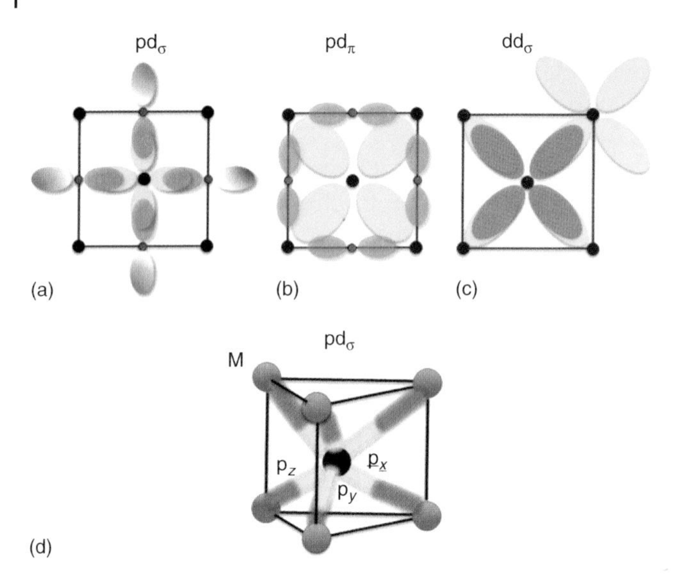

Figure 2.9 Schematic representation of (a) pd$_\sigma$ bonds between M te$_g$ and nonmetal 2p orbitals in the (100) plane of the NaCl structure, (b) pd$_\sigma$ bonds between M t$_{2g}$ and nonmetal 2p orbitals and (c) dd$_\sigma$ bonds between M–M t$_{2g}$ orbitals. In all cases, the X atoms are located at the center of each edge. (d) pd$_\sigma$ bonds between d orbitals of M and p orbitals of A atoms in the MAX phases.

When the experimental and theoretical results of the binary 3d-transition-metal carbides and nitrides are analyzed (Haglund *et al.*, 1991), striking regularities in the variation of the cohesive energies E_{coh} with z_{av} were found (Table 2.5). In their paper, Hagelund *et al.* plotted their results versus z_{av}. A better overall correlation, however, is obtained when their results – both experimental and theoretical – are plotted as a function of r_s/r_B instead (Figure 2.10). From this plot, it is obvious that higher r_s/r_B values result in higher cohesive energies, or more stable compounds.

Table 2.5 Summary of experimental and theoretical cohesive energies (kJ mol^{-1}) of select MX compounds. Also listed are the lattice parameters *a* (Å), r_s/r_B, and z_{av}. All data except r_s/r_B were taken from Fernandez-Guillermet, Haglund, and Grimvall (1992) and Haglund *et al.* (1991).

MX	z_{av}	E_{coh} (theoretical)	E_{coh} (experimental)	*a* (Å)	r_s/r_B	MX	z_{av}	E_{coh} (theoretical)	E_{coh} (experimental)	*a* (Å)	r_s/r_B
TiC	4.0	882	690	4.31	1.59	TiN	4.5	840	646	4.26	1.52
VC	4.5	841	669	4.19	1.49	VN	5.0	761	602	4.16	1.42
CrC	5.0	718	559	4.13	1.42	CrN	5.5	613	496	4.17	1.38
ZrC	4.0	896	765	4.72	1.74	ZrN	4.5	854	726	4.60	1.63
NbC	4.5	878	796	4.49	1.59	NbN	5.0	793	723	4.41	1.51

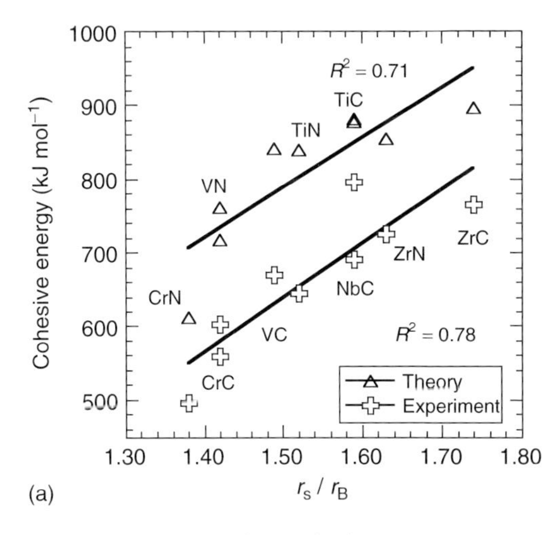

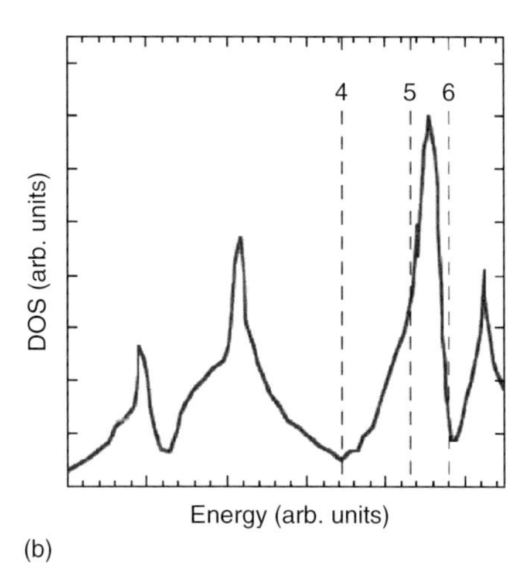

Figure 2.10 (a) Dependence of cohesive energies, both theoretical and experimental, of the MX compounds on r_s/r_B. Least squares fit of the results yielded the lines shown with the R^2 values indicated on plot. (b) Rigid band model for MX compounds. The dotted lines indicate the positions of the DOS curve at which the average valence charge z_{av} takes the values 4.0, 5.0 and 6.0 electrons per atom (Table 2.3) (Haglund *et al.*, 1991). In (a), only the MX chemistries that exist for the MAX phases are included.

Note that higher r_s/r_B values imply a *lower* density of the "free" electrons. In other words, the *lower* the free electron density, the more stable the compound. One explanation for this result is to invoke Heisenberg's uncertainty principle, which states that confining electrons to smaller volumes increases their kinetic energy, which in turn would reduce E_{coh}. A similar pattern would be expected for the MAX phases. Some evidence for this state of affairs is presented below when the surface energies of various MAX phases are compared (see section 2.8).

A complementary, and arguably more sophisticated, view of the same problem – that may iron out some of the scatter in Figure 2.10a – is to follow the arguments of Haglund *et al.* (1991). They adopted a rigid band model, wherein the density of states, DOS, versus energy plot is fixed and rigid, and the cohesive energy of the crystal is partially determined by the location of E_F (Figure 2.10b). As the number of electrons increases, they are pushed into higher energy orbitals that tend to become nonbonding/antibonding, destabilizing the crystal. This phenomenon will henceforth be referred to as *destabilization* on account of having "too many electrons."

2.4.4 Ionic Bonds

The ionic contribution to the bonding results from a partial charge transfer from the M to the X atoms or vice versa. For example, FLMTO calculations on TiC and

TiN concluded that the following charge transfers occur in TiC: Ti, −0.36e; C + 0.43e; in TiN; Ti: −0.32e; N, +0.46e (Neckel, 1990).

2.4.5
Metallic Bonds

For most of the early transition-metal carbides and nitrides, the DOS at E_F, that is, $N(E_F)$, is not insignificant (Figure 2.10b), which is why again for the most part – TiC being a notable exception – the MX phases are excellent conductors of electricity and heat.

2.4.6
The MAX Phases

The first *ab initio* calculation on a MAX phase was published in 1998 on Ti_3SiC_2 (Medvedeva *et al.*, 1998). Since then, the number of such papers has risen dramatically, with many groups around the world tackling the problem. (For a partial list of who is doing what, the reader is referred to Tables 3.1, 3.5, and 4.5.) Not surprisingly, there are many similarities between the electronic structures of the MAX phases and their corresponding binaries. There are, however, some important differences.

In the following, the electronic structures of the MAX phases are reviewed starting with the 211 phases. Before doing so, however, it is important to note that by far the three most important conclusions of the theoretical calculations, valid for most MAX phases, are:

- Similar to the MX compounds, the p–d interactions between the M and X atoms are quite strong. The states located approximately −2 to −5 eV below E_F result from hybridization between the M d orbitals and the X 2p orbitals, leading to strong covalent bonds.
- The DOS at E_F is dominated by the d orbitals of the M atoms.
- For the most part, the interactions between the d electrons of the M atoms and p electrons of the A atoms, around −1 eV relative to E_F, are weaker than those between the M and X atoms.

We now look at specific cases that support these conclusions, starting with the 211 phases.

2.4.7
211 Phases

To gain an understanding of the electronic structure of the 211 phases, it is instructive to refer to some typical results. Figure 2.11a,b plots the partial DOS for Ti_2AlC and Ti_2AlN, respectively. Figure 2.11c compares the total DOS for the two compounds. Figure 2.12 plots the local DOS for Zr_2InC. Figure 2.13 compares the DOS of Ti_2AlC, V_2AlC, and Cr_2AlC. Note that in these, and all other such plots, the location of E_f is at zero energy.

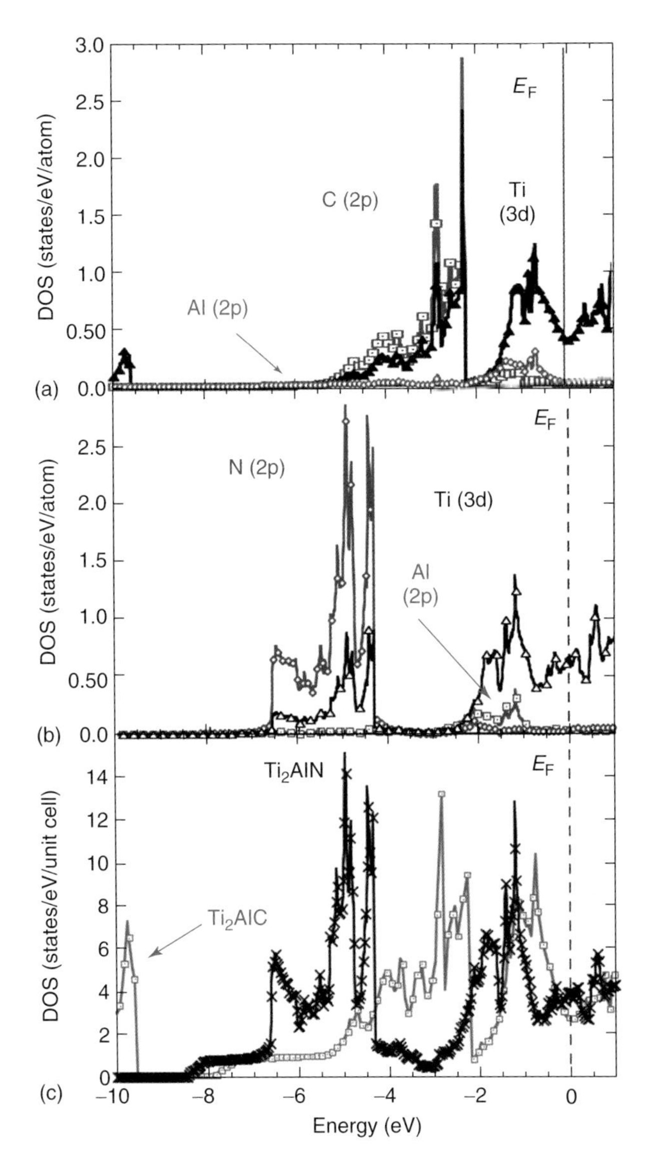

Figure 2.11 Density of states: (a) Ti_2AlC (local), (b) Ti_2AlN (local), (c) Ti_2AlC (total), and Ti_2AlN (total) (Hug, Jaouen, and Barsoum, 2005).

Referring to these figures the following general conclusions can be drawn:

- The p–d interactions between the M and X atoms are quite strong, and comparable to the ones in the MX phases. The states located approximately −2 to −5 eV below E_F result from hybridization between the M d orbitals and the X 2p orbitals, leading to strong covalent bonds. Since they occur at lower energies, it is reasonable to conclude that the Ti–N bonds are stronger than the Ti–C bonds.

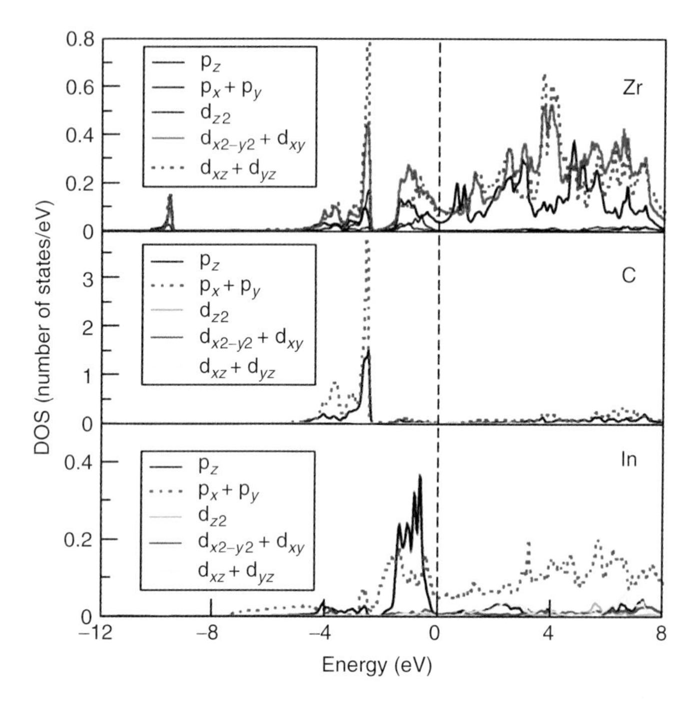

Figure 2.12 Local density of states for Zr, In, and C in Zr_2InC (Jurgens *et al.*, 2011).

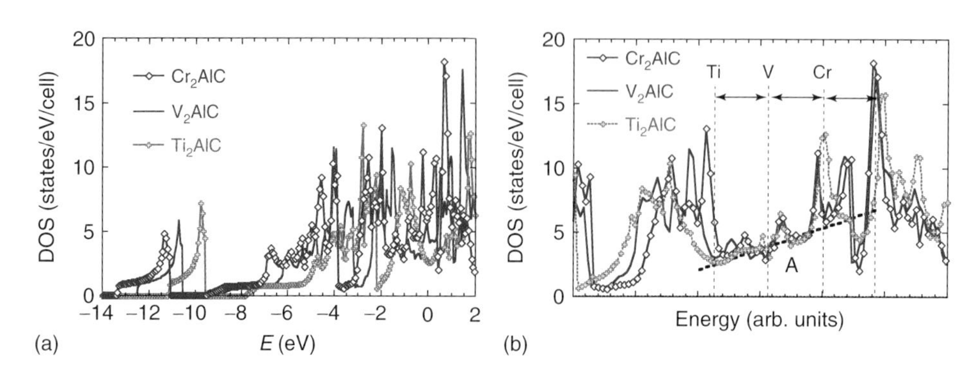

Figure 2.13 (a) Total density of states for Ti_2AlC, V_2AlC, and Cr_2AlC. (b) Same as (a), but now the various curves are shifted to align their various features. The resulting master curve – which only covers the −4 to +2 eV range – can be used as a rigid band model for the MAX phases: that is, it is the equivalent of Figure 2.10b for the MX phases. The Fermi level location for each compound is denoted by a vertical dashed line. The dashed line labeled (A) shows that, if the small fluctuations in the DOSs are ignored, the latter increases more or less linearly with increasing number of electrons.

- The bonds that form between the M and A atoms are pd_σ and are most probably arranged in space as shown in Figure 2.9d. For the most part, the interactions between the d electrons of the M and p electrons of the A atoms, around −1 eV relative to E_F, are weaker than those between the Ti and C atoms. This conclusion is valid even when the A element is much larger than Al. For example, in Zr_2InC, the In d electrons do not appear to play a role in bonding (Figure 2.12).
- There are little to no A–A or X–X bonds.
- In *all* cases, $N(E_F)$ is dominated by the d orbitals of the M atoms.
- When the total DOS of Ti_2AlC, V_2AlC, and Cr_2AlC (Figure 2.13) are compared, the utility of a rigid band model becomes apparent. This is best seen by shifting the various curves in Figure 2.13a to align their features. When so shifted (Figure 2.13b), the equivalence between this picture and that of Figure 2.10b for the MX compounds should be apparent. Note that Cr_2AlC falls near a local maximum.

2.4.8
Density of States at E_F, $N(E_F)$

In general, the theoretical $N(E_F)$ increases with increasing n_{val} (Figure 2.14a). Least squares fitting of all the data points – *except* that of Ti_2SC – shown in Figure 2.14a yields an R^2 value of 0.85. Interestingly, when the theoretical $N(E_F)$ values are plotted versus z_{av} (not shown), the R^2 value is 0.64: that is, the fit is worse than when plotted against n_{val}. This is important because it signifies that it is

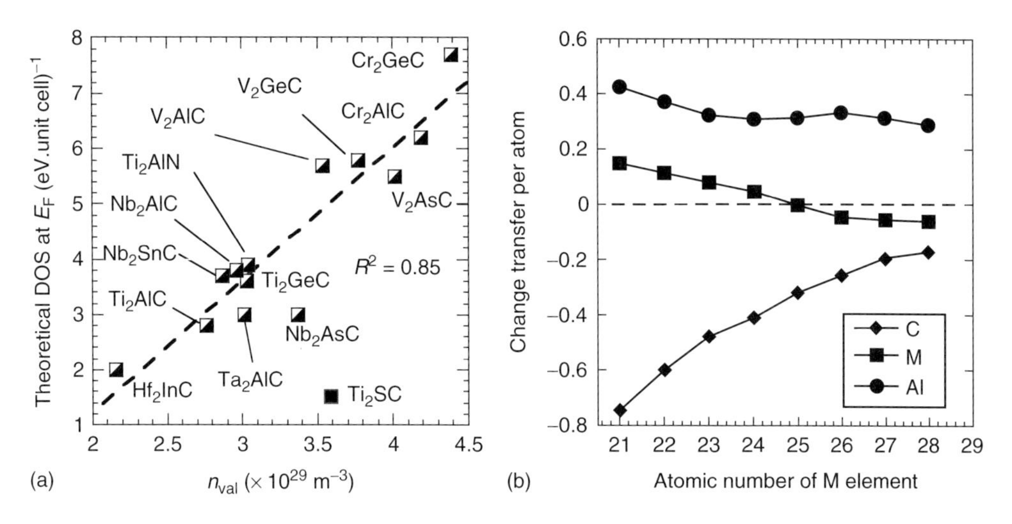

Figure 2.14 (a) Effect of n_{val} on $N(E_F)$ calculated from density functional theory. The R^2 value of all data points, except Ti_2SC, is 0.85. (b) Charge transfer in electrons/atom for the M_2AlC phases as a function of the atomic number of the M element. For the MAX phases that exist – that is, up to Cr – there is a net charge transfer from the Al to the C layers (Hug, Jaouen, and Barsoum 2005).

the concentration of valence electrons *per unit volume*, rather than their average number, that is crucial in determining $N(E_F)$.

The results shown in Figure 2.14a are consistent with those shown in Figure 2.13b, since the dashed line, labeled A in Figure 2.13b, increases more or less linearly with increasing number of electrons. In other words, the simple rigid band model, shown in Figure 2.13b, can more or less explain the linear relationship between $N(E_F)$ and n_{val} (Figure 2.14a).

As noted above, and throughout this book, Ti_2SC is a clear outlier. It is tempting to attribute this observation to its anomalously low c parameter (Figure 2.6c) and the intricacies of bonding – which are probably more covalent and less metallic. It is important to note that the Ti_2SC samples are almost certainly substoichiometric in S, which would shift the location of the Ti_2SC point to the left and thus closer to the rest of the points in Figure 2.14. The same is also probably true of Nb_2AsC.

These comments notwithstanding, DFT calculations for Ti_2SC have all shown that E_F for Ti_2SC falls near a pseudogap (Figure 2.15). The same calculations show that the Ti 3d–S 2p hybridization is shifted to −4.0 eV relative to E_F (Figure 2.15),

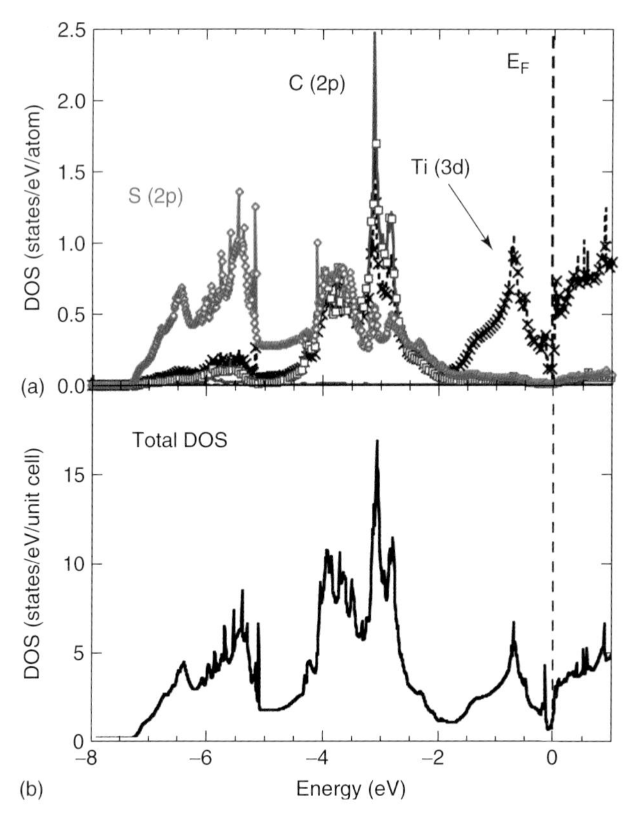

Figure 2.15 (a) Partial and (b) total density of states for Ti_2SC. Note that in this case E_F falls near a pseudogap.

which is lower than the typical Ti–A energies, and closer to the Ti 3d–C 2p hybridization, which suggests that Ti–S bonds in Ti_2SC are stronger than those in most other Ti-containing phases, a result that was anticipated in 2000, based on its low c lattice parameter alone (Barsoum, 2000).

2.4.9
Ionic Component

The charge transfer occurring in the MAX phases has, to date, not been systematically studied. Various researchers have explored various combinations of elements. For example, Hug, Jaouen, and Barsoum (2005) explored the transfer of charge in the M_2AlC phases and found that, in general, there was charge transfer from the Al toward the C; the M elements remained more or less neutral (Figure 2.14b). Note that the values plotted are per atom.

2.4.10
312 Phases

Figure 2.16 plots the partial and total DOS for Ti_3SiC_2. The general conclusions reached in the preceding text for the 211 phases apply here as well and will not be repeated. The partial and total DOS of Ti_3AlC_2 and Ti_3GeC_2 are quite similar to those for Ti_3SiC_2 and are thus not shown.

The theoretical $N(E_F)$ of select 312 phases increases with increasing n_{val}, more or less linearly, with an R^2 value of about 0.76 (Figure 2.17a). A more accurate description of the results, however, suggests that, below $\approx 3.25 \times 10^{29}$ m^{-3}, $N(E_F)$ is a weak function of n_{val}. Above that threshold, $N(E_F)$ increases at a much higher rate. This result can be reconciled with a rigid band model of the DOS. E_F for the MAX phases with n_{val} of $\approx 3.25 \times 10^{29}$ m^{-3} or less, fall in a region where the DOS lies more or less in a shallow valley (He *et al.*, 2010). For those with $n_{val} > 3.25 \times 10^{29}$ m^{-3}, the DOS rises at a faster rate.

In terms of charge transfer, here again, the Ti atoms remain more or less neutral, and about one electron per atom is transferred from the A to the C layers (Figure 2.17b). The transferred charge in this case is not a strong function of composition. Note that this conclusion applies only to the Ti_2AC phases.

2.4.11
413 Phases

The total DOS curves for V_4AlC_3, Nb_4AlC_3, and Ta_4AlC_3 are plotted in Figure 2.18a. The overlap between the C 2p and M d electrons is shown in Figure 2.18b; those of the Al 3p and M d electrons and their location relative to E_F are depicted in Figure 2.18c. Here again, the general conclusions reached for the 211 and 312 phases apply; there are no surprises.

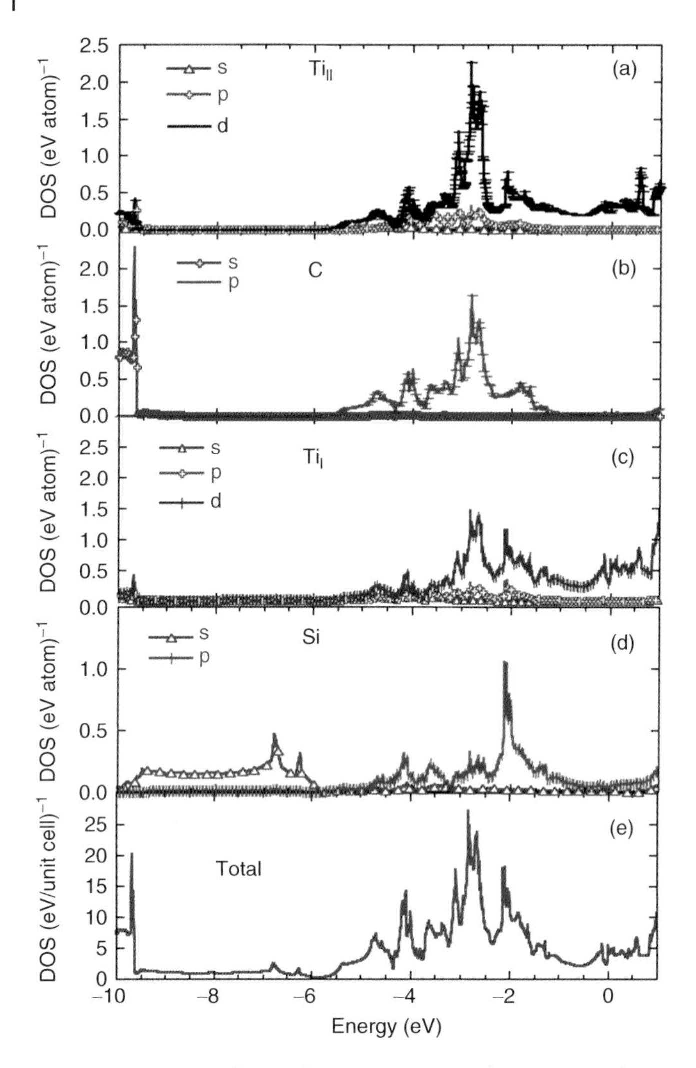

Figure 2.16 Density of states for Ti_3SiC_2, (a) Ti_{II}, (b) C, (c) Ti_I, (d) Si, and (e) total.

In terms of charge transfer, in the case of Ti_4AlN_3, each N atom per formula unit gains ≈0.5 electrons. Concomitantly, each of the Al, Ti_I, and Ti_{II} atoms per formula unit loses ≈0.5, 0.15, and 0.3 electrons, respectively (Jaouen, 2012).

DFT calculations, similar to others, are based on certain assumptions and simplifications. A reasonable question at this time is how good are they, a question that is not trivial to answer. As discussed in the next chapter, these calculations are quite powerful in predicting the elastic constants and Raman frequencies. In the remainder of this section, a few techniques are highlighted that directly or indirectly shed light on the DFT results and their accuracy. More importantly, many of these techniques, - such as X-ray photoelectron spectroscopy (XPS),

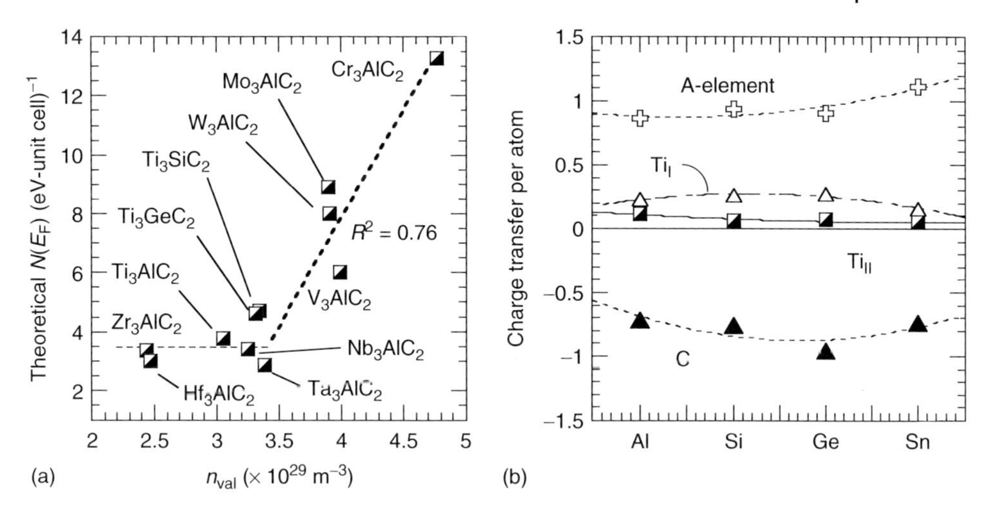

Figure 2.17 (a) Effect of n_{val} on $N(E_F)$ calculated from density functional theory for select 312 phases. The values for the Al-containing 312 phases were taken from He *et al.* (2010). Note most of these phases do not exist. (b) Charge transfer in Ti_3AlC_2, Ti_3SiC_2, Ti_3GeC_2, and Ti_3SnC_2 phases. The lines are guides to the eye (Jaouen, 2012).

X-ray absorption spectroscopy (XAS), X-ray emission spectroscopy (XES), electron energy loss spectroscopy (EELS), perturbed angular correlation (PAC), and nuclear magnetic resonance (NMR) – that are used to shed light on the chemical surroundings of the various atoms, can also be used to characterize these materials.

This section is by no means comprehensive and is included for the sake of completeness. The interested reader should consult the original references cited.

2.4.12 X-Ray Photoelectron Spectroscopy, XPS

Kisi *et al.* (1998) and Medvedeva *et al.* (1998) reported on the XPS studies of the same MAX phase, namely, Ti_3SiC_2. Kisi *et al.* found that the core levels, in general, had lower binding energies than in the related carbides TiC and SiC. They also concluded that all three species were exceptionally well screened, which they attributed to its high electrical conductivity.

To further highlight the chemical similarities of the MX bonds in the ternaries and binaries, refer to Table 2.6 in which the binding energies of the various elements obtained from XPS are summarized. Not surprisingly, the binding energies of the C atoms in the 312 structures are quite comparable to each other and to those in stoichiometric TiC. The same is true for the binding energies of the N atoms in $Ti_4AlN_{2.9}$ and stoichiometric TiN.

Stoltz, Starnberg, and Barsoum (2003) were the first to use synchrotron radiation in the study of XPS of Ti_3SiC_2. To recapitulate the DFT calculations for this

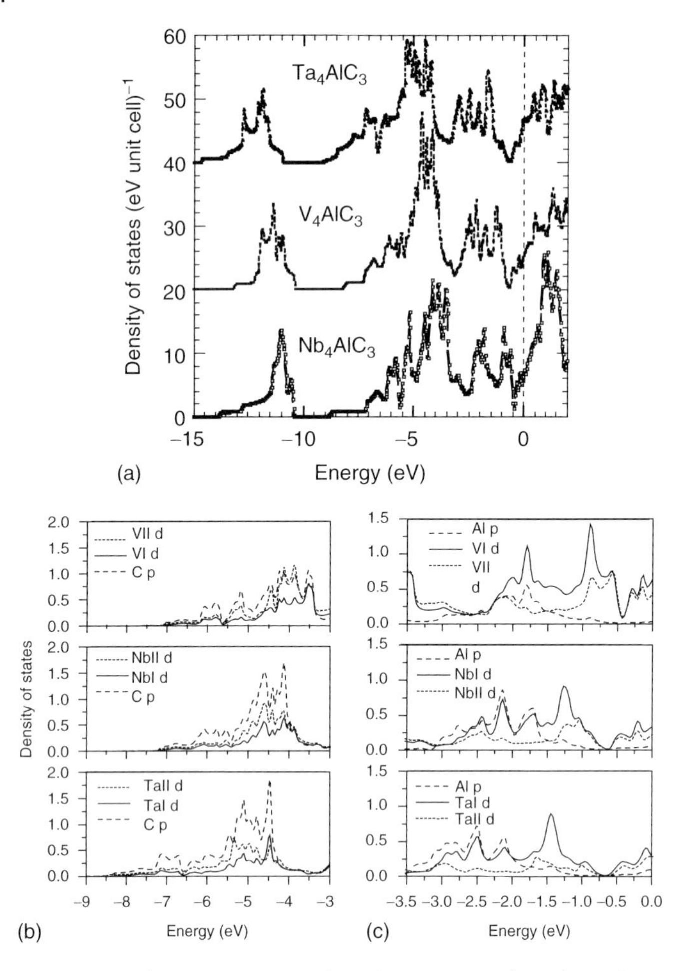

Figure 2.18 Electronic structure of M_4AlC_3 (M = V, Nb, and Ta) phases. (a) Total density of states (DOS). (b) Partial DOS showing M–C overlap. (c) Partial DOS showing M–Al overlap (Du *et al.*, 2009).

compound (Figure 2.16b), the bands at and above the E_F are mainly of Ti 3d character but with a significant Si 3p contribution. The range 1–5 eV below the E_F is the most mixed, with Ti 3d, Si 3p, and C 2p orbitals as the main contributors. In contrast, the range 5–9 eV below the Fermi edge is almost exclusively of Si 3s character, while C 2s dominates in the lowest part, 9–12 eV below E_F. Figure 2.19a compares the experimental valence band spectra measured by Stotlz *et al.* with those calculated by Ahuja *et al.* (2000). While there are some discrepancies, it is reasonable to conclude that the agreement is good, and suggests that state-of-the-art band structure calculations for Ti_3SiC_2, and by extension the other MAX phases, are indeed relevant and useful.

Table 2.6 Binding energies for select MAX phases and related compounds.

	A	M (2p)	C/N (1s)	References
			Carbides	
Ti_3SiC_2	98.5/102.3	454.0	281.0	Kisi *et al.* (1998)
Ti_3SiC_2	99.5	454.9	282.1	Medvedeva *et al.* (1998)
Ti_3SiC_2	98.9	453.7/459.8	281.6	Riley *et al.* (2002)
Ti_3SiC_2	98.91/99.52	454.5/460.6	281.8	Stoltz, Starnberg, and Barsoum (2003)
Ti_3AlC_2	71.9 (2p)	454.5	281.5	Myhra, Crossley, and Barsoum (2001)
Ti_2InC	443.6 (3d)	458.7	281.1	Barsoum, Crossley, and Myhra (2002)
TiC	—	454.4	281.6	Briggs and Seah (1990)
V_2GeC	28.6 (3d)	512.5 (2p)	282.0	Barsoum, Crossley, and Myhra (2002)
VC	—	512.8	—	(Wagner *et al.*, 2003)
Nb_2AsC	40.5 (3d)	202.1 (3d)	281.5	Barsoum, Crossley, and Myhra (2002)
NbAs	40.8 (2p)	—	—	Briggs and Seah (1990)
Hf_2InC	70.9 (2p)	13.0 (4f)	281.7	Barsoum, Crossley, and Myhra (2002)
			Nitrides	
Ti_4AlN_3	72.0	454.7	396.9	Myhra, Crossley, and Barsoum (2001)
TiN	—	455.5	396.7	(Wagner *et al.*, 2003)

2.4.13 X-Ray Absorption and Emission Studies of MAX-Phases

The electronic structure and chemical bonding can also be probed using XAS, XES, and resonant inelastic X-ray scattering (RIXS). These studies include electronic structure investigations of the spectral shapes of the A layers, namely, Al, Si, and Ge, in respectively Ti_3SiC_2, and Ti_3GeC_2 – which are strongly modified by hybridization with the neighboring Ti atoms in comparison to, say, the corresponding pure elements (Magnuson *et al.*, 2005). In that paper, and despite the fact that SXE is a site-selective spectroscopy, the Ti_I and Ti_{II} atoms were not experimentally resolved because of their negligible energy difference.

When Ti_2AlC was compared to TiC (Magnuson *et al.*, 2006b), three different types of bond regions in the spectra were identified: (i) relatively weak M 3d–A 3p bonds, (ii) M 3d–C 2p bonds, and (iii) M 3d–C 2s bonds. In agreement with the DFT calculations, the latter two were stronger and lower in energy relative to E_F. This was also the case for Ti_4SiC_3 (Magnuson *et al.*, 2006a). When Ti_2AlC, Ti_2AlN, and TiN were compared, the electronic structure and chemical bonds were found to be significantly different in the nitrides than in Ti_2AlC, with more

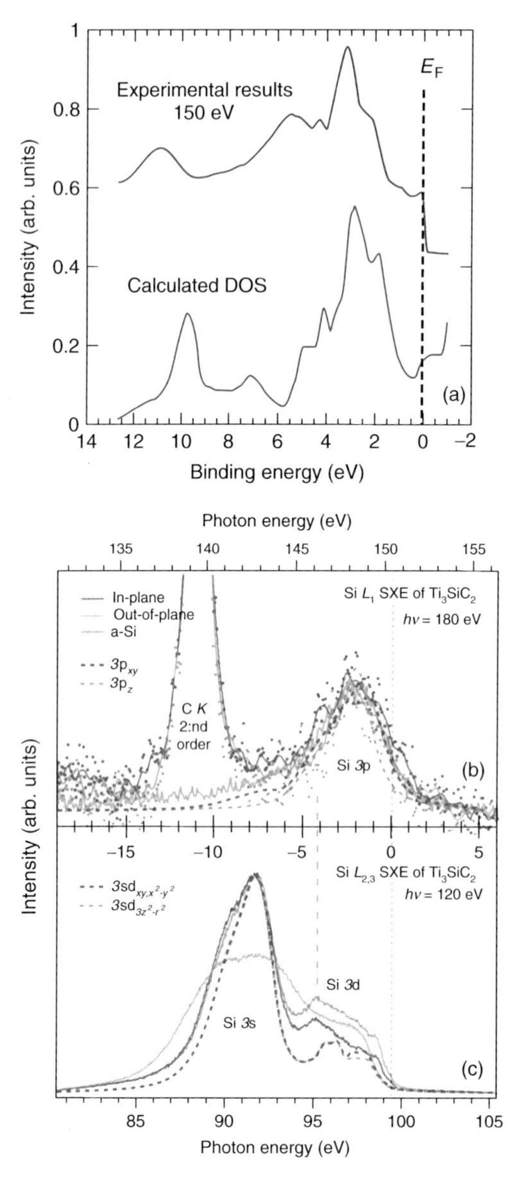

Figure 2.19 (a) Comparison of the valence band spectrum determined from XPS to that calculated by Ahuja *et al.* (2000). The DOS curve was convoluted with a Gaussian of 0.35 eV FWHM (full-width at half-maximum; Stoltz, Starnberg, and Barsoum, 2003). (b) Si L_1 SXE spectra of Ti_3SiC_2 in comparison to pure amorphous Si (a-Si). Dashed curves correspond to calculated $3p_{xy}$ and $3p_z$ spectra. (c) Si $L_{2,3}$ SXE spectra in comparison to a-Si. Dashed curves correspond to calculated $3sd_{xy}$,$_{x2-y2}$, and $3sd_{3z2}$ spectra (Magnuson *et al.*, 2012).

separated and deeper bond regions leading to stronger bonds (Magnuson *et al.*, 2007).

The anisotropy in the electronic structure of V_2GeC was also investigated (Magnuson *et al.*, 2008), demonstrating that it was possible to obtain information about the symmetry and anisotropy of the electronic structure using polarization-dependent Ge M_1 and Ge M_{23} X-ray spectroscopy. The anisotropy in the spectra, due to the different in-plane and out-of-plane bonding orbitals of V_2GeC, was found to be large and sensitive to the local coordination of the V atoms in plane and out of plane.

Quite recently, by employing angular-dependent XAS and RIXS, it was also possible to probe the electronic states of thin, epitaxial Ti_3SiC_2 films at grazing and near normal incidence angles using synchrotron radiation (Magnuson *et al.*, 2012). As shown in Figure 2.19b,c, it is possible to differentiate the out-of-plane p_z and d_{3z2} states from the in-plane p_{xy}, d_{xy}, d_{x2-y2} states. As discussed in Chapter 5, these measurements shed light on the origin of the anisotropy of the Seebeck coefficients in Ti_3SiC_2.

While the agreement between theory and experiment is not perfect, in general it is fair to conclude the XAS and XES results are consistent with those predicted from DFT calculations.

2.4.14
Electron Energy Loss Spectroscopy, EELS

Hug, Jaouen, and Barsoum (2005) were the first to use EELS to shed light on the electronic structure of Ti_2AlC, Ti_2AlN, Nb_2AlC, and TiNbAlC. The lattice parameters and relaxations obtained from the *ab initio* calculations were in good agreement with those deduced from the analysis of the experimental data. For the solid solution TiNbAlC, the most salient result was that the basal planes were corrugated at the atomic level.

By analyzing the C K edge of the EELS spectra, Lin *et al.* (2006) concluded that in Cr_2AlC, a sigma bond existed between the Cr and C atoms, in agreement with the DFT calculations. More recently, EELS was used to probe the anisotropy in dielectric properties of the MAX phases (Mauchamp *et al.*, 2010).

2.4.15
Perturbed Angular Correlation

In this technique, radioactive ^{111}In atoms were implanted in the MAX phases and their decay was used to extract the local electric field at which the embedded atoms were residing. (Jürgens *et al.*, 2007, 2010, 2011). In a series of papers, Jürgens *et al.* have convincingly shown that the implanted ^{111}In atoms end up, not too surprisingly, on the A sites. Here again, good agreement was found between DFT predictions and the experimental results (Jürgens *et al.*, 2011).

2.4.16 Nuclear Magnetic Resonance

As far as we are aware, there is only one paper on NMR of the MAX phases. Lue, Lin, and Xie (2006) carried out ^{27}Al NMR spectroscopy on Ti_2AlC, V_2AlC, and Cr_2AlC. They found that the sign of the isotropic Knight shift changed from positive for Ti_2AlC and V_2AlC to negative for Cr_2AlC, and attributed it to the enhancement of hybridization with increasing number of valence electrons. Universally, long relaxation times were found.

2.5 To Be or Not to Be

For many of those carrying out research on the MAX phases, whether experimental or theoretical, almost inevitably a time comes when the question arises as to why some of them exist and others do not. There are clear compositional and structural gaps. For example, V_2PC and Nb_2PC exist, but not Ti_2PC; the same is true of As-containing phases. Ti_2SC and Nb_2SC exist, but not V_2SC. Similarly, Ti_2SiC and Ti_4SiC_3 do not exist, but Ti_3SiC_2 does and is the *only* Si-containing bulk MAX phase; Ti_2AlN and Ti_4AlN_3 exist, but *not* Ti_3AlN_2. Conversely, Ti_2AlC and Ti_3AlC_2 exist, but not Ti_4AlC_3. The same is true of the $Nb_{n+1}AlC_n$, compounds, and so on.

The only theoretical methodology available at this time to answer the question of being is to carry out careful *ab initio* calculations. Of these there are two types. The first, and easier method, calculates whether a compound is stable on an absolute scale. For example, in Cover *et al.*'s (2009) comprehensive study on 240 M_2AX phases, 17 were found to be intrinsically unstable in that they tended to dissociate into nonhexagonal structures. The unstable phases were V_2SN, Cr_2SN, and a host of Mo- and W-containing MAX phases.

The second, more difficult and time-consuming, approach is to show that a given MAX phase is *more* stable than *all* other competing phases. When done properly, the question to be or not to be answers itself. This approach was applied to select MAX phase recently, and showed excellent agreement with the observed phases. For example, Keast, Harris, and Smith (2009) studied the stability of various phases in five systems (Ti–Al–C, Ti–Si–C, Ti–Al–N, Ti–Si–N, and Cr–Al–C) for $n = 1–3$. Their results are summarized in Table 2.7. More recently, Dahlqvist, Alling, and Rosén (2010) reached similar conclusions and added the results for the V–Al–C system. In both studies and with one exception – Ti_4SiC_3 – the *ab initio* calculations predicted the existence of the phases known to exist and vice versa.

In their comprehensive examination, Cover *et al.* (2009) concluded: "These analyses have, as yet, all failed to reveal any correlations or systematic behavior. In all cases the differences between all of these quantities is very subtle. This would suggest that verification and prediction of phase formation in this set of alloys will

Table 2.7 Summary of DFT calculations for the existence or absence of various MAX phases. X denotes that the phase does not exist. All predictions are correct except Ti_4SiC_3 that is predicted to exist, but does not.

	M_2AX	M_3AX_2	M_4AX_3	References
Ti–Si–N	X	X	X	Keast, Harris, and Smith (2009)
Ti–Si–C	X	√	√	Keast, Harris, and Smith (2009)
Ti–Al–C	√	√	X	Keast, Harris, and Smith (2009)
Ti–Al–N	√	X	√	Keast, Harris, and Smith (2009)
Cr–Al–C	√	X	X	Keast, Harris, and Smith (2009)
V–Al–C	√	√	√	Dahlqvist, Alling, and Rosén (2010)

always require full DFT calculations": a sobering conclusion, but maybe one that is not too bleak if one looks at the problem from the point of view of n_{val}.

2.5.1 Average Number of Valence Electrons, Too Many Electrons, and Covalency

One of the leitmotivs of this book is that the MAX phases can be roughly separated into four subclasses. Referring to Figure 2.6c, they are the following:

- Low *c*-parameter phases, whose *c* parameter is between 11 and 12 Å. These compounds have relatively high n_{val} values and yet are quite stable. They include Ti_2SC, and most probably the P- and some of the As-containing MAX phases.
- Phases whose *c*-lattice parameters lie on the almost vertical line shown in Figure 2.6c, but have values of $n_{val} < 3.25 \times 10^{29}$ m^{-3}. These include most of the Al- and the Ga-containing MAX phases.
- Phases, whose *c*-lattice parameters lie on the almost vertical line shown but with values of $n_{val} > 3.25 \times 10^{29}$ m^{-3}. These MAX phases, in essence, have *too many electrons* and thus have borderline stability. This borderline stability manifests itself, as discussed in Chapters 3–5, sometimes quite dramatically on their properties. The borderline-stable compounds are Cr_2GeC, Cr_2AlC, and, possibly, V_2AsC.
- The MAX phases, which include Hf, Ta, Pb, In, Tl, and Sn. In these phases, the size of the atoms and their mass appear to overshadow other considerations.

In subsequent chapters, and for some properties, this classification will come in handy.

2.6 Distortion of Octahedra and Trigonal Prisms

The *z*-coordinate is important because its deviation from the canonical position is a measure of the distortion of the M–X octahedra. For example, for the

211 phases the distortion parameter α_r is given by Hug, Jaouen, and Barsoum (2005):

$$\alpha_r = \frac{d_1}{d_2} = \frac{\sqrt{3}}{2\sqrt{4z_M^2\left(\frac{c}{a}\right)^2 + \frac{1}{12}}} \tag{2.18}$$

where d_1 and d_2 are defined in Figure 2.20a and z_M is the internal coordinate of the M atoms (see Figure 2.20a). For a perfect cubic octahedron, $\alpha_r = 1$. On the basis of Eq. (2.18), if the M atoms are sitting on their canonical positions, then the c/a ratio that would yield an undistorted octahedron is 4.9.

Similarly, referring to Figure 2.20b, a measure of the distortion of an ideal trigonal prism is given by Hug, Jaouen, and Barsoum (2005) as

$$p_r = \frac{d_{M-M}}{d_{M-A}} = \frac{1}{\sqrt{\frac{1}{3} + \left(\frac{1}{4} - z_M\right)^2 \frac{c}{a}}} \tag{2.19}$$

Here again, an ideal trigonal prism will have $p_r = 1$.

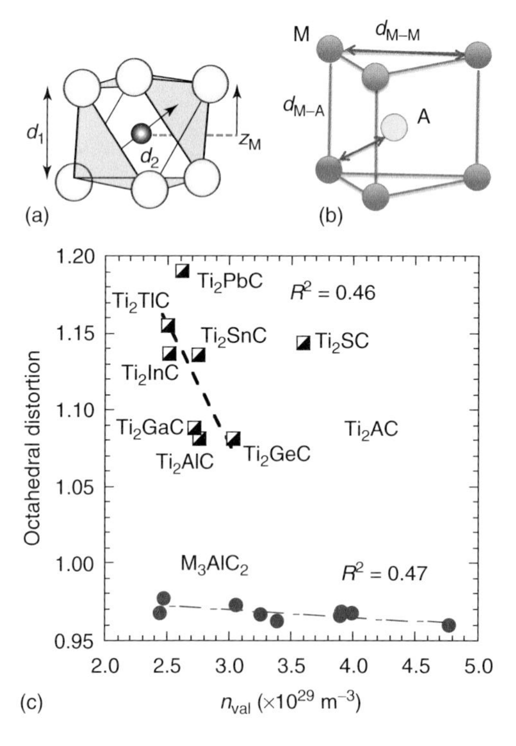

Figure 2.20 (a) Schematic showing definition of d_1, d_2, and z_M in Eq. (2.18). (b) Same as (a) but showing definition of d_{M-M} and d_{M-A} in Eq. (2.19). (c) Evolution of α_r with n_{val} for all known Ti_2AC phases (Hug, 2006) and M_3AlC_2 (He *et al.*, 2010). Most of the latter do not exist. Note anomalous nature of Ti_2SC in (c).

Using *ab initio* calculations, Hug (2006) showed that, at least for the Ti_2AC phases, both a_r and p_r increased as the valence of the A element increased. The two distortions were coupled and tended to mirror each other.

Figure 2.20c replots the octahedral distortions α_r determined by DFT calculations as a function of n_{val} for both the Ti_2AC (Hug, 2006) and M_3AlC_2 (He *et al.*, 2010) phases. This side-by-side comparison shows the following: (i) In both cases, increasing n_{val} decreases α_r. However, for the 211 phases, increasing n_{val} results in less distortion, (i.e. α_r approaches unity) while for the 312 phases the octahedra become more distorted; (ii) Somewhat surprisingly, the A-group element has a much higher effect on α_r than the M elements. (iii) Ti_2SC is an outlier and was not used to calculate the R^2 value of 0.47 for the dashed line shown. The same R^2 value was obtained for the M_3AlC_2 phases. In this case there are no outliers.

Here again, the anomalous nature of Ti_2SC manifests itself. This compound possesses one of the highest z_M values of all the 211 phases (Hug, 2006), and thus its octahedral distortion is large relative to its value of n_{val}. On the basis of Figure 2.20b, had Ti_2SC not been anomalous, the distortion of its octahedra would have been minimal. It is worth noting here that the large distortion is consistent with the presence of strong Ti-Ti bonds across the A-layer that would tend to increase d_1 at the expense of d_2.

Whether, and how, these distortions influence properties remains an open question. At this time, there have been no systematic studies on the role of distortions on properties. This comment notwithstanding, it is reasonable to assume that such distortions would have an effect on thermal expansion and its anisotropy at the very least.

2.7 Solid Solutions

In addition to the list shown in Table 2.1, the number of possible permutations and combinations of solid-solutions is obviously quite large; roughly a quarter of the periodic table can, in principle, come into play. Solid solutions form on the M sites, the A sites, and the X sites, and combinations thereof. For example, a continuous solid solution $Ti_2AlC_{0.8-x}N_x$, where $x = 0$–0.8, occurs at 1490 °C (Pietzka and Schuster, 1996). We confirmed the existence of this solid solution and also synthesized and characterized $Ti_3Al(C_{0.5},N_{0.5})_2$ and $Ti_2Al(C_{0.5},N_{0.5})$ (Manoun *et al.*, 2007a; Radovic, Ganguly, and Barsoum, 2008; Scabarozi *et al.*, 2008). More recently, Cabioch *et al.* (2012) explored the phases in the $Ti_2Al(C_xN_{(1-x)})_y$ system for x from 0 to 1.0 and $0.7 < y < 1.0$. They showed no or little substoichiometry in C when $y = 1$. For $y < 1.0$, at least 20% vacancies were present on the X sites. It is worth noting here that those samples were not necessarily at equilibrium.

Similarly, a continuous solid solution $Ti_3Si_xGe_{1-x}C_2$, $x = 0$–1 exists (Ganguly, Zhen, and Barsoum, 2004). On the M sites, the following solid solutions are known to exist: $(Nb,Zr)_2AlC$, $(Ti,V)_2AlC$, $(Ti,Nb)_2AlC$, $(Ti,Cr)_2AlC$, $(Ti,Ta)_2AlC$, $(V,Nb)_2AlC$, $(V,Ta)_2AlC$, $(V,Cr)_2AlC$ (Salama, El-Raghy, and Barsoum, 2002; Schuster, Nowotny, and Vaccaro, 1980), $(Ti,Hf)_2InC$ (Barsoum *et al.*, 2002b), and $(Ti,V)_2SC$ (Nowotny, Schuster, and Rogl, 1982). More recently, the following phases have been also reported to exist: $Ti_3(Al,Sn)C_2$ and $V_2Ga_{1-x}Al_xC$ (Etzkorn, Ade, and Hillebrecht, 2007b) and $Cr_2(Al_xGe_{1-x})C$ systems (Cabioch *et al.*, 2013).

Yu, Li, and Sloof (2010) found that the amount of Si that can be dissolved into Cr_2AlC was limited to ≈3.5 at% for a limiting composition of $Cr_2Al_{0.96}Si_{0.13}C$.

There is little solubility between B and C in the Ge-containing ternaries (Kephart and Carim, 1998). The generality of this result, and whether it applies to other MAX phases, remains an outstanding question at this time.

Vegard's law applies quite well to most solid-solution compositions explored to date. In the $V_2Ga_{1-x}Al_xC$ (Etzkorn, Ade, and Hillebrecht, 2007b) and $Cr_2(Al_xGe_{1-x})C$ systems (Cabioch *et al.*, 2013), not only the a and c, but also the z parameters follow Vegard's law. Note that the list of compounds listed in this section is by no means exhaustive.

2.8 Defects

Similar to all solids, the MAX phases have point, line, and surface defects. In this section, we review what is known about vacancies, stacking faults (SFs) and the surface energies of the MAX phases. Dislocations are dealt with separately in Chapter 8.

2.8.1 Vacancies

Vacancies are a thermodynamic necessity and, as in all solids, they exist in the MAX phases. Their role can also be important. For example, it is well established in the MX literature that vacancies on the X sites are strong scatterers of both electrons and phonons. As discussed in Chapters 3–5, not surprisingly, the same is true of the MAX phases.

Experimentally, direct evidence for vacancies in the MAX phases is scarce. The first suggestion for the presence of vacancies was our report that the actual chemistry of the first Ti_4AlN_3 sample made was actually $Ti_4AlN_{2.9}$, that is, it was N-deficient (Barsoum *et al.*, 1999c; Rawn *et al.*, 2000). This study, in turn, motivated Music, Ahuja, and Schneider (2005) to use *ab initio* calculations to model vacancies in Ti_4AlN_3. Music *et al.* identified the N 2a sites in $Ti_4AlN_{2.75}$ as the most probable vacancy sites based on the energetics of vacancy formation. The first *ab initio* study of defects in the MAX phases, however, was by Medvedeva *et al.* (1998).

Du *et al.* (2008) also used first principles to examine the formation of C vacancies in Ta_4AlC_3 and found that the introduction of vacancies decreased the stability of the crystal. They identified the 2a site – that is, the C layer at the center of the unit cell in Figure 2.1c – as the most probable location for the C vacancies. This conclusion is in accordance with the results shown in Figure 2.8c.

Using DFT calculations, Liao, Wang, and Zhou (2008) showed that in Ti_2AlC the Ti vacancies, V_{Ti}, had higher energies of formation than the Al vacancies, V_{Al}, or C vacancies, V_C. The energies of migration were calculated to be 0.83, 2.38, and 3.00 eV, respectively, for V_{Al}, V_{Ti}, and V_C, none of which is in the least bit surprising given the known propensity of the A-group elements to selectively react with their surroundings (Chapter 7).

A weakness of much of the work on the MAX phases, including some of our own, is a less than careful accounting of chemistry. The difficulty of quantifying C in EDX and the difficulty of carrying out careful wavelength dispersive spectroscopy (WDS), measurements are two obvious impediments. For the most part, researchers assume that if they obtain clean XRD diffractograms, it follows that the ratios of the elements in the MAX phases are those given by the formal chemistries of the compounds. This is not always the case; in some instances knowing the exact chemistry was key to understanding properties. A good example is the aforementioned study on Ti_4AlN_3, where careful WDS measurements showed it to be N-deficient, with an actual chemistry of $Ti_4AlN_{2.9}$. The effect of this deficiency on properties is discussed in Chapters 3–5.

In general, it is fair to say that little is known about vacancies, where they occur, their energetics, and so on. A related question is the extent of nonstoichiometry of the MAX phases. As a general rule of thumb, if the lattice parameters of a MAX phase vary significantly from report to report – assuming the latter were properly measured – then it is reasonable to assume that this phase exists over a wider range of stoichiometry than, say, a phase in which the lattice parameters do not vary from report to report. It so happens that one of the most stoichiometric MAX phases is Ti_3SiC_2. The lattice parameters reported for this compound appear to be immune to both time and geography. On the other side of the spectrum are Ti_4AlN_3, Ti_2SnC, and Nb_2AlC.

In the Nb–Al–C system – based on small changes in lattice parameters (a = 3.10–3.11 Å; c = 13.86–13.92 Å) – Schuster and Nowotny (1980) postulated the existence of a small homogeneous region consistent with the formula $Nb_{2-x}Al_{1+x}C_{1-x/2}$. To explore this facet, over a dozen different samples centered around the Nb_2AlC composition were hot isostatically pressed (HIPed), at 1600 °C for 8 h (Salama, El-Raghy, and Barsoum, 2002). When the lattice parameters of these samples were measured, a wide scatter (a = 3.096–3.126 Å; c = 13.804–13.88 Å) was observed. Furthermore, WDS and EDS analyses established that these samples were not in equilibrium because various grains had differing compositions, when, based on the phase rule, they should all have been identical. When the samples were further annealed in Ar at 1600 °C for 48 h, many of the samples, especially those that were relatively Al-rich, were found to have visibly reacted with impurities in the Ar atmosphere. The XRD diffractograms did not indicate the emergence of new phases.

However, when the lattice parameters were remeasured, they were indistinguishable from each other. The average and standard deviations of the remeasured a and c parameters were, respectively, 3.107 ± 0.001 and 13.888 ± 0.001 Å. Since the most likely reaction during annealing is the preferential loss of Al from the sample, it was concluded that the aforementioned lattice parameters were those of Al-poor samples. The homogeneity range of Nb_2AlC thus remains an open question at this time and requires more work. Note that the loss of Al led to an increase in the c lattice parameter.

Similarly, by carefully mapping the Nb–Sn–C ternary phase diagram at 1300 °C (Figure 7A4.a), it was concluded that Nb_2SnC was not a line compound but existed over compositions that ranged from a Nb:Sn:C ratio of 54:22:24 at.% when in equilibrium with NbC and Nb_3Sn to 49:28:23 at% when in equilibrium with NbC and Sn (Barsoum *et al.*, 2002a). If one assumes the Nb and Sn sublattices to be fully occupied, these chemistries translate to $Nb_{2.2}Sn_{0.8}C_{0.93}$ and $Nb_{1.92}Sn_{1.08}C_{0.93}$, respectively. Interestingly enough, these chemistries suggest some substitutions of Nb on Sn sites and vice versa.

A more recent example pertains to Ti_3GeC_2 thin films. Buchholt *et al.* (2012) showed that with a chemistry of $Ti_3Ge_{0.8}C_2$, it is more likely than not that Ti_3GeC_2 is substoichiometric in Ge. There are other examples. In general, this is a fruitful area of research that has not been much explored.

2.8.2
Surface Energies

Before reviewing the surface energies of the MAX phases, it is important to define a few terms. Herein, the energy needed to create a surface is termed the *cleavage energy*, γ_{CL}, and is defined as

$$\gamma_{CL} = \frac{E^n_{slab} - n^* E_{bulk}}{2S_o} \tag{2.20}$$

where E^n_{slab} is the total energy of the n layer surface slab and E_{bulk} is the energy of the bulk; S_o is the surface area created. In some of the theoretical work, after the surfaces are created, they are relaxed. The relaxed values of γ_{CL} will be referred to herein as γ_{0001}.

It is important to note that some equate the relaxed γ_{CL} values to the surface energies, which is only partially correct. A more nuanced approach has to take into consideration that when a MAX phase surface is created, two asymmetrical surfaces result, each with a different energy. To illustrate this point, consider the six possible surfaces – shown schematically in Figure 2.21a–f and their notations – that can form when a 312 phase is fractured. For example, when cleavage occurs between the M and A layers, two surfaces are created: $A(M_I)$ (Figure 2.21b) and $M(X_I)$ (Figure 2.21a). The sum of these two surface energies is the cleavage energy given by Eq. (2.20) and listed in Table 2.8.

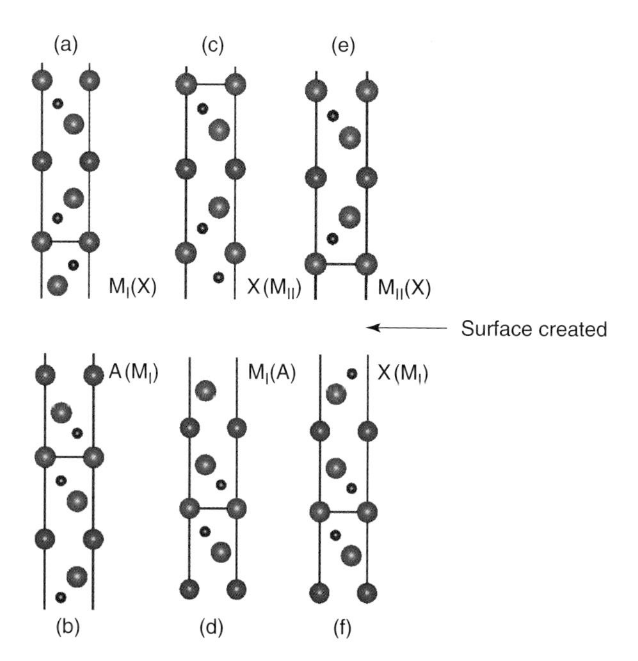

Figure 2.21 Structure models and notations for (0001) surfaces in M_3AX_2: (a) $M_I(X)$, (b) $A(M_1)$, (c) $X(M_{II})$, (d) $M_I(A)$, (e) $M_{II}(X)$, and (f) $X(M_I)$.

2.8.3
Surface Energies of the 211 Phases

Music *et al.* (2006) were the first to calculate γ_{CL} and γ_{0001} of the basal planes in Ti_2AlC, V_2AlC, and Cr_2AlC (Table 2.8). Two years later, Wang, Wang, and Zhou, (2008) reproduced their results; the agreement between the two studies is good. In 2007, Music *et al.* extended the work to other MAX phases (Music *et al.*, 2007). A perusal of the results in Table 2.8 shows that in all cases, the M–A bonds are roughly 50% weaker than the M–X bonds.

As a first approximation, the cleavage energies of a given (*hkl*) plane in any structure are a measure of the strengths of the bonds broken to create the surface. It is thus instructive to plot γ_{0001} when the M–A bonds are ruptured as a function of n_{val} (Figure 2.22). There are several ways to look at these results. The first is to accept that Cr_2GeC is an outlier and draw a line through the remaining points. Least squares fit of such an exercise (inset in Figure 2.22a) yields a straight line with a positive slope and a respectable R^2 value of $\approx$0.7. The second is to focus on the effect of the M element, for a given A, where it is obvious that the relationship is highly dependent on the M atom (Figure 2.22a). For the M_2AlC phases, γ_{0001} increases more or less linearly with increasing n_{val} (solid black line in Figure 2.22a). For the M_2GaC phases, γ_{0001} initially increases before plateauing off (dotted line in Figure 2.22a). In contradistinction, γ_{0001} for the M_2GeC compositions decreases with increasing n_{val}. The third approach is to focus on the effects of the A-group

Table 2.8 Unrelaxed and relaxed cleavage energies of basal planes in select MAX phases. On the basis of these results, it is clear that the surface relaxations are relatively small. Also listed are γ_{111} of MX surfaces estimated as $\sqrt{3}\ \gamma_{100}$. The latter are taken from Hugosson *et al.* (2004).

Solid	Bond	γ_{CL}; unrelaxed ($J\ m^{-2}$)	γ_{0001}; relaxed ($J\ m^{-2}$)	References
		312 phases		
Ti_3SiC_2	Ti_I-Si	1.6	—	Fang *et al.* (2006)
	Ti_I-C	3.1	—	Fang *et al.* (2006)
	$Ti_{II}-C$	3.6	—	Fang *et al.* (2006)
	Ti_I-Si	2.9	—	Zhang and Wang (2007)
	Ti_I-C	6.3	—	Zhang and Wang (2007)
	$Ti_{II}-C$	5.1	—	Zhang and Wang (2007)
	Ti_I-Si	2.8	—	Medvedeva and Freeman (2008)
	Ti_I-C	5.9	—	Medvedeva and Freeman (2008)
	$Ti_{II}-C$	4.9	—	Medvedeva and Freeman (2008)
Ti_3AlC_2	Ti_I-Al	2.1	—	Zhang and Wang (2007)
	Ti_I-C	6.4	—	Zhang and Wang (2007)
	$Ti_{II}-C$	5.7	—	Zhang and Wang (2007)
		211 phases		
Ti_2AlC	Ti–Al	2.0	2.0	Music *et al.* (2006)
	Ti–C	6.2	5.3	Music *et al.* (2006)
TiC	Ti–C	—	4.7	Hugosson *et al.* (2004)
V_2AlC	V–Al	2.4	2.4	Music *et al.* (2006)
	V–C	6.0	5.4	Music *et al.* (2006)
VC	V–C	—	4.8	Hugosson *et al.* (2004)
Cr_2AlC	Cr–Al	2.5	2.4	Music *et al.* (2006)
	Cr–C	5.8	5.4	Music *et al.* (2006)
CrC	Cr–C	—	4.6	Hugosson *et al.* (2004)
Ti_2GaC	Ti–Ga	—	2.0	Music *et al.* (2007)
V_2GaC	V–Ga	—	2.3	Music *et al.* (2007)
Cr_2GaC	Cr–Ga	—	2.3	Music *et al.* (2007)
Ti_2GeC	Ti–Ge	—	2.3	Music *et al.* (2007)
V_2GeC	V–Ge	—	2.1	Music *et al.* (2007)
Cr_2GeC	Cr–Ge	—	1.6	Music *et al.* (2007)

element (Figure 2.22b). For the Ti_2AC phases, γ_{0001} increases with increasing n_{val}. For V_2AC and Cr_2AC, on the other hand, γ_{0001} decreases with increasing n_{val}.

Focusing on the M–X bonds, comparable dependencies on n_{val} are found (Figure 2.23). If the outlier Cr_2GeC is ignored, then γ_{M-X} is a weak function of the M element (horizontal dashed line in Figure 2.23). Note that the differences between γ_{M-X} for the Al- and Ga-containing phases is small, which is not surprising

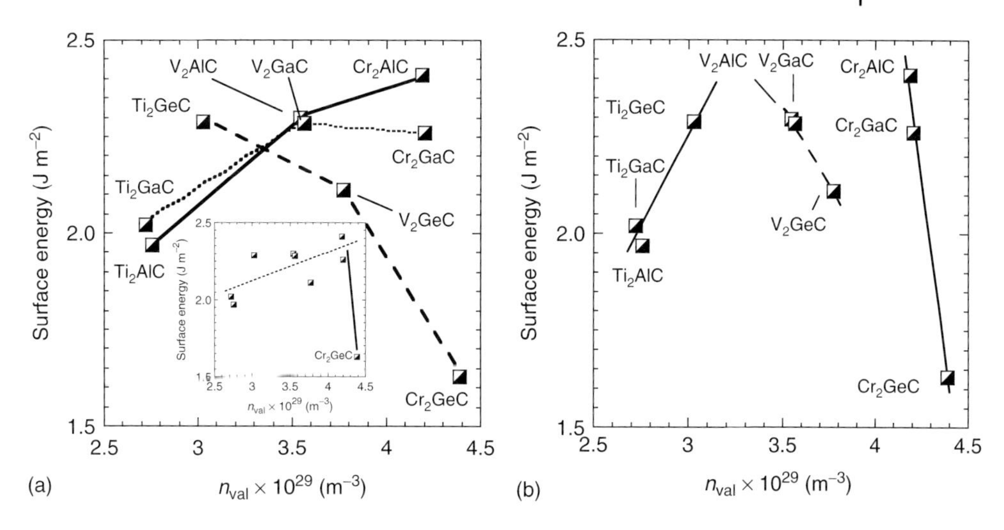

Figure 2.22 Effect of n_{val} on γ_{0001} with A terminations, focusing on (a) the effect of M element and (b) effect of A element. (Source: Data taken from Music *et al.* 2007). Inset in (a) plots the surface energy of all the MAX phases. Note anomalous nature of Cr_2GeC.

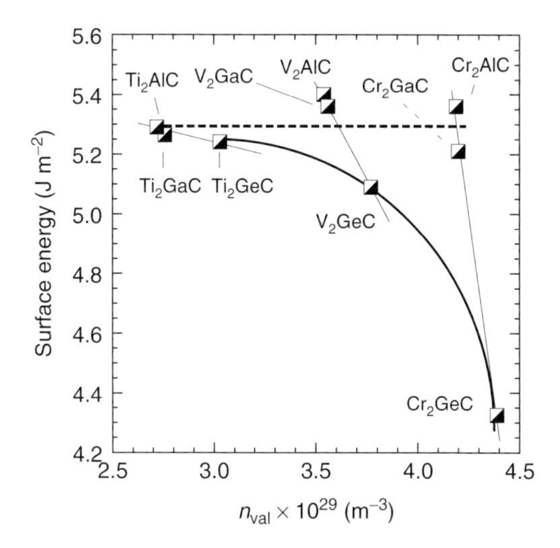

Figure 2.23 Effect of n_{val} on γ_{0001} with X terminations. Lines are guides to the eyes. (Source: Data taken from Music *et al.* 2007.)

since they are both group-13 elements. For the M_2GeC phases, on the other hand, going from Ti to Cr significantly decreases γ_{M-X} (solid arc in Figure 2.23).

Taken together, the results shown in Figures 2.22 and 2.23 suggest that, in general, increasing n_{val} tends to destabilize the structure. Note that the relative changes in Figure 2.23 are smaller than those shown in Figure 2.22. These results also clearly confirm that Cr_2GeC is quite anomalous. In the following chapters, the anomalous nature of this and other Cr-containing phases will be evidenced further. Needless to add, these conclusions only apply to the phases plotted in the figures. Whether such conclusions apply to other MAX phases must await further work.

In general, and as done repeatedly in this book, it is always instructive to compare MAX phase properties to their respective MX cousins. The γ_{100} surface energies of the MX compounds have been calculated (Hugosson *et al.*, 2004). Since it can be easily shown that $\gamma_{111} = \sqrt{3}\gamma_{100}$, and the latter for the MC compounds were calculated, it follows that γ_{111} for TiC, VC, and CrC are, respectively, 4.7, 4.8, and 4.6 J m^{-2}. These values are roughly 15% lower than the corresponding bonds in the MX phases. This is an important result because it suggests that the A-group element has a relatively small effect on the MX bonds.

2.8.4
Surface Energies of the 312 Phases

Fang *et al.* (2006) were the first to calculate γ_{0001} for the basal planes of Ti_3SiC_2. Zhang and Wang (2007) and Medvedeva and Freeman (2008) repeated the calculations. The results of Fang *et al.* are roughly 50% lower than those of Zhang and Wang or Medvedeva and Freeman (Table 2.8).

Zhang and Wang calculated the energies of the surfaces defined in Figure 2.21a–f. Their results for both Ti_3SiC_2 and Ti_3AlC_2 are summarized in Table 2.9, together with more recent results by Orellana and Gutiérrez (2011) on Ti_3SiC_2. Here again, for reasons unclear, the two sets of results differ substantially. However, given that the sum of any two complementary values should equal the values listed in Table 2.8, the results of Orellana and Gutiérrez for the M–X surfaces are too high. For example, γ_{M-X} of the M_{II}–X surface is 10.6 J m^{-2}, a value that is simply too high.

Regardless of which values are more accurate, as in the 211 phases, and in all cases, the Ti–C bonds are roughly 50% stronger than the Ti–A bonds. The C atoms in M_3AC_2 that are bonded to the Ti_I atoms (columns 4 and 5 in Table 2.9) are stronger than the Ti_{II}–C bonds in the middle of the unit cell (columns 6 and 7 in Table 2.9). This is not a surprising result, as the Ti_I–C bonds are significantly shorter than the Ti_{II}–C bonds (Table 2.3). The other important conclusion is that the Ti–Si bonds are roughly 50% stronger than the Ti–Al bonds.

Table 2.9 Surface energies (J m^{-2}) of the surfaces shown in Figure 2.21a–f. Orellana and Gutiérrez's results are for a relaxed 3 × 3 surface; the others are for relaxed 1 × 1 surfaces.

	$A(M_I)$	$M_I(X)$	$M_I(A)$	$X(M_{II})$	$M_{II}(X)$	$X(M_I)$	References
Ti_3SiC_2	1.2	1.7	0.7	5.6	1.0	4.1	Zhang and Wang (2007)
Ti_3SiC_2	1.8	4.0	1.9	8.5	2.5	8.4	Orellana and Gutiérrez (2011)
Ti_3AlC_2	0.6	1.5	0.9	5.5	1.0	3.7	Zhang and Wang (2007)

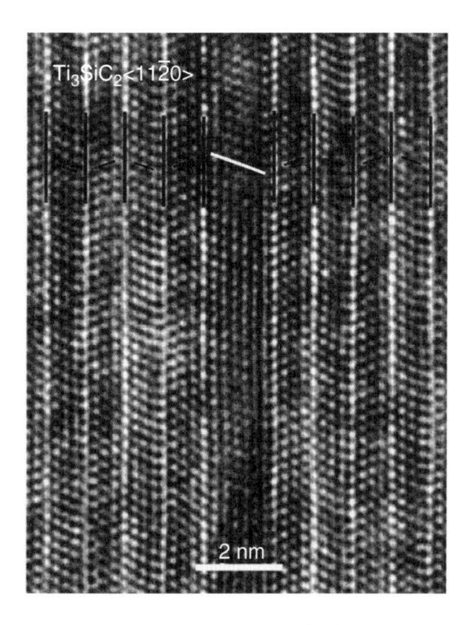

Figure 2.24 HRTEM image of a basal plane SF in Ti_3SiC_2. The viewing direction is [11$\bar{2}$0] (Kooi *et al.*, 2003).

2.8.5
Stacking Faults

The first report on stacking faults, SFs, in a MAX phase was by Farber *et al.* (1998) who showed that they exist in Ti_3SiC_2. Later, using HRTEM, Kooi *et al.* (2003) showed that at the core of the basal plane SF (Figure 2.24) the Si mirror plane is missing and the remaining thin lamella consist of six close-packed Ti planes, with five C planes in between showing identical stacking as the {111} TiC planes. This interpretation is emphasized in Figure 2.24 by inserting vertical black lines on the Si planes. The thin lamella with three Ti and two C basal planes are indicated by the black lines; the region around the SF core is indicated by the white line. It is important to emphasize that these basal plane SFs *must* be growth defects because they could never originate from a simple shearing operation.

2.9
Summary and Conclusions

The MAX phases – of which over 60 exist – are layered hexagonal with $P6_3/mmc$ symmetry and two formula units per unit cell. Most of the phases have the 211 composition. Their densities range from a low of about 4 Mg m^{-3} to a high of almost 15 Mg m^{-3}. The 211 phases only exist in one polymorph; the 312 in two; and the 413 in three. The overwhelming majority exist in the α–polymorph.

The M–X bonds in the binaries are quite comparable in energy and strengths to those in the ternaries. For the $n > 1$ phases, the M–X bonds are shorter when they are closer to the A layers.

The fact that the Drude and Sommerfeld models are useful and can explain some, but not all, property trends is consistent with the fact that the electrons in these phases can be considered "free." To understand property and structural trends, the useful metrics to use are r_s/r_B and/or the average number n_{val} of formal valence electrons per unit volume.

According to DFT calculations, the bonds in the MAX phases exhibit covalent, ionic, and metallic character. The DOS at E_F is substantial and consistent with their high electrical and thermal conductivities. The rigid band model is applicable to the MAX phases and explains the more or less linear increase in $N(E_F)$ with increasing n_{val}. Similar to the MX binaries, there is strong orbital overlap between the p orbitals of the X atoms and the d orbitals of the M atoms as well as the d–orbitals of the M atoms with thermselves; the latter dominate the DOS at E_F. In addition, in the MAX phases, the p orbitals of the A atoms and the d orbitals of the M atoms overlap. The energy of the latter is higher than those of the p orbitals of the X atoms and d orbitals of the M atoms.

In some cases, charge is transferred from the A to the X layers.

Increasing n_{val} beyond a threshold of $\approx 3.5 \times 10^{29}$ m^{-3} destabilizes the structure by, presumably, pushing the electrons into antibonding orbitals.

In several Al-containing MAX phases, and in Ti_3SiC_2, the M–X bond strengths are roughly double those of the M–A bonds. Whether that applies to other MAX phases awaits further studies.

2.A Appendix

Bond distances and distortions in the M_3AX_2 and M_4AX_3 phases

The M_3AX_2 Phases

In the 312 phases, there are two polymorphs (α and β) and two free parameters; z_M is the height of the M_I atom in the unit cell and z_X is the height of the X_I atoms in the unit cell (see Figure 2.1b). The M atoms that are bonded to the A atoms are labelled M_I; M atoms that are only bonded to X atoms are labelled M_{II}.

Bond Distances

The same expressions are valid for the α- and β-phases.

$$R_{M_I-A} = \sqrt{\frac{a^2}{3} + c^2\left(\frac{1}{4} - z_M\right)^2} \qquad (2.A.1)$$

$$R_{M_{II}-X} = \sqrt{\frac{a^2}{3} + c^2 z_X^2} \qquad (2.A.2)$$

$$R_{M_I-X} = \sqrt{\frac{a^2}{3} + c^2 (z_M - z_X)^2} \qquad (2.A.3)$$

$$R_{M-M} = a \text{ (for M atoms belonging to the same basal plane)} \tag{2.A.4}$$

$$R_{M_I-M_{II}} = \sqrt{\frac{a^2}{3} + c^2 z_M^2} \text{ (between M atoms out of the same basal planes)} \tag{2.A.5}$$

Distortions

$$O_d = \frac{d_1}{d_2} = \frac{2z_M}{d_2} = \frac{a\frac{\sqrt{3}}{2}}{\sqrt{R_{M_I-M_{II}} - \frac{a^2}{4}}} = \frac{\sqrt{3}}{2\sqrt{z_M^2\left(\frac{c}{a}\right)^2 + \frac{1}{12}}} \tag{2.A.6}$$

$$P_d = \frac{R_{M-M}}{R_{M_I-A}} = \frac{1}{\sqrt{\frac{1}{3} + \left(\frac{1}{4} - z_M\right)^2\left(\frac{c}{a}\right)^2}} \tag{2.A.7}$$

$$S_d = \frac{R_{M-M}}{R_{M_I-M_{II}}} = \frac{1}{\sqrt{\frac{1}{3} + \left(\frac{c}{a}\right)^2 z_M^2}} \tag{2.A.8}$$

The M_4AX_3 Phases

For the 413 phases, three polymorphs (α, β and γ) and three free parameters exist; z_{M_I} and $z_{M_{II}}$ are the heights of the M_I and M_{II} atoms in the unit cell and z_X is the height of the X_I atoms in the unit cell (see Figure 2.1c). M atoms that are bonded to the A atoms are labelled M_I; M atoms that are only bonded to X atoms are labelled M_{II}; X atoms that are bonded to the M_I atoms are labelled X_I; X atoms that are only bonded to M_{II} atoms are labelled X_{II}.

Bond distances

The same expressions are valid for α and β phases:

$$R_{M_I-A} = \sqrt{\frac{a^2}{3} + c^2\left(\frac{1}{4} - z_{M_I}\right)^2} \tag{2.A.9}$$

$$R_{M_{II}-X_{II}} = \sqrt{\frac{a^2}{3} + c^2 z_{M_2}^2} \tag{2.A.10}$$

$$R_{M_I-X_I} = \sqrt{\frac{a^2}{3} + c^2\left(z_{M_1} - z_X\right)^2} \tag{2.A.11}$$

$$R_{M_{II}-X_I} = \sqrt{\frac{a^2}{3} + c^2\left(z_{M_2} - z_X\right)^2} \tag{2.A.12}$$

$$R_{M_{II}-X_2} = \sqrt{\frac{a^2}{3} + c^2 z_{M_2}^2} \tag{2.A.13}$$

$$R_{M-M} = a \text{ (for M atoms is the same basal plane)} \tag{2.A.14}$$

$$R_{M_I-M_{II}} = \sqrt{\frac{a^2}{3} + c^2\left(z_{M_2}^2 - z_{M_1}^2\right)} \tag{2.A.15}$$

$$R_{M_{II}-M_{II}} = \sqrt{\frac{a^2}{3} + 4c^2 z_{M_2}^2} \text{ (between } M_{II} \text{ atoms not in the same basal planes)} \tag{2.A.16}$$

References

Ahuja, R., Eriksson, O., Wils, J.M., and Johansson, B. (2000) Electronic structure of Ti_3SiC_2. *Appl. Phys. Lett.*, **76**, 2226–2228.

Ashcroft, N.W. and Mermin, N.D. (1976) *Solid State Physics*, Saunders College Publishing, Philadelphia.

Banerjee, R., Ahuja, R., and Fraser, H.L. (1996) Dimensionally-induced structural transformations in Ti-Al multilayers. *Phys. Rev. Lett.*, **76**, 3778.

Barsoum, M.W. (2000) The $M_{n+1}AX_n$ phases: a new class of solids; thermodynamically stable nanolaminates. *Prog. Solid State Chem.*, **28**, 201–281.

Barsoum, M.W., Crossley, A., and Myhra, S. (2002) Crystal-chemistry from XPS analysis of carbide-derived $M_{(n+1)}AX_{(n)}$ (n=1) nano-laminate compounds. *J. Phys. Chem. Solids*, **63**, 2063–2068.

Barsoum, M.W., Ganguly, A., Siefert, H.J., and Aldinger, F. (2002a) The 1300 °C isothermal section in the Nb-Sn-C ternary phase diagram. *J. Alloys Compd.*, **337**, 202–207.

Barsoum, M.W., Golczewski, J., Siefert, H.J., and Aldinger, F. (2002b) Fabrication and electrical and thermal properties of Ti_2InC, Hf_2InC and $(Ti,Hf)_2AlC$. *J. Alloys Compd.*, **340**, 173–179.

Barsoum, M.W., El-Raghy, T., Farber, L., Amer, M., Christini, R., and Adams, A. (1999a) The topotaxial transformation of Ti_3SiC_2 to form a partially ordered cubic $TiC_{0.67}$ Phase by the diffusion of Si into molten cryolite. *J. Electrochem. Soc.*, **146**, 3919–3923.

Barsoum, M.W., El-Raghy, T., Rawn, C.J., Porter, W.D., Wang, H., Payzant, A., and Hubbard, C. (1999b) Thermal properties of Ti_3SiC_2. *J. Phys. Chem. Solids*, **60**, 429.

Barsoum, M.W., Farber, L., Levin, I., Procopio, A., El-Raghy, T., and Berner, A. (1999c) High-resolution transmission electron microscopy of Ti_4AlN_3, or $Ti_3Al_2N_2$ revisited. *J. Am. Ceram. Soc.*, **82**, 2545–2547.

Barsoum, M.W., Rawn, C.J., El-Raghy, T., Procopio, A.T., Porter, W.D., Wang, H., and Hubbard, C.R. (2000) Thermal properties of Ti_4AlN_3. *J. Appl. Phys.*, **83**, 825.

Beckmann, O., Boller, H., and Nowotny, H. (1968) Neue H-phasen. *Monatsh. Chem.*, **99**, 1580.

Beckmann, O., Boller, H., Nowotny, H., and Benesovsky, F. (1969) Einige Komplexcarbide und -nitride in den Systemen Ti-(Zn,Cd,Hg)-(C,N) und Cr-Ga-N. *Monatsh. Chem.*, **100**, 1465.

Boller, H. and Nowotny, H. (1966) Ronthenographische untersuchungen in system: V-As-C. *Monatsh. Chem.*, **97**, 1053.

Boller, H. and Nowotny, H. (1968) Die Kristallstruktur von V_2PC and V_5P_3N". *Monatsh. Chem.*, **99**, 672.

Briggs, D. and Seah, M.P. (eds) (1990) *Pratical Surface Analysis*, John Wiley & Sons, Ltd, Chichester.

Buchholt, K., Eklund, P., Jensen, J., Lu, J., Ghandi, R., Domeij, M., Zetterling, C.M., Behan, G., Zhang, H., Lloyd Spetz, A. *et al.* (2012) Growth and characterization of epitaxial Ti_3GeC_2 thin films on 4H-SiC(0001). *J. Cryst. Growth*, **343**, 133–137.

Cabioch, T. (2012) Private communication.

Cabioch, T., Eklund, P., Mauchamp, V., and Jaouen, M. (2012) Structural investigation of substoichiometry and solid solution effects in $Ti_2Al(C_x,N_{1-x})_y$ compounds. *J. Eur. Ceram. Soc.*, **32**, 1803–1811.

Cabioch, T., Eklund, P., Mauchamp, V., Jaouen, M., and Barsoum, M.W. (2013) Tailoring of the thermal expansions of the MAX phases in the $Cr_2(Al_{1-x},Ge_x)C_2$ system. *J. Eur. Ceram. Soc.*, **33**, 897–904.

Calais, J.L. (1977) Band structure of transition metal compounds. *Adv. Phys.*, **26**, 847.

Cottrell, A. (1995) *Chemical Bonding in Transition Metal Carbides*, Institute of Materials, London.

Cover, M.F., Warschkow, O., Bilek, M.M., and McKenzie, D.R. (2009) A comprehensive survey of M_2AX phase elastic properties. *J. Phys.: Condens. Matter*, **21**, 305403.

Dahlqvist, M., Alling, B., and Rosén, J. (2010) Stability trends of MAX phases from first principles. *Phys. Rev. B*, **81**, 220102.

Dregia, S.A., Banerjee, R., and Fraser, H.L. (1998) Polymorphic phase stability in thin multilayers. *Scr. Mater.*, **39**, 217.

Du, Y.L., Sun, Z.-M., Hashimoto, H., and Tian, W.B. (2008) First-principles study of carbon vacancy in Ta_4AlC_3. *Mater. Trans.*, **49**, 1934–1936.

Du, Y.L., Sun, Z.M., Hashimoto, H., and Tian, W.B. (2009) Bonding properties and bulk modulus of M_4AlC_3 (M = V, Nb, and Ta) studied by first-principles calculations. *Phys. Status Solidi*, **246**, 1039–1043.

Dubois, S., Cabioch, T., Chartier, P., Gauthier, V., and Jaouen, M. (2007) A new ternary nanolaminate carbide: Ti_3SnC_2. *J. Am. Ceram. Soc.*, **90**, 2642–2644.

Eklund, P., Palmquist, J.-P., Howing, J., Trinh, D.H., El-Raghy, T., Hogberg, H., and Hultman, L. (2007) Ta_4AlC_3: phase determination, polymorphism and deformation. *Acta Mater.*, **55**, 4723.

Etzkorn, J., Ade, M., and Hillebrecht, H. (2007a) Ta_3AlC_2 and Ta_4AlC_3 - single-crystal investigations of two new ternary carbides of tantalum synthesized by the molten metal technique. *Inorg. Chem.*, **46**, 1410–1418.

Etzkorn, J., Ade, M., and Hillebrecht, H. (2007b) V_2AlC, V_4AlC_{3-x} (x approximate to 0.31), and $V_{12}Al_3C_8$: synthesis, crystal growth, structure, and superstructure. *Inorg. Chem.*, **46**, 7646–7653.

Etzkorn, J., Ade, M., Kotzott, D., Kleczek, M., and Hillebrecht, H. (2009) Ti_2GaC, Ti_4GaC_3 and Cr_2GaC - synthesis, crystal growth and structure analysis of Ga-containing MAX-phases $M_{n+1}GaC_n$ with M = Ti, Cr and n = 1–3. *J. Solid State Chem.*, **182**, 995.

Fang, C.M., Ahuja, R., Eriksson, O., Li, S., Jansson, U., Wilhelmsson, O., and Hultman, L. (2006) General trend of the mechanical properties of the ternary carbides M_3SiC_2. *(M=transition. Phys Rev B*, **74**, 054106.

Farber, L., Barsoum, M.W., Zavaliangos, A., El-Raghy, T., and Levin, I. (1998) Dislocations and stacking faults in Ti_3SiC_2. *J. Am. Ceram. Soc.*, **81**, 1677–1681.

Farber, L., Levin, I., Barsoum, M.W., El-Raghy, T., and Tzenov, T. (1999) High-resolution transmission electron microscopy of some $Ti_{n+1}AX_n$ compounds. *J. Appl. Phys.*, **86**, 2540–2543.

Fernandez-Guillermet, A., Haglund, J., and Grimvall, G. (1992) Cohesive properties of 4d-transition-metal carbides and nitrides in the NaC1-type structure. *Phys. Rev. B*, **45**, 11557.

Ganguly, A., Zhen, T., and Barsoum, M.W. (2004) Synthesis and mechanical properties of Ti_3GeC_2 and $Ti_3(Si_xGe_{1-x})C_2$ (x = 0.5, 0.75) solid solutions. *J. Alloys Compd.*, **376**, 287–295.

Haglund, J., Grimvall, G., Jarlborg, T., and Fernandez-Guillermet, A. (1991) Band structure and cohesive properties of 3d-transition-metal carbides and nitrides with the NaC1-type structure. *Phys. Rev. B*, **43**(14), 400.

He, X., Bai, Y., Zhu, C., and Barsoum, M.W. (2011) Polymorphism of newly-discovered Ti_4GaC_3: a first-principle study. *Acta Mater.*, **59**, 5523–5533.

He, X., Bai, Y., Zhu, C., Sun, Y., Li, M., and Barsoum, M.W. (2010) General trends in the structural, electronic and elastic properties of the M_3AlC_2 phases (M = transition metal): a first-principle study. *Comput. Mater. Sci.*, **49**, 691–698.

Hu, C., Li, F., Zhang, J., Wang, J., Wanga, J., and Zhou, Y.C. (2007a) Nb_4AlC_3: a new compound belonging to the MAX phases. *Scr. Mater*, **57**, 893–896.

Hu, C., Lin, Z., He, L., Bao, Y., Wang, J., Li, M., and Zhou, Y.C. (2007b) Physical and mechanical properties of bulk Ta_4AlC_3 ceramic prepared by an in situ reaction synthesis/hot-pressing method. *J. Am. Ceram. Soc.*, **90**, 2542–2548.

Hug, G. (2006) Electronic structures of and composition gaps among the ternary carbides Ti_2MC. *Phys. Rev. B*, **74**, 184113–184117.

Hug, G., Jaouen, M., and Barsoum, M.W. (2005) XAS, EELS and full-potential augmented plane wave study of the electronic structures of Ti_2AlC, Ti_2AlN, Nb_2AlC and $(Ti_{0.5},Nb_{0.5})_2AlC$. *Phys. Rev. B*, **71**, 24105.

Hugosson, H.W., Eriksson, O., Jansson, U., Ruban, A.V., Souvatzis, P., and Abrikosov, I.A. (2004) Surface energies and work functions of the transition metal carbides. *Surf. Sci.*, **557**, 243–254.

Jaouen, M. (2012) Private communication.

Jeitschko, W. and Nowotny, H. (1967) Die Kristallstructur von Ti_3SiC_2 – Ein Neuer

Komplxcarbid-Typ. *Monatsh. Chem.*, **98**, 329–337.

Jeitschko, W., Nowotny, H., and Benesovsky, F. (1963a) Die H-phasen Ti_2InC, Hf_2InC und Ti_2GeC. *Monatsh. Chem.*, **94**, 1201.

Jeitschko, W., Nowotny, H., and Benesovsky, F. (1963b) Kohlenstoffhaltige ternare Verbindungen (H-phase). *Monatsh. Chem.*, **94**, 672.

Jeitschko, W., Nowotny, H., and Benesovsky, F. (1963c) Kohlenstoffhaltige ternare Verbindungen (V-Ge-C, Nb-Ga-C, Ta-Ga-C, Ta-Ge-C, Cr-Ga-C und Cr-Ge-C). *Monatsh. Chem.*, **94**, 844.

Jeitschko, W., Nowotny, H., and Benesovsky, F. (1963d) Ti_2AlN, eine stickstoffhaltige H-phase. *Monatsh. Chem.*, **94**, 1198.

Jeitschko, W., Nowotny, H., and Benesovsky, F. (1964a) Carbides of formula T_2MC. *J. Less-Common Met.*, **7**, 133–138.

Jeitschko, W., Nowotny, H., and Benesovsky, F. (1964b) Die H-phasen Ti_2TlC, Ti_2PbC, Nb_2InC, Nb_2SnC und Ta_2GaC. *Monatsh. Chem.*, **95**, 431.

Jeitschko, W., Nowotny, H., and Benesovsky, F. (1964c) Die H-phasen: Ti_2CdC, Ti_2GaC, Ti_2GaN, Ti_2InN, Zr_2InN and Nb_2GaC. *Monatsh. Chem.*, **95**, 178.

Jürgens, D., Uhrmacher, M., Gehrke, H.-G., Nagl, M., Vetter, U., Brusewitz, C., Hofsass, H., Mestnik-Filho, J., and Barsoum, M.W. (2011) Electric field gradients at 111In/111Cd probe atoms on A-sites in 211-MAX phases. *J. Phys.: Condens. Matter*, **23**, 505501–505519.

Jürgens, D., Uhrmacher, M., Hofsäss, H., Mestnik-Filho, J., and Barsoum, M.W. (2010) Perturbed angular correlation studies of the MAX phases Ti_2AlN and Cr_2GeC using Ion implanted 111In as probe nuclei. *Nucl. Instrum. Methods Phys. Res., Sect. B*, **268**, 2185–2188.

Jürgens, D., Uhrmacher, M., Hofsäss, H., Roder, J., Wodniecki, P., Kulinska, A., and Barsoum, M.W. (2007) First PAC experiments in MAX-phases. *Hyperfine Interact.*, **178**, 23–30.

Keast, V.J., Harris, S., and Smith, D.K. (2009) Prediction of the stability of the $M_{n+1}AX_n$ phases from first principles. *Phys. Rev. B*, **80**, 214113.

Kephart, J.S. and Carim, A.H. (1998) Ternary compounds and phase equilibria in Ti-Ge-C and Ti-Ge-B. *J. Electrochem. Soc.*, **145**, 3253.

Kisi, E.H., Crossley, J.A.A., Myhra, S., and Barsoum, M.W. (1998) Structure and crystal chemistry of Ti_3SiC_2. *J. Phys. Chem. Solids*, **59**, 1437–1442.

Kooi, B.J., Poppen, R.J., Carvalho, N.J.M., De Hosson, J.T.M., and Barsoum, M.W. (2003) Ti_3SiC_2: a damage tolerant ceramic studied with nanoindentations and transmission electron microscopy. *Acta Mater.*, **51**, 2859–2872.

Kudielka, H. and Rohde, H. (1960) Strukturuntersuchungen an Carbosulfiden von Titan und Zirkon. *Z. Kristallogr.*, **114**, 447.

Lane, N.J., Eklund, P., Lu, J., Spencer, C.B., Hultman, L., and Barsoum, M.W. (2011) High-temperature phase stability of α-Ta_4AlC_3. *Mater. Res. Bull.*, **46**, 1088–1091.

Lane, N.J., Naguib, M., Lu, J., Eklund, P., Hultman, L., and Barsoum, M.W. (2012a) Comment on "$Ti_5Al_2C_3$: a new ternary carbide belonging to MAX phases in the Ti–Al–C system". *J. Am. Ceram. Soc.*, **95**, 1058–1510.

Lane, N.J., Naguib, M., Lu, J., Hultman, L., and Barsoum, M.W. (2012b) Structure of a new bulk $Ti_5Al_2C_3$ MAX phase produced by the topotactic transformation of Ti_2AlC. *J. Eur. Ceram. Soc.*, **32**, 3485–3491.

Liao, T., Wang, J., and Zhou, Y. (2008) Ab initio modeling of the formation and migration of monovacancies in Ti_2AlC. *Scr. Mater.*, **59**, 854–857.

Lin, Z.J., Zhuo, M.J., Zhou, Y.C., Li, M.S., and Wang, J.Y. (2006) Atomic scale characterization of layered ternary Cr_2AlC ceramic. *J. Appl. Phys.*, **99**.

Lue, C.S., Lin, J.Y., and Xie, B.X. (2006) NMR study of the ternary carbides M_2AlC, M = Ti, V, Cr. *Phys. Rev. B*, **73**, 035125.

Lye, R.G. and Logothetis, E.M. (1966) Optical properties and band structure of TiC. *Phys. Rev.*, **147**, 662.

Magnuson, M., Mattesini, M., Li, S., Hoglund, C., Beckers, M., Hultman, L., and Eriksson, O. (2007) Bonding mechanism in the nitrides Ti_2AlN and TiN: an experimental and theoretical investigation. *Phys. Rev. B*, **76**, 195127.

Magnuson, M., Mattesini, M., Van Nong, N., Eklund, P., and Hultman, L. (2012)

The electronic-structure origin of the anisotropic thermopower of nanolaminated Ti_3SiC_2 determined by polarized x-ray spectroscopy. *Phys. Rev. B*, **85**, 195134.

Magnuson, M., Mattesini, M., Wilhelmsson, O., Emmerlich, J., Palmquist, J.-P., Li, S., Ahuja, R., Hultman, L., and Eriksson, O. (2006a) Electronic structure and chemical bonding in Ti_4SiC_3 investigated by soft x-ray emission spectroscopy and first principle theory. *Phys. Rev. B*, **74**, 205102.

Magnuson, M., Wilhelmsson, O., Palmquist, J.-P., Jansson, U., Mattesini, M., Li, S., Ahuja, R., and Eriksson, O. (2006b) Electronic structure and chemical bonding in Ti_2AlC investigated by soft x-ray emission spectroscopy. *Phys. Rev. B*, **74**, 195108.

Magnuson, M., Palmquist, J.P., Mattesini, M., Li, S., Ahuja, R., Eriksson, O., Emmerlich, J., Wilhelmsson, O., Eklund, P., Hogberg, H. *et al.* (2005) Electronic structure investigation of Ti_3AlC_2, Ti_3SiC_2, and Ti_3GeC_2 by soft x-ray emission spectroscopy. *Phys. Rev. B*, **72**.

Magnuson, M., Wilhelmsson, O., Mattesini, M., Li, S., Ahuja, R., Eriksson, O., Högberg, H., Hultman, L., and Jansson, U. (2008) Anisotropy in the electronic structure of V_2GeC investigated by soft x-ray emission spectroscopy and first-principles theory. *Phys. Rev. B*, **78**, 035117.

Manoun, B., Saxena, S.K., El-Raghy, T., and Barsoum, M.W. (2006) High-pressure X-ray study of Ta_4AlC_3. *Appl. Phys. Lett.*, **88**, 201902.

Manoun, B., Saxena, S.K., Hug, G., Ganguly, A., Hoffman, E.N., and Barsoum, M.W. (2007a) Synthesis and compressibility of $Ti_3(Al_{1.0}Sn_{0.2})C_2$ and $Ti_3Al(C_{0.5},N_{0.5})_2$. *J. Appl. Phys.*, **101**, 113523.

Manoun, B., Yang, H., Saxena, S.K., Ganguly, A., Barsoum, M.W., Liu, Z.X., Lachkar, M., and El-Bali, B. (2007b) Infrared spectrum and compressibility of Ti_3GeC_2 to 51 GPa. *J. Alloys Compd.*, **433**, 265–268.

Mauchamp, V., Hug, G., Bugnet, M., Cabioch, T., and Jaouen, M. (2010) Anisotropy of Ti_2AlN dielectric response investigated by ab initio calculations and electron energy-loss spectroscopy. *Phys. Rev. B*, **81**, 035109.

Medvedeva, N.I. and Freeman, A.J. (2008) Cleavage fracture in Ti_3SiC_2 from first-principles. *Scr. Mater.*, **58**, 671–674.

Medvedeva, N., Novikov, D., Ivanovsky, A., Kuznetsov, M., and Freeman, A. (1998) Electronic properties of Ti_3SiC_2-based solid solutions. *Phys. Rev. B*, **58**, 16042–16050.

Music, D., Ahuja, R., and Schneider, J.M. (2005) Theoretical study of nitrogen vacancies in Ti_4AlN_3. *Appl. Phys. Lett.*, **86**, 031911.

Music, D., Sun, Z., Ahuja, R., and Schneider, J.M. (2006) Electronic structure of $M_2AlC(0001)$ surfaces (M = Ti,V, Cr). *J. Phys.: Condens. Matter*, **18**, 8877.

Music, D., Sun, Z., Ahuja, R., and Schneider, D. (2007) Surface energy of $M_2AC(0001)$ determined by density functional theory (M = Ti,V,Cr; A = Al,Ga,Ge). *Surf. Sci.*, **601**, 896–899.

Myhra, S., Crossley, A., and Barsoum, M.W. (2001) Crystal-chemistry of the Ti_3AlC_2 and Ti_4AlN_3 layered carbide/nitride phases — characterization by XPS. *J. Phys. Chem. Solids*, **62**, 811–817.

Neckel, A. (1990) *Electronic Structure of Stoichiometric and Non-Stoichiometric TiC and TiN* Vol. 485, Kluwer Academic Press, Amsterdam.

Nowotny, H. (1970) Strukturchemie einiger Verbindungen der Ubergangsmetalle mit den elementen C, Si, Ge, Sn. *Prog. Solid State Chem.*, **2**, 27–70.

Nowotny, H., Schuster, J.C., and Rogl, P. (1982) Structural chemistry of complex carbides and related compounds. *J. Solid State Chem.*, **44**, 126.

Orellana, W. and Gutiérrez, G. (2011) First-principles calculations of the thermal stability of $Ti_3SiC_2(0001)$ surfaces. *Surf. Sci.*, **605**, 2087–2091.

Palmquist, J.P., Li, S., Persson, P.O.A., Emmerlich, J., Wilhelmsson, O., Hogberg, H., Katsnelson, M.I., Johansson, B., Ahuja, R., Eriksson, O. *et al.* (2004) $M_{(n+1)}AX_{(n)}$ phases in the Ti-Si-C system studied by thin-film synthesis and ab initio calculations. *Phys. Rev. B*, **70**, 165401.

Pietzka, M.A. and Schuster, J. (1994) Summary of constitution data of the system Al-C-Ti. *J. Phase Equilib.*, **15**, 392.

Pietzka, M.A. and Schuster, J.C. (1996) Phase equilibria in the quaternary system Ti-Al-C-N. *J. Am. Ceram. Soc.*, **79**, 2321.

Radovic, M., Ganguly, A., and Barsoum, M.W. (2008) Elastic properties and phonon conductivities of $Ti_3Al(C_{0.5},N_{0.5})_2$ and $Ti_2Al(C_{0.5},N_{0.5})$ solid solutions. *J. Mater. Res.*, **23**, 1517–1521.

Rawn, C.J., Barsoum, M.W., El-Raghy, T., Procopio, A.T., Hoffman, C.M., and Hubbard, C.R. (2000) Structure of Ti_4AlN_{3-x}- a layered $M_{n+1}AX_n$ nitride. *Mater. Res. Bull.*, **35**, 1785–1796.

Riley, D.P., O'Conner, D.J., Dastoor, P., Brack, N., and Pigram, P.J. (2002) Comparative analysis of Ti_3SiC_2 and associated compounds using x-ray diffraction and x-ray photoelectron spectroscopy. *J. Phys. D: Appl. Phys.*, **35**, 1603–1611.

Salama, I., El-Raghy, T., and Barsoum, M.W. (2002) Synthesis and mechanical properties of Nb_2AlC and $(Ti,Nb)_2AlC$. *J. Alloys Compd.*, **347**, 271–278.

Scabarozi, T., Ganguly, A., Hettinger, J.D., Lofland, S.E., Amini, S., Finkel, P., El-Raghy, T., and Barsoum, M.W. (2008) Electronic and thermal properties of $Ti_3Al(C_{0.5},N_{0.5})_2$, $Ti_2Al(C_{0.5},N_{0.5})$ and Ti_2AlN. *J. Appl. Phys.*, **104**, 073713.

Schuster, J. and Nowotny, H. (1980) Investigation of the ternary systems (Zr, Hf, Nb, Ta)-Al-C and studies on complex carbides. *Z. Metallkd.*, **71**, 341.

Schuster, J.C., Nowotny, H., and Vaccaro, C. (1980) The ternary systems: Cr-Al-C, V-Al-C and Ti-Al-C and the behavior of the H-phases (M_2AlC). *J. Solid State Chem.*, **32**, 213.

Schwarz, K. (1987) Band structure and chemical bonding in transition metal carbides and nitrides. *CRC Crit. Rev. Solid State Mater. Sci.*, **13**, 211.

Shechtman, D., van Heerden, D., and Josell, D. (1994) FCC titanium in Ti-Al multilayers. *Mater. Lett.*, **20**, 10.

Stoltz, S.E., Starnberg, H.I., and Barsoum, M.W. (2003) Core level and valence band studies of layered Ti_3SiC_2 by high resolution photoelectron spectroscopy. *J. Phys. Chem. Solids*, **64**, 2321–2328.

Tepper, T., Shechtman, D., Van Heerden, D., and Josell, D. (1997) FCC titanium in Ti/Ag multilayers. *Mater. Lett.*, **33**, 181.

Toth, L. (1967) High superconducting transition temperatures in the molybdenum carbide family of compounds. *J. Less. Comm. Met.*, **13**, 129.

Van Heerden, D., Josell, D., and Shechtman, D. (1996) The formation of FCC titanium in Ti-Al multilayers. *Acta Mater.*, **44**, 296.

Wagner, C.D., Powell, C.J., Allison, J.W., and Rumble, J.R. (2003) Standard Reference Database 20, X-ray Photoelectron Spectroscopy Database Version 2.0, NIST.

Wang, J., Wang, J., and Zhou, Y. (2008) Stable M_2AlC (0001) surfaces (M = Ti, V and Cr) by first-principles investigation. *J. Phys.: Condens. Matter*, **20**, 225006.

Wang, J., Zhou, Y., Lin, Z., and Hu, J. (2008) Ab initio study of polymorphism in layered ternary carbide M_4AlC_3 (M = V, Nb and Ta). *Scr. Mater.*, **58**, 1043–1046.

Wang, Z., Zha, S., and Barsoum, M.W. (2004) Compressibility and pressure induced phase transformation in Ti_3GeC_2. *Appl. Phys. Lett.*, **85**, 3453–3455.

Wang, X., Zhang, H., Zheng, L., Ma, Y., Lu, X., Sun, Y., and Zhou, Y. (2012) $Ti_5Al_2C_3$: a new ternary carbide belonging to MAX phases in the Ti–Al–C system. *J. Am. Ceram. Soc.*, **95**, 1508–1510.

Wolfsgruber, H., Nowotny, H., and Benesovsky, F. (1967) Die Kristallstuktur von Ti_3GeC_2. *Monatsh. Chem.*, **98**, 2401.

Yu, W., Li, S., and Sloof, W.G. (2010) Microstructure and mechanical properties of a $Cr_2Al(Si)C$ solid solution. *Mater. Sci. Eng., A*, **527**, 5997–6001.

Zhang, H.Z. and Wang, S.Q. (2007) First-principles study of Ti_3AC_2 (A = Si, Al) (001) surfaces. *Acta Mater.*, **55**, 4645–4655.

3
Elastic Properties, Raman and Infrared Spectroscopy

3.1
Introduction

For the most part, the $M_{n+1}AX_n$ phases are elastically quite stiff. When combined with the fact that the densities of some of them are relatively low, $\approx$4.1–5 g cm^{-3}, their specific stiffness values can be high (Figure 3.1). For example, the specific stiffness of Ti_3SiC_2 is comparable to that of Si_3N_4, and roughly *three* times that of Ti. When the latter is combined with the fact that the MAX phases are most readily machinable, their specific stiffness values start to encroach on the domain of other more traditional ceramics with comparable specific stiffness values that are not readily machinable (Figure 3.1). Said otherwise, prior to the discovery of the MAX phases, the price paid for high specific stiffness values had been difficulty or lack of machinability. This is no longer necessarily the case.

In this chapter, the elastic – both measured and calculated – properties of the MAX phases are summarized and reviewed. Section 3.2 summarizes the elastic constants obtained from density functional theory (DFT) calculations. Section 3.3 summarizes the Young's moduli E and shear moduli G values. Poisson's ratio is discussed briefly in Section 3.4. Section 3.5 deals with the bulk moduli B. Extrema in elastic properties are examined in Section 3.6. The effect of temperature on elastic properties is discussed in Section 3.7. Raman and infrared spectroscopy are reviewed in Sections 3.8 and 3.9, respectively. Conclusions are summarized in Section 3.10.

3.2
Elastic Constants

Elastic deformation in crystalline solids is a fully reversible and nondissipative process. For hexagonal symmetry, there are five independent elastic constants; they are c_{11}, c_{12}, c_{13}, c_{33}, and c_{44}. To date, the nonavailability of large MAX single crystals has made it difficult to experimentally determine their elastic constants. What can be used until such measurements are available, however, are the results of *ab initio* calculations. Table 3.1 lists the elastic constants of most of the MAX

MAX Phases: Properties of Machinable Ternary Carbides and Nitrides, First Edition. Michel W. Barsoum.

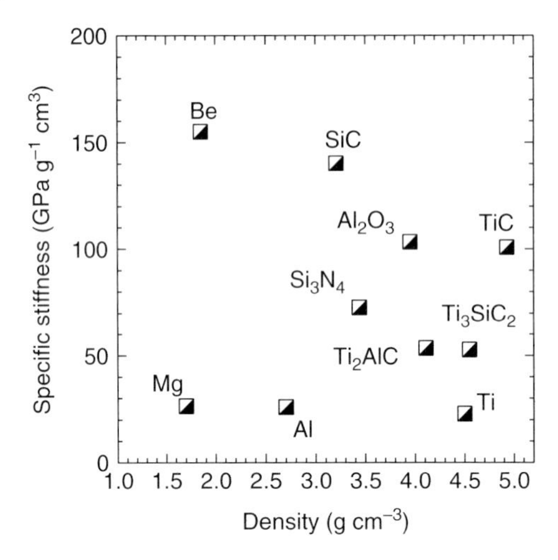

Figure 3.1 Specific stiffness versus density map for select MAX phases and structural ceramics.

phases. At this juncture, it is worth noting that Cover *et al.* (2009) published a comprehensive review on the elastic properties of over 250 M_2AX phases, both those known to exist and many that do not. The reader is referred to that comprehensive paper for more details.

The results listed in Table 3.1 prompt the question: how good are they? In other words, how close are they to the true elastic constants? This question is especially pertinent here since, as noted above, no experimental results exist. The problem is further compounded by the fact that applying a load to the MAX phases, more often than not, results in nonlinear, that is, non-Hookean, elasticity (Chapter 8).

In contrast to most other MAX phases, and for reasons that are not entirely clear, Ti_2SC, especially when the grains are small, deforms in a linear elastic fashion up to a stress of at least 700 MPa (Figure 3.2a). The stress–strain curves were obtained *in situ* in a neutron beam (Shamma *et al.*, 2011). To compare the theoretical and experimental elastic constants the following methodology was used: First, the macroscopic stress–strain plots measured using an extensometer were plotted (gray background in Figure 3.2a). Second, stress–strain curves derived using the elasto-plastic self-consistent (EPSC) approach (Clausen *et al.*, 1999; Clausen, Lorentzen, and Leffers, 1998; Clausen *et al.*, 2008; Turner and Tomé, 1994) were generated. This model treats grains as inclusions in an infinite, homogenous matrix, whose elastic properties are equal to the average polycrystal moduli of all grains in the sample. Because this model averages over a large number of grains, it captures the large sampling volume of the neutron diffraction measurements and is thus ideally suited to analyze the latter. The EPSC model takes elastic anisotropy into account and uses as input the DFT-calculated c_{ij} values. These values for Ti_2SC have been calculated a number of times (Table 3.2) using various assumptions. The first entry in Table 3.2 used a local density approximation (LDA); the others, the generalized gradient approximation (GGA).

Table 3.1 Summary of MAX phases' elastic constants c_{ij} determined from *ab initio* calculations. Also listed are the values of G_v and E_v calculated using Eqs. 3.2 and 3.4, respectively.

	c_{11}	c_{12}	c_{13}	c_{33}	c_{44}	E_v	G_v	Reference
413 Phases								
α-Ti_4AlN_3	420	73	70	380	128	359	153	Holm *et al.* (2002)
α-Ti_4AlN_3	405	94	102	361	160	365	153	Scabarozi *et al.* (2009)
α-Ti_4AlN_3	387	96	96	352	153	349	146	Wang *et al.* (2008)
Ti_4GeC_3	381	96	95	349	148	341	143	Li and Wang (2010)
α-Ta_4AlC_3	437	158	197	416	165	364	143	Du *et al.* (2008a)
α-Ta_4AlC_3	496	154	181	417	200	429	174	Wang *et al.* (2008)
β-Ta_4AlC_3	509	143	156	440	147	404	162	Wang *et al.* (2008)
Ta_4GaC_3	389	84	78	323	131	332	140	He *et al.* (2011)
α-V_4AlC_3	435	121	105	384	168	384	160	Wang *et al.* (2008)
α-V_4AlC_3	458	107	110	396	175	406	171	Li *et al.* (2009)
α-Nb_4AlC_3	413	124	135	328	161	353	144	Wang *et al.* (2008)
312 Phases								
Ti_3SiC_2	365	125	120	375	122	307	123	Holm, Ahuja, and Johansson (2001)
Ti_3SiC_2	355	96	103	357	160	339	142	Wang and Zhou (2004b)
Ti_3GeC_2	355	143	80	404	172	345	144	Finkel *et al.* (2004)
Ti_3AlC_2	361	75	70	299	124	311	132	Sun *et al.* (2004)
Ti_3AlC_2	368	81	76	313	130	320	135	He *et al.* (2010)
Ti_3SnC_2	346	92	84	313	123	300	124	He *et al.* (2010)
211 Phases								
Ti_2AlN	342	56	96	283	123	300	126	Holm *et al.* (2002)
Ti_2AlN	312	69	86	283	127	285	119	Scabarozi *et al.* (2009)
Ti_2AlN	304	68	91	290	135	288	121	Jia and Yang (2010)
Ti_2AlN	315	61	79	299	123	289	122	Du *et al.* (2009a)
Ti_2AlN	311	71	102	298	133	290	120	Bouhemadou, Khenata, and Chegaar (2007)
Ti_2AlN	309	67	90	282	125	282	118	Cover *et al.* (2009)
Ti_2AlC	301	59	55	278	113	272	117	Du *et al.* (2009a)
Ti_2AlC	308	55	60	270	111	274	117	Wang and Zhou (2004a)
Ti_2AlC	321	76	100	318	144	306	128	Sun *et al.* (2004)
Ti_2AlC	307	58	63	284	118	280	120	Bouhemadou, Khenata, and Chegaar (2007)
Ti_2AlC	302	62	58	270	109	268	114	Cover *et al.* (2009)
Ti_2SnC	337	86	102	329	169	332	140	Kanoun, Goumri-Said, and Reshak (2009b)
Ti_2SnC	260	78	70	254	93	226	93	Cover *et al.* (2009)
Ti_2GeC	279	99	95	283	125	257	105	Bouhemadou (2009a)
Ti_2GeC	278	75	96	283	118	257	106	Cover *et al.* (2009)
Ti_2GeC	309	84	105	321	142	293	123	Gui *et al.* (2011)
Ti_2GaC	314	66	59	272	122	283	121	Bouhemadou and Khenata (2007)

(*continued overleaf*)

Table 3.1 (*Continued*)

	c_{11}	c_{12}	c_{13}	c_{33}	c_{44}	E_v	G_v	References
Ti_2GaC	303	66	63	263	101	260	109	Cover *et al.* (2009)
Ti_2SC	335	98	99	362	161	310	127	See Table 3.2
Ti_2SnC	303	84	88	308	121	275	114	Bouhemadou (2008b)
Ti_2CdC	258	68	46	205	33	174	69.6	Bai *et al.* (2010)
Ti_2InC	288	62	53	248	88	241	102	Bai *et al.* (2010)
Ti_2InC	282	65	55	240	86	234	98.0	Cover *et al.* (2009)
Ti_2InN	229	56	106	248	92	208	83.3	Cover *et al.* (2009)
Ti_2PbC	235	90	53	211	66	182	73.2	Cover *et al.* (2009)
V_2AlC	346	71	106	314	151	324	136	Wang and Zhou (2004a)
V_2AlC	338	92	148	328	155	315	128	Sun *et al.* (2004)
V_2AlC	339	71	100	319	148	319	134	Cover *et al.* (2009)
V_2GeC	282	121	95	259	160	277	114	Bouhemadou (2009a)
V_2GeC	311	122	140	291	158	290	116	Scabarozi *et al.* (2009)
V_2GeC	289	122	134	279	142	263	105	Cover *et al.* (2009)
V_2GeC	302	99	132	308	145	278	112	Yang *et al.* (2012)
V_2GaC	343	67	124	312	157	326	136	Bouhemadou and Khenata (2007)
V_2GaC	334	81	111	299	138	302	125	Cover *et al.* (2009)
V_2GaC	329	82	123	306	124	288	117	Shein and Ivanovskii (2010b)
V_2GaN	281	71	142	293	128	263	106	Cover *et al.* (2009)
V_2AsC	334	109	157	321	170	318	128	Scabarozi *et al.* (2009)
V_2AsC	330	124	137	320	161	307	124	Cover *et al.* (2009)
V_2PC	376	113	168	386	204	376	154	Cover *et al.* (2009)
Cr_2AlC	384	79	107	382	147	351	146	Sun *et al.* (2004)
Cr_2AlC	345	67	95	333	153	330	140	Jia and Yang (2010)
Cr_2AlC	365	84	102	369	140	332	138	Cover *et al.* (2009)
Cr_2AlC	340	69	93	472	113	320	132	Du *et al.* (2011)
Cr_2GeC	315	99	146	354	89	249	96.7	Bouhemadou (2009a)
Cr_2GeC	308	147	140	333	74	213	80.5	Cover *et al.* (2009)
Cr_2GaC	312	81	139	325	128	283	114	Cover *et al.* (2009)
Mo_2GaC	294	98	160	289	127	257	101	Cover *et al.* (2009)
Mo_2GaC	306	101	169	302	102	241	93.0	Shein and Ivanovskii (2010b)
Nb_2AlC	310	90	118	289	139	285	116	Cover *et al.* (2009)
Nb_2AlC	341	94	117	310	150	313	129	Sun *et al.* (2004)
Nb_2GaC	309	80	138	262	126	270	108	Cover *et al.* (2009)
Nb_2GaC	374	88	135	310	149	329	135	Bouhemadou and Khenata (2007)
Nb_2GaC	303	81	132	260	110	253	101	Shein and Ivanovskii (2010b)
Nb_2SnC	268	86	119	267	98	226	89.3	Scabarozi *et al.* (2009)
Nb_2SnC	341	106	169	321	183	314	134	Kanoun, Goumri-Said, and Reshak, 2009b)
Nb_2SnC	321	101	99	311	132	277	112	Kanoun, Goumri-Said, and Reshak (2009b)
Nb_2SnC	252	96	129	244	99	209	81.5	Cover *et al.* (2009)
Nb_2SnC	286	91	127	288	100	237	93	Romero and Escamilla (2012)
Nb_2SnC	315	99	141	309	124	189	107	Bouhemadou (2008b)

Table 3.1 (*Continued*)

	c_{11}	c_{12}	c_{13}	c_{33}	c_{44}	E_v	G_v	References
Nb_2InC	291	76	108	267	102	247	99.4	Cover *et al.* (2009)
Nb_2InC	291	77	118	289	57	219	80	Shein and Ivanovskii (2010a)
Nb_2InC	363	103	131	306	148	314	128	Bouhemadou (2008a)
Nb_2AsC	334	104	169	331	167	317	127	Scabarozi *et al.*, 2009
Nb_2AsC	327	108	155	347	162	313	126	Cover *et al.* (2009)
Nb_2PC	369	113	171	316	170	333	134	Cover *et al.* (2009)
Nb_2SC	309	106	159	310	118	259	101	Cover *et al.* (2009)
Hf_2SC	344	116	138	369	175	336	137	Bouhemadou and Khenata (2008)
Hf_2SC	320	91	114	334	148	306	126	Cover *et al.* (2009)
Hf_2SC	311	92	116	329	147	299	122	Fu *et al.* (2010)
Hf_2SnC	318	96	99	301	123	280	114	Kanoun, Goumri-Said, and Reshak (2009b)
Hf_2SnC	330	54	125	292	167	326	138	Kanoun, Goumri-Said, and Reshak (2009b)
Hf_2SnC	240	62	103	236	92	211	84.5	Cover *et al.* (2009)
Hf_2SnC	293	87	101	278	110	254	103	Roumily, Medkour, and Maouche (2009)
Hf_2SnC	311	92	97	306	119	275	112	Bouhemadou (2008b)
Hf_2SnN	240	62	103	236	92	211	84.5	Cover *et al.* (2009)
Hf_2SnN	266	99	138	278	110	230	89.7	Roumily, Medkour, and Maouche (2009)
Hf_2InC	284	69	65	243	91	238	98.7	Scabarozi *et al.* (2009)
Hf_2InC	281	66	61	242	82	230	95.4	Cover *et al.* (2009)
Hf_2SC	320	91	114	334	148	306	126	Cover *et al.* (2009)
Hf_2PbC	241	77	70	222	69	191	81	Cover *et al.* (2009)
Hf_2PbC	245	73	70	230	76	201	76.5	Qian *et al.* (2012)
Ta_2AlC	334	114	130	322	148	303	122	Cover *et al.* (2009)
Ta_2AlC	339	113	130	326	154	311	126	Scabarozi *et al.* (2009)
Ta_2GaC	335	106	137	315	137	294	118	Cover *et al.* (2009)
Ta_2GaC	420	101	146	333	175	374	154	Bouhemadou and Khenata (2007)
Ta_2GaN	333	187	150	364	141	277	107	Cover *et al.* (2009)
Zr_2SnC	225	62	90	224	88	199	80.3	Cover *et al.* (2009)
Zr_2SnC	269	80	107	290	148	275	114	Kanoun, Goumri-Said, and Reshak (2009b)
Zr_2SnC	279	92	97	272	111	252	104	Bouhemadou (2008b)
Zr_2SC	326	103	119	351	160	318	130	Bouhemadou and Khenata (2008)
Zr_2SC	288	94	104	312	135	275	112	Cover *et al.* (2009)
Zr_2InC	251	62	58	215	73	204	84.0	Cover *et al.* (2009)
Zr_2InN	241	71	89	223	85	203	81.4	Cover *et al.* (2009)
Zr_2PbC	219	70	67	206	68	178	71.4	Cover *et al.* (2009)
Zr_2PbC	217	73	71	227	59	169	67	Qian *et al.* (2012)

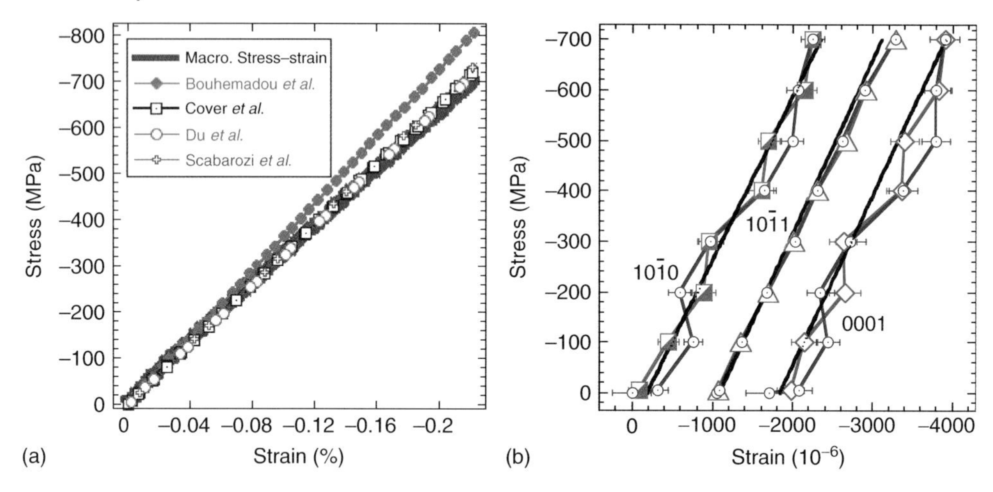

Figure 3.2 (a) Stress–strain curves obtained on Ti_2SC cylinders compressed to 700 MPa (Shamma *et al.*, 2011) using an extensometer (dark gray background) and those calculated from the EPSC model assuming the c_{ij} values listed in Table 3.2 calculated by Scabarozi *et al.* (2009) (open crosses); Cover *et al.* (2009) (open squares); Bouhemadou and Khenata (2008) (solid diamonds); Du *et al.* (2008b) (open circles). (b) Single-peak stress–strain curves on loading (open circles) and unloading (other symbols) for the 0001, 10$\bar{1}$0, and 10$\bar{1}$1 reflections in the longitudinal direction. The EPSC results are plotted as solid lines in the background. The agreement is again quite good.

Table 3.2 Elastic constants c_{ij} (in GPa) for Ti_2SC used in the EPSC model. With the exception of the top entry, which was obtained from local density approximation (LDA), all others used the generalized gradient approximation (GGA). The former resulted in values that were too high. The last row lists the average of entries 3–5. These values are currently recommended for Ti_2SC.

c_{11}	c_{12}	c_{13}	c_{33}	c_{44}	c_{66}	**Comments**
368.0	108.0	123.0	395.0	189.0	130.0	Bouhemadou and Khenata (2008)
319.0	101.0	106.0	351.0	149.0	109.0	Cui *et al.* (2009)
339.0	90.0	100.0	354.0	162.0	125.0	Scabarozi *et al.* (2009)
331.0	98.0	97.0	384.0	157.0	117.0	Cover *et al.* (2009)
336.0	106.0	99.0	348.0	163.0	115.0	Du *et al.* (2008b)
335 ± 4	98 ± 8	99 ± 2	362 ± 19	161 ± 3	119 ± 5	Average of the last three entries

Figure 3.2a compares the effects of stress on strain measured directly using a strain gauge (dark background), with those of the EPSC model using the c_{ij} values listed in Table 3.2 (data points in Figure 3.2a). From these results it is obvious that the values derived by Bouhemadou and Khenata are too high; those of Cui *et al.* (not shown) are slightly too low. The other three studies are closer to each other and to the actual experimental results. The last row in the Table 3.2 averages

the latter, and at this time these values can be considered to be the most reliable. Figure 3.2b compares the single peak longitudinal strains upon loading (open blue circles) and unloading (other symbols) together with those of the EPSC model predictions shown as solid black lines, using the averaged values listed in the last row of Table 3.2. Clearly, the agreement is quite good for the three reflections. The same is true of the transverse strains (not shown) (Shamma *et al.*, 2011).

These results are important for a number of reasons. First and foremost, they are the first evidence that the *ab initio* c_{ij} values are indeed valid. They also confirm that Ti_2SC behaves in a linear elastic fashion and that the EPSC model captures that response adequately. Based on this set of results, it is reasonable to assume that the c_{ij} values obtained from the DFT calculations are accurate to, conservatively, 10%. Note that the values reported herein are, strictly speaking, valid only at 0 K and assume perfect crystals. It thus follows that at least some of the discrepancies between the measured and calculated values can be ascribed to the fact that the experiments were carried out at ambient temperatures and that the Ti_2SC polycrystalline sample was not perfect.

A more serious problem is the discrepancy in the c_{ij} values reported by various investigators using different methods (Table 3.1). Some of these discrepancies arise from the different assumptions made in the *ab initio* calculations. This problem is discussed further below when discussing the bulk moduli B. Until further results, especially on large single crystals, are obtained, the best strategy is to average the various c_{ij} values reported in the literature for a given MAX phase as was done for Ti_2SC (last entry in Table 3.2).

Most layered solids, such as graphite and mica, are elastically quite anisotropic. The MAX phases, as shown in Table 3.1, are mildly so. For example, Holm, Ahuja, and Johansson (2001) predicted c_{33} and c_{11} for Ti_3SiC_2 to be 375 and 365 GPa, respectively. The same is true of some of the M_2AlC phases (Sun *et al.*, 2004; Wang and Zhou, 2004a). With a c_{11} of 308 GPa and a c_{33} of 270 GPa, Ti_2AlC is slightly more anisotropic than Ti_3SiC_2 (Zhou and Sun, 2000).

In short, the little evidence that exists suggests that until direct measurements of the elastic constants are available on large single crystals the c_{ij}s obtained from DFT calculations are accurate enough for most practical purposes.

3.3 Young's Moduli and Shear Moduli

Experimentally, E and G can be calculated from the longitudinal v_L and shear v_S velocities of sound in solids assuming the following relationships:

$$E = v_S^2 \rho \frac{\left(3v_L^2 - 4v_S^2\right)}{\left(v_L^2 - v_S^2\right)} \tag{3.1a}$$

and

$$G = \rho\, v_S^2 \tag{3.1b}$$

where ρ is the density.

Table 3.3 Summary of theoretical and measured longitudinal and shear sound velocities and elastic moduli (in GPa) for Ti_2SC calculated from the c_{ij}s using Eqs. (3.1a) and (3.1b) listed in the last row in Table 3.2. Agreement between theory and experiment is quite good. The density of this sample was measured to be 4.62 Mg m^{-3}.

v_L (m s^{-1})	v_S (m s^{-1})	E_v	G_v	B_v	Comments
8870[a]	5446[a]	328[b]	137[b]	180[b]	Using c_{ij}s listed in last row in Table 3.2
8200[c]	5200[c]	291[a]	125[a]	145[b]	Scabarozi *et al.* (2008)
8180 ± 3[c]	5248 ± 7[c]	293[a]	129[a]	137	Shamma *et al.* (2011)

[a]Calculated using Eqs. (3.1 a and b).
[b]Calculated using Eqs. (3.2–3.4).
[c]Measured values.

To convert the c_{ij}s, listed in Table 3.1 to moduli values, the Voigt shear modulus G_v is calculated assuming:

$$G_v = \frac{1}{15}\left(2c_{11} + c_{33} - c_{12} - 2c_{13}\right) + \frac{1}{5}\left[2c_{44} + \frac{1}{2}\left(c_{11} - c_{12}\right)\right] \tag{3.2}$$

The corresponding Voigt bulk modulus B_v is given by

$$B_v = \frac{2}{9}\left(c_{11} + c_{12} + 2c_{13} + \frac{c_{33}}{2}\right) \tag{3.3}$$

Finally, the E_v values are calculated assuming:

$$E_v = \frac{9G_vB_v}{3B_v + G_v} \tag{3.4}$$

When the values of B_v, G_v, E_v, v_L, and v_S for Ti_2SC computed from Eqs. (3.2–3.4) are compared to those measured (Table 3.3), the agreement is good. Furthermore, at 316 and 327 GPa, the Young's moduli in the $[10\bar{1}0]$ and [0001] directions for Ti_2SC (Figure 3.2b), respectively, calculated from neutron diffraction (ND) results, agree well with those predicted from *ab initio* calculations viz. 324 and 337 GPa, respectively. These results are further evidence that the c_{ij}s calculated from DFT are reasonably accurate. They also confirm that at least some of the MAX phases are more or less elastically isotropic.

Table 3.4 lists the measured – mostly from the velocity of sound, that is, Eq. (3.1) – values of E, G, and corresponding Poisson ratios, ν, of a number of MAX phases.

For the most part, the agreement between the measured and calculated E values is acceptable (see Figure 3.3a). Least squares linear regression of the data results in a correlation coefficient R^2 value of 0.75. On average, however, the theoretical values tend to overestimate the measured values. This is not too surprising since defects such as vacancies and impurities/solid solutions – which are not taken into account in the DFT calculations – can result in a reduction of the elastic properties. The fact that the DFT calculations were carried out at 0 K must also play a role.

Table 3.4 Measured Young's modulus E, shear modulus G, and Poisson ratio ν of select MAX phases calculated from Eqs. (3.1) and (3.8). Also listed are the densities and, when reported, ν_L and ν_S.

Solid	Density (Mg m^{-3})	G (GPa)	E (GPa)	ν	ν_L (m s^{-1})	ν_S (m s^{-1})	Comments and references
				413 Phases			
Ti_4AlN_3	4.7	127	310	0.22	8685	5201	Finkel, Barsoum, and El-Raghy (2000)
Nb_4AlC_3	7.0	127	306	0.20	7000[a]	4270	Du et al. (2009b) and Hu et al. (2008c)
Nb_4AlC_3	7.0	149	365	0.22	7770 (average)	4675 (average)	parallel to basal planes; Hu *et al.* (2011)
Nb_4AlC_3	7.0	153	353	0.15			⊥ to basal planes; Hu *et al.* (2011)
β-Ta_4AlC_3	13.2	132	324	0.23	5400[a]	3150	Hu *et al.* (2007)
				312 Phases			
Ti_3SiC_2	4.5	139	343	0.20	9100	5570	Finkel *et al.* (2004) and Radovic *et al.* (2006)
Ti_3GeC_2	5.6	142	343	0.19	8230	5063	Finkel *et al.* (2004)
$Ti_3(Si,Ge)C_2$	5.02	137	322	0.18	8400	5650	Finkel *et al.* (2004)
Ti_3AlC_2	4.2	124	297	0.20	8880	5440	Radovic *et al.* (2006)
Ti_3AlCN	4.5	137	330	0.21	9092	5514	Radovic, Ganguly, and Barsoum (2008)
				211 Phases			
Ti_2AlC	4.1	118	277	0.19	8500	5400	Hettinger *et al.* (2005)
$Ti_2AlC_{0.5}N_{0.5}$	4.2	123	290	0.18	8670	5407	Radovic, Ganguly, and Barsoum (2008)
Ti_2AlN-a	4.25	120	285	0.18	8553	5328	Radovic, Ganguly, and Barsoum (2008)
Ti_2AlN-b	4.3	112	277	0.16	8700	5100	(Radovic, Ganguly, and Barsoum (2008)
Ti_2SC	4.6	125	290	0.16	8200	5200	Scabarozi *et al.* (2008)
Ti_2SC	4.62	129	293	0.16	8180	5248	See Table 3.2 Shamma *et al.* (2011)
V_2AlC	4.8	116	235	0.20	7000	4913	Hettinger *et al.* (2005)
V_2AlC	—	—	283	—	—	—	Hu *et al.* (2008a)
Cr_2AlC	5.24	105	245	0.20	6200	4450	Hettinger *et al.* (2005)
Cr_2AlC	5.1	116	288	0.24	7550	4770	Tian *et al.* (2007)
Cr_2AlC	—	121	285	0.18	—	—	Lin, Zhou, and Li (2007)
Cr_2AlC	5.2	—	282	—	—	—	Ying *et al.* (2011)
Cr_2GeC	6.9	80	208	0.29	6300	3422	Amini *et al.* (2008)
Nb_2AlC	6.3	117	286	0.21	7165	4306	Hettinger *et al.* (2005)
Nb_2AlC	6.44	—	294	—	6750[a]	—	Zhang *et al.* (2009b)
Nb_2SnC	8.0	—	216	—	5200[a]	—	El-Raghy, Chakraborty, and Barsoum (2000)
Ta_2AlC	11.46	121	292	0.2	5350	3250	Hu *et al.* (2008b)

(*continued overleaf*)

Table 3.4 (*Continued*)

Solid	Density ($Mg\ m^{-3}$)	G (GPa)	E (GPa)	ν	ν_L ($m\ s^{-1}$)	ν_S ($m\ s^{-1}$)	Comments and references
Hf_2SnC	11.2	—	237	—	4600[a]	—	El-Raghy, Chakraborty, and Barsoum (2000)
Zr_2SnC	6.9	—	178	—	5080	—	El-Raghy, Chakraborty, and Barsoum (2000)

[a] Calculated assuming $E = \rho\, v^2{}_L$

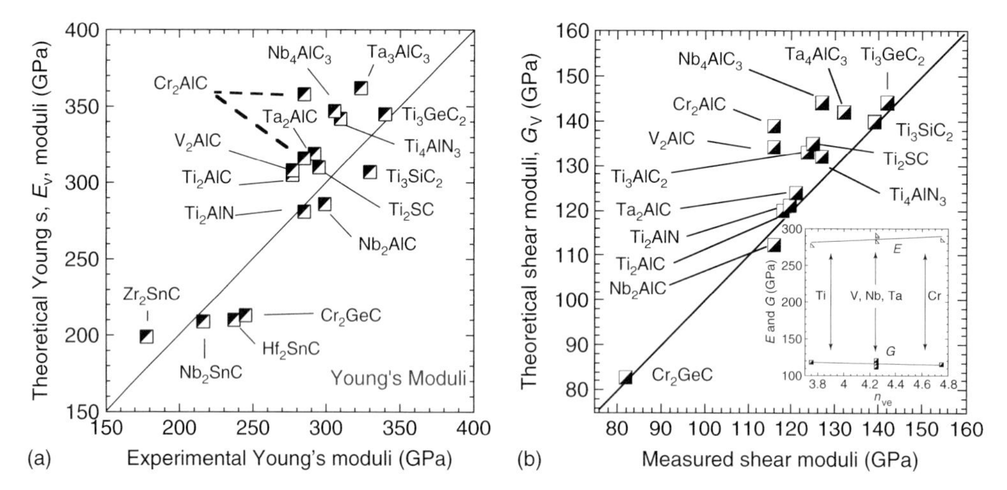

Figure 3.3 Comparison of experimental and theoretical (a) E and (b) G values of select MAX phases. The theoretical values are G_v and E_v calculated from Eqs. (3.2) and (3.4), respectively. Inset in (b) plots the experimental G and E results as a function of z_{av}. No correlation is found. In this plot, the results of Tian *et al.* (2007) and Lin, Zhou, and Li (2007) for Cr_2AlC were averaged. The results of Hettinger *et al.* (2005) for Cr_2AlC appear to be anomalously low and were not included.

This is not to imply that the details of the DFT calculations do not play a role. Recently Du *et al.* (2011) – spurred in part by the large discrepancy between the measured and calculated values of E for Cr_2AlC – recalculated its elastic properties using various approximations. The results showed that the c_{ij} values obtained are strong functions of the assumptions made.

A comparison of the measured and calculated G values (Figure 3.3b) again shows decent agreement between the two sets of results. Least squares linear regression yields $R^2 \approx 0.8$ (Figure 3.3b). As for E, $G_v > G$. As discussed below, the most likely reason for this state of affairs is the presence of defects in general, and point defects in particular.

3.3.1
The Bohm–Staver Relationship

Given that the velocities of sound, v_{ph} in some of the MAX phases are known (Table 3.4), they can be compared with the same values derived from the Bohm–Staver model, which makes use of the Fermi velocity v_F – calculated using Eq. (2.15) – which in turn is based on the Drude model. In that model, v_{ph} for a monoatomic solid with atoms of mass, M, and valence, Z, is given by the Bohm and Staver (1950) relationship

$$v_{Ph} = \sqrt{\frac{Z}{3}\frac{m_e}{M}}\; v_F \tag{3.5}$$

For compounds such as the MAX phases, the average mass, $M_{av,}$ and the average number of *formal* valence electrons in a unit cell, z_{av} are used instead of M and Z respectively. Since for the MAX phases, m_e/M is of the order of 10^{-4}–10^{-5}, z_{av} is of the order of 5 and v_F is of the order of 2.5×10^6 m s^{-1} (Table 2.2), Eq. (3.5) predicts sound velocities of the order of 8000–13 000 m s^{-1}. These values are roughly double those experimentally observed (Table 3.4) suggesting that the v_F values given by Eq. (2.15) are correct to within a factor of 2.

By combining Eqs. (2.15) and (3.5), equating v_{ph} with v_L, and replacing M by M_{av} in gm/mol and Z by z_{av}, the following relationship is recovered:

$$v_L = \frac{9.8 \times 10^5}{r_s/r_B}\sqrt{\frac{z_{av}}{3M_{av}}} \tag{3.6}$$

It follows that a plot of v_L versus $r_B/r_s\left(\sqrt{z_{av}/3M_{av}}\right)$ should yield a straight line, as observed (Figure 3.4a). Similarly, when v_S is plotted versus $r_B/r_s\left(\sqrt{z_{av}/3M_{av}}\right)$,

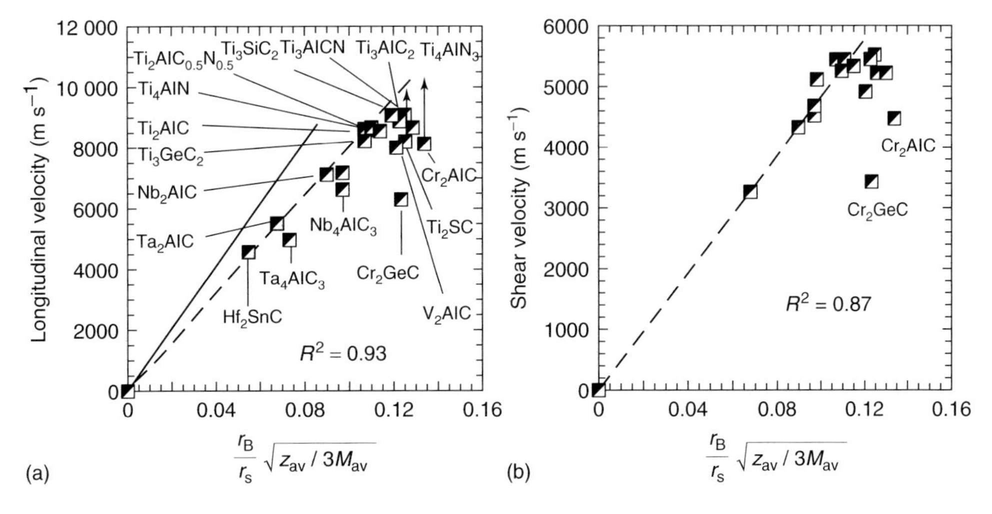

Figure 3.4 Functional dependence of (a) v_L and (b) v_S on $(r_B/r_s)\sqrt{z_{av}/3M_{av}}$. The solid inclined line in (a) is a plot of Eq. (3.6), i.e., has a slope of almost 10^6 with no other adjustable parameters. Note that in all cases the formal charges are used to calculate z_{av}.

again a good correlation is observed (Figure 3.4b). The outliers in Figure 3.4b are the same as Figure 3.4a.

And while the experimental results do not match those predicted by Eq. (3.6) – given by the inclined solid line in Figure 3.4a – they are not too far removed. Least squares linear regression yields $R^2 = 0.93$, implying that the assumptions made in deriving the Bohm–Staver relationship are applicable to the MAX phases. These assumptions are as follows (Ashcroft and Mermin, 1976): (i) the ions are considered point charges, with the *formal* charges assigned to them (Chapter 2); (ii) the forces the conduction electrons exert on the ions are neglected; (iii) electron–electron interactions are neglected; and (iv) the ions are not static, but oscillate with a phonon plasma frequency $\omega_{p,ph}$ given by (Ashcroft and Mermin 1976)

$$\omega_{p,ph} = \sqrt{\frac{z_{av} m_e}{M_{av}}}\, \omega_{p,val} \tag{3.7}$$

where $\omega_{p,val}$ is the electron plasma frequency given by Eq. (2.17).

Returning to Figure 3.4a, the dashed line was deliberately *not* drawn through all the data points, but rather through a select few that also included the origin. Given that defects, especially vacancies, are potent phonon scatterers, it is reasonable to assume that the concentration of defects in the compositions lying on the dashed line are small and/or do not adversely reduce ν_L or ν_S. Conversely, it is reasonable to assume that for the compositions that do *not* fall on the line do not do so because they are defective, nonstoichiometric, contain impurities, or are so anomalous that the simple model breaks down.

The vertical arrows in Figure 3.4a, indicate what the ν_L values would have been had they not been defective and/or anomalous. To illustrate, consider Ti_4AlN_3, Cr_2GeC, Ti_2SC, Ti_3AlCN, V_2AlC, Ta_4AlC_3 and Nb_4AlC_3.

The true composition of the Ti_4AlN_3 sample is $Ti_3AlN_{2.9}$. Given how potent vacancies are in scattering phonons, it is not surprising that ν_L for this composition lies below the dashed line. Similar arguments can be made concerning Ti_2SC, Cr_2AlC, Ta_4AlC_3 and Nb_4AlC_3. All four are probably nonstoichiometric. One signature of nonstoichiometry is scatter in elastic properties. The velocity of sound in Nb_4AlC_3 was measured twice (Hu *et al.*, 2008c, 2011); the first time the velocity was $\approx$6611 m s^{-1} and in the second it was closer to 7700 m s^{-1} in (nominally) identical compositions. The same is true for Cr_2AlC; the first study reported a ν_L of $\approx$7100 m s^{-1} (Hettinger *et al.*, 2005), the next two reported values closer to 8000 m s^{-1} (Lin, Zhou, and Li, 2007; Tian *et al.*, 2007). The simplest explanation for such variations is the presence of a hitherto uncontrolled variable, most likely point defects, on one or more sublattices.

The velocity of sound in the quaternary compound Ti_3AlCN is slightly lower than where it should be probably as a result of it being a solid solution.

The results for Cr_2GeC are the most intriguing. The discrepancy between the experimental values and where, based on the Drude model, ν_L or ν_S for this compound should be, is the largest. In other words, Cr_2GeC – and probably Cr_2AlC – is a true outlier. The reason for this state of affairs is unclear at this time, but *ab initio* calculations suggest the ν_L should be $\approx$6500 m s^{-1}, a value that is not too far from the experimental results. The same is true of ν_S; the DFT calculations predict $\approx$4100 m s^{-1}, which is $\approx$18% higher than the experimentally measured values. It thus appears that not all deviations from the Bohm–Staver expression are due to defects. In some cases, some of the deviation can be accounted for by bonding intricacies, which are only captured by the DFT calculations. Lastly, it is important to note here, again, how increasing z_{av} results in a maximum in properties, not unlike what was observed for the surface energies (Figures 2.22 and 2.23) and other properties discussed in subsequent chapters.

3.3.2
Effect of z_{av}

There are other ways to look at trends in E and G. Figure 3.5a plots the latter as a function z_{av}, for select MAX phases. Two effects – outlined by the solid and dashed lines – can be discerned: the first is the "too many electrons" effect; the other is the size of the A-group element. The former idea, touched upon in Chapter 2, postulates that when z_{av} is too large, electrons are pushed into antibonding orbitals, destabilizing the structure. The abrupt drop in E and G for the Cr_2GeC is probably the best evidence for this hypothesis. Thermally, this is also manifested by one of the highest thermal expansions of all the MAX phases (Chapter 4). It is also

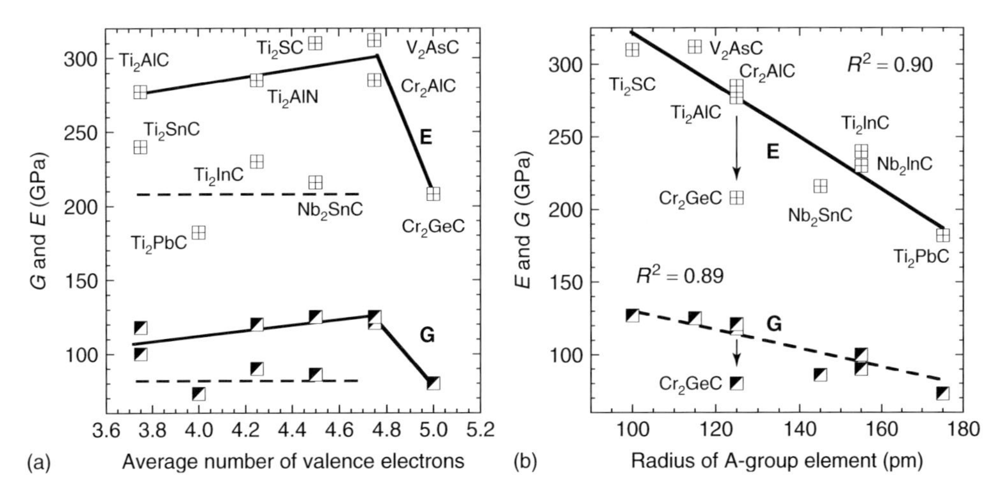

Figure 3.5 Dependencies of E and G on (a) z_{av} and (b) r_A. The values of E and G plotted are those listed in Table 3.1.

consistent with the fact that its density of states (DOS) at E_F is the highest of all MAX phases.

3.3.3
Effect of r_A

The results shown in Figure 3.5a also suggest that the A-group element radius, r_A, also plays an important role. This is better appreciated, and clearly demonstrated, when E and G are plotted against r_A (Figure 3.5b). In both cases, E and G decrease with increasing r_A. When the results for Cr_2GeC, denoted by arrows, are plotted, its anomalous nature is again clear. The R^2 value obtained when this point is not included (shown by the straight lines) is about 0.9; when that point is included, the R^2 value drop to $\approx$0.5.

A number of theoretical papers have predicted systematic changes in E or G as a function of M-group element in the M_2AlC family (Emmerlich *et al.*, 2007; Sun *et al.*, 2003, 2004; Wang and Zhou, 2004a). However, when the experimental G and E results are plotted as a function of z_{av} (inset in Figure 3.3b), no correlation is found.

In general, the effect of X on the elastic properties is weakest. This comment notwithstanding, in one case at least, replacing C with N resulted in a large increase in B (see below).

3.3.4
Anisotropy in Elastic Properties

In Table 3.1, c_{11} and c_{33} for Ti_3SiC_2 are predicted to be comparable. This prediction was indirectly confirmed experimentally (Kooi *et al.*, 2003; Murugaiah *et al.*, 2004). At 302 $\pm$ 5 GPa, the Young's moduli measured – using a Berkovich nanoindenter – of grains whose basal planes were parallel to the applied load were about 10% lower than the 346 $\pm$ 10 GPa of those whose basal planes were normal to the direction of applied load (Kooi *et al.*, 2003).

More recently, Hu *et al.* (2011) measured the velocities of sound in highly textured Nb_4AlC_3 samples and found that the E values parallel and perpendicular to the basal planes were, respectively, 365 and 353 GPa. The respective G values were 149 and 153 GPa. Note that there is a nontrivial difference in the E and G values reported by Hu *et al.* in 2008 and 2011 (Table 3.4). The higher of the two sets of values, namely, the more recent results, should be considered more accurate on the assumption that there are many contributing factors, such as vacancies and other defects, that can reduce a modulus, but few that can augment it. Along the same lines, and in light of more recent results, it appears that the results reported by Hettinger *et al.* (2005) for Cr_2AlC are probably too low. More systemic work on the effects of stoichiometry on the elastic properties of Cr_2AlC and other MAX phases that are not line compounds are needed.

3.4 Poisson's Ratios

Poisson's ratio, ν, is the ratio of the transverse to longitudinal strains when a material is elastically loaded. If the moduli are known, ν can be estimated assuming:

$$\nu = \frac{3B_v - 2G_v}{2(3B_v + G_v)} = \frac{v_L^2 - 2v_S^2}{(2v_L^2 - 2v_S^2)} = \frac{E}{2G} - 1 \tag{3.8}$$

Poisson's ratios for most of the MAX phases are around 0.2 (Table 3.4), which is lower than the 0.3 of Ti and most metals, and closer to the 0.19 of near-stoichiometric TiC.

At 0.215, Poisson's ratio of Ti_2SC measured from the *in situ* ND results is in excellent agreement with the value 0.217 calculated from the EPSC model (Shamma *et al.*, 2011). Hu *et al.* (2011) reported that the Poisson's ratios in their highly oriented Nb_4AlC_3 samples were anisotropic (Table 3.4).

3.5 Bulk Moduli

The bulk modulus B of a solid is a measure of its resistance to hydrostatic pressure. There are two methods by which B can be calculated. If the elastic constants are known, then the Voigt approximation, that is, Eq. (3.3), can be used to calculate B_v. A second method is to calculate B directly from the Birch–Murnaghan equation, henceforth referred to as B_{BM} to differentiate it from B_v given Eq. (3.3). To calculate B_{BM}, P versus V/V_o plots are fitted assuming the Birch–Murnaghan equation (Birch, 1978)

$$P = \frac{3}{2}B_{BM}\left[\left(\frac{V}{V_0}\right)^{-7/3} - \left(\frac{V}{V_0}\right)^{-5/3}\right]\left[1 + \frac{3}{4}\left(B' - 4\right)\left(\left(\frac{V}{V_0}\right)^{-2/3} - 1\right)\right] \tag{3.9}$$

is valid. In Eq. (3.9), V and V_0 are the unit cell volumes at pressure P and 0.1 MPa, respectively. B' is the pressure derivative of B. This equation is also used to determine B from experimental V/V_0 versus P curves obtained from diamond anvil cell experiments (Onodera *et al.*, 1999). Note that, for the vast majority of the MAX phases, B' is ≈ 4.

Table 3.5 compares the bulk moduli of select MAX phases. Note that both B_{BM} and B_v are listed. The scatter seen is typical of DFT calculations and can be traced back to the various assumptions and approximations made (see below). Figure 3.6 plots B_{BM} versus the measured B. Least squares fitting of the data results in an R^2 value of ≈ 0.68. This agreement is not too bad considering that defects and temperature – two factors not typically taken into account in theoretical calculations – are ignored. Along the same lines, there is little correlation (not shown) between the B values predicted from the Drude model (Table 2.2) and the experimental results.

The following factors are currently known to affect the values of B: (i) unit cell volume V_{UC}, (ii) chemistry and the value of n, (iii) defects, and (iv) solid solutions and the puckering of the basal planes. Each factor is briefly discussed below.

Table 3.5 Bulk moduli values of select MAX and MX phases measured directly in an anvil cell (B), calculated from DFT (B_{BM}), and calculated from Eq. (3.6) (B_V). All moduli values are given in GPa.

Solid	B	B_{BM}	B_V	References
		413 Phases		
$Ti_4AlN_{2.9}$	216	185	196	Manoun, Saxena, and Barsoum (2005), and Scabarozi *et al.* (2009)
Nb_4AlC_3	—	247	216	Du *et al.* (2009b), Hu *et al.* (2008c), and Wang *et al.* (2008)
β-Ta_4AlC_3	261 ± 2	266	260	(Manoun *et al.* (2006c), Hu *et al.* (2007), and Du *et al.* (2009b)
V_4AlC_3	—	255	—	Du *et al.* (2009b)
		312 Phases		
Ti_3SiC_2	206	—	190 225 202	Onodera *et al.* (1999), Ahuja *et al.* (2000), Zhou *et al.* (2001), and Finkel *et al.* (2004)
Ti_3GeC_2	179	—	191 198	Manoun *et al.* (2007c), Finkel *et al.* (2004), and Zhou *et al.* (2001)
$Ti_3(Si,Ge)C_2$	183 ± 4	—	—	Manoun *et al.* (2004a)
$Ti_3(Al,Sn_{0.2})C_2$	226 ± 3	—	—	Manoun *et al.* (2007b)
Ti_3AlC_2	156 ± 5	—	157 168 190	Zhang *et al.* (2009a), Zhou *et al.* (2001), and He *et al.* (2010)
Ti_3AlCN	219 ± 4	—	—	Manoun *et al.* (2007b)
		211 Phases		
Ti_2AlC	186	—	138 137 166	Manoun *et al.* (2006a), Sun *et al.* (2003), Cover *et al.* (2009), and Hug (2006)
Ti_2GeC	211	160	152	Phatak *et al.* (2009a), Bouhemadou (2009a), and Cover *et al.* (2009)
Ti_2AlN	169	—	155 163	Manoun *et al.* (2006b), Cover *et al.* (2009), and Holm *et al.* (2002)
Ti_2GaN	189	182	—	Manoun *et al.* (2010) and Bouhemadou (2009b)
Ti_2SC	191	205	177 176	Kulkarni *et al.* (2008b), Cover *et al.* (2009), Bouhemadou and Khenata (2008), and Hug (2006)
Ti_2SnC	152	170	135 167	Manoun *et al.* (2009), Kanoun, Goumri-Said, and Jaouen (2009a), Cover *et al.* (2009), and Hug (2006)
Ti_2InC	148	128	128 137	Manoun *et al.* (2009) and Cover *et al.* (2009)
TiZrInC	131	—	—	Manoun *et al.* (2009)
V_2AlC	201	—	175 203 181	Manoun *et al.* (2006a), Cover *et al.* (2009), Schneider, Mertens, and Music (2006), and Wang and Zhou (2004a)

Table 3.5 (*Continued*)

Solid	B	B_{BM}	B_V	References
V_2GeC	165 201	180	182	Manoun *et al.* (2007a), Bouhemadou (2009a), Cover *et al.* (2009), and Phatak *et al.* (2009a)
Cr_2AlC	166	182	186 200 226	Manoun *et al.* (2006a), Wang and Zhou (2004a), Cover *et al.* (2009), Sun *et al.* (2003), and Du *et al.* (2011)
CrVAlC	219	—	—	Phatak *et al.* (2009b)
Cr_2GeC	182 169	214	200	Manoun *et al.* (2007a), Phatak *et al.* (2008), Bouhemadou (2009a), and Cover *et al.* (2009)
Cr_2GaC	188	—	—	Manoun *et al.* (2010)
Nb_2AlC	208	186	183 173	Manoun *et al.* (2006a), Wang and Zhou (2004a), and Cover *et al.* (2009)
Nb_2AsC	224	—	204	Kumar *et al.* (2005), Cover *et al.* (2009)
Nb_2SnC	180	166	162 206	Manoun *et al.* (2009), Kanoun, Goumri-Said, and Jaouen (2009a), and Cover *et al.* (2009)
Ta_2AlC	251	—	221 191	Manoun *et al.* (2006a), Sun *et al.* (2004), and Cover et al. (2009)
Hf_2SnC	169	165	139	(Manoun *et al.* (2009), Kanoun, Goumri-Said, nd Jaouen (2009a), and Cover *et al.* (2009)
Hf_2InC	—	136	131	He *et al.* (2009) and Cover *et al.* (2009)
Zr_2SC	186	188	165	Bouhemadou and Khenata(2008), Kulkarni *et al.* (2008a), and Cover *et al.* (2009)
Zr_2SnC	—	—	156	Kanoun, Goumri-Said, and Jaouen (2009a)
Zr_2InC	127	131 137	119	Cover *et al.* (2009) and Manoun *et al.* (2004b, 2009)
		MX		
$TiC_{0.96}$	272	—	—	—
VC	200–390	—	—	Neither of the extreme values are credible.
ZrC	206	—	—	Toth (1971)
NbC	296	—	—	Pierson (1996)
HfC	241	—	—	Pierson (1996)
TaC	330	—	—	Lopez-de-la-Torrea *et al.* (2005)

3.5.1 Unit Cell Volume

For a family of isostructural compounds such as the 211 phases, it is reasonable to assume that the smaller the unit cell volume, V_{UC}, the stronger the bonds, and the higher the B values. When all known experimentally measured B values are plotted versus V_{UC} (Figure 3.7a), least squares fit yield a R^2 value of ≈ 0.1. If the Ta_2AlC

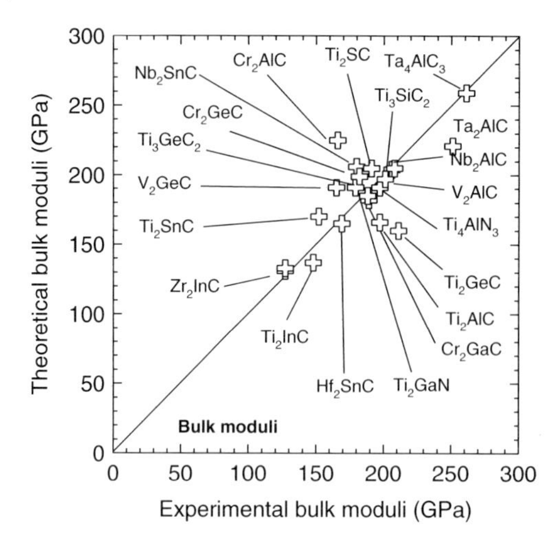

Figure 3.6 Comparison of experimental B and theoretical B_{BM} values of select MAX phases.

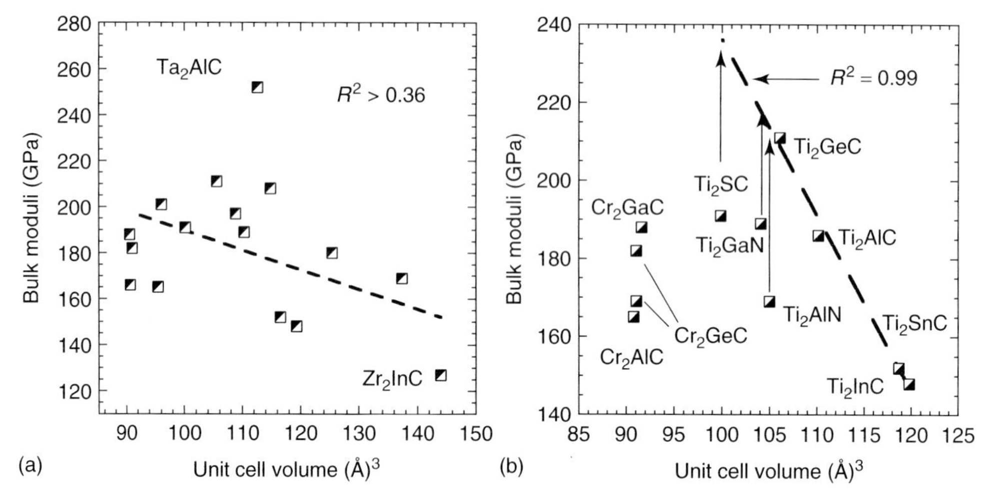

Figure 3.7 Dependence of experimental bulk moduli on (a) V_{UC} of all 211 phases measured. Note the outlier Ta_2AlC. When its value is excluded from the dataset, $R^2 = 0.36$. (b) V_{UC} of all Ti- and Cr-containing phases. In the case of Ti, an excellent correlation ($R^2 = 0.99$) is found for four of the seven phases listed (Manoun *et al.*, 2010). The Cr_2GeC composition was measured twice, once by Manoun *et al.* (2007a) and a second time by Phatak *et al.* (2008).

outlier (labeled in Figure 3.7a) is omitted, *the* R^2 increases to a more respectable, but unremarkable, 0.36.

When the B values of the Ti- and Cr-containing MAX phases are plotted versus V_{UC} (Figure 3.7b), an interesting picture emerges. Four of the Ti-containing phases – Ti_2AlC, Ti_2SnC, Ti_2GeC, and Ti_2InC – fall on a straight (dashed) line, with $R^2 = 0.99$. The other three fall off the line. The discrepancy is most probably due to the presence of vacancies, at least for the Ti_2AlN and Ti_2SC phases.

The *B* values of the Cr-containing ternaries, on the other hand, range from a low of 165 GPa for Cr_2AlC to a high of 188 GPa for Cr_2GaC (Figure 3.7b). These variations indicate that factors other than V_{UC} play a role. This comment notwithstanding, the differences between the various *B* values for the Cr-containing compounds are of the order of 10%. The variations between samples measured by two groups on nominally the same composition – Cr_2GeC – can account for about 5% of that variability. It follows that no strong conclusions can be made concerning the effect of V_{UC} on *B* for the Cr-containing compounds. Note, however, that the B values for the Cr containing phases are significantly lower than where they should have been, had they followed the dashed line in Figure 3.7b.

3.5.2 MAX Phase Chemistry

A number of DFT-based papers have explored the effects of chemistry on *B*. For example, when the DFT results of Hug (2006) are plotted (not shown) versus the electronic configuration of the A-group element for all possible Ti_2AC phases – including ones that do not exist – the general trend is for *B* to increase as the group of the A-element increases from 13 to 15, before decreasing slightly for Ti_2SC. This trend is less apparent in the experimental results, however. For example, the *B* values for Ti_2AlC and Ti_2SC predicted by Hug are 137 and 176 GPa, respectively. At 186 and 191 GPa, the measured values are significantly closer.

Similarly, when B_v for all known 211 phases – calculated by Cover *et al.* (2009) – are plotted versus V_{UC} (not shown), a good correlation is found. However, surprisingly, the data clearly split into two bands: one band included the 211 phases comprising Ti, V, and Cr, that is, row-4 elements; and the other containing row-5 and 6 M elements, with the latter being higher. Why for a given V_{UC} the larger M atoms result in solids with larger *B* values, is a mystery. This, together with the fact that the results of Cover *et al.* do not capture the Ta_2AlC outlier (Table 3.5 or Figure 3.7), is somewhat worrisome.

3.5.3 The Effect of n

Assessing the role of *n* on *B* is more straightforward. Most theoretical papers have shown a one-to-one correspondence between the *B* values of the MAX phases, B_{MAX}, and those of their respective MX binaries, B_{MX}, (Du *et al.*, 2009b; Fang *et al.*, 2006; He *et al.*, 2010; Hug, 2006; Kanoun, Goumri-Said, and Jaouen, 2009a). This is not too surprising given that the latter are comprised of blocks of the former. In general, the B_{MAX}/B_{MX} ratio is ≈ 0.75 (Du *et al.*, 2009b; He *et al.*, 2010; Music *et al.*, 2006). The ratio in the case of the M_2SnC phases is 0.68 (Kanoun, Goumri-Said, and Jaouen, 2009a). The aforementioned studies all compared theoretical values.

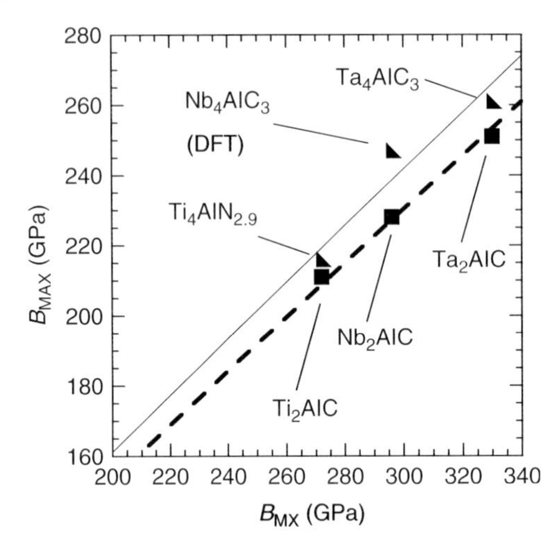

Figure 3.8 Relationship between mostly measured B values of select $M_{n+1}AX_n$ phases as a function of n, and their MX counterparts. The results for Nb_4AlC_3 are from DFT calculations. In both cases, the lines go through the origin. (Source: The values for the MX compounds were taken from Lopez-de-la-Torrea *et al.* (2005).)

Ironically, a major problem with testing this idea is that the B_{MX} values are subject to large variabilities (Lopez-de-la-Torrea *et al.*, 2005).

Gratifyingly, a good correlation can be found between the (mostly) experimental B values of the MAX phases and their MX counterparts (Figure 3.8). Least squares fit yield lines that go through the origin, as they should. The experimental B_{MAX}/B_{MX} ratio is ≈ 0.78, which is consistent with theoretical modeling.

Interestingly, there is *no* correlation between either z_{av} or n_{val} and B.

3.5.4
Defects

An important, but subtle factor, influencing the MAX phases' B values is their stoichiometry, more specifically their vacancy concentration. This effect is best seen in the B values of Ti_2AlN, where theory and experiment show a *decrease* in lattice parameters as C is substituted by N. Given that the lattice parameters shrink, it is not surprising that theory predicts that this substitution should *increase* B. Experimentally, however, B actually *decreases* with increasing N content, as shown by the solid blue lines in Figure 3.9 (Manoun *et al.*, 2007b). The paradox is resolved when it is appreciated that B is a function of vacancies and that the substitution of C by N results in the formation of vacancies on the Al and/or N sites. Note that for the *binary* Ti carbides, replacing C with N atoms results in an increase in B as shown by the three data points labeled I, II, and III in Figure 3.9. Along the same lines, replacing N by Al to form the MAX phases results in a decrease in B, as indicated by the dashed lines in Figure 3.9, and a concomitant decrease in z_{av}. Note that vacancies in the binary carbides also results in a decrease in B (see red X shown in Figure 3.9).

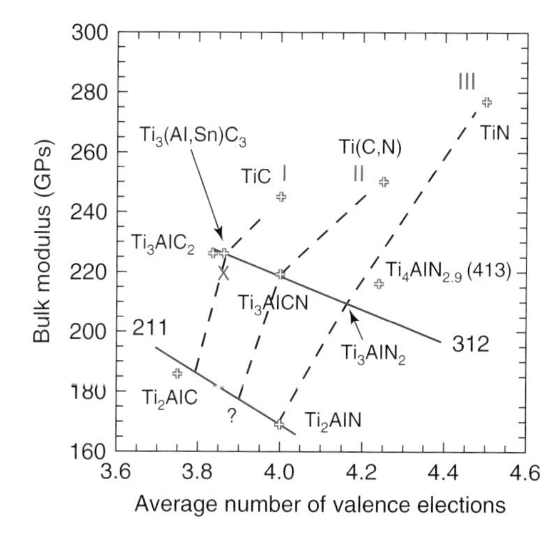

Figure 3.9 Dependence of B on z_{av} for binary TiX and $Ti_{n+1}AlX_n$ compounds (Manoun *et al.*, 2007b). TiX data were taken from Yang *et al.* (2000). Note that the addition of a small amount of Sn does not change the z_{av} from that of Ti_3AlC_2. The datum point marked with a red X was also taken from Yang *et al.* (2000) for a sample of composition $Ti(C_{0.5},N_{0.5})_{0.81}$.

Another example is Ti_2SC. The Ti/S ratio in the sample used to measure B was 2 : 0.96 instead of 2 : 1 (Amini, Barsoum, and El-Raghy, 2007). In other words, it is likely that this sample was substoichiometric in S, which in turn would explain its low B value. According to Figure 3.7b, a stoichiometric Ti_2SC sample should have a B value of $\approx$235 GPa. The same is presumably true for Ti_2AlN in Figure 3.7b. According to these results, it is reasonable to assume that had Ti_2AlN not contained vacancies, its B value would have been closer to 205 GPa. This comment notwithstanding, more systematic and careful work is needed to confirm some of these conjectures.

3.5.5 Solid Solutions and the Puckering of Basal Planes

Another subtle effect on B is the puckering, or corrugation, of the basal planes. Electron energy loss spectroscopy (EELS) results, together with *ab initio* calculations, have shown that in the solid solution TiNbAlC the basal planes are corrugated (Hug, Jaouen, and Barsoum, 2005). This corrugation, in turn, leads to a larger decrease in B – due to a softening along the c-axis – than one would otherwise anticipate based on the lattice parameters and unit cell volumes that decrease monotonically with increasing Nb content (Manoun *et al.*, 2007d).

Note that the formation of solid solutions does not always lead to a softening. For example, Phatak *et al.* (2009b) reported a B value for CrVGeC that was higher than those of both end members. The exact reason for this state of affairs is not clear at this time and awaits theoretical modeling and further experiments.

3.5.6
Anisotropy in Shrinkage

The MAX phases are not cubic, and it follows that the shrinkage along the a or c directions is not equal with increasing hydrostatic pressure P. This is best seen in Figure 3.10a, in which the relative lattice parameters a/a_0 and c/c_0 are plotted as a function of P. To further accentuate the importance of crystallographic direction, Figure 3.10b plots the ratio $(c/c_0)/(a/a_0)$ as a function of P. From these results, one can conclude that the shrinkage along the a direction is higher than that in the c direction for Ti_2AlC and V_2AlC and vice versa for Nb_2AlC and Cr_2AlC. For Ta_2AlC, the shrinkage is more or less the same along both directions (Manoun *et al.* 2006a).

Emmerlich *et al.* (2007) predicted that, with increasing P, the $(c/c_0)/(a/a_0)$ ratio should (i) decrease slightly before ultimately increasing for Ti_2AlC; (ii) increase very slightly for V_2AlC, Nb_2AlC, and Ta_2AlC; and (iii) increase for Cr_2AlC. Comparing these predictions with experimental results (Figure 3.10b) indicates that, while the predictions maybe valid for Ti_2AlC, Ta_2AlC, and Cr_2AlC, they are invalid for V_2AlC and Nb_2AlC. Another problem is that, in the 0–60 GPa range, the theoretical calculations predict that the $(c/c_0)/(a/a_0)$ ratio would range from 0.992 to 1.02. Experimentally that range is 0.97–1.01 (Figure 3.10b). At this time, the reasons for these discrepancies are unclear, but like much else in this chapter, they probably stem from two sources: the inherent assumptions made in the DFT calculations, and the fact that real solids contain defects and that the measurements are carried out at ambient temperatures and not at 0 K.

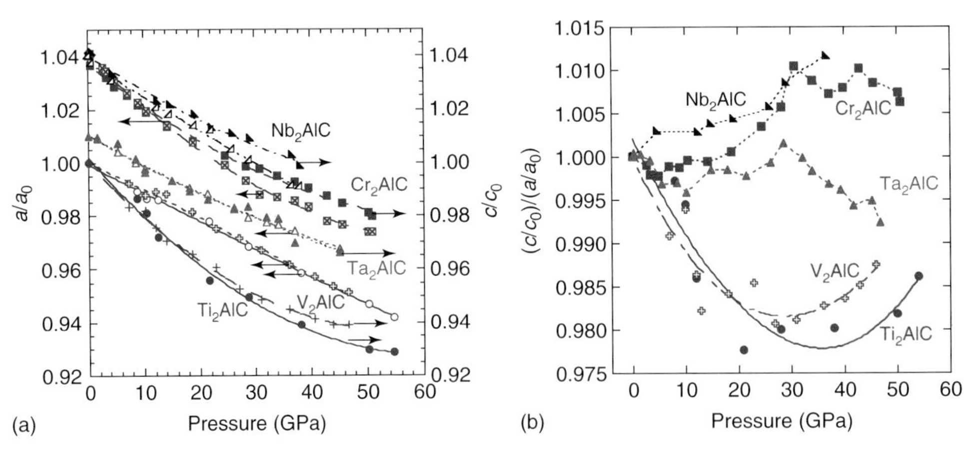

Figure 3.10 Pressure dependencies of (a) a/a_0 left axis and c/c_0 right axis and (b) their ratio for select M_2AlC phases. Curves are separated vertically for clarity; the symbol colors and those of the letters representing the various compounds are matched. The contraction along the a direction with pressure is greater than along the c direction for Nb_2AlC and Cr_2AlC; the opposite it true for Ti_2AlC and V_2AlC; Ta_2AlC is unique in that the variations along the two orthogonal directions are almost identical. The lines are guides to the eye (Manoun *et al.* 2006a).

3.5.7
Critical Analysis of Elastic Properties Predictions

At this point, using measured *B* values to validate *ab initio* calculations or vice versa is an exercise that is fraught with pitfalls for the following reasons:

1) The predicted *B* values depend on assumptions made in the DFT codes. The GGA used in CASTEP calculations tends to "underbind." In contrast, the LDA, used to obtain B_{BM} tends to "overbind," that is, gives smaller lattice parameters. As consequence of this overbinding, an overestimation of B can be expected, and indeed for the most part $B_{BM} > B_V$ (Table 3.5). For more details and examples, the interested reader is referred to a 2009 paper by Bouhemadou (2009b).
2) The predicted *B* values depend on the potentials chosen (Music *et al.*, 2006).
3) The predicted *B* values are influenced by experimental results. In general, the calculated *B* values are closer to the experimental results when the latter are known. For example, Zhou and coworkers predicted the *B* values for Ti_3AlC_2 to be 190 GPa (Zhou *et al.*, 2001) and 187 GPa (Wang and Zhou, 2003). More recently, B was measured for Ti_3AlC_2 to be 156 ± 5 GPa (Zhang *et al.*, 2009a). In the same paper, *ab initio* calculations predicted a *B* value of 157 GPa.
4) The predicted *B* values typically assume a temperature of 0 K and perfect crystals.

Based on a careful comparison of theoretical and experimental results discussed above, the following general conclusion can be reached: the LDA approximation is better at predicting the *B* values of the MAX phases than the GGA approximation and should be the one to use. In some cases – most notably the work of Kanoun, Goumri-Said, and Jaouen (2009a) who used the scheme of Wu and Cohen (2006) for exchange and correlation – the GGA approximation yields good agreement with experiment. Whether this agreement is coincidental or not awaits more work. Most other GGA approximations yield *B* values that are too low.

The opposite is probably true of the c_{ij}s, where the LDA approximation is worse at predicting the values than the GGA approximation. It is crucial to note that this conclusion is quite tenuous at this time and only based on the results for Ti_2SC (Table 3.2).

3.6
Extrema in Elastic Properties

Based on the foregoing discussion, it is now possible to outline strategies one could employ to engineer the elastic properties of the MAX phases. To obtain very stiff solids, small atoms should be used, especially for the A-group elements, while keeping the total number of electrons in the system to a minimum. It is presumably this combination that endows V_2PC with the *highest* predicted values of E_v and G_v of all 211 MAX phases explored (Cover *et al.*, 2009). Another approach is to combine the aforementioned ideas with the highest *n* value possible. This is presumably why $Ti_3SiC_2$and Ti_3GeC_2 have some of the highest E and G values reported to date.

To obtain MAX phases with low elastic properties one should (i) keep away from $n > 1$ phases; (ii) use 211 phases with the largest atoms, especially r_A; and (iii) have the largest average number of electrons possible. Note that destabilizing the structure is less effective than increasing the size of the A atoms, which explains why Ti_2PbC has one of the lowest E and G values (Figure 3.5b). However, if density and/or toxicity are issues, then destabilizing the structure is a good alternative. In all cases, adding defects should further reduce the elastic constants.

Similar strategies can be employed to engineer solids with low B values. For example, at 127 GPa, the B value of Zr_2InC (Manoun *et al.*, 2004b) is the lowest reported to date for any MAX phase. At the other extreme is Ta_4AlC_3; at 260 GPa, its B value is the highest experimentally reported to date (Manoun *et al.*, 2006c); Ta_2AlC comes in second with a B value of 251 GPa. It follows that the strategy for maximizing B is to choose a MAX phase for which the corresponding MX phase has the highest B value, together with maximizing the value of n.

3.7 Effect of Temperature on Elastic Properties

In general, and for most solids, increasing the temperature T reduces the elastic moduli. Typically, there are two regimes: a low T regime in which the moduli are weak functions of T; and a second, higher temperature regime wherein the moduli slowly decrease with increasing T. Each is discussed separately below.

3.7.1 Low Temperatures (4–300 K)

Figure 3.11 plots G – determined from ultrasound – for Ti_3SiC_2, Ti_3AlC_2, and $Ti_4AlN_{2.9}$ in the 4–300 K temperature regime (Finkel, Barsoum, and El-Raghy, 2000). In this case, the moduli increase linearly from room temperature to about 125 K; below about 100 K they are more or less temperature independent. A least squares fit of the results shown in Figure 3.11a for $T > 125$ K for Ti_3SiC_2, Ti_3AlC_2, and $Ti_4AlN_{2.9}$, respectively, yield

$$\begin{aligned} \frac{G}{G_{RT}} &= 1 - 1.4 \times 10^{-4}(T - 298), \quad T>125 \text{ K} \\ \frac{G}{G_{RT}} &= 1 - 1.2 \times 10^{-4}(T - 298), \quad T>125 \text{ K} \\ \frac{G}{G_{RT}} &= 1 - 1.5 \times 10^{-4}(T - 298), \quad T>125 \text{ K} \end{aligned} \tag{3.10}$$

where G_{RT} is the shear modulus at room temperature. The temperature dependences of E of these compounds are also comparable (Finkel, Barsoum, and El-Raghy, 2000).

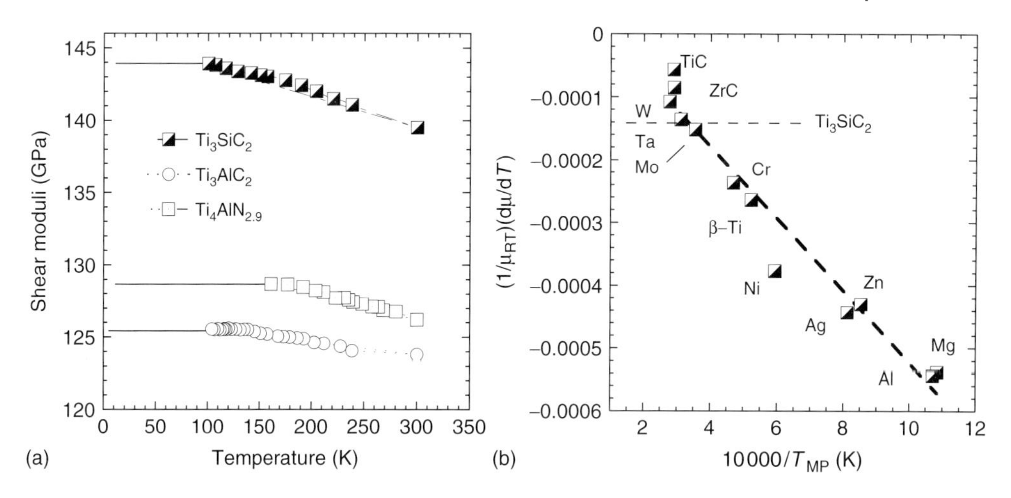

Figure 3.11 (a) Dependence of G on T in the 4–300 K range for Ti_3SiC_2, Ti_3AlC_2, and $Ti_4AlN_{2.9}$. (b) $d(\mu/\mu_{RT})/dT$ versus $1/T_M$, where T_M is the melting point. Also plotted as a horizontal line is the value of $d(\mu/\mu_{RT})/dT$ for Ti_3SiC_2 (Finkel, Barsoum, and El-Raghy, 1999 and 2000).

Typically, the T dependence of G of metals and simple binary compounds can be represented by

$$\frac{\mu(T)}{\mu_{RT}} = 1 - K\frac{(T-300)}{T_M} \tag{3.11}$$

where K is a constant of the order 0.5 and T_M is the melting point in degrees kelvin. Based on Eq. (3.11), a plot of $d(\mu/\mu_{RT})/dT$ versus $1/T_M$ should yield a straight line, as observed (Figure 3.11b). To check the validity of this expression for the MAX phases is problematic because they do not melt congruently but decompose peritectically. If, however, one assumes that Eq. (3.11) applies to the MAX phases, then based on the results plotted in Figure 3.11b the "equivalent" melting point for Ti_3SiC_2 can be estimated to be 2600 °C (Finkel, Barsoum, and El-Raghy, 1999).

3.7.2
High Temperatures, 300–1500 K

Not only are many MAX phases quite stiff at room temperature but, as importantly, their elastic properties are relatively weak functions of temperature. This is best seen in Figure 3.12a,b. For example, at 1273 K, the G and E values of Ti_3AlC_2 are ≈88% of their room temperature values (Finkel, Barsoum, and El-Raghy, 2000; Radovic *et al.*, 2006). In that respect, their resemblance to the MX binaries is notable. A perusal of Figure 3.12 quickly establishes that the rates at which the moduli drop are weak functions of chemistry, at least for the Al- and Si-containing phases listed. Such low temperature dependences are usually associated with ceramic materials, or refractory metals such as Mo or W.

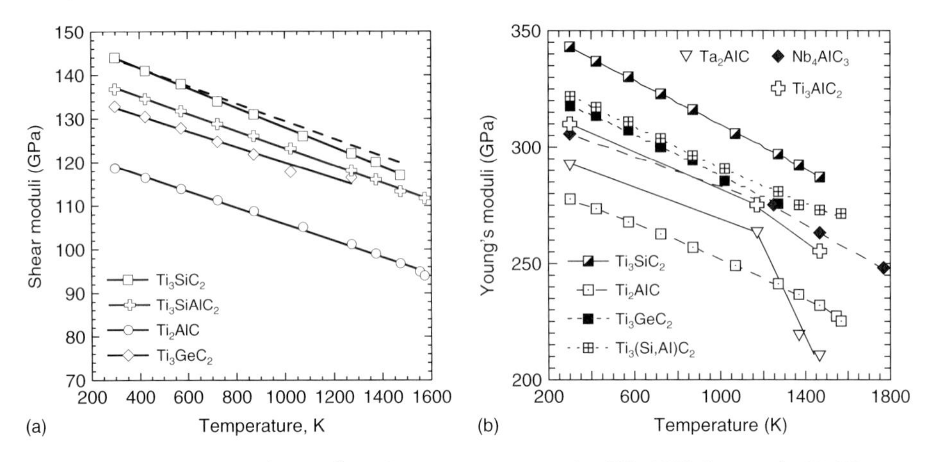

Figure 3.12 Dependence of (a) G on temperature in the 300–1500 K range for Ti_3SiC_2, Ti_3AlC_2, Ti_3SiAlC_2 and Ti_3GeC_2 (Radovic *et al.*, 2006). (b) Dependence of E on temperature for Ti_3SiC_2, Ti_3AlC_2, Ti_3GeC_2, $Ti_3(Si_{0.5}Al_{0.5})C_2$, and Ti_2AlC (Radovic *et al.*, 2006), Ti_3AlC_2 (Wang and Zhou, 2009), Nb_4AlC_3 (Hu *et al.*, 2008c), and Ta_2AlC (Hu *et al.*, 2008b).

It important to point out here that measuring the elastic moduli of the MAX phases is fraught with pitfalls. If care is not taken during the measurements to keep the applied stresses quite low, plastic deformation can occur, resulting in moduli that are too low. A good example of this problem are the results plotted for Ta_2AlC in Figure 3.12b; the change in slope at $\approx$1200 K is most probably an artifact of the measurements.

Such drops are typically *not* observed when the moduli are measured by ultrasound (Radovic *et al.*, 2006). When the latter technique is used, the moduli decrease linearly with temperature across a wide temperature range. Direct ultrasound or resonant ultrasound spectroscopy (RUS), techniques are thus preferred to others. As a bonus, RUS can simultaneously measure the damping of the ultrasound waves. This aspect is discussed in more detail in Chapter 8.

3.8 Raman Spectroscopy

The various Raman-active vibrational modes for a number of MAX phases have been measured and calculated using DFT (Amer *et al.*, 1998; Lane *et al.*, 2012; Leaffer *et al.*, 2007; Mercier *et al.*, 2011; Presser *et al.*, 2012; Spanier *et al.*, 2005; Wang *et al.*, 2005).

At this juncture, it is useful to discuss the Raman modes of each family – starting with the 211 phases – separately. Before doing so, it is instructive to briefly recall what affects the vibration of a bond. The basic equation for the frequency of a bond, assuming harmonic oscillation, is given by

$$\omega = \sqrt{\frac{k_S}{m}} \quad (3.12)$$

where k_S is the bond stiffness and m is the mass of the atoms involved. For a two-body system, m can be represented by the reduced mass, and therefore

$$\omega = \sqrt{\frac{k_S}{m_{red}}} = \sqrt{k_S \left(\frac{1}{m_1} + \frac{1}{m_2}\right)} \quad (3.13)$$

where m_1 and m_2 are the masses of the two atoms bonded together. It follows that for bonds of the same stiffness values, a plot of ω versus $1/\sqrt{m_{red}}$ should yield a straight line.

3.8.1 The 211 Phases

Table 3.6 summarizes the Raman frequencies for select 211 phases. Figure 3.13a shows a high-quality Raman spectrum of a V_2AlC single crystal in which four modes are clearly visible (Spanier *et al.*, 2005). These modes – depicted in Figure 3.13b–e – only involve the M and A atoms; the C atoms do not play a small role. The highest energy mode ω_4 involves atomic vibrations along [0001]; the other three involve vibrations in the basal planes. Mode ω_3 is noteworthy since it occurs at almost the identical frequency – $\approx$270 cm^{-1} – in both Ti_2AlC and Ti_3AlC_2 (see below). This result is consistent with the assignment of the mode to the relative motion of the M and A atoms along the *a* direction (compare Figures 3.13d and 3.15d).

Since all the modes in the 211 phases involve the M and A atoms (Figure 3.13b–e), Spanier *et al.* (2005) showed that the relationship between the various phonon energies and the reduced masses of the M and A atoms – more specifically, $\sqrt{(2/m_M) + (1/m_A)}$ where m_M and m_A are the atomic masses of M and A atoms, respectively – is linear (Figure 3.14a).

3.8.2 The 312 Phases

The Raman spectra of Ti_3SiC_2, Ti_3AlC_2, and Ti_3GeC_2 are compared in Figure 3.15a. Also shown (lowest dashed curve in Figure 3.15a) is the Raman spectrum of a Ti_3AlC_2 sample in which the Al layers were selectively etched. The importance of the latter results is discussed below. Table 3.7 summarizes the results for select 312 phases. For Ti_3SiC_2, Ti_3AlC_2, and Ti_3GeC_2, the four modes at 190–200, 279–297, 625–631, and 664–678 cm^{-1} exist for all of them and are thus quite reliable and are to be considered characteristic peaks of these phases (Presser *et al.*, 2012). In Table 3.7, the modes with a high degree of certainty are highlighted in gray.

Figure 3.15b–g schematically depict the Raman modes corresponding to those listed in Table 3.7. The highest modes in the 312 and 413 phases (see below) are

Table 3.6 Summary of experimental and theoretical Raman peaks (in cm^{-1}) for select 211 phases. Highlighted peaks are the most reliable and there is little doubt of their existence. Whether the peaks that are not highlighted also belong to the MAX phases listed is questionable and requires more work.

Phase	ω_1		ω_2		ω_3		ω_4		References
	experiment	theory	experiment	theory	experiment	theory	experiment	theory	
Ti_2AlN	149	147	234	228	—	224	367	360	Presser *et al.* (2012)
Ti_2AlC	150	151	262	256	268.1	270	365	366	Spanier *et al.* (2005)
	—	136	266	266	—	266	359	358	Wang *et al.* (2005)
V_2AlC	158	145	240	244	257	248	361	361	Spanier *et al.* (2005)
	157	—	239	—	257	—	364	—	Presser *et al.* (2012)
Ti_2InC	159	66	262	258	292	261	368	363	Leaffer *et al.* (2007)
Cr_2AlC	151	160	246	261	—	271	339	358	Leaffer *et al.* (2007)
	—	168	—	263	—	269	—	352	Wang *et al.* (2005)
Nb_2AlC	149	144	211	211	190	193	263	251	Spanier *et al.* (2005)
Hf_2InC	—	67	139	136	139	136	194	192	Leaffer *et al.* (2007)
V_2GeC	138	111	218	254	257	260	299	328	Leaffer *et al.* (2007)
Cr_2GeC	—	108	230	262	244	241	301	352	Leaffer *et al.* (2007)
V_2AsC	151	132	280	280	234	234	—	348	Leaffer *et al.* (2007)
Nb_2AsC	136	130	236	230	205	197	283	279	Leaffer *et al.* (2007)
Ta_2AlC	118	115	188	185	137	132	199	199	Leaffer *et al.* (2007)
Ta_2AlC	117	114	186	182	—	130	197	203	Lane *et al.* (2012)

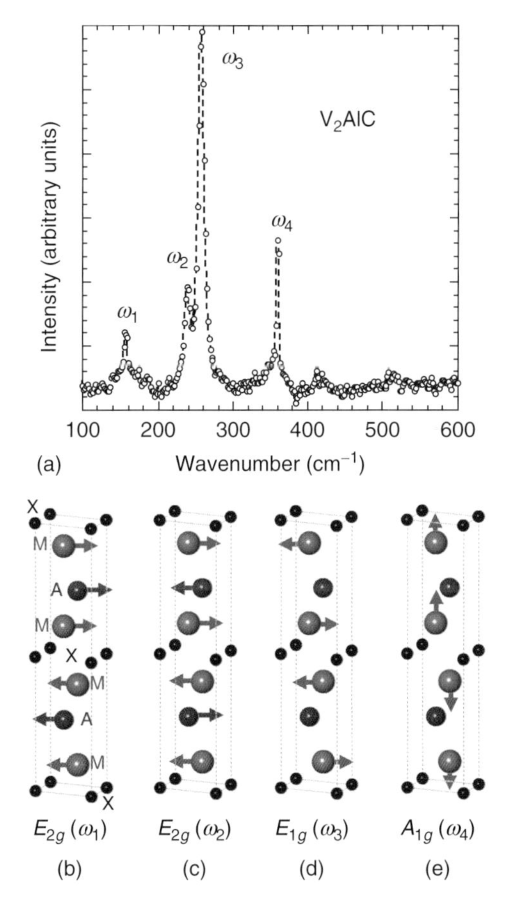

Figure 3.13 (a) High-quality first-order Raman spectra of a V_2AlC single crystal (Spanier *et al.*, 2005). (b–e) Schematics of the atomic displacements associated with the four Raman-active modes in the 211 phases (Spanier *et al.*, 2005). The modes are labeled to correspond to the ones listed in Table 3.6.

due to the vibrations of the X atoms relative to other X atoms within the MX layers. It is for this reason that modes ω_5 and ω_6 in Table 3.7 are *not* observed in the 211 phases, confirming their nature.

The strongest evidence that the lower modes, namely, ω_1, ω_2, and ω_3, involve the A-group elements can be seen in Figure 3.15a, in which the Raman spectra of Ti_3AlC_2 before (solid line) and after (dashed line) the Al was preferentially extracted while maintaining the hexagonal structure (Naguib *et al.*, 2011). From these results, it is obvious that removing the A-group element eliminates these modes, confirming their assignment. The loss of Al also apparently merges the two higher energy modes and shifts them to slightly lower frequencies.

Figure 3.16a plots the energies of the Raman modes ω_1, ω_2, and ω_3 for Ti_3SiC_2, Ti_3GeC_2, and Ti_3AlC_2 as a function of the reduced mass of the atoms participating

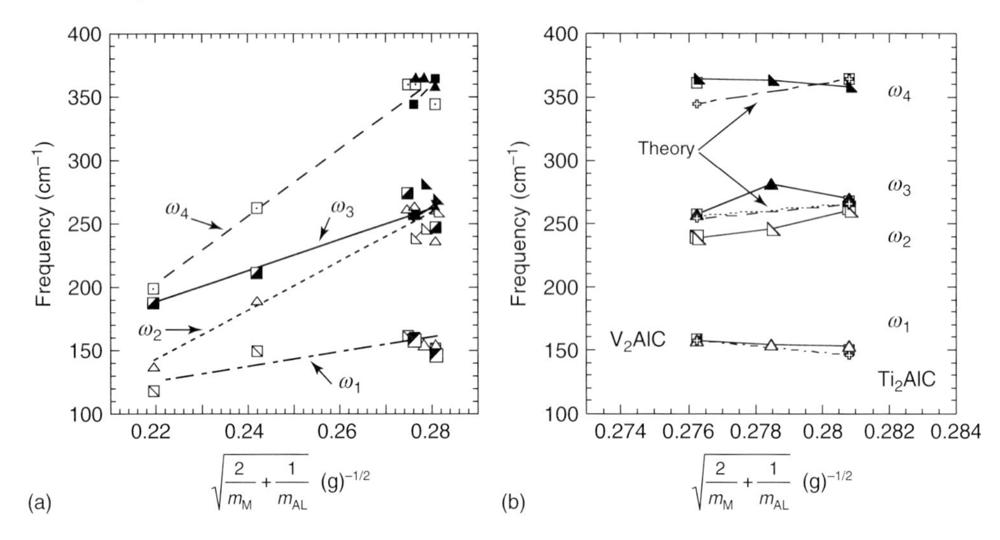

Figure 3.14 Functional dependence of Raman frequencies for the 211 phases on reduced mass in (a) the 0.21–0.29 $g^{-1/2}$ range. (b) Same as (a) but in the narrower range relevant to the end members Ti_2AlC, TiVAlC, and V_2AlC. The frequencies calculated from DFT (Table 3.6) are denoted by crosses and dashed lines (Presser *et al.*, 2012).

in those modes. The latter is given by $\sqrt{(2/m_{Ti}) + (1/m_A)}$, where m_{Ti} and m_A are the atomic masses of Ti and the A-group element, respectively. In this figure, the experimental results are plotted as squares joined by solid lines and the *ab initio* predictions as crosses joined by dashed lines.

Modes ω_5 and ω_6 are plotted in Figure 3.16b. Strictly speaking, if only first nearest neighbors were to be considered, then the reduced masses for modes ω_5 and ω_6 for the three compounds would have been identical, since they all involve *only* the Ti and C atoms (Fig. 3.15f and g). From the results shown in Figure 3.16b, it is clear that this is not the case, which is why the energies are plotted versus the same reduced masses as for the Ti–A vibration modes, that is, $\sqrt{(2/m_{Ti}) + (1/m_A)}$. As in Figure 3.16a, the experimental results are plotted as squares joined by solid lines and the *ab initio* predictions as crosses joined by dashed lines. From the results shown in Figure 3.16, the following conclusions can be reached.

- The relationship between reduced mass and mode energies is neither monotonic nor linear.
- Gratifyingly, theory predicts *all* the trends of the experimental modes as a function of reduced mass.
- In all cases, the theoretical energies are lower than the measured ones. The agreement between theory and experiment depends on the modes. For modes ω_1, ω_3, ω_5, and ω_6, theory underestimates experiment by less than ≈10%. For ω_3 of Ti_3SiC_2, theory underestimates experiment by ≈25%.

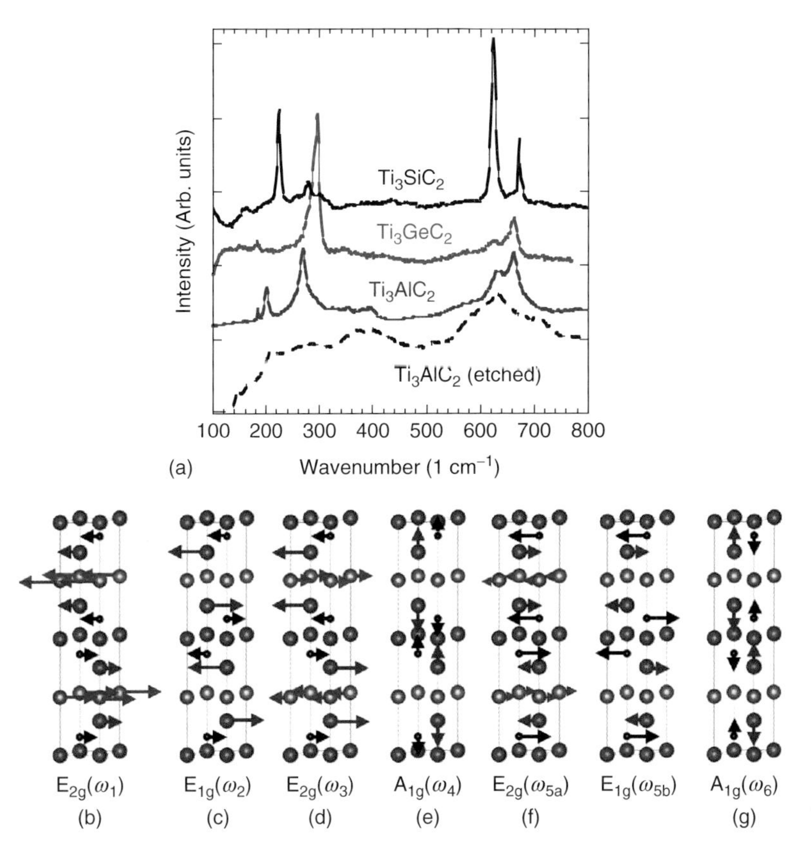

Figure 3.15 (a) First-order Raman spectra of Ti_3SiC_2, Ti_3GeC_2, and Ti_3AlC_2 before and after HF etching. The HF treatment selectively dissolves the A layers (Chapter 7) (Presser *et al.*, 2012). (b–g) Schematics of the atomic displacements associated with the six Raman-active modes of the 312 phases. The modes are labeled to correspond to the ones listed in Table 3.7. The lengths of the arrows is proportional to the amplitude of the vibrations (N. Lane, 2013).

- Despite the fact that modes ω_5 and ω_6 only involve vibrations of the C atoms (Figure 3.15f and g), as shown in Figure 3.16b, the A-group element still affects their Raman vibrational frequencies and cannot be ignored.

3.8.3 The 413 Phases

Until recently, the only 413 phase for which Raman spectra existed was Ti_4AlN_3 (Spanier *et al.*, 2005). More recently, we reported on the Raman spectra of Ta_4AlC_3 (Lane *et al.*, 2012). In the same paper, the Raman active modes were sketched and predicted from first-principles calculations using DFT. In this case, there are 10 modes ($3A_{1g} + 3E_{1g} + 4E_{2g}$), as shown schematically in Figure 3.17a–j.

Table 3.7 Summary of Raman peaks (in cm^{-1}) for select 312 phases.

Phase	ω_1 E_{1g}	ω_2 E_{2g}	ω_3 E_{2g}	ω_4 A_{1g}	ω_5 E_{1g} and E_{2g}	ω_6 A_{1g}	References
Ti_3SiC_2	—	185	224	278	625	673	Mercier *et al.* (2011)
	159	228	281	312	631	678	Amer *et al.* (1998)
	159	226	279	301	625	673	Spanier *et al.* (2005)
	111[a]	194[a]	208[a]	265[a]	610 and 610[a]	664[a]	Wang and Zhou (2004b)
	145[a]	217[a]	253[a]	301[a]	590 and 622[a]	657[a]	Spanier *et al.* (2005)
	129[a]	190[a]	222[a]	273[a]	610 and 611[a]	666[a]	Presser *et al.* (2012)
Ti_3AlC_2	—	183	201	270	632	663	Presser *et al.* (2012)
	125[a]	182[a]	197[a]	268[a]	620 and 621[a]	655[a]	Presser *et al.* (2012)
Ti_3GeC_2	—	183	—	297	627	664	Presser *et al.* (2012)
	87[a]	180[a]	182[a]	268[a]	610 and 611[a]	658[a]	Presser *et al.* (2012)

[a] Predicted modes from DFT calculations.
Highlighted peaks are the most reliable and there is little doubt of their existence. Whether the peaks that are not highlighted also belong to the MAX phases listed is questionable and requires more work.

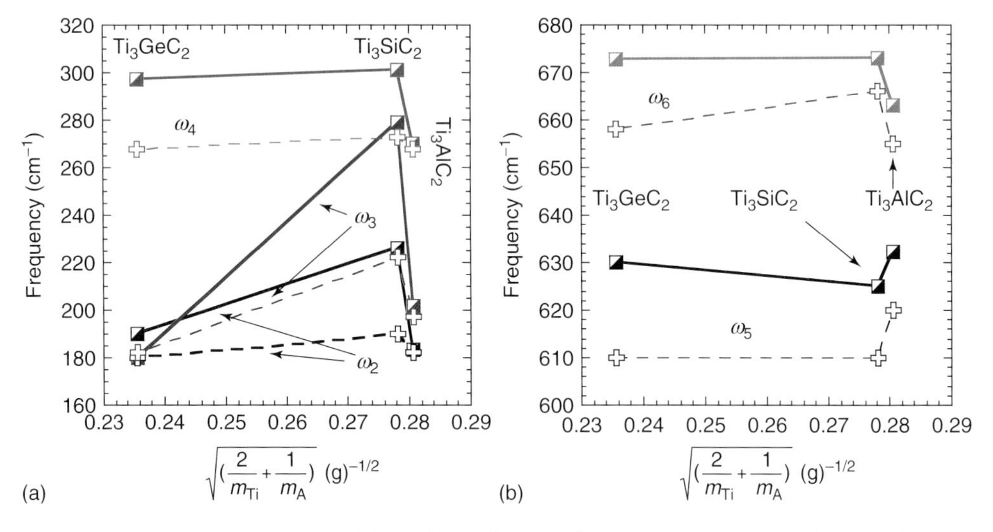

Figure 3.16 Functional dependence of Raman frequencies (a) ω_2, ω_3, and ω_4 and (b) ω_5 and ω_6, on the reduced masses of the 312 phases. The squares denote experimental values; crosses and dashed lines those calculated in Table 3.7 (Presser *et al.*, 2012).

The Raman spectra of Ta_4AlC_3 and Ti_4AlN_3 are shown in Figure 3.18a,b, respectively. The red markers represent the calculated wavenumbers. The results are also listed in Table 3.8. In general, the agreement between theory and experiment is good. Also included in Table 3.8 are the predicted wavenumbers for Nb_4AlC_3 and β-Ta_4AlC_3.

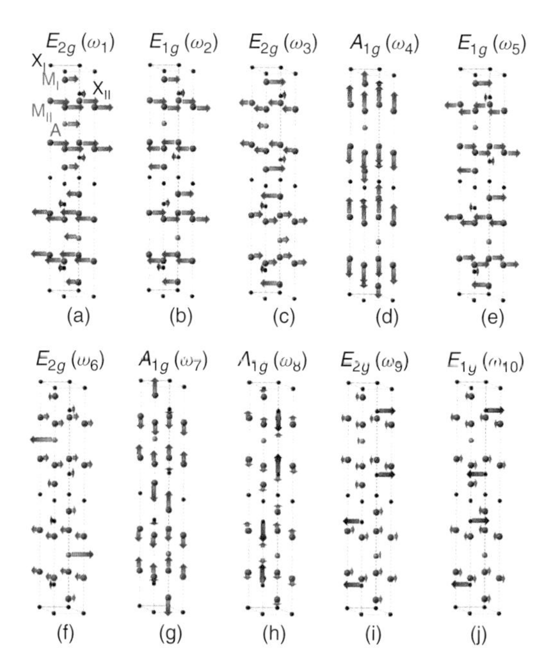

Figure 3.17 (a–j) Schematics of the atomic displacement associated the 10 Raman-active modes in the 413 phases (Lane *et al.* 2012).

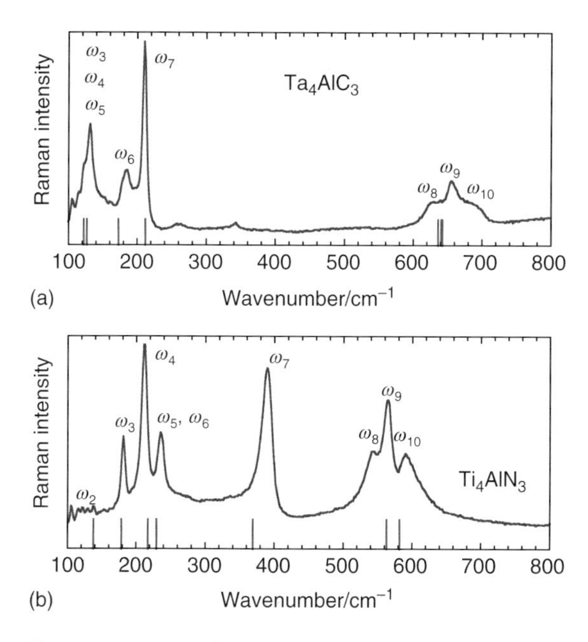

Figure 3.18 First-order Raman spectra of (a) Ta_4AlC_3 and (b) Ti_4AlN_3. The red markers represent the calculated wavenumbers (Lane *et al.* 2012).

Table 3.8 Experimental and theoretical wavenumbers ω (in cm^{-1}), symmetry assignment, of the Raman modes of select 413 phases (Lane *et al.*, 2012). Values in parentheses are from Spanier *et al.* (2005).

Mode	Irreducible	Ti_4AlN_3		Nb_4AlC_3	α-Ta_4AlC_3		β-Ta_4AlC_3
		ω_{expt}	ω_{calc}	ω_{calc}	ω_{expt}	ω_{calc}	ω_{calc}
ω_1	E_{2g}	—	95	82	—	65	44
ω_2	E_{1g}	132 (132)	138	106	—	80	68
ω_3	E_{2g}	181 (181)	179	158	115	123	120
ω_4	A_{1g}	211 (208)	219	169	123	127	125
ω_5	E_{1g}	236 (235)	229	176	132	128	123
ω_6	E_{2g}	—	229	182	188	173	179
ω_7	A_{1g}	387 (386)	370	291	211	212	201
ω_8	A_{1g}	546 (539)	563	626	629	637	637
ω_9	E_{2g}	563 (563)	582	610	657	641	627
ω_{10}	E_{1g}	596 (592)	581	612	685	643	627

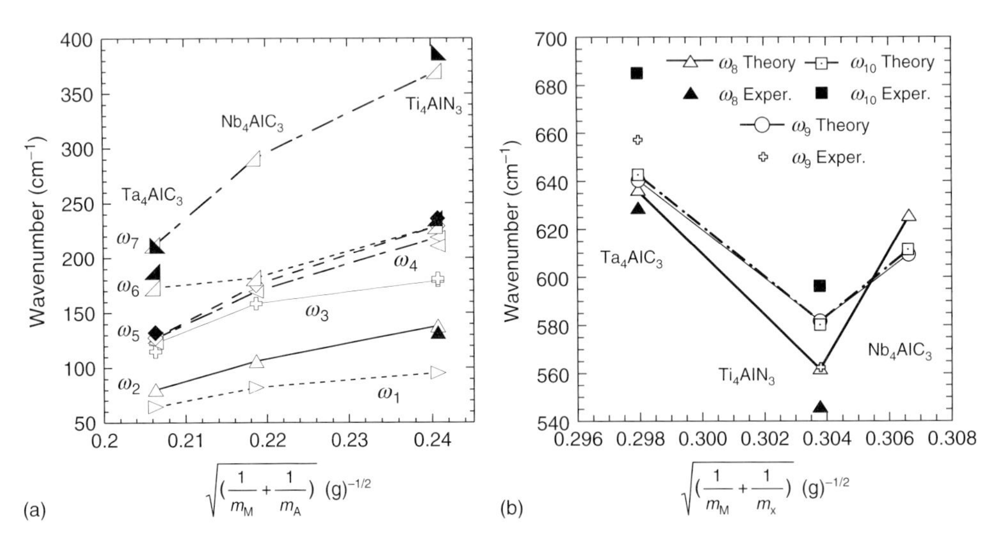

Figure 3.19 Plots of ω versus $\sqrt{1/m_{red}}$ for the 413 phases, where m_{red} is the reduced mass for (a) $\omega_1 - \omega_7$ using the reduced mass for the M and A atoms and (b) $\omega_8 - \omega_{10}$ using the reduced masses for the M and X atoms. Open symbols and the lines correspond to calculated values; solid symbols represent experimental values (Lane *et al.* 2012).

When modes $\omega_1-\omega_7$ are plotted versus $1/\sqrt{m_{red}}$ where m_{red} is that of the M and A atoms (Figure 3.19a), the plots are close to linear with a positive slope, indicating that mass differences between the M atoms are responsible for at least some of the differences in frequencies. The experimental data are shown as solid symbols; the open symbols and the lines correspond to calculated values. On the contrary, when the $\omega_8-\omega_{10}$ modes are plotted versus the reduced mass of the M and A atoms (not shown), the overall slope is negative and the results are quite nonlinear. Since modes $\omega_8-\omega_{10}$ are vibrational modes involving primarily the X atoms, this result is not surprising. However, even when using the masses of the M and X atoms, the results are also nonlinear (Figure 3.19b). More importantly, the overall slope is again negative, which implies that in this case bond stiffness is more a function of chemistry than mass. Furthermore, the results imply the following order for the bond stiffness values: Ta–C >Nb–C > Ti–N. This is fully consistent with what is known about the M–X bonds, as evidenced, indirectly, by the fact that the melting points of TaC, NbC, and TiN are 3983, 3600, and 2949 °C, respectively (Toth, 1971).

3.9 Infrared Spectroscopy

In contrast to Raman spectroscopy, very little work has been carried out on the infrared (IR) spectroscopy of the MAX phases. As far as we are aware, only one study – on Ti_3GeC_2 – exists (Manoun *et al.*, 2007c). Figure 3.20 reproduces the results.

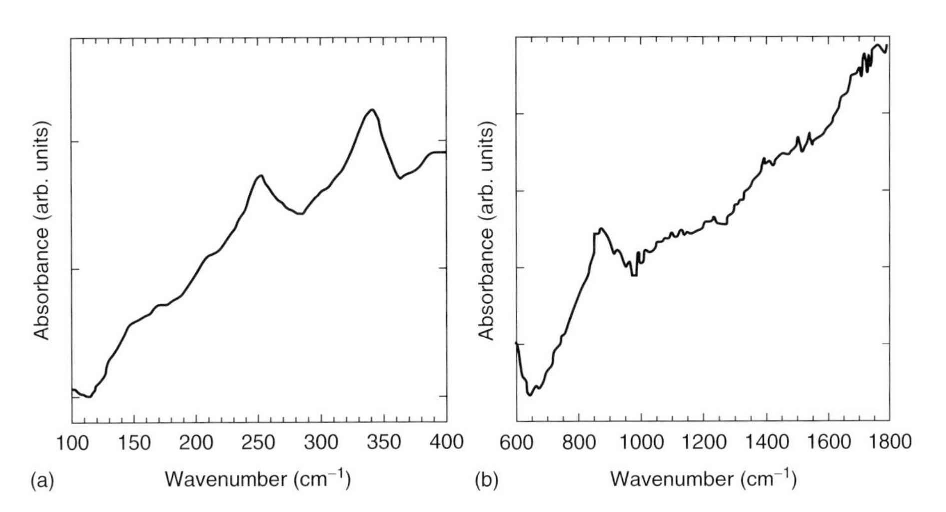

Figure 3.20 Synchrotron IR spectrum of Ti_3GeC_2 in (a) mid-IR and (b) far-IR region. Four of the five predicted modes are observed. The peaks observed at 210.4, 252.3, and 340.5 cm^{-1} are close to those calculated for Ti_3SiC_2. The far-IR peak observed around 875 cm^{-1} is higher than that predicted for Ti_3SiC_2 at 607 cm^{-1}, and may be due to multiphonon absorptions (Manoun *et al.*, 2007c).

3.10
Summary and Conclusions

With Young's and shear moduli of the order of 340 and 140 GPa, respectively, some of the MAX phases have exceptional specific stiffness values, especially for readily machinable solids.

Depending on the chemistry, the longitudinal velocities range from a low of $\approx$4000 m s^{-1} for Hf_2SnC to a high of $\approx$9000 m s^{-1} for the Ti-containing phases. The shear velocities range from $\approx$3000 m s^{-1} to values of the order of 5500 m s^{-1}. For the most part, the sound velocities, and concomitantly the elastic constants, are well described by the Bohm–Staver equation. Plots or ν_S and ν_L versus $r_B/r_S\sqrt{z_{av}/3M_{av}}$ yield straight lines that pass through the origin, with high R^2 values. Deviations from that line are most likely due to the presence of point defects. In the case of the Cr-containing phases, on the other hand, the discrepancies are probably related to factors that the Bohm–Staver equation does not capture.

For the most part, the agreement between theory and experiment is acceptable. When the values obtained in the latter are lower than those predicted, the discrepancy can, in some cases, be traced to the presence of vacancies.

When the various elastic constants are compared with those predicted from DFT calculations, the following conclusion can be made: the LDA approximation is better at predicting the B values than the GGA approximation. The opposite is true for the c_{ij}s.

Like other refractory materials, the elastic moduli of the MAX phases decrease slowly with increasing T – roughly $\approx$5% for every 1000 °C.

There are generally two types of Raman active modes in the MAX phases: (i) low-energy modes–observed in all MAX phases–that range from $\approx$50 to 300 cm^{-1} and are due to the vibrations of the A and M atoms, and (ii) higher energy modes – only observed in the 312 and 413 phases–that range from about 550 to 650 cm^{-1}, which are due to the vibrations of the X atoms. At this time, it is reasonable to conclude that first-principles calculations of the MAX phases' Raman modes agree with the experimental results, in some cases, quite well.

References

Ahuja, R., Eriksson, O., Wils, J.M., and Johansson, B. (2000) Electronic structure of Ti_3SiC_2. *Appl. Phys. Lett.*, **76**, 2226–2228.

Amer, M., Barsoum, M.W., El-Raghy, T., Weiss, I., LeClair, S., and Liptak, D. (1998) Raman spectrum of Ti_3SiC_2. *J. Appl. Phys.*, **84**, 5817.

Amini, S., Barsoum, M.W., and El-Raghy, T. (2007) Synthesis and mechanical properties of fully dense Ti_2SC. *J. Am. Ceram. Soc.*, **90**, 3953–3958.

Amini, S., Zhou, A., Gupta, S., DeVillier, A., Finkel, P., and Barsoum, M.W. (2008) Synthesis and elastic and mechanical properties of Cr_2GeC. *J. Mater. Res.*, **23**, 2157–2165.

Ashcroft, N.W. and Mermin, N.D. (1976) *Solid State Physics*, Saunders College Publishing, Philadelphia, PA.

Bai, Y., He, X., Li, M., Sun, Y., Zhu, C., and Li, Y. (2010) Ab initio study of the bonding and elastic properties of Ti_2CdC. *Solid State Sci.*, **12**, 144–147.

Birch, F. (1978) Finite strain isotherm and velocities for single-crystal and polycrystalline NaCl at high pressures and 300 K. *J. Geophys. Res. B*, **83**, 1257.

Bohm, D. and Staver, T. (1950) *Phys. Rev.*, **84**, 836.

Bouhemadou, A. (2008a) Calculated structural and elastic properties of M_2InC (M = Sc, Ti, V, Zr, Nb, Hf, Ta). *Mod. Phys. Lett. B*, **22**, 2063–2076.

Bouhemadou, A. (2008b) Prediction study of structural and elastic properties under pressure effect of M_2SnC (M = Ti, Zr, Nb, Hf). *Physica B-Condens. Matter*, **403**, 2707.

Bouhemadou, A. (2009a) Calculated structural, electronic and elastic properties of M_2GeC (M = Ti, V, Cr, Zr, Nb, Mo, Hf, Ta and W). *Appl. Phys. A*, **96**, 959–967.

Bouhemadou, A. (2009b) Structural, electronic and elastic properties of MAX phases M_2GaN (M = Ti, V and Cr). *Solid State Sci.*, **11**, 1875–1881.

Bouhemadou, A. and Khenata, R. (2007) Prediction study of structural and elastic properties under the pressure effect of M_2GaC, M = Ti, V, Nb, Ta. *J. Appl. Phys.*, **102**, 043528.

Bouhemadou, A. and Khenata, R. (2008) Structural, electronic and elastic properties of M_2SC (M = Ti, Zr, Hf) compounds. *Phys. Lett. A*, **372**, 6448–6452.

Bouhemadou, A., Khenata, R., and Chegaar, M. (2007) Structural and elastic properties of Zr_2AlX and Ti_2AlX (X = C and N) under pressure effect. *Eur. Phys. J. B*, **56**, 209–215.

Clausen, B., Lorentzen, T., Bourke, M.A.M., and Daymond, M.R. (1999) Lattice strain evolution during uniaxial tensile loading of stainless steel. *Mater. Sci. Eng., A*, **259**, 17–24.

Clausen, B., Lorentzen, T., and Leffers, T. (1998) Self-consistent modelling of the plastic deformation of FCC polycrystals and its implications for diffraction measurements of internal stresses. *Acta Mater.*, **46**, 3087–3098.

Clausen, B., Tomé, C.N., Brown, D.W., and Agnew, S.R. (2008) Reorientation and stress relaxation due to twinning: modeling and experimental characterization for Mg. *Acta Mater.*, **56**, 2456–2468.

Cover, M.F., Warschkow, O., Bilek, M.M., and McKenzie, D.R. (2009) A comprehensive survey of M_2AX phase elastic properties. *J. Phys.: Condens. Matter*, **21**, 305403.

Cui, S., Feng, W., Hu, H., Feng, Z., and Liu, H. (2009) Hexagonal Ti_2SC with high hardness and brittleness: a first-principles study. *Scr. Mater.*, **61**, 576–579.

Du, Y.L., Sun, Z.-M., Hashimoto, H., and Barsoum, M.W. (2009a) Theoretical investigations on the elastic and thermodynamic properties of $Ti_2AlC_{0.5}N_{0.5}$ solid solution. *Phys. Lett. A*, **374**, 78–82.

Du, Y.L., Sun, Z.M., Hashimoto, H., and Tian, W.B. (2009b) Bonding properties and bulk modulus of M_4AlC_3 (M = V, Nb, and Ta) studied by first-principles calculations. *Phys. Status Solidi*, **246**, 1039–1043.

Du, Y.L., Sun, Z.-M., Hashimoto, H., and Barsoum, M.W. (2011) Electron correlation effects in the MAX phase Cr_2AlC from first principles. *J. Appl. Phys.*, **109**, 063707.

Du, Y.L., Sun, Z.M., Hashimoto, H., and Tian, W.B. (2008a) Elastic properties of Ta_4AlC_3 studied by first-principles calculations. *Solid State Commun.*, **147**, 246–249.

Du, Y.L., Sun, Z.M., Hashimoto, H., and Tiana, W.B. (2008b) First-principles study on electronic structure and elastic properties of Ti_2SC. *Phys. Lett. A*, **372**, 5220–5223.

El-Raghy, T., Chakraborty, S., and Barsoum, M.W. (2000) Synthesis and characterization of Hf_2PbC, Zr_2PbC and M_2SnC (M = Ti, Hf, Nb or Zr). *J. Eur. Ceram. Soc.*, **20**, 2619.

Emmerlich, J., Music, D., Houben, A., Dronskowski, R., and Schneider, J.M. (2007) Systematic study on the pressure dependence of M_2AlC phases (M = Ti, V, Cr, Zr, Nb, Mo, Hf, Ta, W). *Phys. Rev. B*, **76**, 224111.

Fang, C.M., Ahuja, R., Eriksson, O., Li, S., Jansson, U., Wilhelmsson, O., and Hultman, L. (2006) General trend of the mechanical properties of the ternary carbides M_3SiC_2 (M = transition metal). *Phys. Rev. B*, **74**, 054106.

Finkel, P., Barsoum, M.W., and El-Raghy, T. (1999) Low temperature dependence of the elastic properties of Ti_3SiC_2. *J. Appl. Phys.*, **85**, 7123–7126.

Finkel, P., Barsoum, M.W., and El-Raghy, T. (2000) Low temperature dependencies of the elastic properties of Ti_4AlN_3 and $Ti_3Al_{1.1}C_{1.8}$ and Ti_3SiC_2. *J. Appl. Phys.*, **87**, 1701–1703.

Finkel, P., Seaman, B., Harrell, K., Hettinger, J.D., Lofland, S.E., Ganguly, A., Barsoum, M.W., Sun, Z., Li, S., and Ahuja, R. (2004) Low temperature elastic, electronic and transport properties of $Ti_3Si_{1-x}Ge_xC_2$ solid solutions. *Phys. Rev. B*, **70**, 085104.

Fu, H.Z., Teng, M., Liu, W.F., Ma, Y., and Gao, T. (2010) The axial compressibility, thermal expansion and elastic anisotropy of Hf_2SC under pressure. *Eur. Phys. J. B*, **78**, 37–42.

Gui, S., Feng, W., Hu, H., Lv, Z., Zhang, G., and Gong, Z. (2011) First-principles studies of the electronic and elastic properties of Ti_2GeC. *Solid State Commun.*, **151**, 491–494.

He, X., Bai, Y., Li, Y., Zhu, C., and Li, M. (2009) Ab initio calculations for properties of MAX phases Ti_2InC, Zr_2InC, and Hf_2InC. *Solid State Commun.*, **149**, 564–566.

He, X., Bai, Y., Zhu, C., and Barsoum, M.W. (2011) Polymorphism of newly-discovered Ti_4GaC_3: a first-principle study. *Acta Mater.*, **59**, 5523–5533.

He, X., Bai, Y., Zhu, C., Sun, Y., Li, M., and Barsoum, M.W. (2010) General trends in the structural, electronic and elastic properties of the M_3AlC_2 phases (M = transition metal): a first-principle study. *Comput. Mater. Sci.*, **49**, 691–698.

Hettinger, J.D., Lofland, S.E., Finkel, P., Palma, J., Harrell, K., Gupta, S., Ganguly, A., El-Raghy, T., and Barsoum, M.W. (2005) Electrical transport, thermal transport and elastic properties of M_2AlC (M = Ti, Cr, Nb and V) phases. *Phys. Rev. B*, **72**, 115120.

Holm, B., Ahuja, R., and Johansson, B. (2001) Ab initio calculations of the mechanical properties of Ti_3SiC_2. *Appl. Phys. Lett.*, **79**, 1450.

Holm, B., Ahuja, R., Li, S., and Johansson, B. (2002) Theory of ternary layered system Ti-Al-N. *J. Appl. Phys.*, **91**, 9874–9877.

Hu, C., He, L., Liu, M., Wang, X., Wang, J., Li, M., Bao, Y., and Zhou, Y. (2008a) In situ reaction synthesis and mechanical properties of V_2AlC. *J. Am. Ceram. Soc.*, **91**, 4029–4035.

Hu, C., He, L., Zhang, J., Bao, Y., Wang, J., Li, M., and Zhou, Y. (2008b) Microstructure and properties of bulk Ta_2AlC ceramic synthesized by an in situ reaction/hot pressing method. *J. Eur. Ceram. Soc.*, **28**, 1679–1685.

Hu, C., Li, F., He, L., Liu, M., Zhang, J., Wang, J., Bao, Y., Wang, J., and Zhou, Y. (2008c) In situ reaction synthesis, electrical and thermal, and mechanical properties of Nb_4AlC_3. *J. Am. Ceram. Soc.*, **91**, 2258–2263.

Hu, C., Lin, Z., He, L., Bao, Y., Wang, J., Li, M., and Zhou, Y.C. (2007) Physical and mechanical properties of bulk Ta_4AlC_3 ceramic prepared by an in situ reaction synthesis/hot-pressing method. *J. Am. Ceram. Soc.*, **90**, 2542–2548.

Hu, C., Sakka, Y., Nishimura, T., Guo, S., Grasso, S., and Tanaka, H. (2011) Physical and mechanical properties of highly textured polycrystalline Nb_4AlC_3 ceramic. *Sci. Technol. Adv. Mater.*, **12**, 044603.

Hug, G. (2006) Electronic structures of and composition gaps among the ternary carbides Ti_2MC. *Phys. Rev. B*, **74**, 184113 (184117pp).

Hug, G., Jaouen, M., and Barsoum, M.W. (2005) XAS, EELS and full-potential augmented plane wave study of the electronic structures of Ti_2AlC, Ti_2AlN, Nb_2AlC and $(Ti_{0.5},Nb_{0.5})_2AlC$. *Phys. Rev. B*, **71**, 24105.

Jia, G.-Z. and Yang, L.-J. (2010) Ab initio calculations for properties of Ti_2AlN and Cr_2AlC. *Physica B-Condens. Matter*, **405**, 4561–4564.

Kanoun, M.B., Goumri-Said, S., and Jaouen, M. (2009a) Steric effect on the M site of nanolaminate compounds M_2SnC (M = Ti, Zr, Hf and Nb). *J. Phys.: Condens. Matter*, **21**, 045404–045406.

Kanoun, M.B., Goumri-Said, S., and Reshak, A.H. (2009b) Theoretical study of mechanical, electronic, chemical bonding and optical properties of Ti_2SnC, Zr_2SnC, Hf_2SnC and Nb_2SnC. *Comput. Mater. Sci.*, **47**, 491–500.

Kooi, B.J., Poppen, R.J., Carvalho, N.J.M., De Hosson, J.T.M., and Barsoum, M.W.

(2003) Ti_3SiC_2: a damage tolerant ceramic studied with nanoindentations and transmission electron microscopy. *Acta Mater.*, **51**, 2859–2872.

Kulkarni, S.R., Phatak, N.A., Saxena, S.K., Fei, Y., and Hu, J. (2008a) High pressure structural behavior and synthesis of Zr_2SC. *J. Phys.: Condens. Matter*, **20**, 135211.

Kulkarni, S.R., Vennila, R.S., Phatak, N.A., Saxena, S.K., Zha, C.S., El-Raghy, T., Barsoum, M.W., Luo, W., and Ahuja, R. (2008b) Study of Ti_2SC under compression up to 47 GPa. *J. Alloys Compd.*, **448**, L1–L4.

Kumar, R.S., Rekhi, S., Cornelius, A.L., and Barsoum, M.W. (2005) Compressibility of Nb_2AsC to 41 GPa. *Appl. Phys. Lett.*, **86**, 111904.

Lane, N.J., Naguib, M., Presser, V., Hug, G., Hultman, L., and Barsoum, M.W. (2012) First-order Raman scattering of the MAX phases Ta_4AlC_3, Nb_4AlC_3, Ti_4AlN_3 and Ta_2AlC. *J. Raman Spectrosc.*, **43**, 954–958.

Lane, N. (2013) PhD, Drexel University.

Leaffer, O.D., Gupta, S., Barsoum, M.W., and Spanier, J.E. (2007) On the Raman scattering from selected M_2AC compounds. *J. Mater. Res.*, **22**, 2651–2654.

Li, C.-W. and Wang, Z. (2010) First-principles study of structural, electronic, and mechanical properties of the nanolaminate compound Ti_4GeC_3 under pressure. *J. Appl. Phys.*, **107**, 123511.

Li, C., Wang, B., Li, Y., and Wang, R. (2009) First-principles study of electronic structure, mechanical and optical properties of V_4AlC_3. *J. Phys. D: Appl. Phys.*, **42**, 065407.

Lin, Z.J., Zhou, Y.C., and Li, M.S. (2007) Synthesis, microstructure, and property of Cr_2AlC. *J. Mater. Sci. Technol.*, **23**, 721–746.

Lopez-de-la-Torrea, L., Winkler, B., Schreuer, J., Knorr, K., and Avalos-Borja, M. (2005) Elastic properties of tantalum carbide (TaC). *Solid State Commun.*, **134**, 245–250.

Manoun, B., Amini, S., Gupta, S., Saxena, S.K., and Barsoum, M.W. (2007a) On the compression behavior of Cr_2GeC and V_2GeC to quasi-hydrostatic pressures of 50 GPa. *J. Phys.: Condens. Matter*, **19**, 456218.

Manoun, B., Saxena, S.K., Hug, G., Ganguly, A., Hoffman, E.N., and Barsoum, M.W. (2007b) Synthesis and compressibility of $Ti_3(Al_{1.0}Sn_{0.2})C_2$ and $Ti_3Al(C_{0.5},N_{0.5})_2$. *J. Appl. Phys.*, **101**, 113523.

Manoun, B., Yang, H., Saxena, S.K., Ganguly, A., Barsoum, M.W., Liu, Z.X., Lachkar, M., and El-Bali, B. (2007c) Infrared spectrum and compressibility of Ti_3GeC_2 to 51 GPa. *J. Alloys Compd.*, **433**, 265–268.

Manoun, B., Zhang, F., Saxena, S.K., Gupta, S., and Barsoum, M.W. (2007d) On the compression behavior of $(Ti_{0.5},V_{0.5})_2AlC$ and $(Ti_{0.5},Nb_{0.5})_2AlC$ to quasi-hydrostatic pressures above 50 GPa. *J. Phys.: Condens. Matter*, **19**, 246215.

Manoun, B., Gulve, R.P., Saxena, S.K., Gupta, S., Barsoum, M.W., and Zha, C.S. (2006a) Compression behavior of M_2AlC (M = Ti, V, Cr, Nb, and Ta) phases to above 50 GPa. *Phys. Rev. B*, **73**, 024110.

Manoun, B., Saxena, S.K., Barsoum, M.W., and El-Raghy, T. (2006b) X-ray high-pressure study of Ti_2AlN and Ti_2AlC. *J. Phys. Chem. Solids*, **67**, 2091–2094.

Manoun, B., Saxena, S.K., El-Raghy, T., and Barsoum, M.W. (2006c) High-pressure X-ray study of Ta_4AlC_3. *Appl. Phys. Lett.*, **88**, 201902.

Manoun, B., Kulkarni, S., Pathak, N., Saxena, S.K., Amini, S., and Barsoum, M.W. (2010) Bulk moduli of Cr_2GaC and Ti_2GaN up 50 GPa. *J. Alloys Compd.*, **505**, 328–331.

Manoun, B., Leaffer, O., Gupta, S., Hoffman, E., Saxena, S.K., Spanier, J., and Barsoum, M.W. (2009) On the compression behavior of Ti_2InC, $(Ti_{0.5}, Zr_{0.5})_2InC$, and M_2SnC (M = Ti, Nb, Hf) to quasi-hydrostatic pressures up to 50 GPa. *Solid State Commun.*, **149**, 1978.

Manoun, B., Saxena, S.K., and Barsoum, M.W. (2005) High pressure study of Ti_4AlN_3 to 55 GPa. *Appl. Phys. Lett.*, **86**, 101906.

Manoun, B., Saxena, S.K., Gulve, R., Ganguly, A., Barsoum, M.W., and Zha, S. (2004a) Compression of $Ti_3Si_{0.5}Ge_{0.5}C_2$ to 53 GPa. *Appl. Phys. Lett.*, **84**, 2799–2801.

Manoun, B., Saxena, S.K., Gulve, R., Liermann, H.P., Hoffman, E.L., Barsoum, M.W., Zha, S., and Hug, G. (2004b) Compression of Zr_2InC to 52 GPa. *Appl. Phys. Lett.*, **85**, 1514–1516.

Mercier, F., Chaix-Pluchery, O., Ouisse, T., and Chaussende, D. (2011) Raman scattering from Ti_3SiC_2 single crystals. *Appl. Phys. Lett.*, **98**, 081912.

Murugaiah, A., Barsoum, M.W., Kalidindi, S.R., and Zhen, T. (2004) Spherical nanoindentations in Ti_3SiC_2. *J. Mater. Res.*, **19**, 1139–1148.

Music, D., Sun, Z., Ahuja, R., and Schneider, J.M. (2006) Coupling in nanolaminated ternary carbides studied by theoretical means: the influence of electronic potential approximations. *Phys. Rev. B*, **73**, 134117.

Naguib, M., Kurtoglu, M., Presser, V., Lu, J., Niu, J., Heon, M., Hultman, L., Gogotsi, Y., and Barsoum, M.W. (2011) Two dimensional nanocrystals produced by exfoliation of Ti_3AlC_2. *Adv. Mater.*, **23**, 4248–4253.

Onodera, A., Hirano, H., Yuasa, T., Gao, N.F., and Miyamoto, Y. (1999) Static compression of Ti_3SiC_2 to 61 GPa. *Appl. Phys. Lett.*, **74**, 3782–3784.

Phatak, N.A., Kulkarni, S.R., Drozd, V., Saxena, S.K., Deng, L., Fei, Y., Hu, J., Luo, W., and Ahuja, R. (2008) Synthesis and compressive behavior of Cr_2GeC up to 48 GPa. *J. Alloys Compd.*, **463**, 220–225.

Phatak, N.A., Saxena, S.K., Fei, Y., and Hu, J. (2009a) Synthesis and structural stability of Ti_2GeC. *J. Alloys Compd.*, **474**, 174–179.

Phatak, N.A., Saxena, S.K., Fei, Y., and Hu, J. (2009b) Synthesis of a new MAX compound $(Cr_{0.5}V_{0.5})_2GeC$ and its compressive behavior up to 49 GPa. *J. Alloys Compd.*, **475**, 629–634.

Pierson, H.O. (1996) *Handbook of Refractory Carbides and Nitrides*, Noyes Publications, Westwood, NJ.

Presser, V., Naguib, M., Chaput, L., Togo, A., Hug, G., and Barsoum, M.W. (2012) First-order Raman scattering of the MAX phases: Ti_2AlN, $Ti_2AlC_{0.5}N_{0.5}$, Ti_2AlC, $(Ti_{0.5}V_{0.5})_2AlC$, V_2AlC, Ti_3AlC_2 and Ti_3GeC_2. *J. Raman Spectrosc.*, **43**, 168–172.

Qian, X., Wang, N., Li, Y., Zhou, Y., Wu, H., Li, Y., and He, X. (2012) First-principle studies of properties of ternary layered M_2PbC (M = Ti, Zr and Hf). *Comput. Mater. Sci.*, **65**, 377–382.

Radovic, M., Ganguly, A., and Barsoum, M.W. (2008) Elastic properties and phonon conductivities of $Ti_3Al(C_{0.5},N_{0.5})_2$ and $Ti_2Al(C_{0.5},N_{0.5})$ solid solutions. *J. Mater. Res.*, **23**, 1517–1521.

Radovic, M., Ganguly, A., Barsoum, M.W., Zhen, T., Finkel, P., Kalidindi, S.R., and Lara-Curzio, E. (2006) On the elastic properties and mechanical damping of Ti_3SiC_2, Ti_3GeC_2, $Ti_3Si_{0.5}Al_{0.5}C_2$ and Ti_2AlC in the 300–1573 K temperature range. *Acta Mater.*, **54**, 2757–2767.

Romero, M. and Escamilla, R. (2012) First-principles calculations of structural, elastic and electronic properties of Nb_2SnC under pressure. *Comput. Mater. Sci.*, **55**, 142–146.

Roumily, A., Medkour, Y., and Maouche, D. (2009) Elastic and electronic properties of Hf_2SnC and Hf_2SnN. *Int. J. Mod. Phys. B*, **22**, 5155–5161.

Scabarozi, T.H., Amini, S., Finkel, P., Leaffer, O.D., Spanier, J.E., Barsoum, M.W., Drulis, M., Drulis, H., Tambussi, W.M., Hettinger, J.D. *et al.* (2008) Electrical, thermal, and elastic properties of the MAX-phase Ti_2SC. *J. Appl. Phys.*, **104**, 033502–033505.

Scabarozi, T.H., Amini, S., Leaffer, O., Ganguly, A., Gupta, S., Tambussi, W., Clipper, S., Spanier, J.E., Barsoum, M.W., Hettinger, J.D. *et al.* (2009) Thermal expansion of select MAX phases measured by high temperature X-ray diffraction and dilatometry. *J. Appl. Phys*, **105**, 013543.

Schneider, J., Mertens, R., and Music, D. (2006) Structure of V_2AlC studied by theory and experiment. *J. Appl. Phys.*, **99**, 013501.

Shamma, M., Presser, V., Clausen, B., Brown, D., Yeheskel, O., and Barsoum, M.W. (2011) On the response of Ti_2SC to stress studied by in situ neutron diffraction and the elasto-plastic self-consistent approach. *Scr. Mater.*, **65**, 573–576.

Shein, I.R. and Ivanovskii, I.V. (2010a) Ab initio calculation of the electronic structure, fermi surface, and elastic properties of the new 7.5K superconductor Nb_2InC. *JETP Lett.*, **91**, 410–414.

Shein, I.R. and Ivanovskii, I.V. (2010b) Structural, elastic, electronic properties and Fermi surface for superconducting Mo_2GaC in comparison with V_2GaC and Nb_2GaC from first principles. *Physica C*, **470**, 533–537.

Spanier, J.E., Gupta, S., Amer, M., and Barsoum, M.W. (2005) First-order Raman scattering from the $M_{n+1}AX_n$ phases. *Phys. Rev. B*, **71**, 12103.

Sun, Z., Ahuja, R., Li, S., and Schneider, J.M. (2003) Structure and bulk modulus of M_2AlC (M = Ti, V and Cr). *Appl. Phys. Lett.*, **83**, 899–901.

Sun, Z., Li, S., Ahuja, R., and Schneider, J.M. (2004) Calculated elastic properties of M_2AlC (M = Ti, V, Cr, Nb and Ta). *Solid State Commun.*, **129**, 589–592.

Tian, W., Wang, P., Zhang, G., Kan, Y., and Li, Y. (2007) Mechanical properties of Cr_2AlC ceramics. *J. Am. Ceram. Soc.*, **90**, 1663–1666.

Toth, L.E. (1971) *Transition Metal Carbides and Nitrides*, Academic Press, New York.

Turner, P.A. and Tomé, C.N. (1994) A study of residual stresses in Zircaloy-2 with rod texture. *Acta Mater.*, **42**, 4143–4153.

Wang, J.Y. and Zhou, Y.C. (2003) First-principles study of equilibrium properties and electronic structure of $Ti_3Si_{0.75}Al_{0.25}C_2$ solid solution. *J. Phys.: Condens. Matter*, **15**, 5959–5968.

Wang, J. and Zhou, Y. (2004a) Dependence of elastic stiffness on electronic band structure of nanolaminate M_2AlC (M = Ti,V,Nb and Cr) ceramics. *Phys. Rev. B*, **69**, 214111.

Wang, J.-Y. and Zhou, Y.-C. (2004b) Polymorphism of Ti_3SiC_2 ceramic: first-principles investigations. *Phys. Rev. B*, **69**, 144108.

Wang, J. and Zhou, Y. (2009) Recent progress in theoretical prediction, preparation, and characterization of layered ternary transition-metal carbides. *Annu. Rev. Mater. Res.*, **39**, 415–443.

Wang, J., Zhou, Y., Lin, Z., and Hu, J. (2008) Ab initio study of polymorphism in layered ternary carbide M_4AlC_3 (M = V, Nb and Ta). *Scr. Mater.*, **58**, 1043–1046.

Wang, J., Zhou, Y., Lin, Z., Meng, F., and Li, F. (2005) Raman active modes and heat capacities of Ti_2AlC and Cr_2AlC ceramics: first principles and experimental investigations. *Appl. Phys. Lett.*, **86**, 101902.

Wu, Z. and Cohen, R.E. (2006) More accurate generalized gradient approximation for solids. *Phys. Rev. B*, **73**, 235116.

Yang, Z.-J., Guo, Y.-D., Linghu, R.-F., and Yang, X.-D. (2012) First-principles calculation of the lattice compressibility, elastic anisotropy and thermodynamic stability of V_2GeC. *Chin. Phys. B*, **21**, 036301.

Yang, Q., Lengauer, W., Koch, T., Scheerer, M., and Smid, I. (2000) Hardness and elastic properties of $Ti(C_xN_{1-x})$,$Zr(C_xN_{1-x})$ and $Hf(C_xN_{1-x})$. *J. Alloys Compd.*, **309**, L5–L9.

Ying, G., He, X., Li, M., Du, S., Han, W., and He, F. (2011) Effect of Cr_7C_3 on the mechanical, thermal, and electrical properties of Cr_2AlC. *J. Alloys Compd.*, **509**, 8022–8027.

Zhang, H., Wu, X., Nickel, K.G., Chen, J., and Presser, V. (2009a) High-pressure powder x-ray diffraction experiments and ab initio calculation of Ti_3AlC_2. *J. Appl. Phys.*, **106**, 013519.

Zhang, W., Travitzky, N., Hu, C., Zhou, Y., and Greil, P. (2009b) Reactive hot pressing and properties of Nb_2AlC. *J. Am. Ceram. Soc.*, **92**, 2396–2399.

Zhou, Y.C. and Sun, Z.M. (2000) Electronic structure and bonding properties of layered machinable Ti_2AlC and Ti_2AlN ceramics. *Phys. Rev. B*, **61**, 12570–12573.

Zhou, Y.C., Sun, Z., Wang, X., and Chen, S. (2001) Ab initio geometry optimization and ground state properties of layered ternary carbides, Ti_3MC_2 (M = Al,Si and Ge). *J. Phys.: Condens. Matter*, **13**, 10001–10010.

4
Thermal Properties

4.1
Introduction

Some of the MAX phases are quite refractory and are therefore candidates for high-temperature applications. Before such applications are developed, however, it is important to understand their thermal properties, including their thermal conductivities, heat capacities, and thermal expansion coefficients (TECs) among others. In this chapter, these properties are discussed and an attempt to understand the physics behind them is made.

Section 4.2 deals with the thermal conductivities and their temperature dependencies. In this section, the thesis that the MAX phases are good thermal conductors because they are good electrical conductors is developed. The role of the A-group element and defects on scattering phonons is highlighted. When solids are heated, their atoms vibrate and the amplitude of these vibrations are characterized by atomic displacement parameters (ADPs). Section 4.3 reviews what is known about the latter, both theoretically and experimentally. Section 4.4 deals with one of the most important thermodynamic properties of any solid, that is, their heat capacities and their temperature dependencies. When solids are heated, they expand and the latter is characterized by a TEC. Section 4.5 summarizes the thermal expansion results and their anisotropies. The penultimate section, Section 4.6, deals with the thermal stabilities of the MAX phases. Section 4.7 summarizes the chapter.

4.2
Thermal Conductivities

Typically, the total thermal conductivity κ_{th} of a solid, is the sum of its electronic κ_{e} and phononic κ_{ph} thermal conductivity contributions, that is,

$$\kappa_{\mathrm{th}} = \kappa_{\mathrm{e}} + \kappa_{\mathrm{ph}} \tag{4.1}$$

For a metallic conductor, κ_{e} can be estimated from the Wiedmann–Franz law:

$$\kappa_{e} = \frac{LoT}{\rho} \tag{4.2}$$

MAX Phases: Properties of Machinable Ternary Carbides and Nitrides, First Edition. Michel W. Barsoum.

where *Lo* is the Lorenz number, whose value is 2.45×10^{-8} W·Ω K^{-2}. Thus knowing ρ as a function of T, it is straightforward to calculate κ_e as a function of T.

At this point, it is useful to separate the discussion into two temperature regimes: low (4–300 K) and high (300–1200 K). The former sheds light on the physics of thermal conductivity; the latter is of more practical import.

4.2.1
Low Temperatures

The MAX phases are good thermal conductors. In many cases, they are better thermal conductors than their corresponding transition metals at all temperatures. Figure 4.1a plots the values of κ_{th} as a function of T in the 4–300 K temperature range for select Al-containing MAX phases and Ti_3SiC_2; the corresponding values of κ_{ph} are plotted in Figure 4.1b.

Figure 4.2a plots κ_{th} as a function of T for a number of $Ti_{n+1}AlX_n$ phases; Figure 4.2b plots their respective κ_{ph}s. Figure 4.3a,b, respectively, plot the temperature dependencies of κ_{th}, κ_e, and κ_{ph} for Ti_2SC (Scabarozi *et al.*, 2008b) and Ti_2GeC (Barsoum *et al.*, 2011). Ti_2SC exhibits one of the highest room-temperature thermal conductivities amongst the MAX phases.

Table 4.1 summarizes the room-temperature values of κ_{th}, κ_e, and κ_{ph} of roughly 20 MAX phases, including some solid solutions. Also included are the corresponding parameters for near-stoichiometric TiC, TiC_x, and NbC_x for comparison. From these results, and those shown in Figures 4.1–4.3, it is reasonable to conclude the following:

1) The MAX phases are good thermal conductors because they are good electrical conductors.

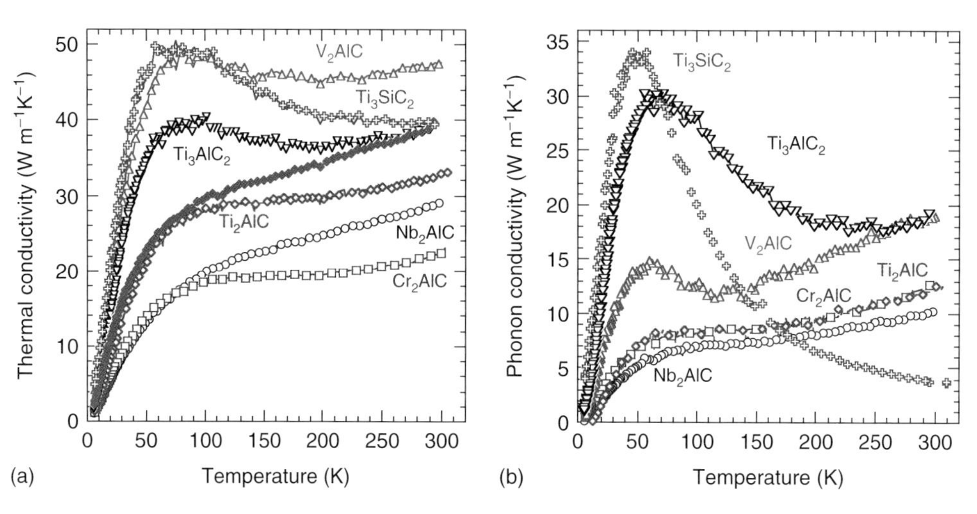

Figure 4.1 Temperature dependencies of (a) κ_{th} and (b) κ_{ph} for select Al-containing MAX phases (Hettinger *et al.*, 2005). The plots are color coded for clarity's sake.

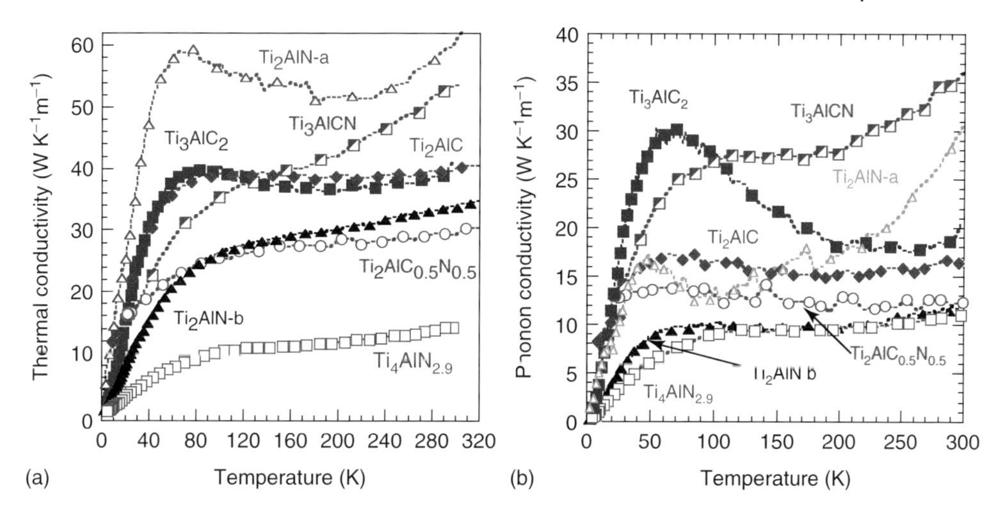

Figure 4.2 Temperature dependencies of (a) κ_{th} and (b) κ_{ph} of select $Ti_{n+1}AlX_n$ phases (Scabarozi *et al.*, 2008a). The plots are color coded for clarity's sake.

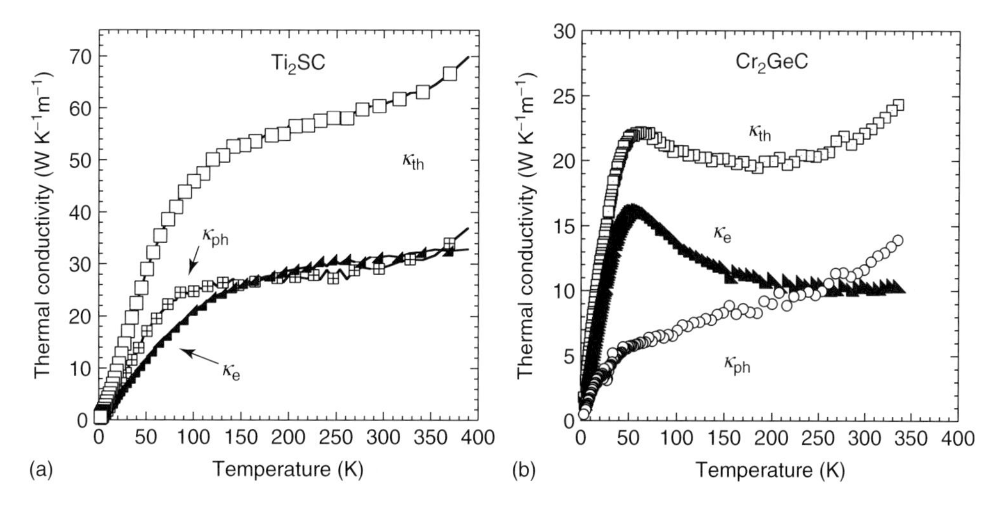

Figure 4.3 Temperature dependencies of κ_{th}, κ_e, and κ_{ph} for (a) Ti_2SC (Scabarozi *et al.*, 2008b) and (b) Cr_2GeC (Barsoum *et al.*, 2011).

2) In general, for non-S- or Al-containing MAX phases, $\kappa_{ph} \ll \kappa_e$ (Table 4.1).
3) The S- and Al-containing MAX phases are good phonon conductors (Table 4.1). At 36 W m^{-1} K^{-1}, κ_{ph} of Ti_3AlCN at room temperature is one of the highest reported for a MAX phase to date. The phonon conductivity of Ti_2SC is also quite high (Figure 4.3a).
4) With a few notable exceptions, as discussed below, a correlation exists between the quality of the crystal, as measured by the residual resistivity ratio (RRR),

Table 4.1 Summary of room-temperature total (κ_{th}), phonon (κ_{ph}), and electronic (κ_e) thermal conductivities (in W m^{-1} K^{-1}) for select MAX phases. Also included for comparison are those for near-stoichiometric TiC_x and NbC_x. The percent contributions of κ_{ph} and κ_e to κ_{ph} are shown in parentheses.

Compound	κ_{th}	κ_e	κ_{ph}	References
			413 phases	
$Ti_4AlN_{2.9}$	12.0	2.8 (23%)	9.2 (77%)	Barsoum *et al.* (2000c)
Nb_4AlC_3	13.5	9.6 (70%)	3.9 (30%)	Hu *et al.* (2008b)
Ta_4AlC_3	38.4	19.0 (50%)	19.0 (50%)	Hu *et al.* (2007b)
			312 phases	
Ti_3SiC_2	34.0	33.0 (97%)	≈1.0 (3%)	Barsoum *et al.* (1999)
	40.0	36.2 (90%)	3.8 (10%)	Finkel *et al.* (2004)
$Ti_3Si_{0.5}Ge_{0.5}C_2$	39.0	38.0 (97%)	1.0 (3%)	Finkel *et al.* (2004)
Ti_3AlC_2	40.0	21.0 (52%)	19.0 (42%)	Scabarozi *et al.* (2008a)
Ti_3GeC_2	38.0	38.0 (100%)	—	Finkel *et al.* (2004)
			211 phases	
Ti_2AlC	33.0	20.5 (62%)	12.5 (38%)	Hettinger *et al.* (2005)
	46.0	20.0 (43%)	26.0 (57%)	Barsoum, Ali, and El-Raghy (2000a)
Ti_2AlN a	60.0	29.0 (49%)	31.0 (51%)	Scabarozi *et al.* (2008a)
b	34.0	23.0 (67%)	11.0 (33%)	
Ti_3AlCN	53.4	18.3 (34%)	36.0 (66%)	Scabarozi *et al.* (2008a)
$Ti_2AlC_{0.5}N_{0.5}$	29.3	16.9 (58%)	12.4 (42%)	Scabarozi *et al.* (2008a)
Ti_2SC	60.0	31.0 (52%)	29.0 (48%)	Scabarozi *et al.* (2008b)
V_2AlC	48.0	29.0 (61%)	19.0 (39%)	Hettinger *et al.* (2005)
Cr_2AlC	23.0	9.0 (39%)	14.0 (61%)	Hettinger *et al.* (2005)
Cr_2AlC	17.8	10.2 (58%)	7.6 (42%)	Tian *et al.* (2006)
Cr_2AlC	14.5	12 (83%)	2.5 (17%)	Zhou, Mei, and Zhu (2009)
Cr_2AlC	15.2	12.7 (81%)	2.5 (19%)	Ying *et al.* (2011)
Cr_2GeC	22.0	11.5 (52%)	10.5 (48%)	Barsoum *et al.* (2011)
Nb_2AlC	29.0	19.0 (66%)	10.0 (34%)	Hettinger *et al.* (2005)
	23.0	23.0 (>100%)[a]		Barsoum *et al.* (2002b)
	20.0			Zhang *et al.* (2009)
TiNbAlC	16.6	9.4 (56%)	7.2 (43%)	Barsoum *et al.* (2002b)
Ta_2AlC	28.4	28.3 (100%)	—	Hu *et al.* (2008a)
Nb_2SnC	17.5	17.5 (100%)	—	Barsoum, El-Raghy, and Chakraborty (2000b)
Ti_2InC	26.5	26.5 (100%)	—	Barsoum *et al.* (2002a)
TiHfInC	20.0	20.0 (100%)	—	Barsoum *et al.* (2002a)
Hf_2InC	26.5	26.5 (100%)	—	Barsoum *et al.* (2002a)
Ti_2SC	60.0	30.0 (50%)	30.0 (50%)	Scabarozi *et al.* (2008b)
TiC_x	33.5	12.0 (36%)	21.5 (64%)	Taylor (1961)
$TiC_{0.96}$	14.4	7.3 (50%)	7.1 (50%)	Lengauer *et al.* (1995)
NbC_x	14	21[a]	—	Pierson (1996)

[a]Implies $Lo < 2.45 \times 10^{-8}$ W Ω K^{-2}.

and κ_{ph} (Figure 4.4). The RRR is the ratio of the resistivity at RT to that at 4 K (Chapter 5).

5) The thermal conductivities of the MAX phases are comparable to those of their binary MX counterparts.

Stiff, lightweight solids with high Debye temperatures are typically good phonon conductors. Given the rigidity of some of the MAX phases (Table 3.2), the fact that κ_{ph} is suppressed in many of them is somewhat surprising. To understand why, one must look at two factors: the rattler effect, and the influence of point defects. And while it is not always easy to deconvolute the two, they are discussed separately below.

4.2.2 The Rattler Effect

Atoms that vibrate about their equilibrium position significantly more than other atoms in a given structure are known as *rattlers* and have been shown to be potent phonon scatterers (Keppens *et al.*, 1998; Sales *et al.*, 1999). Many A elements, especially those with atomic numbers >S, tend to "rattle" in the MAX structures (Barsoum, 2000). Probably, the most convincing evidence for the rattler conjecture is to compare κ_{ph} for the isostructural compounds Ti_3SiC_2 and Ti_3AlC_2. Based on their RRR value – which is higher for Ti_3SiC_2 – and the maximum peak heights of the curves shown in Figure 4.1b, it is reasonable to conclude that the Ti_3SiC_2 sample is less defective than its Ti_3AlC_2 counterpart. And yet at room temperature, κ_{ph} for the latter is ≈5 times higher than that for the former (Figure 4.1b and Table 4.1). Given the similarities of their elastic properties, molecular weights, and Debye temperatures, it is obvious that in this structure Si is a much more potent phonon scatterer than Al. Along the same lines, κ_{ph} for Ti_3GeC_2 is essentially nil (Table 4.1), which suggests that Ge is even more of a rattler than Si. The same is also true of most MAX phases with A elements heavier than S. This is presumably why κ_{ph} for Ti_2InC, Hf_2InC, Nb_2SnC, and so on, are vanishingly small (Table 4.1). Evidence that the A atoms vibrate more than others in the MAX phases is presented in the next section.

The situation for Al is more ambiguous. For reasons that are not entirely clear the Al atoms appear to be better bound, and thus act less as rattlers, which partially explains why κ_{ph} is not negligible in these phases. The fact that the Al is the lightest A element in the MAX phases is also not coincidental.

Another exception to this rule is Ti_2SC. Because of its low c/a ratio (Table 2.1 and Figure 2.6c), it was anticipated that the Ti–S bonds would be exceptionally strong (Barsoum, 2000). Consequently, and despite a low RRR value, at $\approx 30\ W\,m^{-1}\,K^{-1}$ its κ_{ph} is one of the highest reported to date (Table 4.1 and Figure 4.3a). The only higher value is that of Ti_3AlCN (Radovic, Ganguly, and Barsoum, 2008). Note that Ti_2SC, Ti_3AlCN, and possibly Ti_3AlC_2 are apparent outliers in Figure 4.4. The main reason is believed to be the strengths of the M–A bonds in these phases (Scabarozi *et al.*, 2008a, b). The same should presumably be true of the M_2PC and other M_2SC phases.

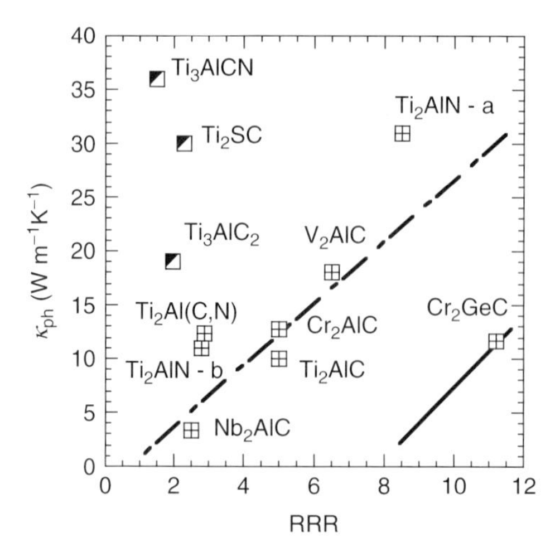

Figure 4.4 Dependence of room-temperature phonon conductivity on the residual resistivity ratio (RRR) of select MAX phases. Clearly, Ti_2SC, Ti_3AlC_2, Cr_2GeC, and $Ti_2AlC_{0.5}C_{0.5}$ are outliers.

4.2.3
Effect of Defects

It is not easy to quantify defect concentrations, especially point defects, in solids. A good measure of the quality of a crystal, however, is its RRR. Low RRR values are associated with low-quality, defective crystals, and vice versa. It is thus not surprising that κ_{ph} scales with RRR (Figure 4.4). This is not unique to the MAX phases but is also true of the MX compounds (Lengauer *et al.*, 1995; Taylor, 1961). A good example of the effect of defects is to compare κ_{ph} of two nominally identical Ti_2AlN samples – "a" and "b" listed in Table 4.1 and plotted in Figure 4.4 – with RRR values of 8.5 and 2.8, respectively. The corresponding κ_{ph} values are 31 and 11 W m^{-1} K^{-1} (Radovic, Ganguly, and Barsoum, 2008). The same is true of $Ti_4AlN_{2.9}$; despite it being quite stiff, the presence of vacancies – presumably on the N sites – resulted in quite low RRR and concomitantly quite low κ_{ph} values (Table 4.1).

A fruitful area of research is to try and fabricate and characterize some of the Al-containing MAX phases with as few defects as possible. The benefit of such a strategy can be seen in Figure 4.3b, where the results for a Cr_2GeC sample are shown. At 11.5, the RRR ratio for this sample is quite high, which resulted in a room temperature κ_{ph} of about 10 W m^{-1} K^{-1}. This value however, is significantly lower than what it would have been had the A-group element been Al, which presumably would have fallen on the dashed line in Figure 4.4. An even better example is sample a-Ti_2AlN; at 31 W m^{-1} K^{-1}, its room temperature κ_{ph} is one of the highest recorded to date which, in turn, explains why its κ_{th} is also one of the highest reported to date (Figure 4.2a).

On the basis of the results shown in Table 4.1, it is reasonable to conclude that the effects of solid solutions on κ_{ph} are mild. For example, κ_{ph} for TiNbAlC is not much different than that of the end members Ti_2AlC and Nb_2AlC. The same is true when κ_{ph} values of $Ti_2Al(C_{0.5}N_{0.5})$ and its end members are compared. Tian, Sun, and Hashimoto (2009) measured κ_{th} of $(Cr_{1-x}V_x)_2AlC$ as a function of x and showed that, as x was increased from 0 to 0.25 to 0.5, κ_{th} decreased, respectively, from $\approx$19 to 15 to 18 W m^{-1} K^{-1}, confirming the weak dependence of κ_{th} on x. The mild decrease in κ_{th} can be attributed to solid-solution scattering.

4.2.4 High Temperatures

The temperature dependences of κ_{th} in the 300–1200 K temperature range for a number of MAX phases are shown in Figure 4.5. To a good first approximation, the κ_{th} values vary linearly with T. For four compounds – $Ti_4AlN_{2.9}$, Nb_2AlC, TiNbAlC, and Ta_4AlC_3 – κ_{th} increases with increasing T. For all others the opposite is true. To better understand the slopes of the curves, we follow the arguments made by Williams (1966) to explain similar results for the MX phases. As discussed in Chapter 5, at $T > 100$ K, the electrical resistivity ρ of the MAX phases can be well described by

$$\rho = aT + b \tag{4.3}$$

where $a > 0$. Combining Eqs. (4.2) and (4.3), it follows that

$$\frac{d\kappa_e}{dT} = \frac{Lob}{(aT + b)^2} \tag{4.4}$$

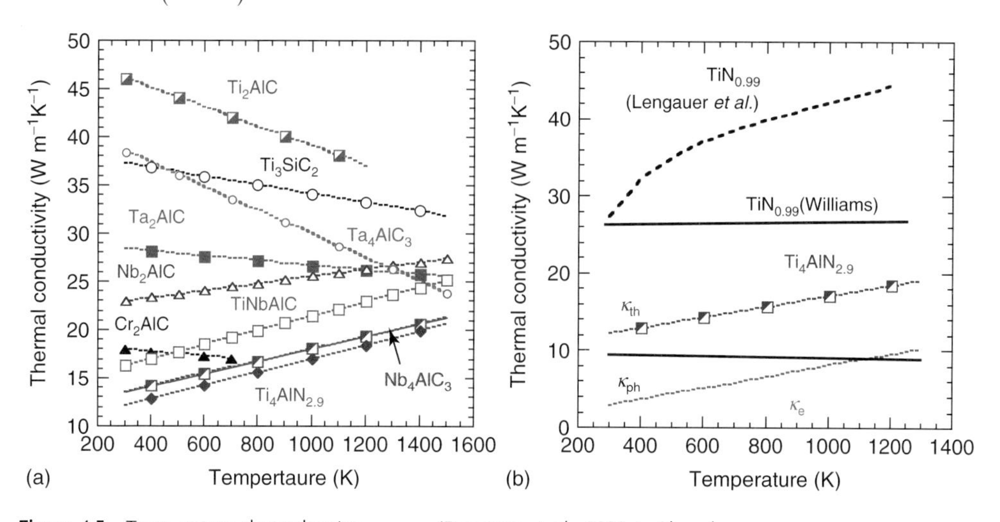

Figure 4.5 Temperature dependencies of κ_{th} in the 300–1200 K temperature range for (a) Ti_3SiC_2 (Barsoum *et al.*, 1999), Ta_2AlC (Hu *et al.*, 2008a), Nb_4AlC_3 (Hu *et al.*, 2008b), Cr_2AlC (Tian *et al.*, 2006), $Ti_4AlN_{2.9}$ (Barsoum *et al.*, 2000c), Ta_4AlC_3 (Hu *et al.*, 2007b), and (b) $Ti_4AlN_{2.9}$ (Barsoum *et al.*, 2000c). Also shown are the results for near-stoichiometric $TiN_{0.99}$ (Williams, 1966) and (Lengauer *et al.*, 1995). The contributions of κ_{ph} and κ_e to κ_{th} of $Ti_4AlN_{2.9}$ are shown below the latter in (b). In both figures the results and labels are color coded.

Table 4.2 Summary of thermal conductivities of select MAX phases in the 300–1300 K temperature range. The percent contributions of κ_{ph} and κ_e to κ_{th} are shown in parentheses.

Compound	κ_{th} (W m^{-1}·K^{-1})		300 K		1300 K		References
	300	1300	κ_e	κ_{ph}	κ_e	κ_{ph}	
				413 phases			
$Ti_4AlN_{2.9}$	12	20	3 (25%)	9 (75%)	10 (50%)	10 (50%)	Barsoum *et al.* (2000c)
Ta_4AlC_3	38	26	19 (50%)	19 (50%)	20 (77%)	6 (23%)	Hu *et al.* (2007b)
Nb_4AlC_3	13	20	10 (77%)	3 (23%)	13 (65%)	7 (35%)	Hu *et al.* (2008b)
				312 phases			
Ti_3SiC_2	37	33	33 (90%)	≈4 (10%)	32 (97%)	≈1 (3%)	Barsoum *et al.* (1999)
				211 phases			
Ti_2AlC	46	36	20 (43%)	26 (57%)	20 (55%)	16 (45%)	Barsoum *et al.* (2002b)
TiNbAlC	17	24	9 (56%)	8 (43%)	20 (85%)	4 (62%)	Barsoum *et al.* (2002b)
Nb_2AlC	23	27	16[a] (77%)	7[a] (32%)	20 (74%)	>4 (26%)	Barsoum *et al.* (2002b)
Nb_2SnC	17	30	12[a] (72%)	5[a] (28%)	25 (82%)	5 (18%)	Barsoum *et al.* (2000b)
Ta_2AlC	28	26	28 (100%)	—	26 (100%)	—	Hu *et al.* (2008a)
Cr_2AlC	19	16[b]	10	9	12	4[b]	Tian *et al.* (2006)
ZrSC	≈38	≈30	—	—	—	—	Opeka *et al.* (2011)
				MX phases			
TiC_x	33.5	39	12 (36%)	7.0 (50%)	24 (66%)	15 (38%)	Lengauer *et al.* (1995)
$TiC_{0.96}$	14.4	33.4	7.35 (50%)	21.5 (64%)	—	—	Taylor (1961)
$TiN_{0.99}$	27.4	45.3	29.4[c]	—	—	—	Lengauer *et al.* (1995)
NbC_x	14	—	21[c]	—	—	—	Pierson (1996)

[a] Assuming $Lo = 1.5 \times 10^{-8}$ W Ω K^{-2}. See text.
[b] Extrapolated.
[c] Implies $Lo < 4.5 \times 10^{-8}$ W Ω K^{-2}.

Hence $d\kappa_e/dT$ *must* have the same sign as b. It follows that solids with low RRR values, that is, those with positive b values, will, in turn, result in positive $d\kappa_e/dT$ values – as observed – and vice versa. Note that, when b is close to zero, the temperature dependence of κ_{th} is identical to that of κ_{ph}. In Table 4.2, two compounds, Ti_2AlC and Ta_4AlC_3, have $b \approx 0$, and in both cases, increasing T decreases κ_{ph}, which is not very surprising (see next section).

On the basis of the results shown in Figure 4.5, and summarized in Table 4.2, the following points are pertinent:

- The highest value of κ_{th} belongs to Ti_2AlC because of its high κ_{ph}. Increasing the temperature to 1300 K from room temperature reduces κ_{ph} by about 20%. The second highest values are those for Ti_3SiC_2. This is true despite the fact that κ_{ph}

is negligible at all temperatures. The relatively high κ_{th} values must thus stem from the high quality of the sample.

- If one assumes *Lo* to be 2.45×10^{-8} WΩ K^{-2}, the impossible result that $\kappa_e > \kappa_{th}$ is obtained for all Nb-containing compounds listed in Table 4.2, including NbC. Furthermore, given that it is unlikely that κ_{ph} of TiNbAlC is greater than that of Nb_2AlC, the value for the latter must be greater than that of the former, that is, >7 W m^{-1} K^{-1}. Using that value for Nb_2AlC reduces *Lo* to $\approx 1.5 \times 10^{-8}$ W·Ω K^{-2}, which is the value assumed for all the Nb-containing compounds listed in Table 4.2.

The primary purpose of the preceding exercise is not to precisely determine κ_{ph} but rather to make the case that for the Nb-containing phases κ_{ph} is small, both in relation to κ_e and in absolute terms. However, given the low thermal expansions, relatively high Debye temperatures, and high stiffness values of Nb_2AlC, this is a surprising result. This is particularly true since it has been shown that, at least at 10 K, the atomic displacements in Nb_2AlC was significantly lower than those in Ti_2AlC (Hug, Jaouen, and Barsoum, 2005). It is thus incorrect to conclude that the lower thermal conductivities of Nb_2AlC are related to the rattler effect. A more likely explanation is that the Nb_2AlC sample is significantly more defective than the Ti_2AlC sample. This is especially true in this case, since the stoichiometries of the Nb-containing phases were $Nb_{2.00}Al_{0.91}C_{0.89}$ and $Ti_{0.94}Nb_{1.06}Al_{0.93}C_{0.94}$ (Barsoum *et al.*, 2002b). It is also consistent with the fact that at 2.8, the RRR of the Nb_2AlC sample reported in Figure 4.4 was quite low. These comments notwithstanding, it is hereby acknowledged that more work is needed to sort out some of these issues, especially the role of nonstoichiometry on κ_{ph}.

To summarize this section: for most of the MAX phases with atoms heavier than S, κ_{ph} is negligible because of the rattler effect, an effect that is more potent when the A-group element is heavy. For Ge-containing phases and presumably others, the polycrystal has to be of exceptional quality for κ_{ph} not to be negligible. For the Al-, S-, and P-, and, probably As-containing phases, κ_{ph} cannot be neglected because these atoms are better bonded in the structure than the rest. Not all MAX phases with light A elements are good phonon conductors, however. Like for the MX phases, point defects are potent phonon scatterers.

An important measure of electron and phonon scattering are the ADPs, discussed in the next section.

4.3
Atomic Displacement Parameters

When solids are heated, their atoms vibrate. The amplitude of these vibrations is a function of the mass of the vibrating atoms, the bonds holding them in place, and their local atomic arrangement. The ADP or U_{ij} is a measure of the square of the amplitude of an atom's vibration. Experimentally, there are several methods to determine the U_{ij} values. Probably the most common is Rietveld analysis of

X-ray diffraction (XRD) and neutron diffraction (ND) diffractograms, typically carried out as a function of temperature.

In structure refinement, the Debye–Waller factor (T) – that accounts for the thermal motion correction to the structure factor due to anisotropic thermal motion–is expressed as

$$T = \exp\left(-2\pi^2\left(U_{11}h^2a^{*2} + U_{22}k^2b^{*2} + U_{33}l^2c^{*2} + 2U_{23}klb^*c^* + 2U_{13}lhc^*a^* + 2U_{12}hka^*b^*\right)\right) \quad (4.5)$$

where a^*, b^*, and c^* are the edges of the unit cell in the reciprocal space associated with the x^*, y^*, and z^* axes, respectively (Larson and Dreele, 2004). For hexagonal structures, due to site symmetry, U_{23} and U_{13} are both equal to 0 and the equation simplifies to

$$T = \exp\left(-2\pi^2\left(U_{11}h^2a^{*2} + U_{22}k^2b^{*2} + U_{33}l^2c^{*2} + 2U_{12}hka^*b^*\right)\right) \quad (4.6)$$

Furthermore, for the hexagonal crystal system, the anisotropic U_{ij}s may be converted to an approximate isotropic temperature factor U_{eq}, given by

$$U_{eq} = \frac{1}{3}\left(U_{11} + U_{22} + U_{33}\right) \quad (4.7)$$

In what follows, the values of U_{eq} are first reviewed, followed by a more detailed exposition of ADP anisotropies.

4.3.1 Isotropic Atomic Displacement Parameters

Table 4.3 summarizes the room-temperature U_{eq} values for the unique atoms in select MAX phases. The penultimate column lists the average for all the atoms in a structure, $U_{eq,av}$. Also listed in Table 4.3 for comparison are values for a number of near-stoichiometric MX phases (Nakamura and Yashima, 2008). In the latter, the authors assumed $U_M = U_X$. Figure 4.6 plots the $U_{eq,av}$ values for MX and select Al- and Si-containing MAX phases. A perusal of the results listed in Table 4.3 and plotted in Figure 4.6, indicates the following:

1) Replacing the C atoms in MX with an A atom results in a substantial increase in $U_{eq,av}$ (Figure 4.6).
2) The Ti- and Ta-containing MAX phases have relatively low $U_{eq,av}$ values, compared to the V-containing ones. A comparison with the corresponding values for the MX compounds makes it clear that at least part of that increase is because $U_{eq,av}$ of VC is significantly higher than those of the other MX compounds. Whether this is related to weaker bonds or to the presence of vacancies is unclear at this time.
3) The Ta–Al phases have lower $U_{eq,av}$ values than those of the Ta–Ga ones.
4) The $U_{eq,av}$ values for the 312 phases are all relatively low. Whether this is a characteristic of this structure or the chemistries listed is unclear at this time. More work is needed.

Table 4.3 Summary of room-temperature U_{eq} (Å^2) for the unique atoms in select MAX phases. The penultimate column lists the average for all the atoms in the structure, $U_{eq,av}$. Also shown are the values for near-stoichiometric MX compounds. The equation used to calculate the values of $U_{eq,av}$ in some of our previous papers was incorrect. The values listed here use the correct equation, viz. Eq. 4.7.

Atom	M_I	M_{II}	A	X_I	X_{II}	$U_{eq,av}$	References
				413 phases			
$Ti_4AlN_{2.9}$	0.0030(6)	0.0033(7)	0.0053(8)	0.0055(3)	0.0039(4)	0.0042	Barsoum *et al.* (2000c)
$V_4AlC_{2.7}$	0.0092(2)	0.0120(2)	0.0114(3)	0.0087(6)	0.009(2)	0.010	Etzkorn, Ade, and Hillebrecht (2007b)
$V_4AlC_{2.67}$ [a]	0.0122(2)	0.0123(2)	0.0154(2)	0.0127(6)	0.014(1)	0.013	Etzkorn, Ade, and Hillebrecht (2007b)
Ta_4AlC_3	0.0070(2)	0.0074(2)	0.009(2)	0.007(3)	0.012(6)	0.008	Etzkorn, Ade, and Hillebrecht (2007a)
Ta_4GaC_3	0.0117(8)	0.0124(8)	0.014(1)	0.001(1) [b]	0.018(3)	0.012	Etzkorn *et al.* (2009)
				312 phases			
Ti_3SiC_2	0.0063(6)	0.0030(9)	0.0093(6)	0.0040(3)	—	0.0079	Barsoum *et al.* (1999)
Ti_3SiC_2	0.0064 [c]	0.0061 [c]	0.0113 [c]	0.0077 [c]	—	0.0057 [c]	Lane, Vogel, and Barsoum (2010)
Ti_3SiGeC_2	0.0033(1)	0.0037(1)	0.0077(1)	0.0042(2)	—	0.0044	Yang *et al.* (2006)
Ti_3GeC_2	0.0047 [c]	0.0087 [c]	0.0300 [c]	0.0061 [c]	—	0.0081 [c]	Lane, Vogel, and Barsoum (2010)
Ta_3AlC_2 [d]	0.0084(2)	0.0082(2)	0.0143(2)	0.01(3)	—	0.0059	Etzkorn, Ade, and Hillebrecht (2007a)
				211 phases			
Ti_2AlN	0.0045 [c]	—	0.0090 [c]	0.0075 [c]	—	0.0070 [c]	Lane, Vogel, and Barsoum (2011)
Cr_2GeC	0.0071	—	0.0077 [c]	0.0061 [c]	—	0.0071 [c]	Lane, Vogel, and Barsoum (2011)
Ti_2GaC	0.016(4)	—	0.0128(6)	0.007(3)	—	0.013	Etzkorn *et al.* (2009)

(continued overleaf)

Table 4.3 (*Continued*)

Atom	M_I	M_{II}	A	X_I	X_{II}	$U_{eq,av}$	References
V_2AlC	0.0113(4)	—	0.0136(6)	0.012(2)	—	0.012	Etzkorn, Ade, and Hillebrecht (2007b)
Cr_2GaC	0.0095(3)	—	0.0123(4)	0.011(1)	—	0.010	Etzkorn *et al.* (2009)
				MX phases			
TiC	0.0015	—	—	0.0015	—	0.0015	Nakamura and Yashima (2008)
VC	0.0058	—	—	0.0058	—	0.0058	Nakamura and Yashima (2008)
NbC	0.0019	—	—	0.0019	—	0.0019	Nakamura and Yashima (2008)
TaC	0.0013	—	—	0.0013	—	0.0013	Nakamura and Yashima (2008)
HfC	0.0052	—	—	0.0052	—	0.0052	Nakamura and Yashima (2008)
ZrC	0.0042	—	—	0.0042	—	0.0042	Nakamura and Yashima (2008)

[a] In this compound the vacancies are ordered.
[b] To be consistent with all other measurements, the values for X_I and X_{II} were reversed.
[c] Extrapolated values from slightly higher temperatures.
[d] Actual chemistry: $Ta_3Al_{0.96}Sn_{0.04}C_2$.

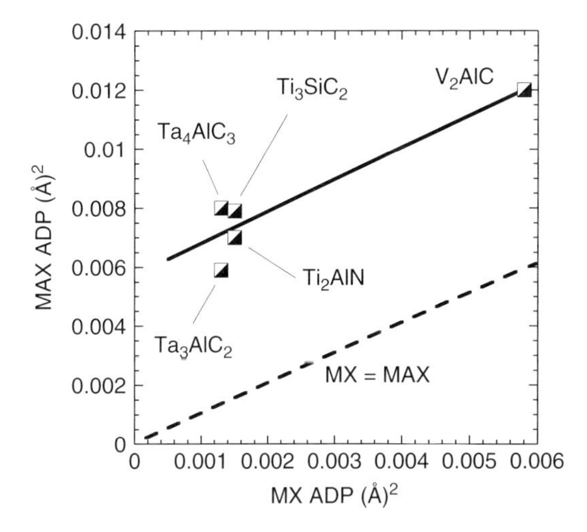

Figure 4.6 Comparison of room temperature $U_{eq,av}$ values for select MX and Al- and Si-containing MAX phases. Dashed line depicts boundary where the $U_{eq,av}$ values of the MAX and MX phases are equal. The MX values were taken from Nakamura and Yashima (2008).

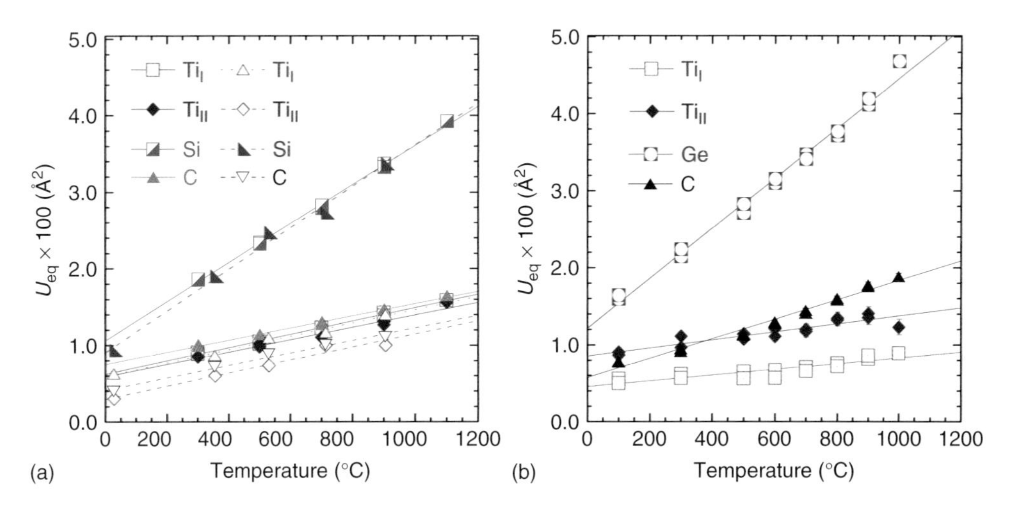

Figure 4.7 Temperature dependence of isotropic U_{ij}s of atoms in (a) Ti_3SiC_2 and (b) Ti_3GeC_2 during both heating and cooling (Lane, Vogel, and Barsoum, 2010). In (a), the results of Lane *et al.* (solid lines) are compared with those of Barsoum *et al.* (1999) (dashed lines).

Making use of recent Rietveld analysis of high-temperature ND results, we explored the vibrations of the four unique atoms in Ti_3SiC_2 and Ti_3GeC_2 (Lane, Vogel, and Barsoum, 2010). Not surprisingly, when the U_{eq} values of all atoms are plotted as a function of temperature (Figure 4.7a,b), they increase approximately linearly with T, with the greatest increases occurring for the A atoms in both

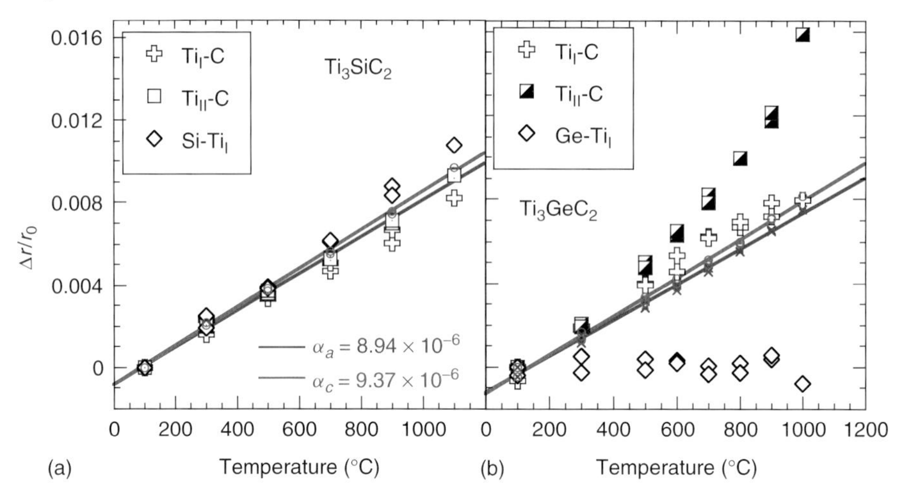

Figure 4.8 Temperature dependence of the thermal expansion of the bonds in (a) Ti_3SiC_2 and (b) Ti_3GeC_2 during both heating and cooling (Lane, Vogel, and Barsoum, 2010). Note that while the TECs of the lattice parameters a and c shown in red and green, respectively, are quite similar, the thermal strains on the various bonds are quite different indeed.

materials. These results are important because they support the hypothesis that the A-group element is indeed a rattler. Given that the two compounds are isostructural and the chemical similarities of Si and Ge, these results were unsurprising. The similarities also extend to the thermal expansions of the lattice parameters that are almost identical as depicted by the colored lines in Figure 4.8a,b.

However, the reasons why the TECs along a and c directions are so similar are quite different in each of the compounds. According to Figure 4.8b, the Ti–Ge bond does not appear to expand with increasing T, while the Ti_{II}–C bond appears to increase the most with increasing T. To understand this surprising result, one needs to look at the anisotropies of the ADPs in some more detail.

4.3.2
Anisotropic Atomic Displacement Parameters

Figure 4.9a,b, respectively, plot the temperature dependencies of U_{11} – which for hexagonal symmetry is equal to U_{22} – and U_{33} of the A atoms in five MAX phases. Note that at all temperatures, the ratio U_{11}/U_{33} is > 1, which implies that the vibration amplitudes of the A atoms along the basal planes are greater than those along the [0001] direction. Furthermore, for the most part, the ADPs of the A elements are larger than those of the M or X atoms, again confirming that the A-group elements can act as rattlers, but not to the same extent. For example, the rattling of Ge and Si atoms in the 312 structures are higher than the others, notably Ge in Ti_2GeC. The latter is consistent with the fact that some phonon conductivity is observed in Ti_2GeC but not in Ti_3GeC_2.

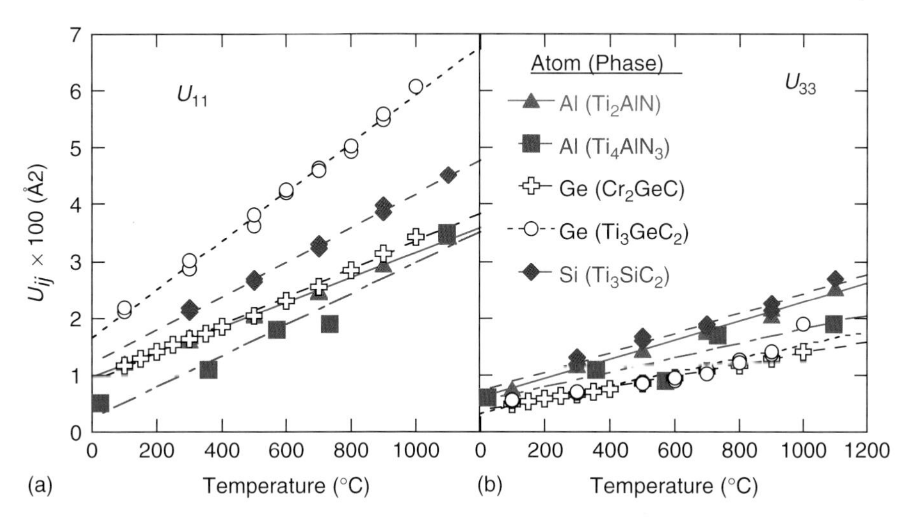

Figure 4.9 Temperature dependence of (a) $U_{11} = U_{22}$ and (b) U_{33} determined from Rietveld analysis of HTND for the A-group elements in Ti_2AlN (red triangles), $Ti_4AlN_{2.9}$ (green squares), Cr_2GeC (black crosses), Ti_3GeC_2 (open circles), and Ti_3SiC_2 (blue diamonds). The results and labels are color-coded (Lane, Vogel, and Barsoum, 2011).

We now return to Ti_3SiC_2 and Ti_3GeC_2. When the details of the thermal expansions of these two compounds, which appear to be identical in Figure 4.8, are looked into in more detail, a very different picture emerges. This is best seen in Figure 4.10a–d, where the T dependencies of the U_{ij}s for the four unique atoms in Ti_3SiC_2 (shown in red) and Ti_3GeC_2 (shown in black) are compared. The results for the Si/Ge atoms are more comparable (Figure 4.10c), but clearly the Ge atoms vibrate more vigorously and *anisotropically* than the Si atoms. The U_{ij}s of the Ti_I, Ti_{II}, and C atoms were also different, not only in magnitude but, more importantly, in their temperature dependencies and relative values (Lane, Vogel, and Barsoum, 2010).

To better appreciate these differences, the temperature dependences of the *ratios* of the thermal motions in the a and c directions, that is, U_{11}/U_{33}, are plotted for Ti_3SiC_2 and Ti_3GeC_2 in Figure 4.11a,b, respectively. While both the Si and Ge atoms tend to vibrate more along the basal planes than along the c-axis (i.e., $U_{11}/U_{33} > 1$), the extent of this anisotropic motion is much greater for Ge, with U_{11}/U_{33} ratios that are more than twice those of Si (compare Figure 4.11a,b). Note that the error bars associated with the ADPs of the Ti_{II} atoms in Ti_3GeC_2 increase dramatically and suddenly between 400 and 500 °C (Figure 4.11b). This is not an experimental artifact however, because (i) the error bars for the same atoms in Ti_3SiC_2 are quite low (Figure 4.8b) and (ii) the error bars at 100 and 300 °C for Ti_3GeC_2 are also quite low. What is occurring here is unclear, but could be related to the large increase in damping observed (Chapter 8) in Ti_3GeC_2 at around the same temperatures (Radovic *et al.*, 2006).

To further accentuate the differences, the thermal ellipsoids at 900 °C for Ti_3SiC_2 and Ti_3GeC_2 are compared in Figure 4.12a,b, respectively. A major difference

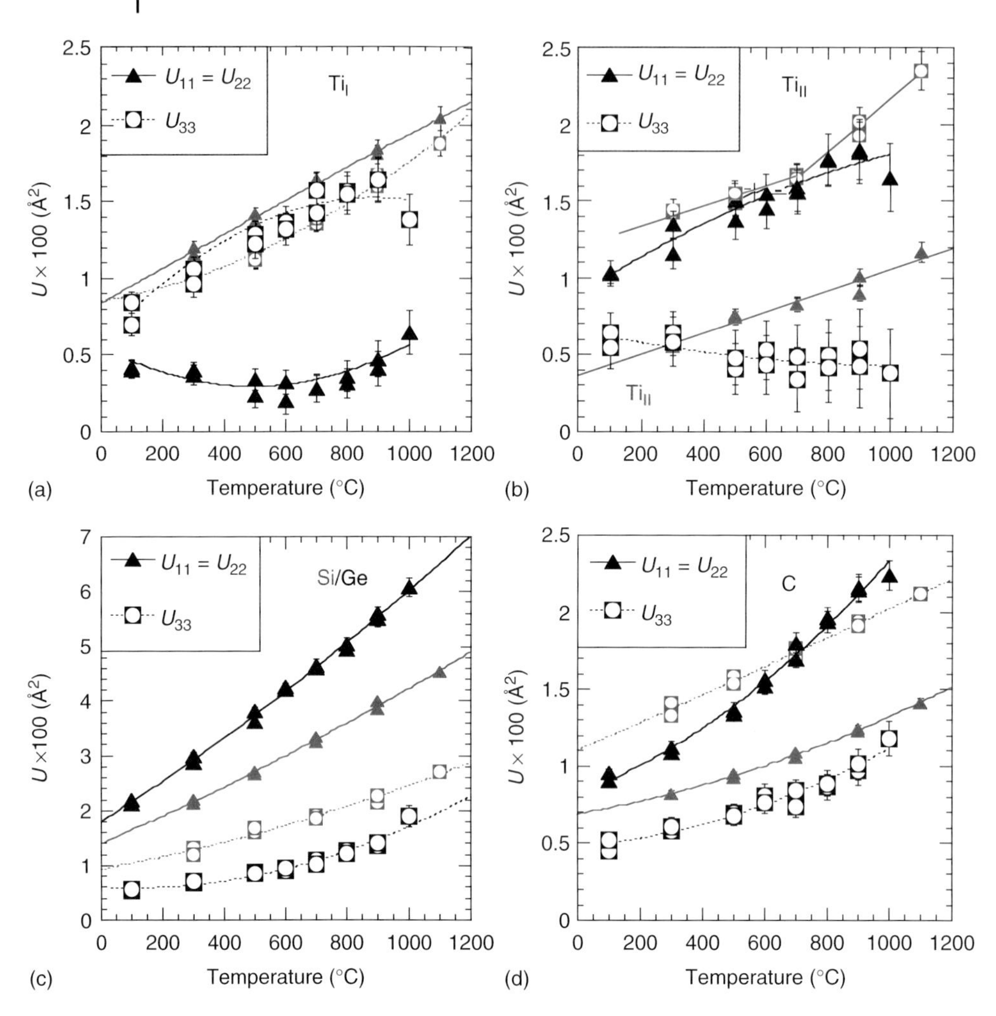

Figure 4.10 Temperature dependence of anisotropic U_{ij}s in Ti_3SiC_2 (red) and Ti_3GeC_2 (black) during heating and cooling for (a) Ti_I, (b) Ti_{II}, (c) Si/Ge, and (d) C. Note change in scale of *y*-axis in (c) (Lane, Vogel, and Barsoum, 2010). The lines are guides to the eyes.

between the two structures is the extent to which the Ge ellipsoids are flattened relative to those of Si. Note that the Ti_{II} atoms in Ti_3GeC_2 also flatten along the basal planes with increasing temperature. The Ti_I atoms, on the other hand, tend to vibrate normal to the basal planes.

4.3.3 Phonon Density of States

To better understand how, or why, a bond can apparently not increase in length with increasing temperature (Figure 4.8b), the details of the vibrations need to be looked into more carefully. Togo *et al.* (2010) calculated the partial phonon

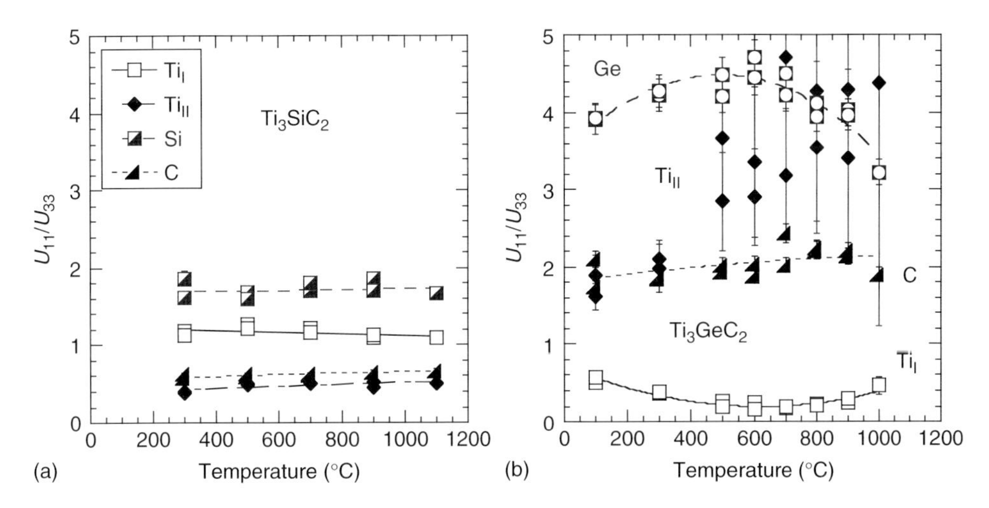

Figure 4.11 The U_{11}/U_{33} ratios determined from Rietveld analysis of HTND results for the various atoms in (a) Ti_3SiC_2 and (b) Ti_3GeC_2 during both heating and cooling (Lane, Vogel, and Barsoum, 2010).

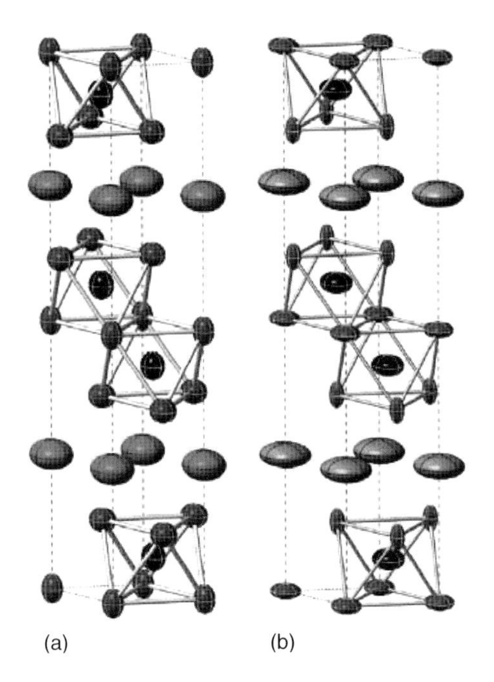

Figure 4.12 Unit cells and thermal ellipsoids (99% probability) at 900 °C of atoms in (a) Ti_3SiC_2 and (b) Ti_3GeC_2 (Lane, Vogel, and Barsoum, 2010).

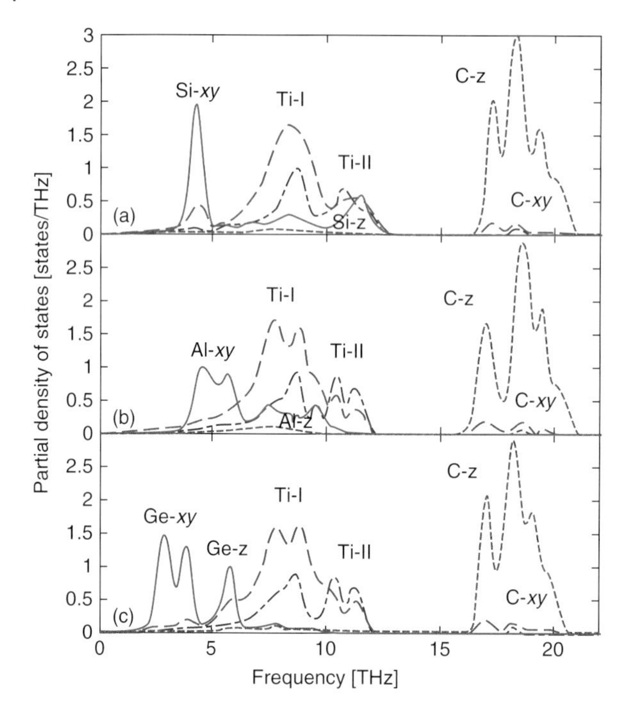

Figure 4.13 Phonon partial density of states of (a) Ti_3SiC_2, (b) Ti_3AlC_2, and (c) Ti_3GeC_2 (Togo *et al.*, 2010). The dashed-dotted, dashed, and dotted curves denote those of Ti_I, Ti_{II}, and C, respectively; the solid curves depict those of Si, Al, and Ge. The *xy* and *z* labels indicate the atomic vibrations parallel and perpendicular to the basal planes, respectively.

density of states (DOS) of Ti_3SiC_2, Ti_3AlC_2, and Ti_3GeC_2. From their results (Figure 4.13), it is possible to conclude the following:

1) The lowest energy vibrations are those associated with the A elements, confirming once more their rattling nature.
2) For Ti_3AlC_2 and Ti_3SiC_2, the vibrations of the A layers along the basal planes occur at significantly lower energies than those along [0001]. The same is true for Ti_3GeG_2, but now the in-plane and out-of-plane vibrations energies are closer together. These results partially explain why the A atoms tend to vibrate more in, than out of, the basal planes (Figure 4.12).
3) The C-atom vibrations occur at the highest energies, a conclusion that is in total accord with the Raman results presented in Chapter 3.
4) The Ti atoms' vibrations are in between those of the A and C elements.

Togo *et al.* (2010) also presented evidence that the motion of Ti and A elements were correlated (Figure 4.14). Piecing the various clues together, we postulated that the anomalous response of the Ti_I–Ge and Ti_{II}–C bonds shown in Figure 4.8b could be related to a *correlated* motion of the Ti_I and Ge atoms away from each other such that the average Ti_I–Ge distance – labeled r_3 in Figure 4.15 – as

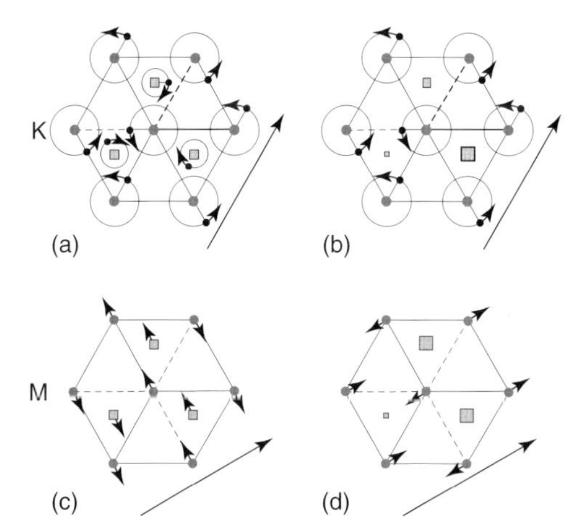

Figure 4.14 Atomic motions of the lower two localized bands. (a,b) at the K-point. (c,d) at the M-point. The circles and squares depict the A and Ti_I atoms, respectively. The size of the square represents the Ti_I atom position along the *c*-axis. It is shown larger when the Ti atoms move closer to the A layers. The small arrows attached to the atoms show the directions of motions at the same moment. The long arrows denote the directions of the wave vectors (Togo *et al.*, 2010).

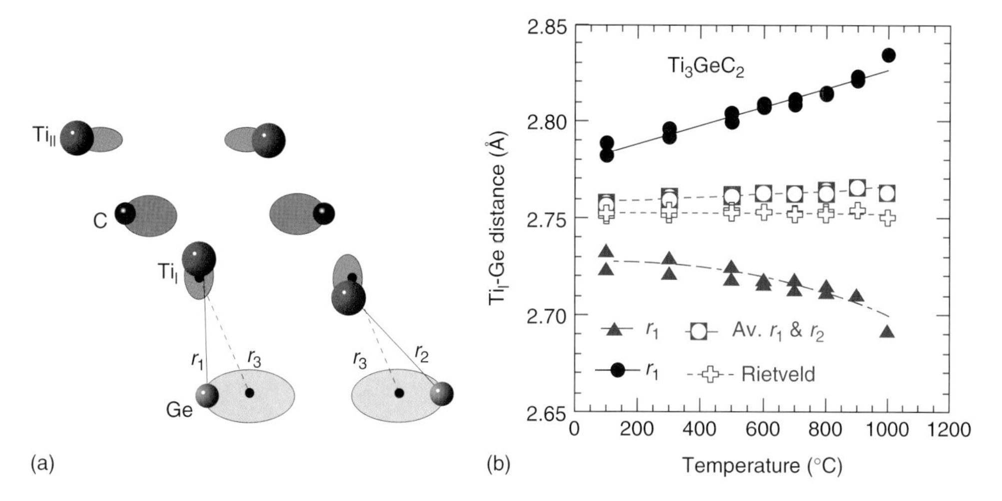

Figure 4.15 a) Schematic of interatomic distances in Ti_3GeC_2 (a) 99% probability thermal ellipsoids at 1000 °C showing Ti_I–Ge instantaneous positions (solid lines, r_1 and r_2) and interatomic distance between time-averaged positions (dashed lines, r_3). (b) Temperature dependence of the Ti_I–Ge bond showing the minimum (r_1, green triangles) and maximum (r_2, black circles) interatomic distances for the 50% probability thermal ellipsoids. Note the average of the two (red squares), is roughly equal to the distances determined by the time- and space-averaged positions obtained from Rietveld analysis (blue crosses). (Lane, Vogel, and Barsoum, 2010).

measured from Rietveld analysis remains more or less constant. The instantaneous bond distance r_2, on the other hand, behaves normally in that it expands with increasing T, at more or less the same rate as the Ti–Si bond, as it should. Note that, because the other instantaneous distance r_1 shrinks with increasing T, the average of r_1 and r_2 is a weak function of T and is quite comparable to that measured from Rietveld analysis (compare open circles and crosses in Figure 4.15b).

What happens in the case of Ti_3SiC_2 is more straightforward and normal in that all the bonds increase linearly with increasing temperatures (Figure 4.8a).

To summarize, the ADPs of the MAX phases depend on the atoms. In general the A atoms tend to vibrate with higher amplitudes than the M or X atoms, indicating that they are more weakly bound in the structure thus confirming their rattler nature. The vibrations of the A elements also tend to occur more along the basal planes than normal to them.

4.4 Heat Capacities

One of the more fundamental thermal properties of any compound is its heat capacity at constant pressure c_p. It is only by knowing c_p can a compound's free energy be calculated as a function of T. At low temperatures, c_p can be used to determine the DOS at the Fermi level, $N(E_F)$, and thus shed important light on the metallicity, or lack thereof, of a given structure. Low-temperature c_p measurements can also be used to determine the Debye temperature θ_D, which in turn is fundamental in understanding c_p. The following section focuses on the low-temperature results; section 4.4.2 summarizes the high-temperature results and combines both the high and low T results. Note that, throughout, c_p is assumed to be equal to the heat capacity at constant volume, that is c_v.

4.4.1 Low-Temperature Heat Capacity

In the 2–10 K temperature range, c_p monotonically increases with increasing T. When replotted as c_p/T versus T^2 (Figure 4.16), a linear fit representing

$$\frac{c_p}{T} = \gamma + \beta T^2 \tag{4.8}$$

prevails, confirming that

$$c_p \approx c_v = \gamma T + \beta T^3 \tag{4.9}$$

as is the expected behavior for metallic-like conducting solids. The parameters γ and β are the coefficients of the electronic and lattice heat capacities, respectively. At this juncture, it is useful to separate the discussion into two sections: one dealing with γ, and the other with θ_D.

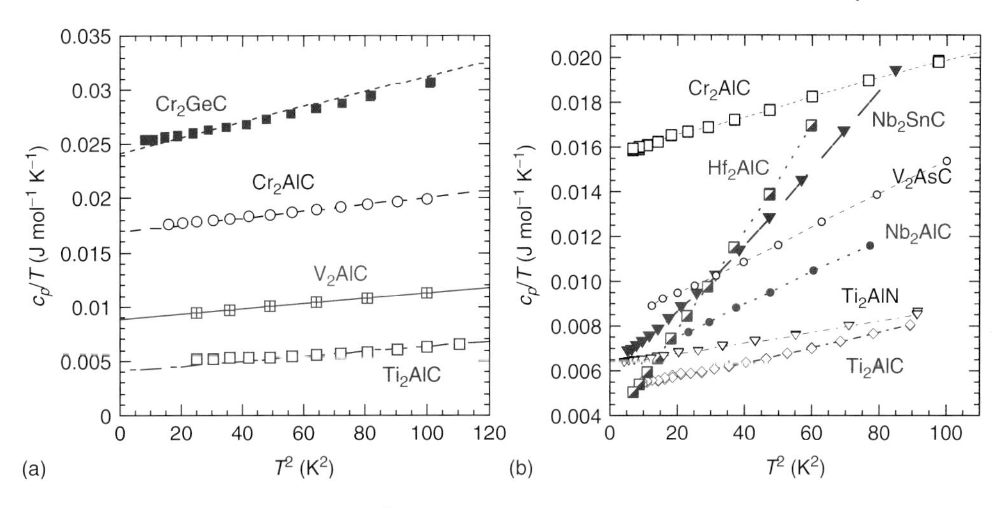

Figure 4.16 (a,b) Plots of c_p/T versus T^2 for select MAX phases. In all cases, straight lines are obtained confirming that Eq. (4.9) is the operative relationship for these compounds. The results and labels are color coded. See Table 4.4 for sources.

4.4.2 Coefficients of Electronic Heat Capacities and the DOS at E_F

Following the free electron model, γ can be related to $N(E_F)$ by

$$N\left(E_F\right) = \frac{3\gamma}{\pi^2 k_B^2} \tag{4.10}$$

The γ values measured to date (Table 4.4) range from a high of 26 mJ K^{-2} mol^{-1} for Cr_2GeC to a low of 3.4 mJ K^{-2} mol^{-1} for Hf_2InC. Also listed in Table 4.4 are the values of $N(E_F)$ calculated from Eq. (4.10), as well as those calculated from *ab initio* or $N_{DFT}(E_F)$.

Since $N(E_F)$ is dominated by the d–d orbitals of the M atoms (Chapter 2), it is not surprising that it depends on the latter. This is best seen in Figure 4.17a, where $N(E_F)$ for the 211 phases is plotted as a function of the transition metal. The values for the Ti-containing MAX phases cluster around 3–5 (eV unit cell)$^{-1}$, those for V vary between 7 and 17 (eV unit cell)$^{-1}$, and those for Cr are the highest and range from 15 to 22 (eV unit cell)$^{-1}$. When the same results are plotted as a function of the number of d electrons of the M element (not shown) the scatter is greater, especially for group 5. This scatter notwithstanding, least squares fitting of the results yields a R^2 value of 0.6, showing a decent correlation.

When $N_{DFT}(E_F)$ is plotted instead (not shown), the R^2 value is ≈0.52. It's worth noting that a similar trend – that is, both increasing DOS *and* increasing scatter with increasing valence electron concentration (VEC) – is predicted from DFT calculations of the M_3AlC_2 phases (He *et al.*, 2010). These comments notwithstanding, it would be incorrect to conclude that the A-group element does not affect the DOS. When $N_{DFT}(E_F)$ is plotted versus the electronic configuration of the A-group element (not shown) for select Ti_2AC phases, a maximum occurs when

Table 4.4 Summary of γ values derived from low temperature c_p measurements, the corresponding $N(E_F)$ values calculated from Eq. (4.10), and those calculated using DFT, $N_{DFT}(E_F)$. Also listed are values for near-stoichiometric TiC and TiN for comparison.

Solid	γ (mJ mol^{-1} K^{-2})	$N(E_F)$ (eV unit cell^{-1})	N_{DFT} (E_F) (eV unit cell^{-1})	References
			413 phases	
$Ti_4AlN_{2.9}$	8.1	6.9	0.0[a]	Ho *et al.* (1999) and Holm *et al.* (2002)
			312 phases	
Ti_3AlC_2	4.5	3.8	3.7	Ho *et al.* (1999) and Zhou *et al.* (2001)
Ti_3SiC_2	6.4	5.0	4.7	Drulis *et al.* (2004), Ho *et al.* (1999), and Zhou *et al.* (2001)
Ti_3GeC_2	6.4	5.4	4.6	Drulis *et al.* (2005), Finkel *et al.* (2004), and Zhou *et al.* (2001)
			211 phases	
Ti_2AlC	4.6	3.9	2.8	Drulis *et al.* (2006), Hug and Fries (2002), Lofland *et al.* (2004)
$Ti_2AlC_{0.5}N_{0.5}$	6.0	5.1	4.8	Drulis *et al.* (2007) and Du *et al.* (2009)
Ti_2AlN	5.9	5.0	3.9	Drulis *et al.* (2007), Hug and Fries (2002), and Lofland *et al.* (2004)
Ti_2GeC	4.8	4.1	3.6	Hug (2006), Lofland *et al.* (2004), and Zhou *et al.* (2000)
Ti_2SC	3.8	3.2	1.5	Drulis *et al.* (2008), Du *et al.* (2008), and Hug (2006)
V_2AlC	9.1	7.5	5.7	Drulis *et al.* (2006), Lofland *et al.* (2004), and Schneider, Mertens, and Music (2006)
V_2GeC	19.0	16.1	5.1	T.H. Scabarozi and S.E. Lofland (unpublished) and Bouhemadou (2009)
V_2AsC	11.7	9.9	5.8	Halilov, Singh, and Papaconstantopoulos (2002), Lofland *et al.* (2004), and Lofland *et al.* (2006)
Cr_2GeC	26.1	22.0	7.7	Bouhemadou (2009) and Drulis *et al.* (2008)
Cr_2AlC	16.2	14.6	6.2	Drulis *et al.* (2006) and Lofland *et al.* (2004)
Nb_2AlC	6.0	5.1	3.8	Hug, Jaouen, and Barsoum (2005) and Lofland *et al.* (2004)
Nb_2SnC	5.7	4.8	3.7	Kanoun, Goumri-Said, and Reshak (2009) and Lofland *et al.* (2004)
Nb_2SnC	3.15	2.65	3.7	Barsoum *et al.* (2000b)
Nb_2AsC	4.7	4.0	3.0	Halilov, Singh, and Papaconstantopoulos (2002), Lofland *et al.* (2004), and Lofland *et al.* (2006)

Table 4.4 (*Continued*).

Solid	γ (mJ mol^{-1} K^{-2})	$N(E_F)$ (eV unit cell^{-1})	N_{DFT} (E_F) (eV unit cell^{-1})	References
Hf_2InC	3.4	2.9	2.0	Lofland *et al.* (2006)
Ta_2AlC	7.1	6.0	3.0	Drulis *et al.* (2008) and Schneider, Music, and Sun (2005)
			MX phases	
$TiC_{0.97}$	<1	0.1–0.5	0.3	Toth (1971)
TiN	2.5–3.3	4–6	0.9	Toth (1971)

[a]The Fermi level in Ti_4AlN_3 falls in a small gap.

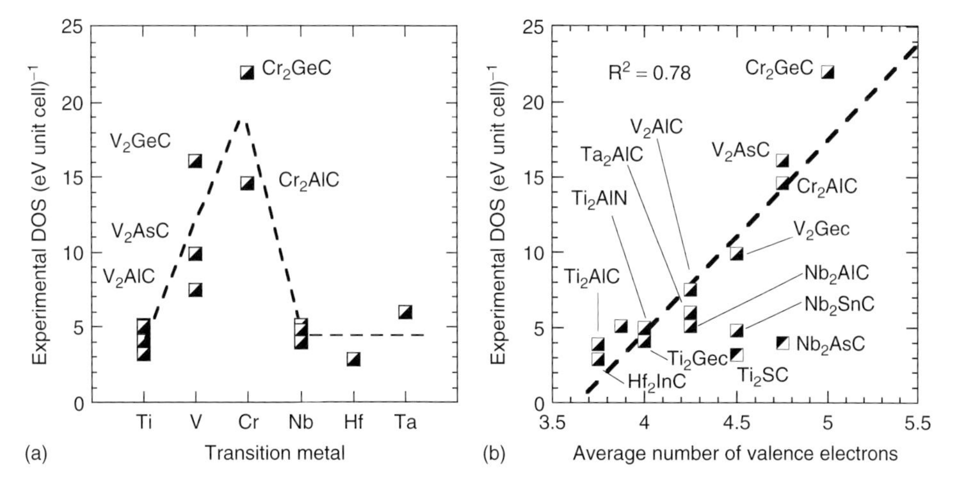

Figure 4.17 Plot of $N(E_F)$ versus (a) M atoms and (b) z_{av}, where it is obvious that $N(E_F)$ increases with the latter. If Nb_2AsC and Ti_2SC are excluded from the plot, least squares fit result in an R^2 value is 0.78. Unlabeled square in b is that for a solid solution between Ti_2AlC and Ti_2AlN.

the A-group element belongs to group 15 (Hug, 2006). The effect of the A elements, however, at least for the Ti-containing MAX phases, is weaker than that of the M elements.

A more illuminating plot is shown in Figure 4.17b, where $N(E_F)$ is plotted versus the average number of valence electrons per formula unit, z_{av}. If Nb_2AsC and Ti_2SC are excluded from the plot, the R^2 value is 0.78. This correlation is to be considered excellent given the wide variations in chemistries and atomic weights. The rationale for excluding Nb_2AsC and Ti_2SC is their significantly lower c-parameter values which probably leads to a localization of electrons between the A and M layers, reducing the metallicity of the bonds.

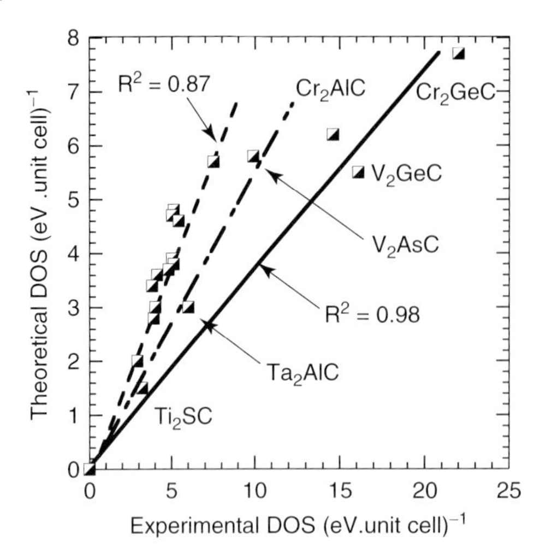

Figure 4.18 Plot of $N_{DFT}(E_F)$ versus $N(E_F)$. In all cases, $N(E_F) > N_{DFT}(E_F)$. Least squares fitting of all the results, except those for V_2GeC, Ta_2AlC and Ti_2SC, Cr_2AlC, Cr_2GeC, and V_2AlC, yield the dashed line on the left. The best fit through Cr_2AlC, Cr_2GeC, and V_2GeC, is given by the solid line. The dashed-dotted line in the center is the fit for V_2AsC, Ta_2AlC and T_2SC. All lines are forced through the origin.

When columns 3 and 4 in Table 4.4 are plotted (Figure 4.18) in all cases, $N(E_F) > N_{DFT}(E_F)$. Following work done in the 1960s and 1970s on pure transition metals, this enhancement in DOS at E_F can be related to the electron–phonon coupling constant λ_{ep}. The latter can be estimated by comparing $N_{DFT}(E_F)$ to the measured $N(E_F)$ calculated from Eq. (4.10), if one assumes (Lofland *et al.*, 2006)

$$N_{\mathrm{DFT}}\left(E_F\right) = \left(\frac{1}{1+\lambda_{\mathrm{ep}}}\right) N\left(E_F\right) \tag{4.11}$$

A perusal of the results shown in Figure 4.18 shows that a single straight line cannot adequately fit all the data points. A more realistic approach is to use three straight lines, one for Cr_2AlC, Cr_2GeC, and V_2GeC, one for Ti_2SC, Ta_2AlC and V_2AsC and one for the rest. Least squares fitting of the dashed line – forced through the origin – in Figure 4.18 yields a slope of 0.73, from which λ_{ep} is calculated to be ≈0.4. The slope of the solid line – also forced through the origin – is 0.36, which yields a λ_{ep} of ≈1.8, or more than four times that of the majority of the other 211 MAX phases. The third line, in between the other two, has a slope of 0.56, which implies a λ_{cp} of 0.78. Why this is the case is not clear at this time.

The fact that for the most part λ_{ep} is not a function of atomic mass is important and consistent with the results of Maksimov (1969) who rigorously showed that, in the harmonic approximation, λ_{ep} does not depend on atomic mass.

At ≈0.4, the value of λ_{ep} determined herein is in good agreement with the corresponding values reported for the early transition metals. For example, McMillan reported the following values for λ_{ep} for Ti, V, Zr, Nb, Hf, and Ta, respectively: 0.38, 0.6, 0.41, 0.82, 0.34, and 0.65 (McMillan, 1968). Sanborn *et al.* (1989) reported similar values. The latter also concluded that the anisotropy in electrical conductivity for these transition metals is quite mild.

Note that λ_{ep} can also be estimated from changes in electrical resistivities with temperature. The agreement between the two sets of values was found to be acceptable (Lofland *et al.*, 2006).

4.4.3 Debye Temperatures

One of the more important thermal property of a solid is its Debye temperature θ_D. The latter can be derived in a variety of ways. If derived from low-temperature c_p results, that is, $\theta_{D,cp}$, then for a monoatomic solid $\theta_{D,cp}$ is related to β by

$$\theta_{D,cp} = \left(\frac{12\pi^4 RN_m}{15\beta}\right)^{1/3} \tag{4.12}$$

where N_m is the number of modes. Once β is experimentally determined, $\theta_{D,cp}$ is calculated from Eq. (4.12). The MAX phases are not monoatomic, which complicates the analysis. As a first approximation we can assume the MAX phases to be monoatomic, with an atom whose mass is the average of the atoms in the unit cell. In other words, the assumption that $N_m = 3n_f$, where n_f is the number of atoms per formula unit, is made.

As shown in Table 4.5, the θ_D values range from a low of 330 K for Hf_2InC to a high of 775 K for $Ti_4AlN_{2.9}$. As expected, in addition to bond strengths, the average atomic mass has a large effect on θ_D, which is why, for example, the θ_Ds of Hf_2InC and Nb_2SnC are some of the lowest reported to date. The effect of mass on β is apparent in Figure 4.16b; the Hf- and Sn-containing compounds clearly have the higher βs, and thus lower θ_Ds.

For the most part, however, the θ_Ds of the MAX phases – especially the Al-containing ones – are quite high. The θ_Ds for the 312 and 413 phases are significantly higher than those of Ti or Ti_3Al, and comparable to that of TiN. This must reflect the partial, but significant, covalent and ionic character of the Ti–Ti, Ti–C, and Ti–N bonds in the ternaries. It should be further noted that at about 430 K, the θ_D value for pure Ti is already amongst the higher ones for pure metals.

A single θ_D representation for an anisotropic multi-element material is an oversimplification. Indeed, Drulis *et al.* (2004, 2005, 2006, 2007, 2008) showed that for most MAX phases one Debye temperature was not sufficient to describe the entire c_p versus T range. They proposed that, in the 20–80 K temperature range, c_v be given as the sum of three terms:

Table 4.5 Summary of θ_D for select MAX phases assuming $N = 3n_f$, where n_f is the number of atoms per formula unit. Also listed are values for select MX compounds.

Compound	$\theta_{D,cp}$ c_p	$\theta_{D,ac}$ Acoustic	$\theta_{D,ht}$ HTD	References
				413 phases
$Ti_4AlN_{2.9}$	775	762	700	Barsoum *et al.* (2000c) and Ho *et al.* (1999)
				312 phases
Ti_3SiC_2	615–715	784	608	Barsoum *et al.* (1999), Drulis *et al.* (2004), Finkel, Barsoum,
	663	813		and El-Raghy (2000), Ho *et al.* (1999), and Radovic *et al.* (2006)
$Ti_3(SiAl)C_2$	—	795	—	Radovic *et al.* (2006)
$Ti_3(SiGe)C_2$	724	728	—	Finkel *et al.* (2004)
Ti_3GeC_2	637/670	725/728	485	Drulis *et al.* (2005), Finkel *et al.* (2004), and Radovic *et al.* (2006)
Ti_3AlC_2	765	758	—	Finkel, Barsoum, and El-Raghy (2000), Ho *et al.* (1999),
Ti_3AlCN	685	795	—	Radovic, Ganguly, and Barsoum(2008) and Scabarozi *et al.* (2008a)
				211 phases
Ti_2AlC	746	732/741	—	Drulis *et al.* (2007), Hettinger *et al.* (2005), and Radovic *et al.* (2006)
$Ti_2AlC_{0.5}N_{0.5}$	671/724	738	—	Drulis *et al.* (2007), Radovic, Ganguly, and Barsoum (2008), and Scabarozi *et al.* (2008a)
Ti_2AlN	761/692	740	600	Drulis *et al.* (2007), Lofland *et al.* (2004), and Radovic, Ganguly, and Barsoum (2008)
Ti_2GeC	625	728	—	Lofland *et al.* (2006) and Radovic *et al.* (2006)
Ti_2SC	611	—	—	Scabarozi *et al.* (2008b)
V_2AlC	658/681	696	—	Drulis *et al.* (2006), Lofland *et al.* (2004), and (Hettinger *et al.* (2005)
V_2AsC	444	—	—	Lofland *et al.* (2004)
Cr_2AlC	673/648	644	—	Drulis *et al.* (2006), Lofland *et al.* (2004), and Hettinger *et al.* (2005)
Cr_2GeC	467/673	—	532	Drulis *et al.* (2008) and Lofland *et al.* (2004)
Nb_2AlC	540	577	—	Hettinger *et al.* (2005) and Lofland *et al.* (2004)
Nb_2SnC	380/324	—	—	Barsoum *et al.* (2000b) and Lofland *et al.* (2004)
Nb_2AsC	520	—	—	Lofland *et al.* (2006)
Hf_2InC	330	—	—	Lofland *et al.* (2004)
Ta_2AlC	326	—	—	Drulis *et al.* (2008)
				MX
TiC	845	940	740	Houska (1964) and Toth (1971)
TiN	—	636	671	Houska (1964) and Toth (1971)
ZrC	—	—	587	Houska (1964)

$$c_v = \gamma T + N_D c_{v(D)} + N_E c_{v(E)} \tag{4.13}$$

where N_E is the number of Einstein modes, such that $N_E + N_D = 3n$. $c_{v(D)}$ and $c_{V(E)}$ are in turn given by

$$c_V = \gamma T + N_D \frac{R}{x^3} \int_0^x \frac{x^4 e^x}{(e^x - 1)^2} dx + N_E R \left(\frac{1}{2x_E}\right)^2 \csc h^2 \left(\frac{1}{2x_E}\right) \tag{4.14}$$

Here, $x_E = \theta_E/T$, $x = \theta_D/T$ and θ_E is the Einstein temperature. The details and ramifications of these refinements are beyond the scope of this book. The interested reader can consult the original references.

There are two other methods to estimate θ_D. The first is from acoustic measurements, $\theta_{D,ac}$; the second from U_{eq}. For the acoustic method, it is assumed that

$$\theta_{D,ac} = \frac{h}{k_B} \left(\frac{3N}{4\pi}\right)^{1/3} v_M \tag{4.15}$$

where h is Planck's constant, and v_M is the average of the shear, v_S and longitudinal, v_L, sound velocities given by

$$v_M = \left[\frac{3(v_S v_L)^3}{2v_L{}^3 + v_S^3}\right]^{1/3} \tag{4.16}$$

The Debye temperatures calculated thus appear in column 3 in Table 4.5.

The Debye temperature $\theta_{D,th}$ can also be related to the $\overline{U}_{eq}$ values by the following equation:

$$\theta_{D,th} = \frac{6h^2}{8\pi^2 m_{av} k_B \overline{U}_{eq}^2} \left\{\frac{\phi(x)}{x} + \frac{1}{4}\right\} \tag{4.17}$$

where m_{av} is the mean atomic mass. $\phi(x)$ is the Debye integral

$$\phi(x) = \frac{1}{x} \int_0^x \frac{x}{e^x - 1} dx \tag{4.18}$$

where x is θ_D/T. This relationship implicitly assumes an artificial, monoatomic, simple cubic lattice and should thus be used with care. Nevertheless, it is useful for comparison purposes, especially if the structures are similar or related.

The $\theta_{D,th}$ values of Ti_3SiC_2 and $Ti_4AlN_{2.9}$, estimated from the $\bar{U}_{eq}$ averaged over all temperatures, are listed in column 4 in Table 4.5. These values, while still quite high, are ≈12–15% lower than the θ_Ds calculated using other methods. In general, however, the agreement between the θ_D values calculated by the various techniques is acceptable especially given the many simplifying assumptions made in most cases.

4.4.4 High-Temperature Heat Capacity

The temperature dependences of c_p for Ti_3SiC_2, $Ti_4AlN_{2.9}$, and Nb_2SnC in the 300–1300 K temperature range are plotted (open symbols) in Figure 4.19. The results for Ti_2AlC are almost indistinguishable from those of Nb_2SnC and are not shown. In all cases, the c_p asymptotes to the Dulong and Petit limit of $3n_fR$, where $n_f = 4$, 6, and 8 for Nb_2SnC, Ti_3SiC_2, and $Ti_4AlN_{2.9}$, respectively.

4.4.5 Heat Capacity over Extended Temperature Range

Figure 4.20a and b combine the low- and high-temperature c_p data for Cr_2GeC and Ti_2SC, respectively (Drulis *et al.*, 2008). Also plotted are the results of *ab initio* calculations with, and without, the experimentally obtained electronic heat capacity term, that is, γT. For both compounds, the agreement between theory and experiment is excellent up to about 800 K. Beyond that temperature the results diverge. The exact reason for this diversion is unclear but could be due to (i) the invalidity of the $c_v \approx c_p$ approximation at higher Ts, (ii) anharmonic effects that are not accounted for in the DFT calculations, and/or (iii) the incipient dissociation of the phases at higher temperatures.

Based on these results, however, it is reasonable to conclude that in the absence of experimental results, the *ab initio* results, when combined with the experimentally determined γT term can be used with some confidence.

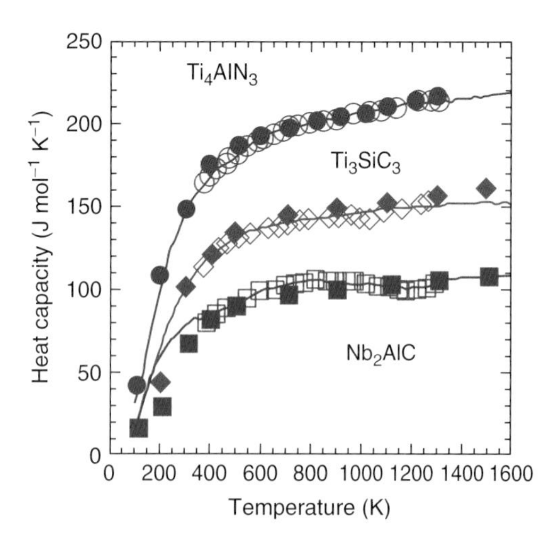

Figure 4.19 High-temperature dependences of c_p in the 0–1500 K temperature range for Ti_4AlN_3, Ti_3SiC_2, and Nb_2SnC (open symbols). The filled symbols are the corresponding c_p values for the respective MX compounds multiplied by the appropriate $(n + 1)$ factor (Eq. 4.19). Solid lines are best fits using the Debye model (Barsoum 2000).

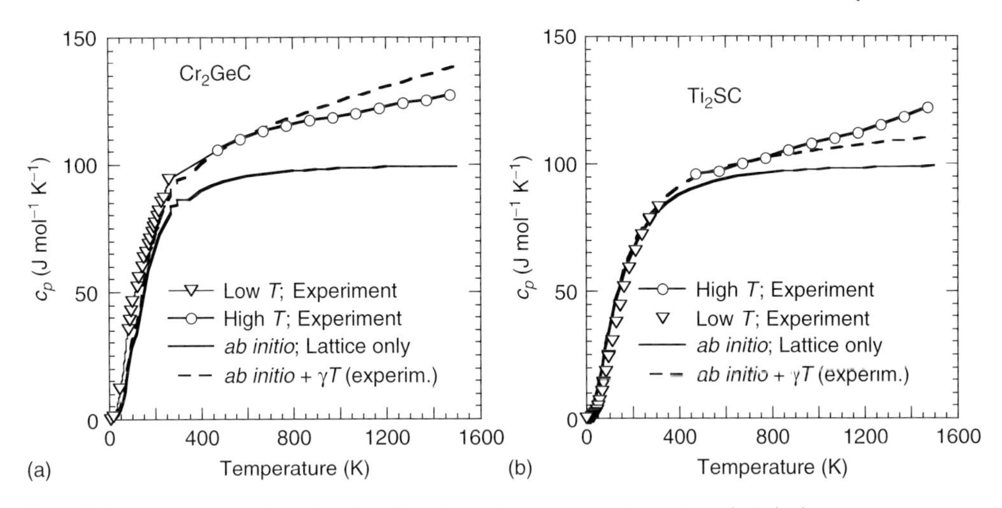

Figure 4.20 Temperature dependencies of c_p at low temperatures (open triangles), high temperatures (open circles), lattice *ab initio* (solid line), and the latter plus γT (dashed line), where γ was experimentally obtained (Table 4.4) for (a) Cr_2GeC and (b) Ti_2SC (Drulis *et al.*, 2008).

Tables 4.A.1 and 4.A.2 list the thermodynamic functions for Ti_3SiC_2 and Ti_3GeC_2 over the 0–300 K temperature range, respectively.

4.4.6
Relationship between c_p of the MAX and MX Phases

Roughly a decade ago (Barsoum, 2000), it was shown that the relationship

$$c_p\left(\mathrm{M}_{n+1}\mathrm{AX}_n\right) = (n+1) \times c_p\,(\mathrm{MX}) \tag{4.19}$$

was a good approximation. In other words, without much loss in accuracy, the A-group element can be replaced by an X atom. This is best seen in Figure 4.19, where the c_ps of select phases are plotted as open symbols. The filled symbols are the corresponding c_p values of their respective MX compounds multiplied by the appropriate $(n + 1)$ factor, that is, Eq. (4.19). The lines represent the best fit of the experimental points using the Debye model. The three sets of results are almost indistinguishable, which implies that, as a first approximation, the entropy content of the binaries and ternaries are comparable. The latter can thus be used until the thermodynamic parameters of the ternaries are independently measured.

This idea can be taken further. Since on average, the bond distances in the MAX and corresponding MX phases are comparable, it follows that, as a first approximation, the standard free energies of formation of the ternaries, ΔG_f° can be assumed to be $(n + 1)$ times those of their corresponding binaries, or

$$\Delta G_f^\circ\left(\mathrm{M}_{n+1}\mathrm{AX}_n\right) = (n+1) \times \Delta G_f^\circ\,(\mathrm{MX}) \tag{4.20}$$

Table 4.6 Comparison of temperature dependence of ΔG_f° (kJ mol^{-1}) for Ti_3SiC_2 and three times ΔG_f° of stoichiometric TiC and difference between them.

Temperature (K)	300	500	700	900	1100	1300	1500	Reference
Ti_3SiC_2	−539.7	−534.7	−529.7	−524.8	−519.8	−514.8	−509.9	Du *et al.* (2000)
3× TiC	−541.2	−534.1	−527.5	−521.1	−514.7	−506.5	−497.5	Chase *et al.* (1985)
Difference (%)	−0.28	0.11	0.43	0.7	1.0	1.65	2.5	—

Only two reports exist in the literature for the experimental determination of ΔG_f° of Ti_3SiC_2 (Du *et al.*, 2000; Sambasivan, 1990). The reliability of the earlier results is suspect however, because they predict an increase in the stability of Ti_3SiC_2 with increasing temperatures. Du *et al.* used a combination of theoretical modeling and experimental observations to conclude that ΔG_f° of Ti_3SiC_2 from the elements is

$$\Delta G_f^\circ \left(\text{J mol}^{-1}\right) = -547,145 + 24.84T \tag{4.21}$$

These results are listed in Table 4.6, together with three times those of TiC in the 300–1500 K temperature range. The two sets of results are quite close, the maximum deviation being 2.5% at 1500 K. Thus, based on this one result, it appears that not only are the entropies of the ternaries and binaries comparable on an absolute scale, but their standard free energies are as well. Whether this is a general conclusion that applies to other ternaries will require further work. These caveats notwithstanding, and until the thermodynamic parameters of the MAX phases are measured, a good approximation is to use the corresponding parameters for the MX binaries multiplied by the appropriate $(n + 1)$ factor.

4.5 Thermal Expansion

One of the consequences of atomic vibration is thermal expansion. For cubic symmetry, the TEC α_{cu} is given by

$$\alpha_{cu} = \frac{c_v \gamma_{Gr}}{3BV_m} \tag{4.22}$$

where V_m is the molar volume, B is the bulk modulus, and γ_{Gr} is the Grüneisen parameter. Since, in general, BV_m is more or less a constant and at high T, the Dulong–Petit law applies, that is, $c_v = 3n_f R$, it follows that γ_{Gr} is a measure of anharmonicity which, in turn, gives rise to thermal expansion.

In a system with hexagonal symmetry, on the other hand, the Grüneisen parameters in the a and c directions, γ_a and γ_c, respectively, are given by

$$\gamma_a = \frac{V_m}{c_v}\left[\left(c_{11} + c_{12}\right)\right]\alpha_a + c_{13}\alpha_c] \qquad (4.23)$$

$$\gamma_c = \frac{V_m}{c_v}\left[2c_{13}\alpha_a + c_{33}\alpha_c\right] \qquad (4.24)$$

where c_{ij}s are the elastic stiffness constants and α_a and α_c are the TECs along the a and c directions, respectively. The average values of γ_c and γ_a for select MAX phases are listed in column 7 of Table 4.7. These values are comparable to those of the MX phases that also range from 1.2 to 2.0 (Scabarozi *et al.* 2009). This comment notwithstanding, there is little correlation between the Grüneisen parameters of a MAX phase and its respective MX phase (see below).

4.5.1 Average Thermal Expansion Coefficients

For hexagonal solids, the average TEC, α_{av}, can be defined as

$$\alpha_{av} = \frac{2\alpha_a + \alpha_c}{3} \qquad (4.25)$$

The average TECs of the MAX phases fall in the range of $\approx$5–15 $\times$ 10^{-6} K^{-1} (column 5 in Table 4.7). With two exceptions, Cr_2GeC and Nb_2AsC, the agreement between the TEC values measured dilatometrically (column 6 in Table 4.7) and those measured using high-temperature diffraction – either XRD or ND – are reasonable. The reason why Cr_2GeC and Nb_2AsC are such outliers is intimately related to their large TEC anisotropies (Cabioch, *et al.* 2013). For both the dilatometric measurements and the high-temperature ND experiments, bulk samples were used. For the XRD results, powders were used, instead. It follows that upon cooling of the bulk samples, the microstrains that develop in the grains reduce both α_a and α_c, as compared to the values measured on powders. Said otherwise, the larger the TEC anisotropies, the larger the difference one would expect between the TECs measured on powders and those measured on bulk samples, whether in high-temperature ND or dilatometrically, as observed.

To shed some light on what affects the TECs, it is instructive to plot α_{av} versus n_{val} (Table 2.2). The resulting plot (Figure 4.21a) is quite intriguing and suggests the presence of two regimes: one in which α_{av} is a weak function of n_{val}; and a second regime, for $n_{val} > 4 \times 10^{29}$ m^{-3}, where α_{av} is a much stronger function of n_{val}. These results, when taken together with the elastic constant results discussed in Chapter 3 (see for e.g., Figure 3.4), are consistent with the idea that, beyond $n_{val} \approx$ 3.5–4 $\times$ 10^{29} m^{-3}, the MAX structure is destabilized. Note that the same conclusion is reached if α_{av} is plotted versus the average number of valence electrons z_{av} per formula unit (Figure 4.21b).

Table 4.7 Summary of α_a, α_c, α_c/α_a, and α_{av} for select MAX phases determined from high temperature XRD and ND. Also included in column 6 are the dilatometric values. Column 7 lists the average Grüneisen parameter. Numbers in parentheses are estimated standard deviations in the last significant figure of the refined parameter.

Compound	α_a (μK^{-1})	α_c (μK^{-1})	α_c/α_a	α_{av} (μK^{-1})	α_{dila} (μK^{-1})	$\langle \gamma \rangle$	References
413 phases							
$Ti_4AlN_{2.9}$	8.3(2)	8.3(9)	1.0	8.3(5)	9.7(2)	1.33	Scabarozi *et al.* (2009)
	9.6(1)	8.8(1)	0.9	9.4(1)			Rawn *et al.* (2000)
Nb_4AlC_3	—	—	—	—	7.2	—	Hu *et al.* (2008b)
Nb_4AlC_3	—	—	—	—	6.7	—	Hu *et al.* (2009)
Ta_4AlC_3	—	—	—	—	8.2 ± 0.3	—	Hu *et al.* (2007b)
312 phases							
Ti_3SiC_2	8.9(1)	10.0(2)	1.1	9.3 (2)	9.1(5)	1.64	Scabarozi *et al.* (2009)
	8.4(1)	9.3(10)	1.1	8.7(4)	9.1(2)		Manoun *et al.* (2005)
	8.6(1)	9.7(1)	1.1	9.1(2)	9.1(2)		Barsoum *et al.* (1999)
	8.9(1)	9.4(1)	1.1	9.2			Lane, Low Vogel, and Barsoum (2010)
Ti_3AlCN	6.0(2)	11.3(2)	1.9	7.8(2)	7.5(5)	—	Scabarozi *et al.* (2009)
Ti_3GeC_2	8.1(2)	9.7(2)	1.2	8.6(2)	7.8(2)	1.46	Scabarozi *et al.* (2009)
Ti_3GeC_2	8.5(1)	9.1(1)	1.1	8.7(1)	—	—	Lane, Vogel, and Barsoum (2010)
Ti_3SiGeC_2	8.8(6)	11.1(3)	1.3	9.6(5)	—	—	Scabarozi *et al.* (2009)
Ti_3AlC_2	8.3(1)	11.1(1)	1.3	9.2(1)		1.34	(Scabarozi *et al.* 2009)
					9.0(2)		Tzenov and Barsoum (2000)
Ti_3AlC_2	8.5	10.2	1.2	9.2	—	—	Pang, Low, and Sun (2010)
211 phases							
Ti_2AlN	10.6(2)	9.75(2)	0.9	10.3(2)	8.8(2)	1.52	Scabarozi *et al.* (2009)
	8.6(2)	7.0(5)	0.8	8.1(5)			Barsoum, (2000)
	$12.75^{\dagger}$	$11.5^{\dagger}$	0.9	9.7			Lane, Vogel, and Barsoum (2011)
	$10.0^{\dagger}$	$9.0^{\dagger}$					Barsoum, Ali, and El-Ragh (2000a)
Ti_2AlCN	8.4(1)	8.8(1)	1.0	8.5(1)	7.9(5)		Scabarozi *et al.* (2009)
Ti_2AlC	7.1(3)	10.0(5)	1.42	8.1(5)	8.8(2)		Barsoum (2000)
	8.6	9.2	1.1	8.8			Lane *et al.* (2013)
Ti_2SnC	—	—	—	—	10	—	El-Raghy, Chakraborty, and Barsoum (2000)
Ti_2GeC	10.3	8.6	0.84	9.7	—	—	N.J. Lane (private communication)
Ti_2InC	—	—	—	—	9.5	—	Barsoum *et al.* (2002a)
Ti_2SC	8.6(1)	8.7(2)	1.0	8.7(1)	9.3(6)	1.40	Scabarozi *et al.* (2009)
	8.5(5)	8.8(2)	1.04	8.6(6)			Kulkarni *et al.* (2009)
V_2AlC	9.1(2)	10.0(7)	1.1	9.4(10)	9.4(5)	1.44	Scabarozi *et al.* (2009)
	9.3(5)	9.5(4)	1.0	9.4(5)			Kulkarni *et al.* (2007)
V_2GeC	6.9(1)	15.8(3)	2.3	9.9(2)	9.4(6)	1.63	(Scabarozi *et al.* 2009)
V_2AsC	7.2(1)	14.0(1)	1.9	9.5(1)	—	1.69	Scabarozi *et al.* (2009)

Table 4.7 (*Continued*)

Compound	α_a (μK^{-1})	α_c (μK^{-1})	α_c/α_a	α_{av} (μK^{-1})	α_{dila} (μK^{-1})	$<\gamma>$	References
Cr_2GeC	12.9(1)	17.6(2)	1.4	14.5(2)	9.5(5)	2.38	Scabarozi *et al.* (2009)
	14.3	17.2	1.2	15.3			Cabioch, *et al.* (2013)
	12.3	14.4	1.2	13.0			Lane, Vogel, and Barsoum (2011)
Cr_2AlC	12.8(3)	12.1(1)	0.9	12.6(2)	12.8(5)	1.99	Scabarozi *et al.* (2009)
	13.3	11.7	0.9	12.8	13.3		Cabioch, *et al.* (2013)
					13		Tian *et al.* (2006)
					13		Lin, Zhou, and Li (2007)
					13		Zhou, Mei, and Zhu (2009)
Nb_2AlC	8.8(2)	6.8(3)	0.8	8.1(2)	7.5(2)	1.56	Scabarozi *et al.* (2009), Barsoum
					8.7(2)		(2000), Barsoum *et al.* (2002b), and
					8.1		Zhang *et al.* (2009)
Nb_2AsC	2.9(1)	10.6(1)	2.6	5.5(1)	7.3(5)	1.40	Scabarozi *et al.* (2009)
Nb_2SnC	6.6(4)	14.5(2)	2.2	9.3(3)	7.8(2)	1.72	Scabarozi *et al.* (2009) and El-Raghy, Chakraborty, and Barsoum (2000)
Hf_2InC	7.2(1)	7.6(2)	1.0	7.3(2)	7.6(2)	1.07	Scabarozi *et al.* (2009) and Barsoum *et al.* (2002a)
Hf_2PbC	—	—	—	—	8.3	—	El-Raghy, Chakraborty, and Barsoum (2000)
Hf_2SnC	—	—	—	—	8.1	—	El-Raghy, Chakraborty, and Barsoum (2000)
Ta_2AlC	—	—	—	—	8.0	1.66	Hu *et al.* (2008a)
Zr_2SnC	—	—	—	—	8.3	—	El-Raghy, Chakraborty, and Barsoum (2000)
Zr_2PbC	—	—	—	—	8.2	—	El-Raghy, Chakraborty, and Barsoum (2000)
Zr_2SC	—	—	—	—	8.8	—	Opeka *et al.* (2011)

4.5.2 Thermal Expansion Anisotropies

When the average Grüneisen parameters of the MAX and MX phases are plotted on the same graph (not shown), there is little correlation between the two, which indirectly implicates the A-group element in the thermal expansion anisotropies observed (Scabarozi *et al.*, 2009). This is further confirmed when the α_c/α_a ratios for the 211 phases are plotted as a function of the A-element group (Figure 4.22a) (Scabarozi *et al.*, 2009). For example, the highest anisotropies were observed for the As-containing ternaries. The same overall dependence is recovered when γ_c/γ_a is plotted versus the group of the A element (Figure 4.22b). This suggests that the anisotropies are in part determined by the elastic constants. It thus follows that, not too surprisingly, α_c/α_a correlates somewhat with c_{13} of the 211 phases (Figure 4.22c).

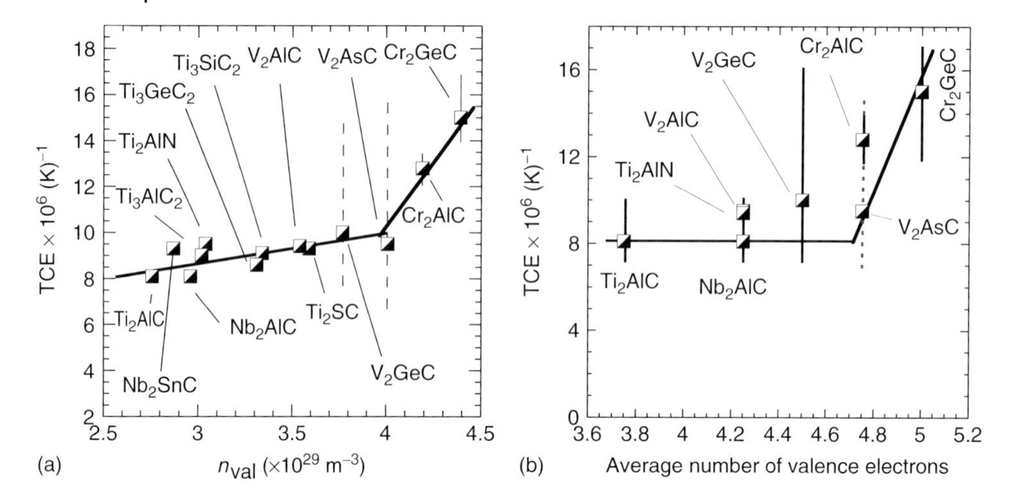

Figure 4.21 (a) Dependence of α_{av} on the total valence electron density n_{val} (Table 2.2). (b) Dependence of α_{av} on z_{av}. Solid and dashed vertical lines outline the maximum and minimum values of α_{av} for a given compound.

Note that, for the most part, $\alpha_c > \alpha_a$, which implies that the M–A bonds are weaker than the M–X bonds. However, since the M–X bonds are already some of the strongest known, the few compounds in which $\alpha_c < \alpha_a$ are noteworthy. They are Nb_2AlC, Cr_2AlC, Ti_2SC, and Ti_2AlN. The case of Cr_2AlC is not too surprising because, as noted above, this phase is close to being unstable which explains why both α_c and α_a are large. The situation for Nb_2AlC, Ti_2SC, and Ti_2AlN, on the other hand, is different since the results suggest that indeed the M–A bonds in those ternaries are stronger or at least more thermally stable, than their M–X counterparts. This is a somewhat astonishing conclusion given the fact that the Ti–X bonds are some of the stronger bonds known.

4.5.3
Effect of Solid Solutions

Solid solutions affect the TEC in various ways. When the TEC of $(Ti_{0.5},Nb_{0.5})_2AlC$, is compared with those of its end members (Barsoum *et al.*, 2002b), the latter are found to be identical at $8.7 \times 10^{-6}\,K^{-1}$. At $8.9 \times 10^{-6}\,K^{-1}$, the TEC for $(Ti_{0.5},Nb_{0.5})_2AlC$ are slightly higher. It follows that, in this case, the solid solution slightly destabilizes the structure at higher temperatures.

Finkel *et al.* measured the TECs of $Ti_3(Ge_x,Si_{1-x})C_2$ solid solutions in the 323–1473 K temperature range (Finkel *et al.*, 2004). They reported that at $9.3 \times 10^{-6}\,K^{-1}$ the TEC for the $x = 0.5$ composition was slightly higher than that of Ti_3SiC_2 ($8.9 \times 10^{-6}\,K^{-1}$). At $7.8 \times 10^{-6}\,K^{-1}$, the TEC of Ti_3GeC_2 was the lowest of the three. Thus, here again, the solid solution slightly destabilizes the structure relative to the end members. It is worth noting that more recent measurements

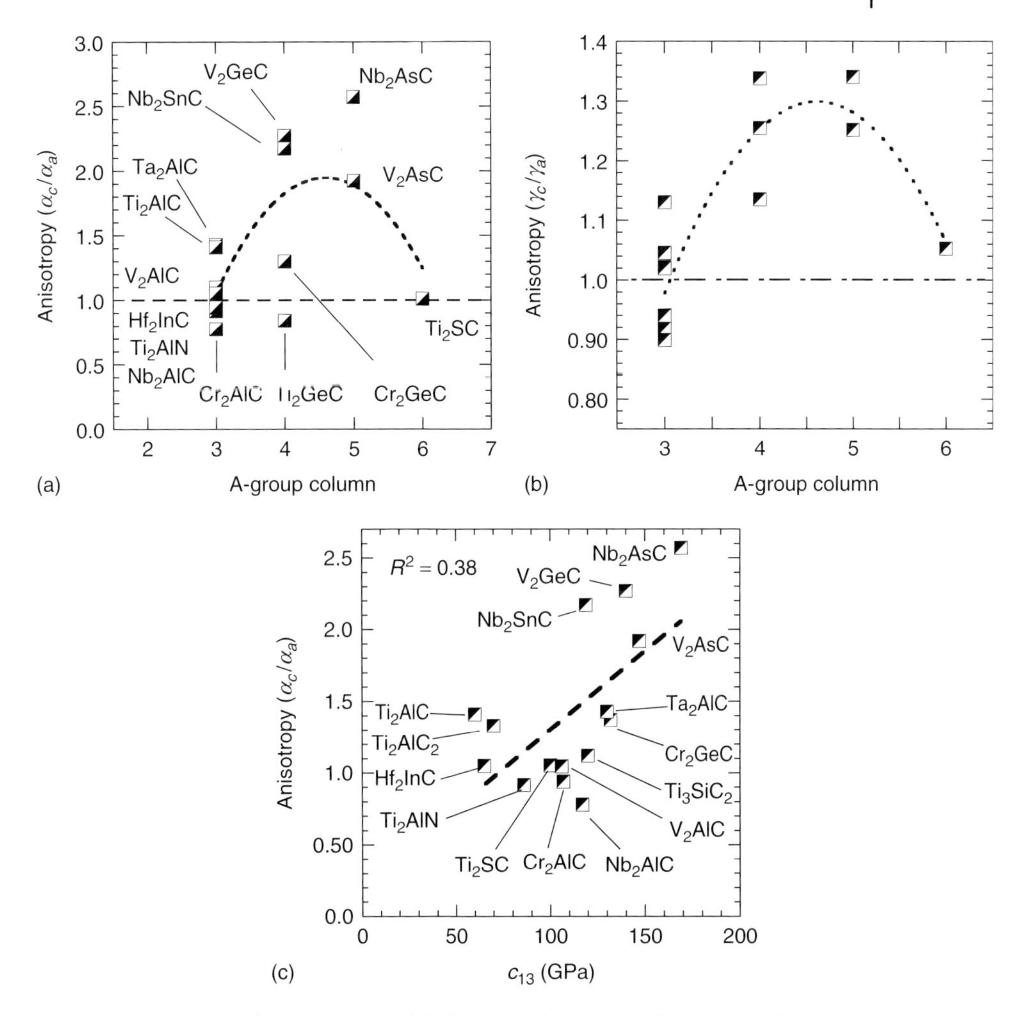

Figure 4.22 Correlation between group of A-element and (a) α_c/α_a, (b) γ_c/γ_a, and (c) c_{13}. Dotted curved lines in (a,b) are guides to the eye. In (c), the dashed line is the least squares fitted line through the results. The R^2 value is 0.38 (Scabarozi *et al.*, 2009).

yielded a TEC value of 8.5×10^{-6} K^{-1} for Ti_3GeC_2 (Lane, Vogel and Barsoum 2010) that is more in line with the other results.

More recently, Cabioch, *et al.* (2013) measured the TECs of the end members and solid-solution compositions in the $Cr_2(Al_x,Ge_{1-x})C$ system using HTXRD in the 25–800 °C temperature range. The results (Figure 4.23) show that, with increasing Al content, α_a remains more or less constant at $\approx 14 \pm 1 \times 10^{-6}$ K^{-1}, whereas α_c, decreases monotonically from $17 \pm 1 \times 10^{-6}$ K^{-1} to about $12 \pm 1 \times 10^{-6}$ K^{-1}. At around the $Cr_2(Al_{0.75},Ge_{0.25})C$ composition, the thermal expansions along the two directions are equal: a crucial property to have as far as minimizing thermal residual stresses are concerned.

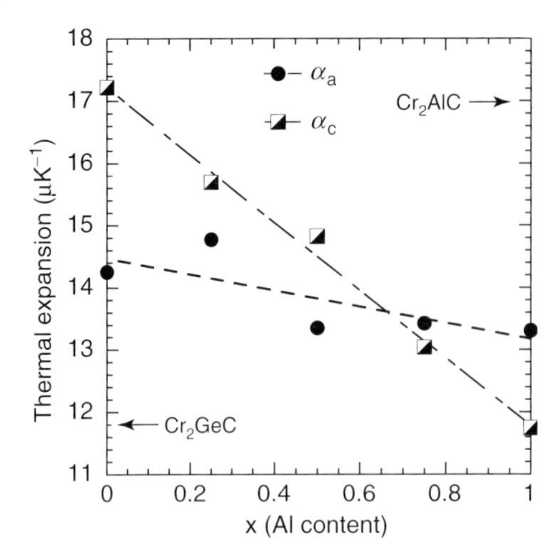

Figure 4.23 Effect of x in $Cr_2(Al_x,Ge_{1-x})C$ on TECs. Increasing the Al content affects α_a less than α_c. Note that at $Cr_2(Al_{0.75},Ge_{0.25})C$, $\alpha_a \approx \alpha_c$ (Cabioch, *et al.* 2013).

4.6
Thermal Stability

The last topic in this chapter deals with the thermal stability of the MAX phases. The MAX phases do not melt congruently but decompose peritectically according to the following reaction:

$$M_{n+1}AX_n \Rightarrow M_{n+1}X_n + A \tag{4.26}$$

Given the chemical stability of the $M_{n+1}X_n$ blocks and the fact that the A layers are relatively loosely held, this result is not too surprising. The decomposition temperatures vary over a wide range: from ≈850 °C for Cr_2GaN (Farber and Barsoum, 1999) to above 2300 °C for Ti_3SiC_2 (Du *et al.*, 2000). The decomposition temperatures of the Sn-containing ternaries range from 1200 to 1400 °C (El-Raghy, Chakraborty, and Barsoum, 2000).

It is important to point out that the stability of the MAX phases, for the most part, is strictly *kinetic* in origin since in most gaseous environments, and especially vacuum, the activity of the A-group element is negligible. Under those circumstances, the free energy change for Eq. (4.26) will *always be negative* at all temperatures. It follows that the main reason why the MAX phases do not decompose at lower temperatures is because they are kinetically stable, that is, the out-diffusion of the A-group element is slow, and/or a protective layer – such as an oxide – forms on the surface, which, in turn, prevents the A-group element from escaping. It is important to emphasize this point because there seems to be some confusion in the literature about this aspect of the MAX phases.

The proper question to ask concerning the chemical stability of the MAX phases is: Under what well-defined activity of its A-group element is a MAX phase stable?

Given that the vast majority of A-group elements are metals, with well-defined melting points that are usually lower than the decomposition temperatures of the MAX phases, it would be useful to define the thermodynamic decomposition temperature to be the temperature at which a MAX phase will decompose in a liquid of its own A-group element. This can only be done, however, if the MAX phase is in equilibrium with its own A liquid. In Chapter 7, a few ternary phase diagrams are shown and, in several cases, the MAX phase is indeed in equilibrium with its own A liquid. As far as we are aware, however, the aforementioned experiment has never been carried out. This is a fruitful area of future research.

In the remainder of this section, what occurs to the MAX phases when they are heated in a vacuum and/or in a low oxygen partial pressure atmosphere is reviewed, with the clear understanding that their stability is kinetic in origin.

4.6.1 Ti_3SiC_2

Jeitschko and Nowotny (1967) were the first to synthesize Ti_3SiC_2 in powder form by a chemical reaction between TiH_2, Si, and graphite at 2000 °C. From that information alone, one has to conclude that Ti_3SiC_2 must be stable at least up to that temperature. However, prior to a series of papers by Barsoum and El-Raghy (1996) and El-Raghy and Barsoum (1999), all previous attempts to produce predominantly single-phase, bulk polycrystalline samples of Ti_3SiC_2 were unsuccessful. For example, some authors reported that Ti_3SiC_2 decomposed at T as low as 1400 °C by dissociating into Si and TiC_x (Lis *et al.*, 1993; Okano, Yano, and Iseki, 1993; Pampuch and Lis, 1995; Racault, Langlais, and Naslain, 1994). In retrospect, it is now clear that the main reason why these early attempts showed such low decomposition temperatures was that by the time the powders were ready to be densified, they were contaminated by impurities, that, in turn, reduced the decomposition temperature.

As noted above, there is little doubt that under the right conditions pure Ti_3SiC_2 is stable up to at least 1700 °C. Typical XRD patterns of Ti_3SiC_2 samples, prepared with high purity Ti, SiC, and graphite powders and annealed at 1600 °C for 4 h, show no TiC_x. The presence of small amounts of impurities, however, can greatly influence the stability of Ti_3SiC_2. For example, additions of as little as 1 at.% Fe- or V to Ti_3SiC_2 resulted in a dramatic decrease in its stability (Tzenov, Barsoum, and El-Raghy, 2000). The presence of Fe results in the formation of a liquid at temperatures as low as 1450 °C. The V goes into solution, forming, most probably, $(Ti,V)_5Si_3C_x$ and stabilizing it at the expense of Ti_3SiC_2. The threshold of Fe or V needed to destabilize the Ti_3SiC_2 is not high (<1 at.%). The destabilization is manifest in the formation of an extensive network of pores when the samples are heated to $T > 1600$ °C, a temperature at which pure Ti_3SiC_2 is stable.

Probably the most convincing evidence for the kinetic nature of the stability is the fact that, when heated in vacuum, thin epitaxial Ti_3SiC_2 films decompose at Ts as low as 1000–1200 °C (Emmerlich *et al.*, 2007). Here, because the barriers to Si loss were low, the decomposition temperature was reduced substantially.

4.6.2 Ti_2AlC

The first successful fabrication of predominantly single-phase samples of Ti_2AlC was carried out by reactively hot-pressing Ti, Al_4C_3, and graphite powders at 1600 °C (Barsoum, Brodkin, and El-Raghy, 1997). This result clearly proves that the free-energy change for the reaction

$$8Ti + Al_4C_3 + C = 4Ti_2AlC \quad (4.27)$$

has to be negative at 1600 °C. Said otherwise, under *certain conditions*, there is no doubt that Ti_2AlC is thermodynamically stable at 1600 °C. In this case upon heating, the Ti in contact with the graphite dies reacted and formed a TiC outer layer that presumably sealed the surface and prevented further loss of Al. Others have shown that Ti_2AlC can be fabricated from a host of other starting materials at relatively high temperatures.

Recently, Spencer *et al.* (2011) have shown that, when commercial Ti_2AlC powders are heated to about 1500 °C, the following reaction occurs:

$$2Ti_2AlC = Ti_3AlC_2 + TiAl_x\,(\text{liq.}) + (1 - x)\,Al\,(\text{liq./vap}), \quad \text{with } x < 1 \quad (4.28)$$

More severe Al loss results in the formation of TiC_y and pure Al. Note that, if the system is "open," then Al evaporates rather than pools up at the grain boundaries. Lastly, we note that, in many cases, the loss of the M element by evaporation cannot be ruled out, especially at the highest temperatures.

This two-step reaction – in which a MAX phase with a lower n, typically 1, converts to one with a higher n before transforming to the $MX_{n/n+1}$ binary – is common and appears to be the preferred mode of reaction whenever a MAX phase exists with more than one n (see below).

4.6.3 Ti_3AlC_2

Pang, Low, and Sun (2010) studied the decomposition of Ti_3AlC_2 into TiC_x at elevated temperatures and surmised the reaction occurred through the continuous sublimation of Al and Ti species in the dynamic environment of high vacuum. The operative reaction was concluded to be

$$Ti_3AlC_2 = 2TiC + Al\,(g) + Ti\,(g) \quad (4.29)$$

Depth profiling of vacuum-annealed Ti_3AlC_2 by grazing synchrotron radiation diffraction revealed a graded composition of TiC_x at the near surface. The rate-limiting step is presumably the evaporation of the Al and Ti atoms.

4.6.4 Ti_2AlN

The first successful fabrication of predominantly single-phase samples of Ti_2AlN was carried out by reactively hot-pressing Ti and AlN, again at 1600 °C (Barsoum,

Brodkin, and El-Raghy, 1997). Here again, this proves that the free-energy change for the reaction

$$2Ti + AlN = Ti_2AlN \qquad (4.30)$$

has to be negative at 1600 °C. Said otherwise, under *certain conditions*, there is no doubt that Ti_2AlN is thermodynamically stable at those high temperatures. Here again, the Ti reacts with the C of the hot pressing dies to form a TiC outer surface that prevents further loss of Al.

In vacuum, Low *et al.* (2011), using *in situ* ND, showed that the reaction

$$Ti_2AlN = 2\,TiN_{0.5} + Al\,(g) \qquad (4.31)$$

was operative at *T*s as low as 1400 °C. For example, when held for 300 min at 1550 °C, 20 wt% of a polycrystalline Ti_2AlN sample decomposed. Consistent with the loss of Al, the samples lost weight after the various anneals. The reaction was also found to occur on the surface of the sample. In that paper, the authors suggested that at higher *T*s the following reaction occurs:

$$Ti_2AlN + 2\,TiN = Ti_4AlN_3 \qquad (4.32)$$

This reaction, however, is thermodynamically impossible. A more likely reaction is

$$3Ti_2AlN = Ti_4AlN_3 + 2\,Al\,(g) + 2\,Ti\,(g) \qquad (4.33)$$

4.6.5 Ti_4AlN_3

Low *et al.* (2011) used *in situ* ND to study the decomposition of Ti_4AlN_3 in vacuum. The operative reaction was reported to be

$$Ti_4AlN_3 = 4\,TiN_{0.75} + Al\,(g) \qquad (4.34)$$

In this case, more than 60 wt% of the sample decomposed after annealing at 1500 °C for 400 min.

4.6.6 Ti_2InC

Heating Ti_2InC in vacuum at ≈800 °C, results in loss of the In and the formation of TiC_x on the surface (Barsoum *et al.*, 2002a).

4.6.7 Ta_2AlC and Nb_2AlC

When heated to 1700 °C in Ar, Ta_2AlC transforms first to Ta_4AlC_3 before ultimately converting to TaC_x (Hu *et al.*, 2007a). A similar reaction occurs when Nb_2AlC is heated in Ar; it transforms to Nb_4AlC_3 before, presumably, converting to NbC_x (Hu *et al.*, 2007a).

4.7
Summary and Conclusions

Thermally, the MAX phases share much in common with their MX counterparts. They are good thermal conductors because they are good electrical conductors. Typically, MAX phases with A elements heavier than S are poor phonon conductors because the combination of the rattling effect of the A-group element together with the enhanced mass, renders the A layers potent phonon scatterers. The $Ti_{n+1}AlC_n$ phases are an exception since the Al atoms appear to behave less as rattlers than the heavier A atoms. The phonon conductivities are also quite sensitive to the presence of defects, most likely point defects.

For the most part, the ADPs increase linearly with increasing temperature and, as expected for refractory solids, are relatively low. All of the MAX phases studied to date indicate that the A layers' atomic displacements are higher than those of the other atoms in the structure, which confirms their rattler nature. The displacements are also anisotropic and tend to favor the basal planes. DFT calculations of the phonon DOS also predict that vibrations of the A elements along the basal planes should occur more readily than along [0001].

As a first approximation, the c_ps of the MAX phases can be assumed to be $(n + 1)$ times the c_p of their corresponding MAX phases. DFT calculations predict the heat capacities over a relatively wide temperature range reasonably well, but only if the experimentally derived values of γ are included.

The Debye temperatures calculated from the low-temperature c_p results agree with those calculated from the velocity of sound and from the ADPs, lending credence to the entire body of work. The discrepancy between the DFT-calculated DOS and those derived from low-temperature heat capacity measurements are most probably due to electron–phonon coupling, λ_{ep}. With the exception of a few compounds, most notably Cr_2AlC and Cr_2GeC, many MAX phases appear to have a λ_{ep} value of $\approx$0.4.

Again, with the exception of Cr_2AlC and Cr_2GeC, the TECs of the MAX phases from room temperature to $\approx$1000 °C fall in the relatively narrow range of $6–10 \times 10^{-6}$ K^{-1}. The Cr-containing phases, on the other hand, have TECs that are higher than those of the rest. These results once more confirm the outlier nature of these two phases.

The thermal decomposition of the MAX phases occurs by the loss of the A element and the formation of higher n-containing MAX phases and/or MX. In most environments, the MAX phases are only kinetically stable. It follows that their decomposition temperatures depend on their morphology – powder versus bulk – state of surface oxidation, or whether a reaction product prevents the escape of the A-group element.

4.A Appendix

Table 4.A.1 Thermodynamic parameters of Ti_3SiC_2.

T (K)	c_p (J mol^{-1} K^{-1})	S_T° (J mol^{-1} K^{-1})	$H_T^\circ - H_0^\circ$ (J mol^{-1})	$(G_T^\circ - H_0^\circ)/T$ (J mol^{-1} K^{-1})
0	0	0	0	0
3.1	0.024	0.012	0.04	0
3.5	0.028	0.015	0.05	0.002
4.02	0.031	0.019	0.06	0.004
5.06	0.04	0.027	0.1	0.008
5.57	0.043	0.031	0.12	0.01
6.06	0.049	0.035	0.14	0.012
6.49	0.053	0.039	0.17	0.013
7.01	0.059	0.043	0.19	0.015
8.13	0.074	0.053	0.27	0.02
9.21	0.09	0.063	0.36	0.024
10.04	0.108	0.072	0.44	0.028
11.1	0.128	0.083	0.56	0.033
11.99	0.152	0.094	0.68	0.037
13.98	0.26	0.135	1.23	0.047
14.99	0.249	0.151	1.46	0.054
20	0.487	0.254	3.28	0.09
25.1	0.909	0.406	6.73	0.138
30.2	1.621	0.632	13.04	0.201
35	2.64	0.94	23.11	0.28
40	4.059	1.382	39.74	0.389
50	7.881	2.676	98.42	0.707
60	12.767	4.531	200.91	1.182
70	18.392	6.914	356.22	1.825
80	24.444	9.762	570.17	2.634
90	30.647	12.999	845.6	3.603
110	42.646	20.328	1580.16	5.963
120	48.315	24.283	2035.09	7.324
130	53.805	28.368	2545.77	8.785
140	59.226	32.554	3110.96	10.333
150	64.621	36.824	3730.2	11.956
160	70	41.166	4403.33	13.646
170	75.345	45.571	5130.1	15.394
180	80.603	50.027	5909.93	17.194
190	85.693	54.522	6741.59	19.04
200	90.505	59.041	7622.87	20.927
210	95.01	63.566	8550.46	22.85
220	99.376	68.088	9522.58	24.803
230	103.527	72.597	10537.25	26.783
240	107.517	77.088	11592.57	28.786

Table 4.A.2 Thermodynamic parameters of Ti_3GeC_2.

T (K)	c_p (J mol^{-1} K^{-1})	S_T° (J mol^{-1} K^{-1})	$H_T^\circ - H_0^\circ$ (J mol^{-1})	$(G_T^\circ - H_0^\circ)/T$ (J mol^{-1} K^{-1})
0	0	0	0	0
3.5	0.026	0.015	0.04	0.002
4	0.03	0.018	0.06	0.004
4.5	0.034	0.022	0.08	0.006
5	0.039	0.026	0.09	0.007
6	0.049	0.034	0.14	0.011
6.5	0.054	0.038	0.16	0.013
7	0.06	0.042	0.19	0.015
8	0.076	0.051	0.26	0.019
9	0.094	0.061	0.34	0.023
10	0.116	0.072	0.45	0.027
11	0.143	0.085	0.58	0.032
12	0.177	0.098	0.74	0.037
13	0.22	0.114	0.94	0.042
14	0.274	0.132	1.18	0.048
15	0.342	0.154	1.49	0.054
20	0.895	0.322	4.5	0.098
25	2.038	0.637	11.65	0.171
30	3.55	1.137	25.5	0.287
35	5.333	1.815	47.61	0.455
40	7.353	2.657	79.23	0.676
50	12.099	4.79	175.71	1.275
60	17.605	7.481	324.18	2.078
70	23.581	10.639	529.84	3.07
80	29.822	14.193	796.7	4.234
90	36.19	18.072	1126.71	5.553
100	42.573	22.215	1520.55	7.01
110	48.889	26.57	1977.94	8.589
120	55.075	31.09	2497.88	10.274
130	61.094	35.737	3078.87	12.054
140	66.922	40.479	3719.11	13.914
150	72.479	45.289	4416.57	15.845
160	77.59	50.131	5167.15	17.837
170	82.348	54.979	5967.05	19.879
180	86.797	59.813	6812.95	21.963
190	90.998	64.619	7702.06	24.082
200	95.016	69.39	8632.23	26.228
210	98.909	74.12	9601.92	28.397
220	102.714	78.809	10610.08	30.582
230	106.439	83.458	11655.89	32.78
240	110.051	88.064	12738.44	34.988
250	113.46	92.627	13856.16	37.202
260	116.511	97.137	15006.3	39.421

References

Barsoum, M.W. (2000) The $M_{n+1}AX_n$ phases: a New class of solids; thermodynamically stable nanolaminates. *Prog. Solid State Chem.*, **28**, 201–281.

Barsoum, M.W., Brodkin, D., and El-Raghy, T. (1997) Layered machinable ceramics for high temperature applications. *Scrip. Met. Mater.*, **36**, 535–541.

Barsoum, M.W. and El-Raghy, T. (1996) Synthesis and characterization of a remarkable ceramic: Ti_3SiC_2. *J. Am. Ceram. Soc.*, **79**, 1953–1956.

Barsoum, M.W., El-Raghy, T., Rawn, C.J., Porter, W.D., Wang, H., Payzant, A., and Hubbard, C. (1999) Thermal properties of Ti_3SiC_2. *J. Phys. Chem. Solids*, **60**, 429.

Barsoum, M.W., Golczewski, J., Siefert, H.J., and Aldinger, F. (2002a) Fabrication and electrical and thermal properties of Ti_2InC, Hf_2InC and $(Ti,Hf)_2AlC$. *J. Alloys Compd.*, **340**, 173–179.

Barsoum, M.W., Salama, I., El-Raghy, T., Golczewski, J., Porter, W.D., Wang, H., Siefert, H., and Aldinger, F. (2002b) Thermal and electrical properties of Nb_2AlC, $(Ti,Nb)_2AlC$ and Ti_2AlC. *Metall. Mater. Trans.*, **33a**, 2779.

Barsoum, M.W., Ali, M., and El-Raghy, T. (2000a) Processing and characterization of Ti_2AlC, Ti_2AlCN and $Ti_2AlC_{0.5}N_{0.5}$. *Metall. Mater. Trans.*, **31A**, 1857–1865.

Barsoum, M.W., El-Raghy, T., and Chakraborty, S. (2000b) Thermal properties of Nb_2SnC. *J. Appl. Phys.*, **88**, 6313.

Barsoum, M.W., Rawn, C.J., El-Raghy, T., Procopio, A.T., Porter, W.D., Wang, H., and Hubbard, C.R. (2000c) Thermal properties of Ti_4AlN_3. *J. Appl. Phys.*, **83**, 825.

Barsoum, M.W., Scabarozi, T.H., Amini, S., Hettinger, J.D., and Lofland, S.E. (2011) Electrical and thermal properties of Cr_2GeC. *J. Am. Ceram. Soc.*, **94**, 4123–4126.

Bouhemadou, A. (2009) Calculated structural, electronic and elastic properties of M_2GeC (M = Ti, V, Cr, Zr, Nb, Mo, Hf, Ta and W). *Appl. Phys. A*, **96**, 959–967.

Cabioch, T., Eklund, P., Mauchamp, V., Jaouen, M., and Barsoum, M.W. (2013). Tailoring of the Thermal Expansions of the MAX Phases in the $Cr_2(Al_{1-x},Ge_x)C_2$ system. *J. Eur. Ceram. Soc.*, **33**, 897–904.

Chase, M.W., Davies, C.A., Downey, J.R., Frurip, D.J., McDonald, R.A., and Syverud, A.N. (1985) JANAF thermodynamic tables third edition. *J. Phys. Chem.*, **14**(Supp. 1).

Drulis, M.K., Czopnik, A., Drulis, H., and Barsoum, M.W. (2004) Low temperature heat capacity and magnetic susceptibility of Ti_3SiC_2. *J. Appl. Phys.*, **95**, 128–133.

Drulis, M.K., Czopnik, A., Drulis, H., Spanier, J.E., Ganguly, A., and Barsoum, M.W. (2005) On the heat capacities of Ti_3GeC_2. *Mater. Sci. Eng., B*, **119**, 159–163.

Drulis, M.K., Drulis, H., Gupta, S., and Barsoum, M.W. (2006) On the heat capacities of M_2AlC (M = Ti, V, Cr) ternary carbides. *J. Appl. Phys.*, **99**, 093502.

Drulis, M.K., Drulis, H., Hackemer, A.E., Ganguly, A., El-Raghy, T., and Barsoum, M.W. (2007) On the low temperature heat capacities of Ti_2AlN and $Ti_2Al(C_{0.5},N_{0.5})$. *J. Alloys Compd.*, **433**, 59–62.

Drulis, M.K., Drulis, H., Hackemer, A.E., Leaffer, O., Spanier, J., Amini, S., Barsoum, M.W., Guilbert, T., and El-Raghy, T. (2008) On the heat capacities of Ta_2AlC, Ti_2SC and Cr_2GeC. *J. Appl. Phys.*, **104**, 23526–23527.

Du, Y., Schuster, J., Seifert, H., and Aldinger, F. (2000) Experimental investigation and thermodynamic calculation of the titanium-silicon-carbon system. *J. Am. Ceram. Soc.*, **83**, 197–203.

Du, Y.L., Sun, Z.-M., Hashimoto, H., and Barsoum, M.W. (2009) Theoretical investigations on the elastic and thermodynamic properties of $Ti_2AlC_{0.5}N_{0.5}$ solid solution. *Phys. Lett. A*, **374**, 78–82.

Du, Y.L., Sun, Z.M., Hashimoto, H., and Tiana, W.B. (2008) First-principles study on electronic structure and elastic properties of Ti_2SC. *Phys. Lett. A*, **372**, 5220–5223.

El-Raghy, T. and Barsoum, M.W. (1999) Processing and mechanical properties of Ti_3SiC_2: part I: reaction path and microstructure evolution. *J. Am. Ceram. Soc.*, **82**, 2849–2854.

El-Raghy, T., Chakraborty, S., and Barsoum, M.W. (2000) Synthesis and characterization of Hf_2PbC, Zr_2PbC and M_2SnC (M = Ti, Hf, Nb or Zr). *J. Eur. Ceram. Soc.*, **20**, 2619.

Emmerlich, J., Högberg, H., Wilhelmsson, O., Jansson, U., Music, D., Schneider, J.M., and Hultman, L. (2007) Thermal stability of MAX-phase Ti_3SiC_2 thin films. *Acta Mater.*, **55**, 1479–1488.

Etzkorn, J., Ade, M., and Hillebrecht, H. (2007a) Ta_3AlC_2 and Ta_4AlC_3 - single-crystal investigations of two new ternary carbides of tantalum synthesized by the molten metal technique. *Inorg. Chem.*, **46**, 1410–1418.

Etzkorn, J., Ade, M., and Hillebrecht, H. (2007b) V_2AlC, V_4AlC_{3-x} (x approximate to 0.31), and $V_{12}Al_3C_8$: synthesis, crystal growth, structure, and superstructure. *Inorg. Chem.*, **46**, 7646–7653.

Etzkorn, J., Ade, M., Kotzott, D., Kleczek, M., and Hillebrecht, H. (2009) Ti_2GaC, Ti_4GaC_3 and Cr_2GaC - synthesis, crystal growth and structure analysis of Ga-containing MAX-phases $M_{n+1}GaC_n$ with M = Ti, Cr and n = 1–3. *J. Solid State Chem.*, **182**, 995.

Farber, L. and Barsoum, M.W. (1999) Isothermal sections in the Cr-Ga-N system in the 650–1000 °C temperature range. *J. Mater. Res.*, **14**, 2560–2566.

Finkel, P., Barsoum, M.W., and El-Raghy, T. (2000) Low temperature dependencies of the elastic properties of Ti_4AlN_3 and $Ti_3Al_{1.1}C_{1.8}$ and Ti_3SiC_2. *J. Appl. Phys.*, **87**, 1701–1703.

Finkel, P., Seaman, B., Harrell, K., Hettinger, J.D., Lofland, S.E., Ganguly, A., Barsoum, M.W., Sun, Z., Li, S., and Ahuja, R. (2004) Low temperature elastic, electronic and transport properties of $Ti_3Si_{1-x}Ge_xC_2$ solid solutions. *Phys. Rev. B*, **70**, 085104.

Halilov, S.V., Singh, D.J., and Papaconstantopoulos, D.A. (2002) Soft modes and superconductivity in the layered hexagonal carbides V_2CAs, Nb_2CAs, and Nb_2CS. *Phys. Rev. B*, **65**, 174519.

He, X., Bai, Y., Zhu, C., Sun, Y., Li, M., and Barsoum, M.W. (2010) General trends in the structural, electronic and elastic properties of the M_3AlC_2 phases (M = transition metal): a first-principle study. *Comput. Mater. Sci.*, **49**, 691–698.

Hettinger, J.D., Lofland, S.E., Finkel, P., Palma, J., Harrell, K., Gupta, S., Ganguly, A., El-Raghy, T., and Barsoum, M.W. (2005) Electrical transport, thermal transport and elastic properties of M_2AlC (M = Ti, Cr, Nb and V) phases. *Phys. Rev. B*, **72**, 115120.

Ho, J.C., Hamdeh, H.H., Barsoum, M.W., and El-Raghy, T. (1999) Low temperature heat capacities of $Ti_3Al_{1.1}C_{1.8}$, Ti_4AlN_3 and Ti_3SiC_2. *J. Appl. Phys.*, **86**, 3609.

Holm, B., Ahuja, R., Li, S., and Johansson, B. (2002) Theory of ternary layered system Ti-Al-N. *J. Appl. Phys.*, **91**, 9874–9877.

Houska, C.R. (1964) Thermal expansion and atomic vibration amplitudes for TiC, TIN, ZrC, ZrN and pure W. *J. Phys. Chem. Solids*, **25**, 359–366.

Hu, C., He, L., Zhang, J., Bao, Y., Wang, J., Li, M., and Zhou, Y. (2008a) Microstructure and properties of bulk Ta_2AlC ceramic synthesized by an in situ reaction/hot pressing method. *J. Eur. Ceram. Soc.*, **28**, 1679–1685.

Hu, C., Li, F., He, L., Liu, M., Zhang, J., Wang, J., Bao, Y., Wang, J., and Zhou, Y. (2008b) In situ reaction synthesis, electrical and thermal, and mechanical properties of Nb_4AlC_3. *J. Am. Ceram. Soc.*, **91**, 2258–2263.

Hu, C., Li, F., Zhang, J., Wang, J., Wang, J., and Zhou, Y. (2007a) Nb_4AlC_3: a new compound belonging to the MAX phases. *Scr. Mater*, **57**, 893–896.

Hu, C., Lin, Z., He, L., Bao, Y., Wang, J., Li, M., and Zhou, Y.C. (2007b) Physical and mechanical properties of bulk Ta_4AlC_3 ceramic prepared by an in situ reaction synthesis/hot-pressing method. *J. Am. Ceram. Soc.*, **90**, 2542–2548.

Hu, C., Sakka, Y., Tanaka, H., Nishimura, T., and Grasso, S. (2009) Low temperature thermal expansion, high temperature electrical conductivity, and mechanical properties of Nb_4AlC_3 ceramic synthesized by spark plasma sintering. *J. Alloys Compd.*, **487**, 675–681.

Hug, G. (2006) Electronic structures of and composition gaps among the ternary carbides Ti_2MC. *Phys. Rev. B*, **74**, 184113–184117.

Hug, G. and Fries, E. (2002) Full-potential electronic structure of Ti_2AlC & Ti_2AlN. *Phys. Rev. B*, **65**, 113104.

Hug, G., Jaouen, M., and Barsoum, M.W. (2005) XAS, EELS and full-potential augmented plane wave study of the electronic structures of Ti_2AlC, Ti_2AlN, Nb_2AlC and $(Ti_{0.5},Nb_{0.5})_2AlC$. *Phys. Rev. B*, **71**, 24105.

Jeitschko, W. and Nowotny, H. (1967) Die kristallstructur von Ti_3SiC_2 - Ein neuer komplxcarbid-Typ. *Monatsh. Chem.*, **98**, 329–337.

Kanoun, M.B., Goumri-Said, S., and Reshak, A.H. (2009) Theoretical study of mechanical, electronic, chemical bonding and optical properties of Ti_2SnC, Zr_2SnC, Hf_2SnC and Nb_2SnC. *Comput. Mater. Sci.*, **47**, 491–500.

Keppens, V., Mandrus, D., Sales, B.C., Chakoumakos, B.C., Dai, P., Coldea, R., Maple, M.B., Gajewski, D.A., Freeman, E.J., and Bennington, S. (1998) Localized vibrational modes in metallic solids. *Nature*, **395**, 876.

Kulkarni, S., Merlini, M., Phatak, N., Saxena, S.K., Artioli, G., Amini, S., and Barsoum, M.W. (2009) Thermal expansion and stability of Ti_2SC in air and inert atmospheres. *J. Alloys Compd.*, **469**, 395–400.

Kulkarni, S.R., Merlini, M., Phatak, N., Saxena, S.K., Artioli, G., Gupta, S., and Barsoum, M.W. (2007) High-temperature thermal expansion and stability of V_2AlC up to 950 degrees C. *J. Am. Ceram. Soc.*, **90**, 3013–3016.

Lane, N.J., Vogel, S.C., and Barsoum, M.W. (2010) High temperature neutron diffraction and the temperature-dependent crystal structures of Ti_3SiC_2 and Ti_3GeC_2. *Phys. Rev. B*, **82**, 174109.

Lane, N.J., Vogel, S.C., and Barsoum, M.W. (2011). Temperature-dependent crystal structures of Ti_2AlN and Cr_2GeC as determined from high temperature neutron diffraction *J. Am. Ceram. Soc.*, **94**, 3473–3479.

Lane, N.J., Vogel, S., Caspi, E., and Barsoum, M.W. (2013). High-temperature neutron diffraction study of Ti_2AlC, Ti_3AlC_2 and $Ti_5Al_2C_3$ *J. Appl. Phys.*, **113**, 183519.

Larson, A.C. and Dreele, R.B.V. (2004) GSAS General Structure Analysis System. Report LAUR 86–748, Los Alamos National Laboratory, University of California, Los Alamos, NM, pp. 122–124.

Lengauer, W., Binder, S., Aigner, K., Ettmayer, P., Guillou, A., Debuigne, J., and Groboth, G. (1995) Solid state properties of group IVb carbonitrides. *J. Alloys Compd.*, **217**, 137–147.

Lin, Z.J., Zhou, Y.C., and Li, M.S. (2007) Synthesis, microstructure, and property of Cr_2AlC. *J. Mater. Sci. Technol.*, **23**, 721–746.

Lis, J., Pampuch, R., Piekarczyk, J., and Stobierski, L. (1993) New ceramics based on Ti_3SiC_2. *Ceram. Int.*, **19**, 91–96.

Lofland, S.E., Hettinger, J.D., Harrell, K., Finkel, P., Gupta, S., Barsoum, M.W., and Hug, G. (2004) Elastic and electronic properties of select M_2AX phases. *Appl. Phys. Lett.*, **84**, 508–510.

Lofland, S.E., Hettinger, J.D., Meehan, T., Bryan, A., Finkel, P., Hug, G., and Barsoum, M.W. (2006) Electron–phonon coupling in MAX phase carbides. *Phys. Rev. B*, **74**, 174501.

Low, I.M., Pang, W.K., Kennedy, S.J., and Smith, R.I. (2011) High-temperature thermal stability of Ti_2AlN and Ti_4AlN_3: a comparative diffraction study. *J. Eur. Ceram. Soc.*, **31**, 159–166.

Maksimov, E.G. (1969) *Zh. Eksp. Teor. Fiz.*, **57**, 1660.

Manoun, B., Saxena, S.K., Liermann, H.-P., and Barsoum, M.W. (2005) Thermal expansion of polycrystalline Ti_3SiC_2 in the 25–1400 °C temperature range. *J. Am. Ceram. Soc.*, **88**, 3489–3491.

McMillan, W.L. (1968) Transition temperature of strong-coupled superconductors. *Phys. Rev.*, **167**, 331.

Nakamura, K. and Yashima, K. (2008) Crystal structure of NaCl-type transition metal monocarbides MC (M = V, Ti, Nb, Ta, Hf, Zr), a neutron powder diffraction study. *Mater. Sci. Eng., B*, **148**, 69–72.

Okano, T., Yano, T., and Iseki, T. (1993) Synthesis and mechanical properties of Ti_3SiC_2. *Trans. Met. Soc. Jpn.*, **14A**, 597.

Opeka, M., Zaykoski, J., Talmy, I., and Causey, S. (2011) Synthesis and characterization of Zr_2SC ceramics. *Mater. Sci. Eng., A*, **528**, 1994–2001.

Pampuch, R. and Lis, J. (1995) Ti_3SiC_2 – A Plastic Ceramic Material, in Advances in Science and Technology, Vol. **3B**, P.

Vincenzini, ed. (Faenza, Techna Srl), pp. 725–732.

Pang, W.-K., Low, I.M., and Sun, Z.-M. (2010) In situ high-temperature diffraction study of the thermal dissociation of Ti_3AlC_2 in vacuum. *J. Am. Ceram. Soc.*, **93**, 2871–2876.

Pierson, H.O. (1996) *Handbook of Refractory Carbides and Nitrides*, Noyes Publications, Westwood NJ.

Racault, C., Langlais, F., and Naslain, R. (1994) Chemically vapor deposition of Ti_3SiC_2 from $TiCl_4$-$SiCl_4$-CH_4-H_2 gas mixtures: part II an experimental approach. *J. Mater. Sci.*, **29**, 5023.

Radovic, M., Ganguly, A., and Barsoum, M.W. (2008) Elastic properties and phonon conductivities of $Ti_3Al(C_{0.5},N_{0.5})_2$ and $Ti_2Al(C_{0.5},N_{0.5})$ solid solutions. *J. Mater. Res.*, **23**, 1517–1521.

Radovic, M., Ganguly, A., Barsoum, M.W., Zhen, T., Finkel, P., Kalidindi, S.R., and Lara-Curzio, E. (2006) On the elastic properties and mechanical damping of Ti_3SiC_2, Ti_3GeC_2, $Ti_3Si_{0.5}Al_{0.5}C_2$ and Ti_2AlC in the 300–1573 K temperature range. *Acta Mater.*, **54**, 2757–2767.

Rawn, C.J., Barsoum, M.W., El-Raghy, T., Procopio, A.T., and Hoffman, C.M. (2000) Thermal properties of Ti_4AlN_{3-x}. *Mater. Res. Bull.*, **35**, 1785–1796.

Sales, B.C., Chakoumakos, B.C., Mandrus, D., and Sharp, J.W. (1999) Atomic displacement parameters and the lattice thermal conductivity of clatharate thermoelectric compounds. *J. Solid State Chem.*, **146**, 528.

Sambasivan, S. (1990) PhD Chemistry, Arizona State University, Tempe, AZ.

Sanborn, B.A., Allen, P.B., and Papaconstantopoulos, D.A. (1989) Empirical electron-phonon coupling constants and anisotropic electrical resistivity in hcp metals., Phys. Rev. B. 40.

Scabarozi, T., Ganguly, A., Hettinger, J.D., Lofland, S.E., Amini, S., Finkel, P., El-Raghy, T., and Barsoum, M.W. (2008a) Electronic and thermal properties of $Ti_3Al(C_{0.5},N_{0.5})_2$, $Ti_2Al(C_{0.5},N_{0.5})$ and Ti_2AlN. *J. Appl. Phys.*, **104**, 073713.

Scabarozi, T.H., Amini, S., Finkel, P., Barsoum, M.W., Tambussi, W.M., Hettinger, J.D., and Lofland, S.E. (2008b) Electrical, thermal, and elastic properties of the MAX phase Ti_2SC. *J. Appl. Phys.*, **104**, 033502.

Scabarozi, T.H., Amini, S., Leaffer, O., Ganguly, A., Gupta, S., Tambussi, W., Clipper, S., Spanier, J.E., Barsoum, M.W., Hettinger, J.D. *et al.* (2009) Thermal expansion of select MAX phases measured by high temperature X-ray diffraction and dilatometry. *J. Appl. Phys*, **105**, 013543.

Schneider, J., Mertens, R., and Music, D. (2006) Structure of V_2AlC studied by theory and experiment. *J. Appl. Phys.*, **99**, 013501.

Schneider, J., Music, D., and Sun, Z. (2005) Effect of the valence electron concentration on the bulk modulus and chemical bonding in Ta_2AC and Zr_2AC, A=Al, Si, and P. *J. Appl. Phys.*, **97**, 066105.

Spencer, C.J., Córdoba, J.M., Obando, N., Sakulich, A., Radovic, M., Odén, M., Hultman, L., and Barsoum, M.W. (2011). Phase evaluation in short Al_2O_3 fiber-reinforced Ti_2AlC matrices during processing in 1300–1500 °C temperature range. *J. Am. Ceram. Soc.*, **94**, 3317–3334.

Taylor, R.E. (1961) Thermal conductivity of TiC at high temperatures. *J. Am. Ceram. Soc.*, **44**, 525.

Tian, W., Sun, Z.-M., and Hashimoto, H. (2009) Synthesis, microstructure and properties of $(Cr_{1-x}V_x)_2AlC$ solid solutions. *J. Alloys Compd.*, **484**, 130–133.

Tian, W., Wang, P., Zhang, G., Kan, Y., Li, Y., and Yan, D. (2006) Synthesis and thermal and electrical properties of bulk Cr_2AlC. *Scr. Mater.*, **54**, 841–846.

Togo, A., Chaput, L., Tanaka, I., and Hug, G. (2010) First-principles phonon calculations of thermal expansion in Ti_3SiC_2, Ti_3AlC_2, and Ti_3GeC_2. *Phys. Rev. B*, **81**, 174301.

Toth, L. (1971) *Transition Metal Carbides and Nitrides*, Academic Press, New York.

Tzenov, N., and Barsoum, M.W. (2000). Synthesis and characterization of $Ti_3AlC_{1.8}$. *J. Am. Ceram. Soc.* **83**. 825–832.

Tzenov, N., Barsoum, M.W., and El-Raghy, T. (2000) Influence of small amounts of Fe and V on the synthesis and stability of Ti_3SiC_2. *J. Eur. Ceram. Soc.*, **20**, 801.

Williams, W. (1966) High temperature thermal conductivity of transition metal cerbides and nitrides. *J. Am. Ceram. Soc.*, **49**, 156.

Yang, H., Manoun, B., Downs, R.T., Ganguly, A., and Barsoum, M.W. (2006) Crystal chemistry of the ternary layered carbide, $Ti_3(Si_{0.43}Ge_{0.57})C_2$. *J. Phys. Chem. Solids*, **67**, 2512–2516.

Ying, G., He, X., Li, M., Du, S., Han, W., and He, F. (2011) Effect of Cr_7C_3 on the mechanical, thermal, and electrical properties of Cr_2AlC. *J. Alloys Compd.*, **509**, 8022–8027.

Zhang, W., Travitzky, N., Hu, C., Zhou, Y., and Greil, P. (2009) Reactive hot pressing and properties of Nb_2AlC. *J. Am. Ceram. Soc.*, **92**, 2396–2399.

Zhou, W.B., Mei, B.C., and Zhu, J.Q. (2009) On the synthesis and properties of bulk ternary Cr_2AlC ceramics. *Mater. Sci. Poland*, **27**, 973–980.

Zhou, Y.C., Dong, H.Y., Wang, X.H., and Chen, S.Q. (2000) Electronic structure of the layered ternary carbides Ti_2SnC and Ti_2GeC. *J. Phys. Condens. Matter*, **12**, 9617.

Zhou, Y.C., Sun, Z., Wang, X., and Chen, S. (2001) Ab initio geometry optimization and ground state properties of layered ternary carbides, Ti_3MC_2 (M = Al,Si and Ge). *J. Phys. Condens. Matter*, **13**, 10001–10010.

5 Electronic, Optical, and Magnetic Properties

5.1 Introduction

Many of the MAX phases are excellent metal-like electrical conductors. Some, such as Ti_3SiC_2 and Ti_3AlC_2, are better conductors than Ti metal itself. Most interestingly and intriguingly, many of the MAX phases whose transport properties have been characterized to date appear to be compensated conductors, wherein not only are the concentrations of electrons and holes roughly equal, but also their mobilities.

Another unique feature of some of these phases, most notably Ti_3SiC_2, is their very low thermoelectric or Seebeck coefficients.

Optically, the MAX phases behave like the good conductors they are, and their optical properties are dominated by delocalized electrons. Magnetically, they are Pauli paramagnets, wherein the susceptibility is, again, determined by the delocalized electrons and is thus not very high, and is temperature independent. Some of the MAX phases have also been shown to be low temperature superconductors.

This chapter reviews the electrical properties of the MAX phases including their conductivities, Hall coefficients, magnetoresistances (MRs), and Seebeck coefficients. Their optical properties are also reviewed and related to their electronic properties. Lastly, their magnetic and superconducting properties are briefly discussed, again in light of their electronic properties.

5.2 Electrical Resistivities, Hall Coefficients, and Magnetoresistances

5.2.1 Electrical Resistivities

The electrical resistivities ρ of the MAX phases, like those of their M and MX counterparts, are metal-like in that they increase linearly with increasing temperature. The resistivities of select MAX phases in the 0–1000 K temperature range are plotted in Figure 5.1a. Also included in the figure are the results for Ti metal. From

MAX Phases: Properties of Machinable Ternary Carbides and Nitrides, First Edition. Michel W. Barsoum.

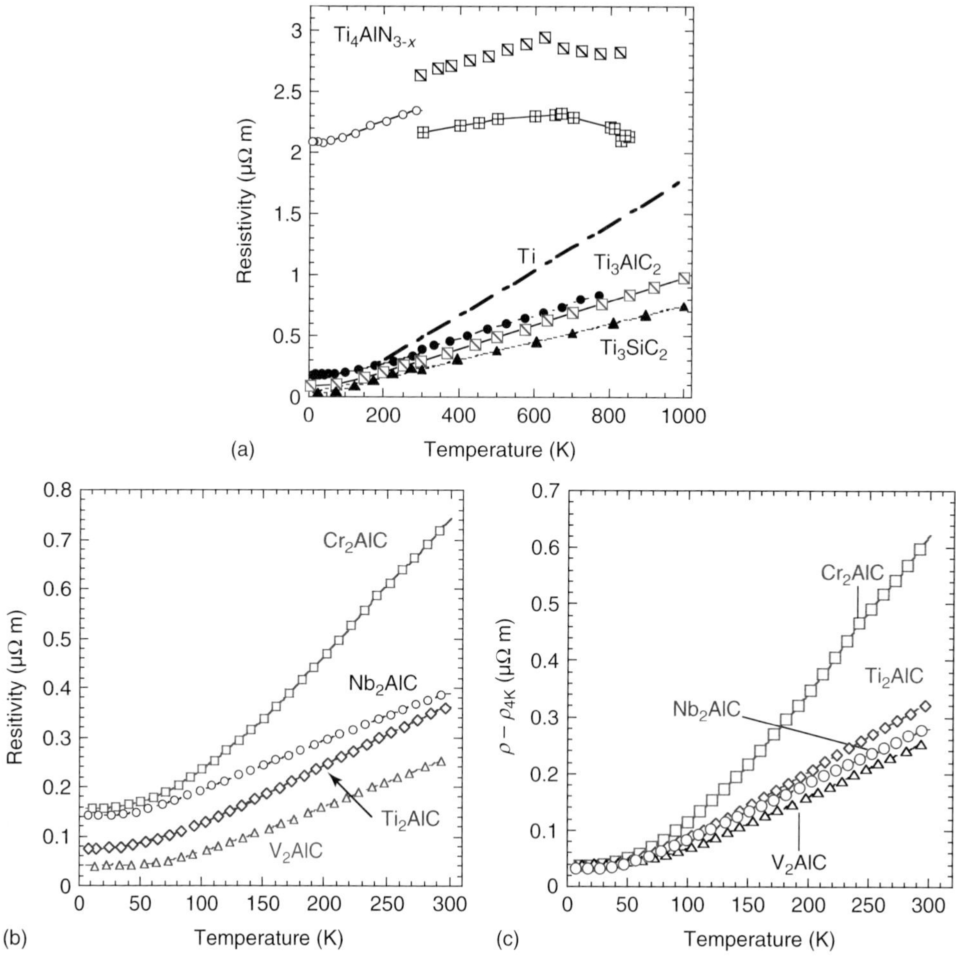

Figure 5.1 Temperature dependence of (a) ρ for select MAX phases in the 0–1000 K range. Results for Ti are from Stanely (1963); for Ti_3AlC_2 (red squares) (Wang and Zhou, 2002) and solid circles; for Ti_4AlN_3 and Ti_3SiC_2 (solid triangles) (Barsoum *et al.*, 2000). (b) ρ for select Al-containing MAX phases. (c) $(\rho - \rho_0)$, where ρ_0 is resistivity at 4 K. The results and their MAX designation are color-coordinated. (Hettinger *et al.*, 2005).

these results it is obvious that Ti_3SiC_2 and Ti_3AlC_2 are better electrical conductors than the latter over the entire temperature regime. The behavior of Ti_4AlN_3 (top results in Figure 5.1a) is unique in that it behaves more as a semimetal.

Figure 5.1b plots ρ versus T for four Al-containing MAX phases, from which it is obvious that $\mathrm{d}\rho/\mathrm{d}T$ is a function of the M element (Hettinger *et al.*, 2005). In Figure 5.1b, the residual resistivity ratios (RRRs) – defined as ρ_{RT}/ρ_{4K}, where ρ_{4K} is the resistivity at 4 K – are different. The RRR values are a measure of the quality of a given sample, with higher RRR values indicating less defective solids. Another way to plot the results of Figure 5.1b is to plot the so-called intrinsic resistivity –

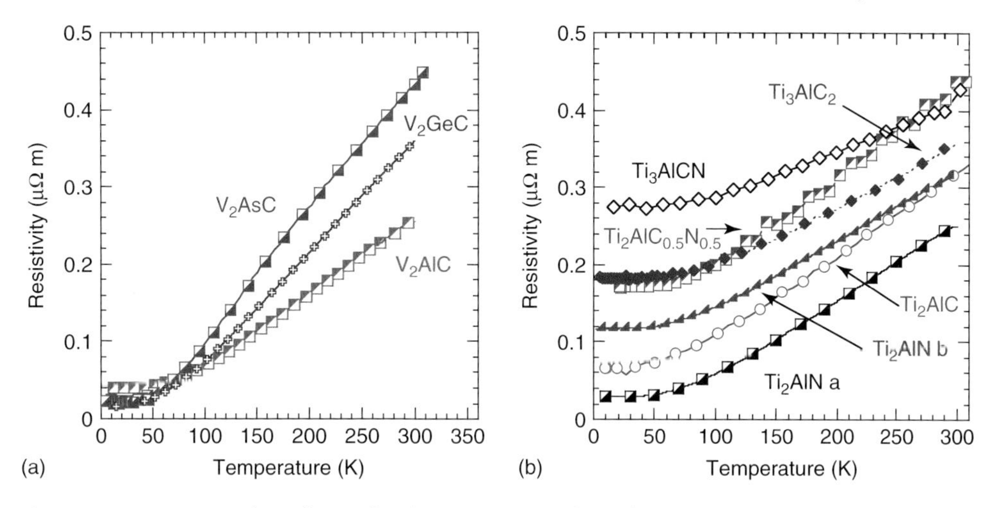

Figure 5.2 Temperature dependence of ρ for (a) V_2AsC, V_2AlC, and V_2GeC and (b) select $Ti_{n+1}AlX_n$ phases (Scabarozi *et al.*, 2008a). The two Ti_2AlN samples are believed to have different defect concentrations. The results and their MAX designations are color-coordinated.

that is, $\rho_i = \rho - \rho_{4K}$ versus T (Figure 5.1c). This plot, even more than Figure 5.1b, shows that $d\rho/dT$ depends on the M element. That is not to imply that the A-group element has no effect. The effect of the latter is best seen when the results for three V-containing phases are compared (Figure 5.2a).

Solid solutions result in solid solution scattering. For example, replacing half the Ti atoms by Nb in Ti_2AlC to form TiNbAlC results in an approximately threefold increase in ρ over the end members (Barsoum *et al.*, 2002b). Replacing 50% of the C by N in the $Ti_{n+1}AlX_n$ compounds also results in solid-solution scattering. Compare Ti_2AlC and Ti_2AlN with $Ti_2AlC_{0.5}N_{0.5}$ in Figure 5.2b.

The effect of replacing all the C or N atoms, on the other hand, is more modest. This is best seen by comparing the results for Ti_2AlC and Ti_2AlN, shown in Figure 5.2b. In this figure, the results for two nominally identical Ti_2AlN samples, a and b, are also plotted, among others; they gave quite different results for reasons discussed below.

The results shown in Figures 5.1 and 5.2 are typical of metallic conductors. Such conduction is quite often represented by the following equation:

$$\rho = \rho_{ref}\left[1 + \alpha_{TCR}\left(T - T_{ref}\right)\right] \tag{5.1}$$

where ρ_{ref} is the resistivity at a reference temperature T_{ref} – usually either 273.15 or 300 K – and α_{TCR} is the temperature coefficient of resistivity. Table 5.1 lists the values of ρ_{RT}, α_{TCR}, RRR, and $d\rho/dT$ for select MAX phases. Note that $d\rho/dT = \rho_{ref} \times \alpha_{TCR}$. The physics behind what determines $d\rho/dT$ is discussed in section 5.2.8.

Based on the results shown in Table 5.1, it is obvious that with the possible exception of Ti_4AlN_3, the MAX phases are good conductors of electricity, with a conductivity that is roughly a tenth that of pure copper. Furthermore, the changes

Table 5.1 Summary of electric resistivities and RRR values of select MAX phases. Also listed in columns 5 and 6 are the α_{TCR} (Eq. (5.1)) and $d\rho/dT$ values, respectively. Note that $d\rho/dT = \rho_{ref} \times \alpha_{TCR}$.

Composition	T (K)	ρ ($\mu\Omega$ m)	RRR	α_{TCR} (K)$^{-1}$	$d\rho/dT$ ($\mu\Omega\cdot$m K^{-1})	References
			413 phases			
Ti_4AlN_4	300	2.64	1.1	0.00034	0.0009	Barsoum *et al.* (2000)
	4	2.35				
α-Ta_4AlC_3	273	0.354	6.0	0.0035	0.0012	Hu *et al.* (2007)
	4	0.064				
Nb_4AlC_3	300	0.75	2.5	0.0025	0.00175	Hu *et al.* (2008b)
	4	0.30				
Nb_4AlC_3	273	0.44	—	0.0030	—	Hu *et al.* (2009)
			312 Phases			
Ti_3SiC_2	300	0.23	—	0.0033	0.00075	Barsoum *et al.* (2000)
	4	—				
Ti_3SiC_2	300	0.23	7.7	0.0028	0.00071	Finkel *et al.* (2001)
	4	≈0.03				
Ti_3AlC_2	300	0.353	1.95	0.00345	0.00078	Scabarozi *et al.* (2008a)
	4	0.18				
	300	0.287	3.2	0.0035	0.00083	Wang and Zhou (2002)
	4	0.09				
Ti_3GeC_2	300	0.28	5.6	0.0032	0.0009	Finkel *et al.* (2004)
	4	0.05				
$Ti_3(Si,Ge)C_3$	300	0.27	3.1	0.0028	0.00075	Finkel *et al.* (2004)
	4	0.084				
Ti_3AlCN	300	0.40	1.5	0.0014	0.00064	Scabarozi *et al.* (2008a)
	4	0.27				
			211 Phases			
Ti_2AlC	300	0.32	4.8	0.0033	0.0011	Scabarozi *et al.* (2008a)
	4	0.067				
Ti_2AlC	300	0.23	2.9	0.0029	0.00067	Wang and Zhou (2002)
	4	0.08 (estimate)				
Ti_2AlC	300	0.39	2.2	0.0023	0.0009	Wang, Hong, and Zhou (2007)
	4	0.18 (estimate)				
$Ti_2AlC_{0.5}N_{0.5}$	300	0.43	1.5	0.0027	0.0012	Scabarozi *et al.* (2008a)
	4	0.167				

Table 5.1 (*Continued*)

Composition	T (K)	ρ ($\mu\Omega$ m)	RRR	α_{TCR} $(K)^{-1}$	$d\rho/dT$ ($\mu\Omega \cdot m\ K^{-1}$)	References
Ti_2AlN-a	300	0.25	8.6	0.0039	0.0010	Scabarozi *et al.* (2008a)
	4	0.029				
Ti_2GeC	300	0.30	3.2	0.0028	0.00084	Scabarozi *et al.* (2008c)
	4	0.094				
Ti_2AlN-b	300	0.343	2.8	0.0026	0.0009	Scabarozi *et al.* (2008a)
	4	0.123				
TiNbAlC	300	0.78	1.3	0.0019	0.00078	Barsoum *et al.* (2002b)
	4	0.6				
Ti_2SC	300	0.52	2.3	0.0027	0.0014	Scabarozi *et al.* (2008b)
	4	0.23				
Ti_2InC	300	0.2	4	0.0030	0.0006	Barsoum *et al.* (2002a)
	4	0.05				
TiHfInC	300	0.27	3.4	0.0026	0.0007	Barsoum *et al.* (2002a)
	4	0.08				
Cr_2GeC	300	0.72	11.2	0.0027	0.0027	Barsoum *et al.* (2011)
	4	0.064				
Cr_2AlC	300	0.74	4.9	0.0036	0.0027	Hettinger *et al.* (2005)
	4	0.15				
Cr_2AlC	300	0.6	2.0	0.0024	0.0014	Ying *et al.* (2011)
	4	0.3 (estimate)				
Cr_2AlC	300	0.625	—	—	—	Zhou, Mei, and Zhu (2009)
Cr_2AlC	300	0.72	2.2	0.0028	0.0020	Tian *et al.* (2006)
	4	0.32				
V_2AlC	300	0.25	6.4	0.004	0.0010	Hettinger *et al.* (2005)
	4	0.04				
V_2AsC	300	1.3	16.5	0.004	0.0053	Lofland *et al.* (2006)
	4	0.08				
V_2GeC	300	0.36	22.5	0.004	0.0014	(see Figure 5.2a) Scabarozi *et al.* (unpublished)
	4	0.016				
Hf_2InC	300	0.19	8.3	0.0033	0.00063	Barsoum *et al.* (2002a)
	4	0.023				
Hf_2InC	300	0.55	2.75	0.0023	0.0013	Lofland *et al.* (2006)
	4	0.2				
Ta_2AlC	273	0.23	8.8	0.0042	0.00097	Hu *et al.* (2008a)
	4	0.026				
Nb_2AlC	300	0.29	2.6	0.0024	0.0007	Hettinger *et al.* (2005)
	4	0.14				
Nb_2SnC	300	0.58	1.93	0.0018	0.001	Lofland *et al.* (2006)
	4	0.30				
Nb_2AsC	300	1.3	18.6	—	0.0052	Lofland *et al.* (2006)
	4	0.07				

of resistance with temperature are relatively mild compared to other metals. For the most part, $d\rho/dT$ for the Ti-containing MAX phases hover around ≈ 0.001. At ≈ 0.02 (Berlincourt, 1959; Collings, 1974), the corresponding value for Ti is 20 times larger. This large difference reflects the strong bonds in the MAX phases relative to those in Ti, despite the fact that Ti is a metal that is prized for its high specific stiffness resulting from its relatively strong bonds.

As noted above, the MAX phases are compensated conductors in that both holes *and* electrons are mobile charge carriers. A two-band conduction model is thus required to understand their electronic transport. In the low magnetic field (B) limit of the two-band model, the following relationships apply:

$$\sigma = \frac{1}{\rho} = e\left(n\mu_n + p\mu_p\right) \tag{5.2}$$

$$\frac{\Delta\rho_B}{\rho_o} = \alpha B^2 = \frac{np\mu_n\mu_p\left(\mu_n + \mu_p\right)^2}{\left(n\mu_n + p\mu_p\right)^2} B^2 \tag{5.3}$$

$$R_H = \frac{\left(\mu_p^2 p - \mu_n^2 n\right)}{e\left(\mu_p p + \mu_n n\right)^2} \tag{5.4}$$

where n and p are the concentration of electrons and holes, and μ_n and μ_p are their mobilities, respectively. σ is the conductivity and R_H is the Hall coefficient. The MR is defined as $\Delta\rho_B/\rho_0 = (\rho_B - \rho_0)/\rho_0$, where ρ_B is the resistivity in the presence of magnetic field B, ρ_0 is the resistivity in the absence of a field and α is the MR coefficient. The importance and usefulness of these relationships will be become clear shortly. Table 5.2 summarizes the electric transport parameters of select MAX phases obtained from Equations 5.2 to 5.4.

5.2.2
Hall Coefficients

To start understanding a solid's electronic transport properties, the charge carrier densities and their mobilities need to be known. For most solids, R_H is used to determine the concentration and sign of the majority charge carriers. Once known, the mobility is simply determined from ρ. The MAX phases, however, are unlike many other metal-like conductors in that their R_H values are quite low – in some cases vanishingly small – and a weak function of temperature (Barsoum, Yoo, and El-Raghy, 2000; Finkel *et al.*, 2003, 2004; Hettinger *et al.*, 2005; Yoo, Barsoum, and El-Raghy, 2000). Sometimes even the sign of R_H changes with increasing T (Figure 5.3a).

The Hall voltage V_H of the MAX phases increases linearly with B (inset in Figure 5.3a). The slope of these lines is proportional to R_H. Figure 5.3b compares the R_H values of a number of Al-containing MAX phases as a function of T, from which it is clear that R_H is indeed small. Note that when R_H is close to zero,

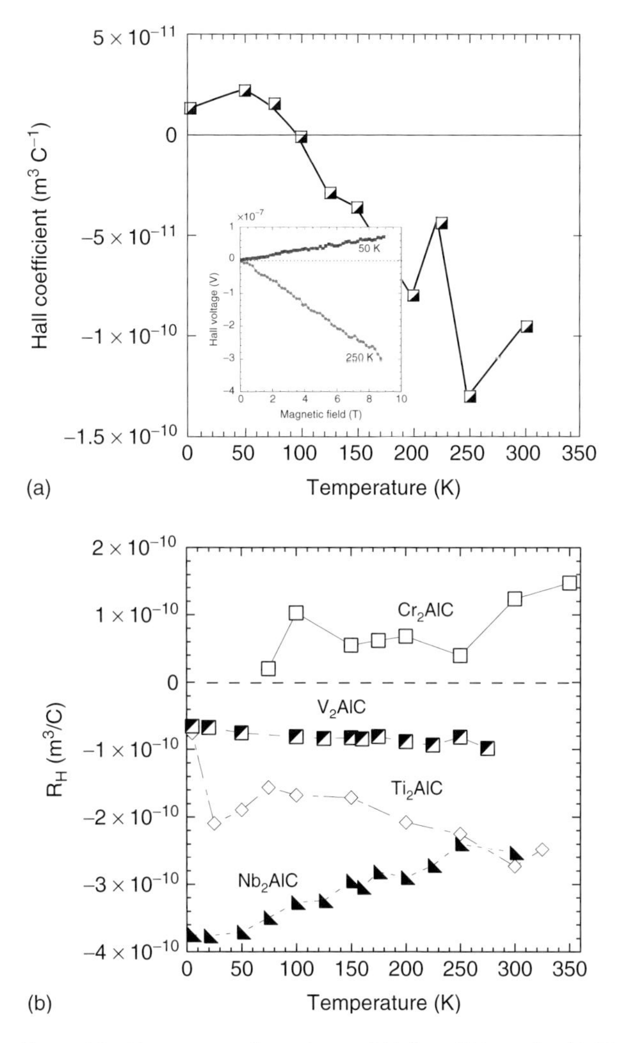

Figure 5.3 Temperature dependence of Hall coefficients for (a) Ti_3AlC_2 (Finkel *et al.*, 2003) and (b) select Al-containing MAX phases (Hettinger *et al.*, 2005). Inset in (a) shows effect of magnetic field, *B*, and temperature on Hall voltage. Note change in Hall voltage sign with increasing temperature.

as is the case here, the signs and magnitudes of R_H are quite sensitive to small variations in the values of the mobilities chosen. For example (taking some typical values from Table 5.2), if $n \approx p \approx 1 \times 10^{27}$ m^{-3}, and if $\mu_p = 0.0074$ m^2 V^{-1}· s^{-1}, then if μ_e changes from 0.0072 to 0.0076 m^2 V^{-1}· s^{-1}, then, according to Eq. (5.4), R_H changes from $+8.5 \times 10^{-11}$ to -8.3×10^{-11} m^3 C^{-1}. Presumably, such small variations in mobilities with T are responsible for the changes in sign of V_H shown in Figure 5.3a. Said otherwise, the change in sign of V_H with increasing T is *strong* evidence that the system is very nearly compensated.

Table 5.2 Summary of electrical transport parameters of select MAX phases calculated from ρ, R_H, and α assuming $n = p$.

Composition	T (K)	ρ (μΩ m)	RRR	R_H ($\times 10^{11}$) $m^3 C^{-1}$	α ($m^4 V^{-2} \cdot s^{-2}$)	$\mu_n = \mu_p = \sqrt{\alpha}$ ($m^2 V^{-1} \cdot s^{-1}$)	$n = p$ ($10^{27} m^{-3}$)	References
				413 Phases				
$Ti_4AlN_{2.9}$	300	2.61	1.1	90 ± 5	0.03×10^{-5}	0.00055	2.2	Finkel *et al.* (2003)
				312 Phases				
Ti_3SiC_2	300	0.22	8.33	38	2.9×10^{-5}	0.0054	2.65	Finkel *et al.* (2001)
	4	0.03		30	206×10^{-5}	0.045	2.30	
Ti_3GeC_2	300	0.28	5.6	−18	15×10^{-5}	0.012	0.90	Finkel *et al.* (2004)
	4	0.05		−2.5	200×10^{-5}	0.045	1.40	
Ti_3AlC_2	300	0.353	1.95	−1.2	3.7×10^{-5}	0.0063	1.40	Scabarozi *et al.* (2008a)
	4	0.181		≈0	15×10^{-5}	0.012	1.41	
	300	0.387	—	−1.2	4×10^{-5}	0.0063	1.27	Finkel *et al.* (2003)
Ti_3AlCN	300	0.40	1.5	17.4	0.65×10^{-5}	0.0025	3.07	Scabarozi *et al.* (2008a)
	4	0.27		33	3.3×10^{-5}	0.0057	2.07	
				211 Phases				
Ti_2AlC	300	0.36	4.9	−27	8.3×10^{-5}	0.009	0.95	Scabarozi *et al.* (2008a)
	4	0.073		−8	0.002	0.045	0.95	
	300	0.36	4.8	−28	20×10^{-5}	0.014	0.65	Hettinger *et al.* (2005)

$Ti_2Al(C_{0.5}N_{0.5})$	300	0.36	2.86	45.6	3.5×10^{-5}	0.0059	1.47	Scabarozi *et al.* (2008c)
	4	0.126		60	22×10^{-5}	0.015	1.67	
Ti_2AlN-a	300	0.25	8.5	−3.9	17×10^{-5}	0.013	0.95	
	4	0.029		6.1	687×10^{-5}	0.083	1.30	
Ti_2AlN-b	300	0.343	2.8	−7	4.8×10^{-5}	0.0069	1.32	
	4	0.123		16	20×10^{-5}	0.014	1.80	
Ti_2SC	300	0.52	2.3	−160	23×10^{-5}	0.015	0.39	Scabarozi *et al.* (2008a)
	4	0.23		−149	120×10^{-5}	0.035	0.39	
Ti_2GeC	300	0.30	3.2	27	5×10^{-5}	0.0071	1.47	Scabarozi *et al.* (2008b)
	4	0.094		16	73×10^{-5}	0.027	1.23	
V_2AlC	300	0.25	6.4	−10	2×10^{-5}	0.0045	2.80	Hettinger *et al.* (2005)
	4	0.04		−6.5	37×10^{-5}	0.019	4.05	
V_2AsC	300	0.64	21	—	—	0.0026	1.87	Lofland *et al.* (2006)
	4	0.03		—	—	—	—	
Cr_2GeC	300	0.72	11.2	−24.3	2×10^{-5}	0.0045	0.97	Barsoum *et al.* (2011)
	4	0.064		0	50×10^{-5}	0.022	2.18	
Cr_2AlC	300	0.74	4.9	15	0.73×10^{-5}	0.0027	1.56	Hettinger *et al.* (2005)
	4	0.15		≈0	5×10^{-5}	0.007	2.95	
Nb_2AlC	300	0.39	2.8	−25	0.9×10^{-5}	0.003	2.67	Hettinger *et al.*, 2005)
	4	0.14		−37	8.9×10^{-5}	0.0094	2.37	
Nb_2AsC	300	1.3	18.6	—	—	0.004	0.60	Lofland *et al.*, (2006)
	4	0.07		—	—	—	—	
Nb_2SnC	300	0.577	1.9	—	—	0.003	1.8	Lofland *et al.* 2006)
	4	0.3		—	—	—	—	
Hf_2InC	300	0.55	2.75	—	—	0.008	1.42	Lofland *et al.* (2006)
	4	0.2		—	—	—	—	

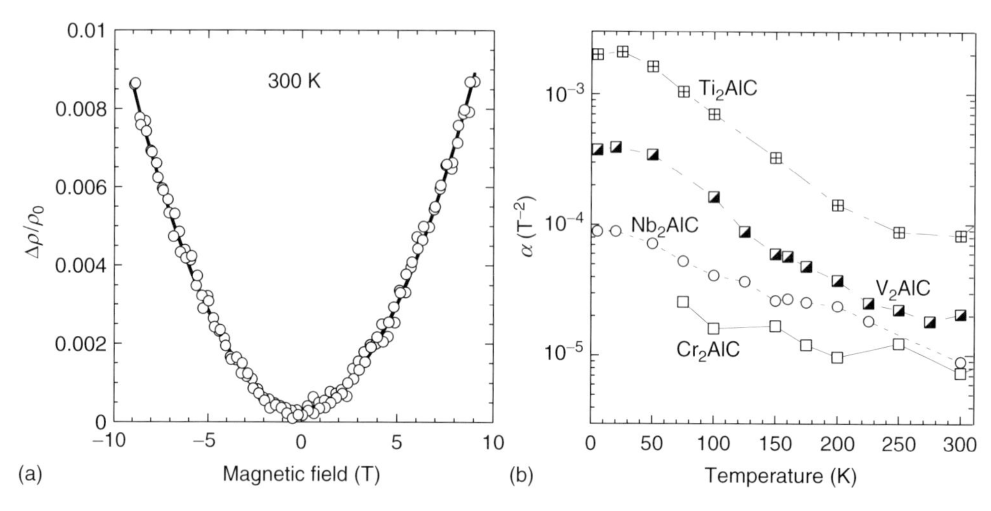

Figure 5.4 (a) Typical magnetic field dependence of resistivity of a Ti_3AlC_2 sample from which the magnetoresistance and Hall coefficients are calculated (Finkel *et al.*, 2003). (b) Semi-logarithmic plot of α as a function of temperature for select M_2AlC phase (Hettinger *et al.*, 2005).

5.2.3
Magnetoresistance

The MAX phases' MRs are positive, parabolic, and nonsaturating (Figure 5.4a). As discussed below, the MR is essentially a measure of the mobility of the electronic carriers, and it is thus not very surprising that it is a strong function of the quality of the crystal as, for example, measured by the RRR. Furthermore, since the mobilities of the electronic carriers are affected by phonon scattering, the MR typically decreases with increasing temperatures. When α is plotted as a function of temperature for a number of Al-containing MAX phases (Figure 5.4b), as expected it decreases with increasing temperature, presumably as a result of the scattering of electrons by phonons.

Like R_H, the Seebeck coefficient Θ of some of the MAX phases can also be vanishingly small over a wide temperature range (Section 5.3).

5.2.4
Compensated Conduction

The combination of small Hall and Seebeck coefficients (see below) and the linearity of V_H with B (inset Figure 5.3a), together with the parabolic, nonsaturating MR (Figure 5.4a), strongly suggest that many of the MAX phases are compensated conductors and thus Eqs. (5.2–5.4) apply. In those relationships, there are four unknowns: n, p, μ_n, and μ_p. Given the small R_H and Seebeck coefficient values, and the nonsaturating and parabolic MR, it is reasonable to assume either $n \approx p$

or $\mu_n = \mu_p$. With either of these assumptions, it is possible to solve for all four unknowns (Scabarozi *et al.*, 2008c).

If one assumes $n = p$, then Eq. (5.2) simplifies to $\mu_n\mu_p = \alpha$. If as a first approximation one further assumes that $\mu_n = \mu_p$, the results shown in Table 5.2 are obtained. Note, however, that assuming $\mu_n = \mu_p$, implies $R_H = 0$. For the most part, it is not difficult to adjust μ_n and μ_p to yield the correct values and sign of R_H. If $\mu_n > \mu_p$, R_H is negative and vice versa. Little is gained by the exercise however, since for the most part $\mu_n \approx \mu_p$. Even in the worst case scenario where R_H is large, such as for Ti_2SC, an exact calculation yields $\mu_n = 0.017\ m^2\ V^{-1}\cdot s^{-1}$ and $\mu_p = 0.014\ m^2\ V^{-1}\cdot s^{-1}$ instead of the value listed in Table 5.2, that is, $0.015\ m^2\ V^{-1}\cdot s^{-1}$. The corrections for all other phases are even smaller.

From the results shown in Table 5.2, it is thus clear that the MAX phases are, indeed, compensated conductors with $n \approx p$ and $\mu_n \approx \mu_p$ and that a two-band model is needed to explain their electronic transport.

Based on the results shown in Table 5.2, the following is apparent.

1) Most of the room-temperature (RT) resistivities fall in the relatively narrow range of 0.2–0.7 µΩ·m (Barsoum, 2000). One notable exception, discussed below, is Ti_4AlN_3.
2) For the most part, $n \approx p$ and $\mu_n \approx \mu_p$. The densities of electronic carriers fall in the relatively narrow – for carrier concentrations – range of $0.3–3 \times 10^{27}\ m^{-3}$. Further, n and p are weak functions of temperature.
3) At 4 K, the less defective samples, as measured by the RRR, have higher mobilities (see Figure 5.5). The 4 K mobilities are also inversely proportional to the density of states (DOS) at the Fermi level, $N(E_F)$. This important point is discussed in more detail below.

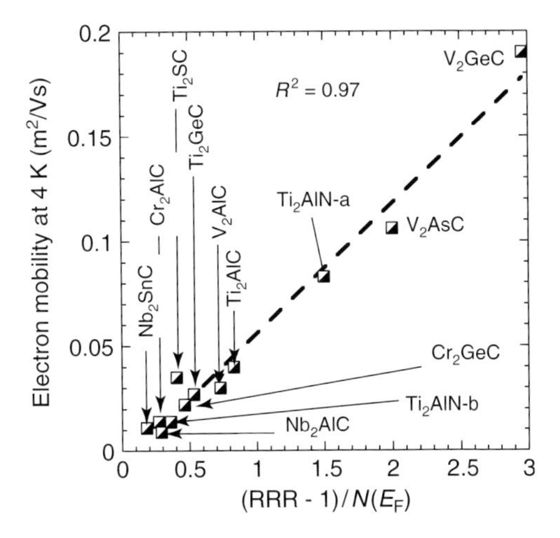

Figure 5.5 Functional dependence of electronic charge mobilities at 4 K on RRR and measured $N(E_F)$ for select 211 phases.

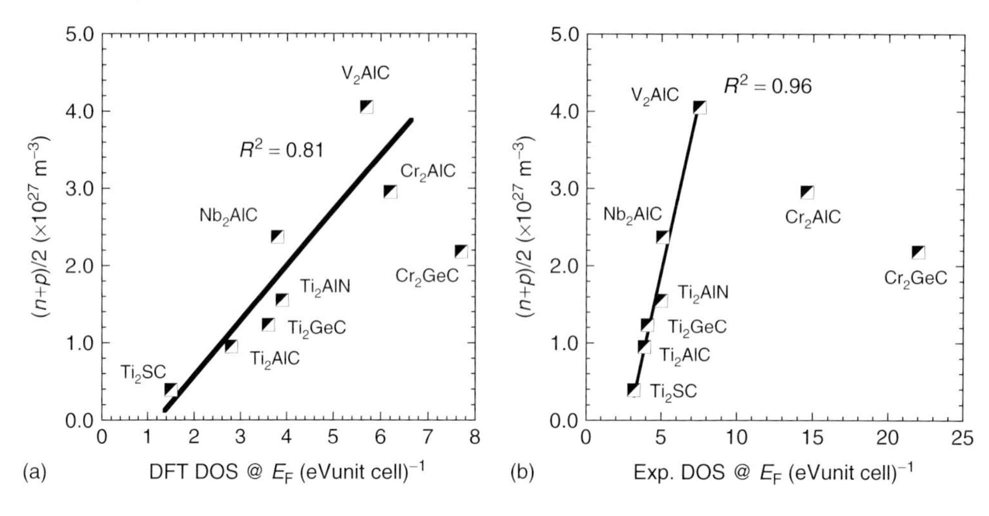

Figure 5.6 Dependence of the average of n and p at 4 K (Table 5.2) on $N(E_F)$. (a) Calculated using DFT. (b) Measured (Table 5.2). In (a), the Cr_2GeC datum point was not included in the least squares which resulted in line shown. In (b), the data points for Cr_2GeC and Cr_2AlC were not included in the least squares fit shown by the line. The R^2 values are indicated on the figures.

4) The room temperature μs fall in the range of 0.008 ± 0.004 m^2 V$^{-1}\cdot$s^{-1}. For reference, μ_n and μ_p in high-quality single-crystalline Si at room temperature are 0.135 and 0.045 m^2 V$^{-1}\cdot$s^{-1}. It follows that μ_n and μ_p in polycrystalline, not particularly defect-free, MAX phases are about one order of magnitude lower than those in intrinsic Si.
5) When the average of n and p listed in Table 5.2 is plotted versus $N(E_F)$ the correlation is weak when either the theoretical (Figure 5.6a) or experimental (Figure 5.6b) $N(E_F)$ values are used. However, if in Figure 5.6a the datum point for Cr_2GeC is excluded, least squares fit yields the line shown with an R^2 value of 0.81. Similarly, if the data points for Cr_2GeC and Cr_2AlC are excluded from the least squares fit in Figure 5.6b, the line shown on the figure can be drawn with an R^2 value of 0.96. Comparing the outliers in Figure 4.18 with those in Figure 5.6b, it is reasonable to assume that the high electron–phonon coupling factor λ_{ep} for Cr_2GeC and Cr_2AlC is maybe, at least partially, responsible for their anomalous nature. This comment notwithstanding, more work is needed in this area. What the results shown in Figure 5.6 confirm, however, is once again the anomalous nature of Cr_2AlC and Cr_2GeC.

To better understand some of the results shown in Figures 5.1–5.6 and Table 5.2, it is important to realize that because the electronic properties of the MAX phases are dominated by the d–d M orbitals (Drulis *et al.*, 2006; Hug, Jaouen, and Barsoum, 2005; Lofland *et al.*, 2004; Medvedeva *et al.*, 1998), it follows that their behavior should be similar to that of their respective transition metals M. It is thus useful to briefly summarize what is known about the latter. The conductivities of transition metals are inversely proportional to $N(E_F)$ (Ashcroft and Mermin, 1976; Mott and Jones, 1936). The RRR is also important. Accordingly, the electron mobilities at 4 K, that is, μ_{4k}, should be *inversely* proportional to $N(E_F)$ and directly proportional to (RRR − 1). That such a correlation exists is shown convincingly in Figure 5.5.

The ternary Ti_4AlN_3 is somewhat unique in that it is nonstoichiometric – its actual chemistry is $Ti_4AlN_{2.9}$ – and also behaves more as a semi-metal than a metal (Figure 5.1a) (Finkel *et al.*, 2003). Consistent with these results are *ab initio* calculations that predict that, if stoichiometric, the Fermi level of this compound falls in a small gap (Holm *et al.*, 2002). Interestingly, when the resistivities of two nominally identical samples were measured, a significant difference between their RT resistivities was found (Figure 5.1). These variations were attributed to slight variations in stoichiometry (Finkel *et al.*, 2003). The high defect concentration also leads to significantly reduced mobilities (Table 5.2). These comments notwithstanding, more work is needed to better understand electronic transport in Ti_4AlN_3.

The same trend can be seen in Ti_2AlN (Table 5.2). The 4 K mobility of sample "a" is approximately seven times higher than that of a nominally identical sample "b." The RRR of the former is approximately six times that of the latter. The difference was ascribed to the presence of electron scattering defects, most probably vacancies (Scabarozi *et al.*, 2008a). A comparison of the n and p values of the two Ti_2AlN samples is instructive for another reason. If one makes the plausible assumption that n and p should be identical in the two samples, it follows that the accuracy of our technique to calculate n and p is of the order of 50%.

5.2.5 Transport Anisotropy

Given the layered nature of the MAX phases, it is not surprising that a number of theoretical papers have predicted strong anisotropic electrical conductivity. Conductivity along the basal planes is typically predicted to be higher than that normal to them. This comes about because the conductivity is assumed to be dominated by the d–d orbitals of the M elements. In general, the experimental results do not support this notion. The few experimentally measured anisotropies show a factor of 2, at most, in the conductivities normal and parallel to the basal planes.

To shed more light on this issue, Scabarozi *et al.* (2008c) measured the transport properties of bulk Ti_2GeC and two highly oriented epitaxial thin films, one of which – film A – was of exceptionally high quality. The transport properties, assuming $n \approx p$, are listed in Table 5.3. Despite the very different natures of the films, as compared to the bulk sample, within the two-band framework, both films and the bulk sample yielded $n \approx p \approx 1.6 \times 10^{27}$ m^{-3}. This result is important since it suggests that our methodology of estimating the values of n and p is robust and applicable to various morphologies.

The MRs of film B and the bulk sample (not shown) were positive, quadratic, and nonsaturating. The 5 K α value for film A is one of the highest reported to date

Table 5.3 Summary of electronic parameters for Ti_2GeC bulk and thin films (Scabarozi *et al.* 2008c).

Sample	ρ_{RT} ($\mu\Omega$ m)	$d\rho/dT$ ($\mu\Omega\cdot$m K^{-1})	RRR	$n \approx p$ (10^{27} m^{-3})	μ_n (5 K) (m^2 V^{-1} s^{-1})	μ_p (5 K) (m^2 V^{-1} s^{-1})
Ti_2GeC bulk[a]	0.30	0.00084	3.2	1.6 ± 0.3	0.027	0.027
Ti_2GeC-A	0.24	0.00097	25.3	3.7 ± 0.5[b]	0.088[b]	0.090[b]
	—	—	—	1.6 ± 0.3[a]	0.100[a]	0.098[a]
Ti_2GeC-B[b]	0.27	0.00100	8.6	1.6 ± 0.3	0.055	0.055

[a] Assuming two bands (see text).
[b] Assuming three bands, electron-like with a carrier density of $4 \pm 2 \times 10^{25}$ m^{-3} and a mobility at 5 K of 0.5 m^2 V^{-1} s^{-1}.
Note that the averaging in carrier density is over the 5–300 K temperature range.

for a MAX phase. It is important to note that, at lower temperatures (<150 K) and high magnetic fields, the MR of film A (not shown) was *not* quadratic. In other words, for sample A, at $T < 150$ K, the conditions for the low B limit were not met for the full range of B used; that is, ρ_{xx} was not quadratic and ρ_{xy} was not linear in B. It was thus necessary to use the general magneto-transport expressions:

$$\sigma_{xx} = \frac{\rho_{xx}}{\rho_{xx}^2 + \rho_{xy}^2} = e\sum \frac{n_i \mu_i}{1 + (\mu_i B)^2} \tag{5.5}$$

$$\sigma_{xy} = \frac{\rho_{xy}}{\rho_{xx}^2 + \rho_{xy}^2} = e\sum \frac{\text{sgn}(n_i)\, B\mu_i^2}{1 + (\mu_i B)^2} \tag{5.6}$$

where $\rho_{xy} = R_H$ B. The inversion of the resistivity tensor to yield the conductivities σ_{xx} and σ_{xy} is allowed here because the measurements were taken in the *ab*-plane of the sample, which ought to be isotropic. From fits to the results to Eqs. (5.5) and (5.6) – at 5 K – it was found that at least three conduction bands were required to achieve reasonable agreement to both σ_{xx} and σ_{xy} simultaneously. From a three-band fit, a feature observed (not shown) at $\approx 0.5T$ in σ_{xy} can be attributed to an electron-like band with a high mobility ($\mu_1 = 0.5 \pm 0.1$ m^2 $V^{-1}\cdot s^{-1}$) yet small carrier concentration ($n_1 = 4 \pm 2 \times 10^{25}$ m^{-3}), and as a result that band actually has minimal contribution to the total conductivity. It follows that at 5 K, the majority of the conductivity occurs in two bands, one electron-like ($\mu_2 = 0.09 \pm 0.01$ m^2 $V^{-1}\cdot s^{-1}$, $n_2 = 3.7 \pm 0.5 \times 10^{27}$ m^{-3}) and one hole-like ($\mu_3 = 0.1 \pm 0.01$ m^2 $V^{-1}\cdot s^{-1}$, $n_3 = 3.7 \pm 0.5 \times 10^{27}$ m^{-3}).

Note that these results do not conflict with those from the other two Ti_2GeC samples. The use of more than two bands to describe the data is unjustified as long as the data can be described by the low-field approximations (R_H linear and α quadratic in B), which was the case for the bulk and film B sample. As noted above, using only the low-B data for sample A, the two-band model yielded $n \approx p \approx 1.6 \times 10^{27}$ m^{-3} (Table 5.3), in excellent accord with the other Ti_2GeC samples (Table 5.3).

However, had the mobility for sample A actually been as large as predicted by the two-band model, the change in conductivity with B (Eq. (5.5)) would have been much greater than that observed. The difference between the results obtained using Eqs. (5.5) and (5.6) and the low-field results (i.e., Eqs. (5.1) and (5.3)) is due to the large mobility of band 1. In the low-field limit

$$\sigma_{xx} \approx e\sum n_i\mu_i[1-(\mu_i B)^2] = \sigma_{xx}(B=0) - e\sum n_i\mu^3 B^2 \quad \text{and}$$

$$\sigma_{xy} = e\sum \text{sgn}\,(n_i)\, n_i B\mu_i^2$$

That is, the magnetoconductance and Hall conductivity are proportional to μ^3 and μ^2, respectively, and the mobilities have a significantly greater impact on the magnetotransport as compared to the carrier densities. Nonetheless, the results on sample A suggest that the assumption that $n \approx p$ is robust and that using only the low-field results provides, at worst, approximate values. Note that it is unlikely that there are additional significant conduction bands in Ti_2GeC since σ_{xx} at 9 T decreased about 40% from the zero-field value (Scabarozi *et al.*, 2008c).

The curves for the intrinsic resistivity values ρ_{in} (not shown) are quite similar for the bulk and the epitaxial thin-film samples, suggesting that the resistivity is nearly isotropic. To estimate the anisotropy, an effective medium approximation for the intrinsic conductivity ($= 1/\rho_{in}$) of the bulk sample, assuming spherical grains (Genchev, 1993) was used. This yields

$$\frac{2\left(\sigma-\sigma_a\right)}{\frac{2}{3}\sigma+\frac{1}{3}\sigma_a} + \frac{\left(\sigma-\sigma_c\right)}{\frac{2}{3}\sigma+\frac{1}{3}\sigma_c} = 0 \tag{5.7}$$

where σ_a and σ_c represent the intrinsic in-plane and out-of-plane conductivities, respectively, and σ is the average conductivity of the bulk sample. Using σ_a values from the films yields $\sigma_c/\sigma_a = 1.6$. Thus, the present results suggest weak anisotropy in the intrinsic mobilities, $\mu_{in} \approx e\tau/m^*$, where τ is the scattering time and m^* is the effective mass of the charge carriers.

The experimental observations that there is conduction by both holes and electrons and that the electronic anisotropy is weak in Ti_2GeC are in conflict with its published band structure (Zhou *et al.*, 2000). The calculations indicate that, for all bands crossing the Fermi energy E_F, with wave vector **k** going from the Γ point to either the K or M point (i.e., in the *ab* plane), $1/m^* \propto \left(\partial^2\varepsilon/\partial k^2\right)_{\varepsilon=E_F} < 0$, where ε is the energy, indicating that conduction is solely by holes. Also, the calculation predicts that along the *c* direction, that is, **k** going from the Γ point to the A point, there are no bands crossing E_F. This in turn suggests a large anisotropy in the conductivity, in contrast to the observed weak anisotropy. In other words, the experimental results indicate a larger out-of-plane conductivity that theory predicts.

In another more recent example, Hu *et al.* (2011) fabricated highly oriented Nb_4AlC_3 samples and measured their RT resistivities along and normal to the basal planes to be, respectively, 1.2 and 2 μΩ m. In other words, here again the anisotropy is weak. In the same paper, the authors reported that the thermal conductivity was higher along the basal planes (21.2 W m^{-1} K^{-1}) than normal to them (14.1 W m^{-1} K^{-1}). Intriguingly, at 0.66, the ratio of the thermal conductivities is slightly higher than the ratio of 0.6 for the electrical conductivities. Whether this means that the phonon conductivities are anisotropic as well, requires further work.

5.2.6
Solid Solutions

Probably one of the more direct methods to increase the resistivities of the MAX phases is to create solid solutions that result in reduced mobilities due to solid solution scattering. However, since $N(E_F)$ is dominated at the d orbitals of the M element, substitutions on the various sites do not impact the resistivity equally. Substitutions on the A sites appear to have little effect on ρ (Finkel *et al.*, 2004). For example, substituting 25% of the Al atoms with Si increased the resistivity from $\approx$0.35 to $\approx$0.37 μΩ·m (Zhou *et al.*, 2006).

Substitutions on the X sites also have a small effect, that is only observed if the concentration of defects – presumably vacancies and/or displaced atoms – in the end members are low and thus do not mask the solid-solution effect.

Consistent with the fact that the Fermi level is dominated by the d–d orbitals of the M atoms, substitution on these can have a more dramatic effect on increasing the resistivities above those of the end members as shown in the Ti–Nb–Al–C system referred to above (Barsoum *et al.*, 2002b). Another good example is shown in Figure 5.7a for a series of solid-solution thin films in the Cr–V–Ge–C system. The residual resistivity peaks around 50 at% (Figure 5.7b) as one would expect from solid-solution scattering (Scabarozi *et al.*, 2012).

5.2.7
Vacancies

Recently Dubois and coworkers (private communication) measured the resistivities of two solid-solution series, namely, $Ti_2AlC_{1-x}N_x$ and $Ti_2Al(C_{1-x}N_x)_y$, with

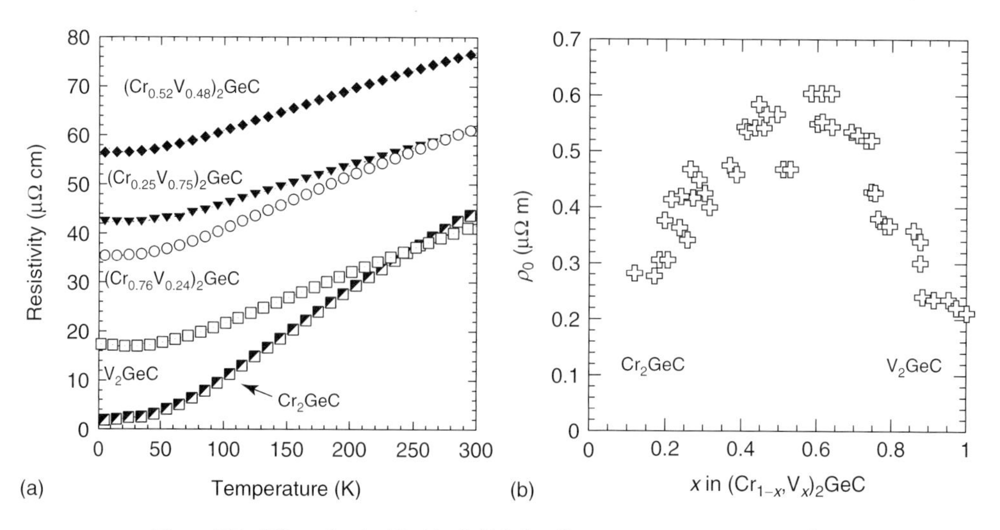

Figure 5.7 Effect of x in $(Cr_x,V_{1-x})_2AlC$ thin films on (a) resistivity as a function of temperature and (b) residual resistivity (Scabarozi *et al.*, 2012).

$y = 0.8$. Not surprisingly, they concluded that the resistivities were a function of both x and y. The effect of y, however, was more important in increasing the resistivities. This result is in total agreement with the fact that vacancies in the MX compounds are strong scatterers of charge carriers. Interestingly, they also showed that, while $d\rho/dT$ was a function of y, it was a weaker function of x. In other words, vacancies on the X-lattice sites appear to impact the electron–phonon coupling factor λ more severely than disorder on the X sites.

5.2.8
Effect of Temperature

In the remainder of this section, the focus shifts to better understanding how temperature affects ρ. One useful method for understanding the RT dc resistivity of metals, proposed by Chakraborty, Pickett, and Allen (1976), makes use of the fact that the electron–phonon interaction determines both conventional superconductivity and normal-state resistivity at high temperatures. In that formalism, ρ is given by

$$\rho = \frac{1}{\varepsilon_0 \tau \omega_R^2} \tag{5.8}$$

where ω_R is the plasma frequency due to conduction electrons alone (see below). The mean free time between collisions, τ, is given by Chakraborty, Pickett, and Allen (1976) as

$$\frac{1}{\tau} = \frac{2\pi\lambda k_B T}{\hbar}\left(1 - \frac{\hbar^2\langle\omega^2\rangle}{12k_B^2T^2}\right) \tag{5.9}$$

where λ is the electron–phonon coupling constant discussed in Chapter 4 and $\langle\omega^2\rangle$ is the renormalized phonon frequency, squared and averaged according to the prescriptions given by McMillan (1968). At 300 K, the second term in brackets in Eq. (5.9) can be safely ignored. Combining Eqs. (5.8, 5.9) and (5.14), it is straight forward to show that

$$\frac{d\rho}{dT} \approx \left(\frac{2\pi k_B}{\hbar}\frac{m_e}{e^2}\right)\frac{\lambda}{(n+p)} \tag{5.10}$$

where m_e is electron rest mass and $n+p$ the number of mobile carriers given in Table 5.2. It follows that a plot of $d\rho/dT$ versus $\lambda/(n+p)$ should yield a straight line, as indeed observed (Figure 5.8). Note that for this plot, λ was equated to λ_{ep} (Eq. (4.11)). In other words, λ_{ep} was assumed to be 1.8 for Cr_2AlC, and Cr_2GeC, 0.56 for V_2AsC, and 0.4 for the rest. Least squares fit of the results shown in Figure 5.8 yield an R^2 value of 0.8, which is an excellent correlation considering all the assumptions made in the model, experimental uncertainties, and, most importantly, the wide ranges of compositions explored. If the average of the three measurements for Cr_2AlC is used instead of the three points shown in Figure 5.8, the R^2 value jumps to 0.9. According to Eq. (5.10), the theoretical slope of a plot of $d\rho/dT$ versus λ/n should be $\approx 2.9 \times 10^{19}$ Ω m^{-2}. At 2.3×10^{18} Ω m^{-2}, the slope of the dashed line shown in Figure 5.8 is about an order of magnitude smaller. The reason for the

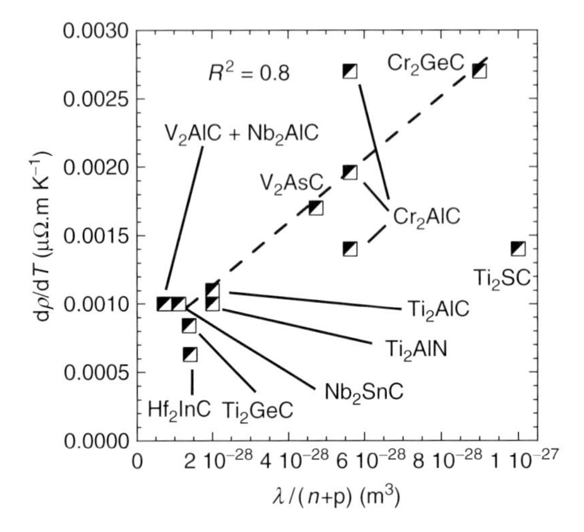

Figure 5.8 Dependence of $d\rho/dT$ (Table 5.1) on $\lambda/(n+p)$. For this plot, λ was equated to λ_{ep} (Eq. (4.11)). For Cr_2AlC and Cr_2GeC, λ_{ep} was assumed to be 1.8; for V_2AsC it was assumed to be 0.56, for the rest it was assumed to be 0.4. Note that $d\rho/dT$ for Cr_2AlC was measured three times as shown. If the average of these three values is used instead of the three points, the R^2 value jumps to 0.9.

discrepancy is unclear at this time. Note that Ti_2SC is an outlier on this figure and was not included in the calculation of the slope.

This discrepancy notwithstanding, the results shown in Figure 5.8 are important for several reasons, amongst which are the following: (i) The assumption that the effective mass of the electrons is the same as the rest mass is a good one. Note that to reconcile the theoretical and experimental slopes one would have to assume that the electron's effective mass to be *less* than m_e in Eq. (5.10). (ii) The λ_{ep} values determined from low-temperature c_p measurements are relevant to phonon scattering at higher temperatures. (iii) For most of the MAX phases with a λ about 0.4, at 300 K, τ – given by Eq. (5.9) – is $\approx 1 \times 10^{-14}$ s, which is a reasonable number indeed. The τ values for Cr_2AlC and Cr_2GeC, on the other hand, are closer to 4.4×10^{-14} s. (iv) λ and n are the important factors in determining $d\rho/dT$. Decoupling the electrons and phonons, and/or increasing the number of charge carriers, should depress $d\rho/dT$ and vice versa.

5.3 Seebeck Coefficients, Θ

As noted above, because of the compensated nature of their conductivity, many of the MAX phases have small Seebeck coefficients Θ. When the Θ values are plotted versus T (Figure 5.9a,b) for select MAX phases, it is clear that, for most of them, Θ is small ($<10\ \mu V\ K^{-1}$) and more or less temperature independent.

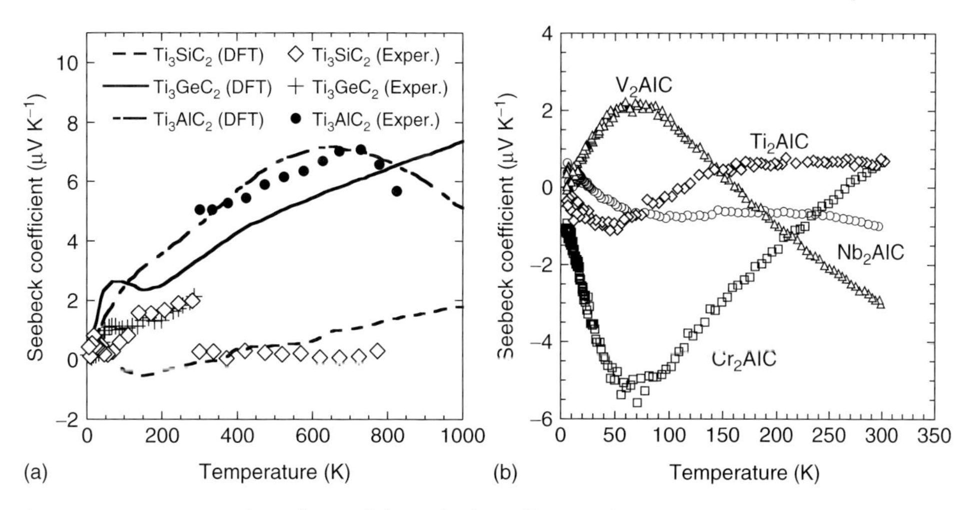

Figure 5.9 Temperature dependence of the Seebeck coefficients of (a) Ti_3SiC_2, Ti_3AlC_2, and Ti_3GeC_2 (Barsoum *et al.*, 2000; Finkel *et al.*, 2004). The theoretical predictions of Chaput *et al.* (2007) are plotted as solid and dashed lines. (b) Ti_2AlC, V_2AlC, Cr_2AlC, and Nb_2AlC (Hettinger *et al.*, 2005).

Figure 5.9a plots Θ for Ti_3SiC_2, Ti_3GeC_2 and Ti_3AlC_2, over an extended temperature range. The Θ value for Ti_3SiC_2 was measured twice: the first time in the 300–1000 K temperature range (Yoo, Barsoum, and El-Raghy, 2000), and the second time in the 0–300 K range (Finkel *et al.*, 2004). In both cases, Θ was quite small and temperature independent. The same is true of other Al-containing MAX phases (Figure 5.9b) and $Ti_3(Si_x,Ge_{1-x})C_2$ solid solutions (not shown) (Finkel *et al.*, 2004).

To explain this unusual behavior, Chaput *et al.*, (2005, 2007) used density functional theory (DFT) to calculate the thermopower of Ti_3SiC_2, Ti_3AlC_2, and Ti_3GeC_2 as a function of temperature. They concluded that the thermopower was negative along the *c*-axis and positive in the basal planes (Figure 5.10). The small experimentally observed value was thus ascribed to compensation between the thermopowers of two nonequivalent crystallographic axes. When the theoretical and experimental results were compared (Figure 5.9a) the agreement was good. To reach these conclusions, Chaput *et al.* plotted the Fermi surfaces of Ti_3SiC_2, Ti_3GeC_2, and Ti_3AlC_2 shown in Figure 5.11a–c, respectively. And while certainly plausible, until quite recently there was no independent experimental verification of this idea. In 2012, Magnuson *et al.*, (2012) showed that the in-plane Seebeck coefficient of Ti_3SiC_2 measured on epitaxial thin films indeed had a positive value of 4–6 $\mu V\ K^{-1}$ as predicted by Chaput *et al.* (Figure 5.10).

In the same paper, Magnuson *et al.* also showed – using a combination of polarized angle-dependent X-ray spectroscopy and DFT – that the DOS of the Ti 3d and C 2p states at E_F in the basal *ab*-plane was about 40% higher than along the *c*-axis. They also confirmed that electron–phonon interactions are important and needed to be taken into account. Positive contribution to the Seebeck coefficient of the element-specific electronic occupations in the basal plane is compensated by 73% enhanced Si 3d electronic states across the laminate plane that give rise to a negative Seebeck coefficient in that direction. Strong phonon vibration modes with 3–4 times higher frequency along the *c*-axis than along the basal *ab*-plane also influence the electronic population and the measured spectra by the asymmetric average displacements of the Si atoms. In other words, this work validated many of the ideas presented in this and previous chapters.

Lastly, we note that the sign of the Seebeck voltage is often used to qualitatively determine the sign of the dominant charge carrier. With this in mind, it may be expected that the Seebeck voltages should roughly reflect the sign and shape of R_H as a function of temperature. On comparing the Hall (Figure 5.3b) and Seebeck coefficients (Figure 5.9b), it is clear there are no obvious correlations between them. Three of the materials have changing signs of the Seebeck voltage with no corresponding changes in R_H, providing additional evidence that these compounds are nearly compensated.

Solids with essentially zero thermopower can, in principle, be used as leads to measure the *absolute* thermopower of other solids. In other words, they could be used as reference materials in thermoelectric measurements.

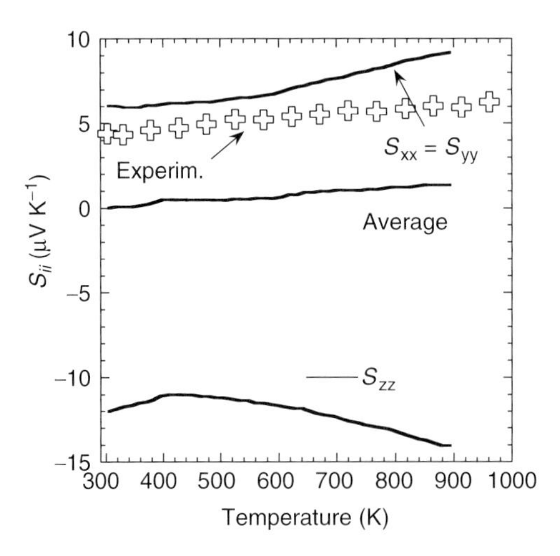

Figure 5.10 Calculated thermoelectric tensor S_{ii}, for Ti_3SiC_2. S_{xx} ($= S_{yy}$) and S_{zz} are the components of the thermoelectric tensor in the basal planes and along [0001], respectively, (Chaput *et al.*, 2005). The average of $S_{xx} = S_{yy}$ and S_{zz} is given by the line labeled "average" that hovers around zero as observed in Figure 5.9a. Data points plotted are measured values for Ti_3SiC_2 along the basal planes (Magnuson *et al.*, 2012). The agreement between them and theory has to be considered good given the weakness of the signal.

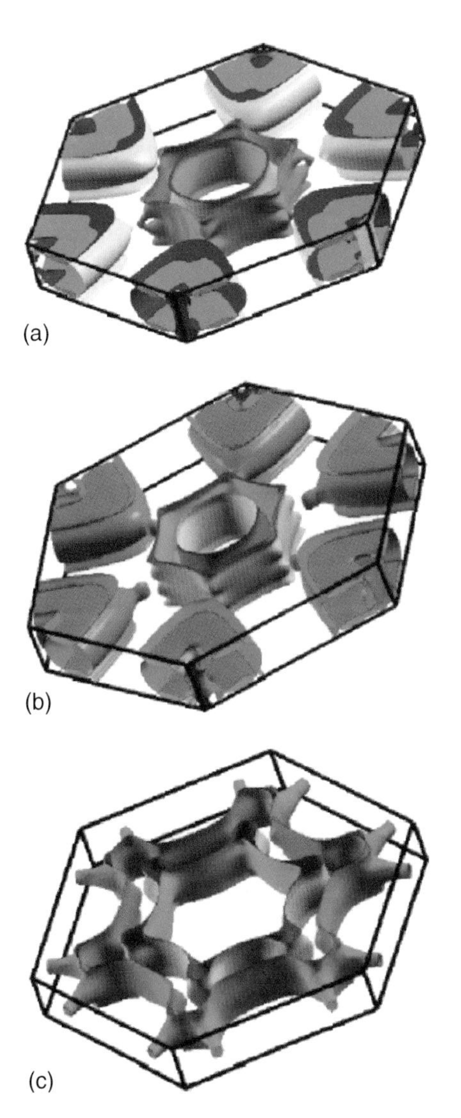

Figure 5.11 Fermi surfaces for (a) Ti_3SiC_2, (b) Ti_3GeC_2, and (c) Ti_3AlC_2 (Chaput *et al.*, 2007).

5.4 Optical Properties

There have been relatively few experimental studies on the optical properties of the MAX phases. In 2008, Li *et al.* reported on the optical response of Ti_3SiC_2 and Ti_4AlN_3 (Li *et al.*, 2008). In the same year, Haddad *et al.*, (2008) reported on the dielectric properties of Ti_2AlC and Ti_2AlN measured by spectroscopic ellipsometry and electron energy loss spectroscopy (EELS). EELS was also used to study the optical response of Ti_3SiC_2 and Ti_3AlC_2 (Hug *et al.*, 2010). More recently, Mendoza-Galván *et al.* reported on the dielectric functions of bulk Ti_2AlN, Ti_2AlC, Nb_2AlC, TiNbAlC, and Ti_3GeC_2 phases using ellipsometry (Mendoza-Galván *et al.*, 2011). In many cases, the experimental response was compared with that predicted from *ab initio* calculations. In the meantime, there has been an increase in the number of theoretical papers dealing with the optical response of the MAX phases (He *et al.*, 2011; Kanoun, Goumri-Said, and Reshak, 2009; Kyriienko and Shelykh, 2011; Li *et al.*, 2009).

Understanding the optical properties is not only important from a basic science point of view but also for a better understanding of the electronic properties. How a material responds to electromagnetic radiation is contained in the complex dielectric function

$$\varepsilon = \varepsilon_1 + i\varepsilon_2 = \left(n_{\text{ref}} + \mathrm{i}\kappa\right)^2 \tag{5.11}$$

where ε_1 and ε_2 are, respectively, the real and imaginary parts of the dielectric function, n_{ref} is the refractive index, and κ is the extinction coefficient or absorption index. Equation (5.11) fully describes the optical properties of a medium at all photon energies $\hbar\omega$, where $\hbar$ is the Plank's constant divided by 2π.

For metal-like conductors such as the MAX phases, the optical response is a function of the delocalized electrons' polarization as well as interband transitions. In the Drude–Lorenz model

$$\varepsilon(\omega) = \varepsilon_\infty - \frac{\omega_p^2}{\left(\omega^2 - \mathrm{i}\omega\gamma_D\right)} + \sum_{j=1} \frac{f_j\omega_{oj}^2}{\omega_{oj}^2 - \omega^2 + \mathrm{i}\Gamma_j\omega} \tag{5.12}$$

and γ_{D} is due to the scattering of electrons which according to the free-electron theory, is the inverse of the conduction electron relaxation time τ. The Lorentz oscillators (third term) describe the interband electronic transitions with energy $\hbar\omega_{oj}$, oscillator strength f_j, and damping or broadening factor Γ_j. Finally, ε_∞ is a background constant >1, due to the contributions of higher energy transitions that are not taken into account by the Lorentz terms.

The Drude term is characterized by the unscreened plasma energy $\hbar\omega_{\text{p}}$ and the damping factor γ_{D}. The plasma frequency ω_{p} is given by

$$\omega_{\text{p}} = \sqrt{\frac{n_{\text{p}}e^2}{\varepsilon_0 m_{\text{e}}}} \tag{5.13}$$

where m_{e}, ε_0, and n_{p} are, respectively, the rest mass of the electron, the permittivity of free space, and the concentration of electrons that *respond* to the optical signal. As discussed below, it is important to emphasize that n_{p} is not a constant but varies with the frequency of the light.

The first question to answer at this stage is: how good is Eq. (5.13). To do so, it is useful to estimate the order of magnitude of ω_{p}, a parameter that is poorly defined in the literature. Here, we define $\omega_{\text{p,val}}$ as the frequency at which *all valence electrons per unit volume*, n_{val}, come into play. In other words, n_{p} is assumed in Eq. (5.13) to be given by n_{val} (Eq. (2.11)), in which the formal number of valence electrons per

Table 5.4 Summary of plasma frequencies of select MAX and MX phases as measured by EELS. Also listed are the values of n determined from transport measurements listed in Table 5.2. (Note that $n < n_{val}$.) Column 4 lists the experimental values $\omega_{p,exp}$ obtained from EELS. Column 5 lists values of $\omega_{p,theo}$ calculated from Eq. (5.13) using n_{val} listed in column 2.

Phase	n_{val} × 10^{28} m^{-3}	n (transport) × 10^{28} m^{-3}	$\hbar\omega_{p,exp}$ (eV)	$\hbar\omega_{p,val}$ (eV)	EELS References for $\hbar\omega_{p,exp}$
			312 Phases		
Ti_3AlC_2	30.2	0.14	20	20.4	Hug *et al.* (2010)
Ti_3SiC_2	33	0.265	21	21.5	Hug *et al.* (2010)
			211 Phases		
Ti_2AlC	27.5	0.095	18.1	19.5	Haddad *et al.* (2008)
Ti_2AlN	30.0	0.095	20	20.5	Haddad *et al.* (2008)
Ti_2GeC	30.3	0.15	20.6	20.5	G. Hug (private communication)
V_2GeC	37.7	—	23	22.8	G. Hug (private communication)
Cr_2AlC	41.9	0.16	21.7	24.0	Mauchamp *et al.* (2012)
Cr_2GeC	43.9	0.1	21.7	24.6	Bugnet (PhD)
			MX phases		
TiC	40.0	—	23.8	23.5	Mirguet/Bugnet
TiN	47.0	—	25.5	25.5	Mirguet/Bugnet
$TiN_{0.8}$	41.7	—	23.0	24.0	Mirguet/Bugnet

atom, z_{av}, is used (Chapter 2). Thus the number of valence electrons for Ti, Zr, and Hf is 4; for V, Nb, and Ta the value is 5; for Al, Ga, and In it is 3; for S it is 6, and so on. Once n_{val} is calculated, it is used in Eq. (5.13) to calculate $\hbar\omega_{p,val}$.

Table 5.4 compares the values of $\hbar\omega_p$ measured by EELS and the $\hbar\omega_{p,val}$ values calculated from Eq. (5.13). When the results are plotted, not only for the MAX phases, but for many other unrelated solids, including the alkali metals (Figure 5.12a), least squares fit yields an $R^2 > 0.96$. This is a crucial result because it implies that (i) if the photon energy is high enough, optical measurements can be used to determine n_{val}, and (ii) the electrons need *not* be free; as long as they are in the valence band, they will respond to the applied field. It follows that the value of $\hbar\omega_{p,val}$ is an excellent anchor, or reference point, without which understanding the relationship between optical phenomena and electrical conductivity is not possible.

It is important to note that the results and conclusions of the aforementioned analysis must be of universal applicability to all metallically conductive solids. The best evidence that this is the case can be seen in Figure 5.12a,

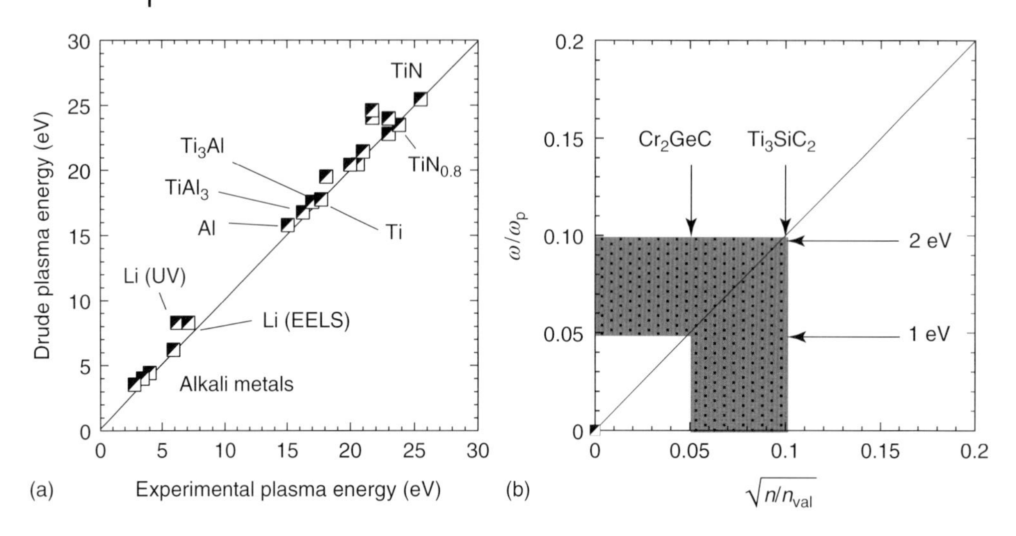

Figure 5.12 (a) Plasma frequency calculated from Eq. (5.13) versus experimental values obtained from EELS measurements (Table 5.4). Also plotted are results for the alkali metals (Born and Wolf, 1964) and other metals and intermetallics. (b) Range of frequencies that have any relationship to transport properties in the MAX phases are of the order of 0.1 of the Drude frequency.

where the theoretical and experimental values of $\hbar\omega_p$ for the alkali metals Li to Cs are plotted. In this case, the experimental frequencies are proportional to the ultraviolet (UV) frequencies at which the alkali metals became *transparent*, that is, strictly from optical measurement. The agreement is very good in this case because in the alkali metals *all* the s valence electrons are activated in the UV range used in the measurements. If a similar experiment were carried out on the MAX phases, only a small fraction of their electrons would be affected.

5.4.1
Relationship of Optical to Transport Properties

In general, ρ is related to the optical parameters by

$$\rho = \frac{\gamma_D}{\varepsilon_0 \omega_R^2} = \frac{m_e}{\tau\left(n+p\right)e^2} \tag{5.14}$$

where ω_R is the plasma frequency due to the conduction electrons solely. Note that using $\omega_{p,val}$ in this equation is incorrect since it would imply that – like in the alkali metals – *all* the valence electrons are conducting, when in fact only a small fraction are (compare columns 2 and 3 in Table 5.4).

In some of the literature on the optical properties of metal-like conductors, an attempt is made to relate plasma frequencies measured in the visible and UV ranges – or, much more frequently fitted – to the transport properties. One strong impetus for this procedure is the aforementioned good agreement obtained for the alkali metals. But as noted above, that agreement had more to do with the coincidence that n_{val} for the alkali metals is significantly smaller than other solids, which results in the plasma frequencies that happen to fall in the UV range (see bottom left in Figure 5.12a).

It is crucial to re-iterate that the n and p values obtained from transport measurements are significantly lower than n_{val} (compare columns 2 and 3 in Table 5.4). This is not too surprising as it is well established that only electrons near E_F are responsible for transport. It follows that, if the values of n and p are introduced into Eq. (5.14), the range of $\hbar\omega_p$ that has any relationship to transport for the MAX phases must fall between 1 and 2 eV (Figure 5.12b). Said otherwise, if the MAX phases' optical properties are determined using radiation with energies that are much higher than 2 eV, it is quite likely that more electrons than those involved in transport are polarized.

5.4.2
Relaxation Times and Mean Free Paths

In calculating the relaxation times τ, instead of using n_{val}, n and p have to be used. In other words, τ is given by

$$\tau = \frac{m_e}{\rho\,(n+p)\,e^2} \tag{5.15}$$

Note that in this case the implicit assumption is made that the relaxation times of the holes and electrons are equal. Referring to Table 5.4, if one assumes $\rho = 0.3\,\mu\Omega\cdot\text{m}$ and $n = p \approx 2 \times 10^{27}\ \text{m}^{-3}$, it follows that $\tau \approx 3 \times 10^{-14}$ s. If n_{val} is used instead, τ values of the order of 10^{-16} s are recovered. Such values are about two orders of magnitude *lower* than those for other metal-like conductors. Note that for a τ value of $\approx 3 \times 10^{-14}$ s, γ_D in Eq. (5.14) should be of the order of 0.02 eV.

Lastly, the mean free path l_m of the charge carriers is given by

$$l_m = v_F \tau \tag{5.16}$$

If one assumes v_F for the MAX phases to be of the order of 2×10^6 m s^{-1} (Table 2.2) and $\tau \approx 3 \times 10^{-14}$ s, then at room temperature l_m is $\approx$600 Å. This value is about an order of magnitude larger than the corresponding values for the relevant transition metals (Sanborn, Allen, and Papaconstantopoulos, 1989), which is not too surprising given the high Debye temperatures of the MAX phases relative to those of the M metals.

The reflectivity spectra of Ti_3SiC_2, Ti_4AlN_3, TiC, and TiN are plotted as a function of energy in Figure 5.13. From these results, it is clear that the reflectivity of the MAX phases are comparable to those of TiC. Surprisingly, the reflectivity of TiN is quite different from the rest; why this is the case is unclear at this time.

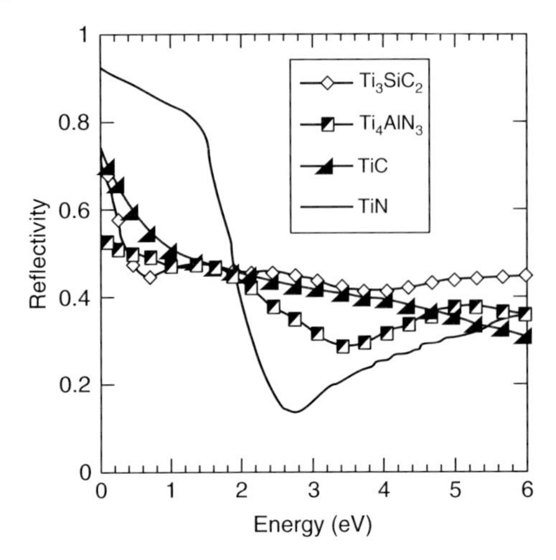

Figure 5.13 Reflectivity spectra of Ti_3SiC_2, Ti_4AlN_3, TiC, and TiN (Li *et al.*, 2008).

5.5 Magnetic Properties

There has not been much work on the magnetic properties of the MAX phases. The little that exists shows them to be Pauli paramagnetic materials (Drulis *et al.*, 2004; Finkel *et al.*, 2001, 2003). Pauli paramagnetic solids are characterized by a magnetization M, which is a linear function of the applied magnetic field intensity H. In that case, the magnetic susceptibility χ is independent of both H and temperature as observed (Figure 5.14).

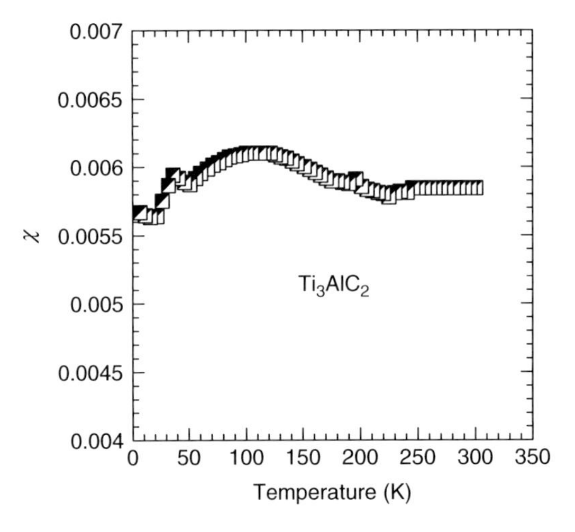

Figure 5.14 Temperature dependence of χ of Ti_3AlC_2 (Finkel *et al.*, 2003). Note χ is small and temperature independent, consistent with Pauli paramagnetism.

For metal-like solids,

$$\chi = \mu_0 \mu_B^2 N(E_F) \tag{5.17}$$

where μ_0 and μ_B are the permeability of free space and the Bohr magneton, respectively. Assuming that nearly all states below E_F are occupied at low temperatures, one can calculate $N(E_F)$ from χ.

The χ of Ti_3SiC_2 was measured twice; the first time the value was 4.1×10^{-6} (Finkel *et al.*, 2001), and the second time it was 4.5×10^{-5} (Drulis *et al.*, 2004). The χ value of Ti_3AlC_2 was measured once (Finkel *et al.*, 2003). From, the results (Figure 5.14), it is obvious that $\chi \approx 1.6 \times 10^{-3}$ and temperature independent. The reasons for the large variations in χ values, in what should have been – based on Eq. (5.17) – quite comparable results is unclear, but most probably reflect the presence of trace amounts of magnetic impurities.

Since the presence of magnetic impurities can only enhance χ, but not reduce it, it is reasonable to assume that the lower χ value is more representative of the pure MAX phases. Thus assuming χ of Ti_3SiC_2 to be 4.1×10^{-6}, from Eq. (5.17) $N(E_F) = 3.8 \times 10^{46}\ J^{-1}\ m^{-3}$, which translates to 0.84 (eV unit cell)$^{-1}$. This value is about 6 times lower than it should be based on the low-temperature c_p measurements (Table 4.4). The reason for the discrepancy is unclear at this time, but the agreement should still be considered good especially since the following simplifications were made: (i) χ contains the diamagnetic component due to the electrons of the atomic core, and (ii) the value of χ can be influenced by electron–electron and electron–ion interactions.

5.6 Superconducting Properties

In 1967, Toth was the first to report on superconductivity in a MAX phase: Mo_2GaC (Toth, 1967). Since then, the following MAX phases have also been claimed to be superconducting at various critical temperatures: Nb_2SC (Sakamaki *et al.*, 1999, 2000), Nb_2SnC (Bortolozo *et al.*, 2006b), Nb_2AsC (Lofland *et al.*, 2006), Ti_2InC (Bortolozo *et al.*, 2007), Nb_2InC (Bortolozo *et al.*, 2006a), and Ti_2InN (Bortolozo *et al.*, 2010). Note that the superconductivity claims made by Bortolozo *et al.* have not been reproduced by others. Moreover, we carefully searched for any evidence of superconductivity in Nb_2SnC and Ti_2InC but found none. We did, however, find a superconducting transition in Nb_2AsC at ≈2 K (Lofland *et al.*, 2006). Sakamaki *et al.* reported a T_c of 5 K in Nb_2SC.

Knowing the electron–phonon coupling constant λ, the critical transition temperature T_c can be estimated from the McMillan (1968) formula

$$T_e = \frac{\theta_D}{1.45} \exp\left[-\frac{1.04(1+\lambda)}{\lambda - \mu^*(1+0.62\lambda)}\right] \tag{5.18}$$

where μ^* is the Coulomb pseudo-potential given by:

$$\mu^* = 0.26 \frac{N_{bs}(E_F)}{1 + N_{bs}(E_F)} \tag{5.19}$$

where $N_{bs}(E_F)$ is expressed in states/eV atom. μ^* is typically 0.1–0.2. For a typical MAX phase, we can assume $\lambda = 0.4$ and a θ_D of, conservatively, 500 K, so the calculated T_c values are, for the most part, <1 K, as observed.

The situation for Cr_2AlC, Cr_2GeC, and V_2AsC is quite different, however, since $\lambda \approx 1.8$ (Chapter 4). Using the latter in Eq. (5.19), and assuming μ^* to be 0.15, and a Debye temperature of 500 K, the predicted T_c values are of the order of 33 K, which is not observed. Why this is the case is not clear at this time.

5.7 Summary and Conclusions

The vast majority of the MAX phases are metal-like conductors in which the resistivity increases linearly with increasing temperature. The electron and hole mobilities are directly proportional to RRR and inversely proportional to $N(E_F)$. The density of charge carriers in these solids is of the order of $1–3 \times 10^{27}$ m^{-3}.

Because the R_H values are quite small – in some cases vanishingly small – and a weak function of temperature, the Seebeck coefficients also hover around zero, and the MR is parabolic and nonsaturating, it is reasonable to conclude that the MAX phases are compensated conductors wherein $n \approx p$ and $\mu_e \approx \mu_p$.

If the results for Cr_2AlC and Cr_2GeC are not included, then an excellent correlation exists between $N(E_F)$, on one hand, and n and p on the other (Fig. 5.6b). If they are included, the correlation is quite weak. Why the Cr-containing solids are outliers is not clear at this time, but probably related to their high electron–phonon coupling.

The value of $d\rho/dT$ is directly proportional to the electron–phonon coupling constant λ, and inversely proportional to n and p.

The Seebeck coefficients of most of the MAX phases are quite small. According to DFT calculations, this stems from the thermopower being negative along the c-axis and positive in the basal planes. The small experimentally obtained Seebeck coefficients can thus be ascribed to the compensation between the thermopowers along two nonequivalent crystallographic axes.

The only solid relationship concerning the optical properties of the MAX phases is between their Drude – where the formal valences of the elements is used – and plasma frequencies measured from EELS. The latter fall in the 20–25 eV range for the MAX phases.

The delocalized electrons give rise to Pauli paramagnetism, where χ is positive, small, and temperature independent.

A few MAX phases have been reported to be superconducting at $T < 5$ K. The McMillan formula predicting T_c does not appear to apply to the MAX phases.

References

Ashcroft, N.W, and Mermin, N.D. (1976) *Solid State Physics*, Saunders College Publishing, Philadelphia, PA.

Barsoum, M.W. (2000) The $M_{n+1}AX_n$ phases: a new class of solids; thermodynamically stable nanolaminates. *Prog. Solid State Chem.*, **28**, 201–281.

Barsoum, M.W., Golczewski, J., Siefert, H.J., and Aldinger, F. (2002a) Fabrication and electrical and thermal properties of Ti_2InC, Hf_2InC and $(Ti,Hf)_2AlC$. *J. Alloys Compd.*, **340**, 173–179.

Barsoum, M.W., Salama, I., El-Raghy, T., Golczewski, J., Porter, W.D., Wang, H., Siefert, H., and Aldinger, F. (2002b) Thermal and electrical properties of Nb_2AlC, $(Ti,Nb)_2AlC$ and Ti_2AlC. *Metall. Mater. Trans.*, **33a**, 2779.

Barsoum, M.W., Scabarozi, T.H., Amini, S., Hettinger, J.D., and Lofland, S.E. (2011) Electrical and thermal properties of Cr_2GeC. *J. Am. Ceram. Soc.*, **94**, 4123–4126.

Barsoum, M.W., Yoo, H.I., and El-Raghy, T. (2000) Electrical conductivity, thermopower and Hall effect of Ti_3AlC_2, Ti_4AlN_3 and Ti_3SiC_2. *Phys. Rev. B*, **62**, 10194.

Berlincourt, T. (1959) *Phys. Rev.*, **114**, 696.

Born, M. and Wolf, E. (1964) *Principle of Optics*, Pergamon Press, New York, NY.

Bortolozo, A.D., Sant'Anna, O.H., da Luz, M.S., dos Santos, C.A.M., Pereira, A.S., Trentin, K.S., and Machado, A.J.S. (2006a) Superconductivity in Nb_2SnC. *Solid State Commun.*, **139**, 57.

Bortolozo, A.D., Sant'Anna, O.H., da Luz, M.S., dos Santos, C.A.M., Pereira, A.S., Trentin, K.S., and Machado, A.J.S. (2006b) Superconductivity in the Nb_2SnC compound. *Solid State Commun.*, **139**, 57.

Bortolozo, A.D., Sant'Anna, O.H., dos Santos, C.A.M., and Machado, A.J.S. (2007) Superconductivity in the hexagonal-layered nanolaminate Ti_2InC compound. *Solid State Commun.*, **144**, 419.

Bortolozo, A.D., Serrano, G., Serquis, A., Rodrigues, D., dos Santos, C.A.M., and Fisk, Z. (2010) Superconductivity at 7.3 K in Ti_2InN. *Solid State Commun.*, **150**, 1364.

Chakraborty, B., Pickett, W.E., and Allen, P.B. (1976) Density of states, optical mass, and dc electrical resistance of Ta, W, Nb, and Mo using Slater-Koster interpolation. *Phys. Rev. B*, **14**, 3227.

Chaput, L., Hug, G., Pecheur, P., and Scherrer, H. (2005) Anisotropy and thermopower in Ti_3SiC_2. *Phys. Rev. B*, **71**, 121104(R).

Chaput, L., Hug, G., Pecheur, P., and Scherrer, H. (2007) Thermopower of the 312 MAX phases Ti_3SiC_2, Ti_3GeC_2, and Ti_3AlC_2. *Phys. Rev. B*, **75**, 035107.

Collings, E.W. (1974) Anomalous electrical resistivity, bcc phase stability, and superconductivity in titanium-vanadium alloys. *Phys. Rev. B*, **9**, 3989.

Drulis, M.K., Czopnik, A., Drulis, H., and Barsoum, M.W. (2004) Low temperature heat capacity and magnetic susceptibility of Ti_3SiC_2. *J. Appl. Phys.*, **95**, 1128–1333.

Drulis, M.K., Drulis, H., Gupta, S., and Barsoum, M.W. (2006) On the heat capacities of M_2AlC (M = Ti, V, Cr) ternary carbides. *J. Appl. Phys.*, **99**, 093502.

Finkel, P., Barsoum, M.W., Hettinger, J.D., Lofland, S.E., and Yoo, H.I. (2003) Low-temperature transport properties of nanolaminates Ti_3AlC_2 and Ti_4AlN_3. *Phys. Rev. B*, **67**, 235108.

Finkel, P., Hettinger, J.D., Lofland, S.E., Barsoum, M.W., and El-Raghy, T. (2001) Magnetotransport properties of the ternary carbide Ti_3SiC_2: Hall effect, magnetoresistance and magnetic susceptibility. *Phys. Rev. B*, **65**, 035113.

Finkel, P., Seaman, B., Harrell, K., Hettinger, J.D., Lofland, S.E., Ganguly, A., Barsoum, M.W., Sun, Z., Li, S., and Ahuja, R. (2004) Low temperature elastic, electronic and transport properties of $Ti_3Si_{1-x}Ge_xC_2$ solid solutions. *Phys. Rev. B*, **70**, 085104.

Genchev, Z.D. (1993) Anisotropic electrical conductivity tensor of granular high-Tc superconductors in an effective-medium theory. *Supercond. Sci. Technol.*, **6**, 523.

Haddad, N., Garcia-Caurel, E., Hultman, L., Barsoum, M.W., and Hug, G. (2008) Dielectric properties of Ti_2AlC and Ti_2AlN MAX phases: the conductivity anisotropy. *J. Appl. Phys.*, **104**, 023531.

He, X., Bai, Y., Chen, Y., Zhu, C., Li, M., and Barsoum, M.W. (2011) Phase stability, electronic structure, compressibility, elastic and optical properties of a newly discovered Ti_3SnC_2: a first-principle study. *J. Am. Ceram. Soc.*, **94**, 3907–3914.

Hettinger, J.D., Lofland, S.E., Finkel, P., Palma, J., Harrell, K., Gupta, S., Ganguly, A., El-Raghy, T., and Barsoum, M.W. (2005) Electrical transport, thermal transport and elastic properties of M_2AlC (M = Ti, Cr, Nb and V) phases. *Phys. Rev. B*, **72**, 115120.

Holm, B., Ahuja, R., Li, S., and Johansson, B. (2002) Theory of ternary layered system Ti-Al-N. *J. Appl. Phys.*, **91**, 9874–9877.

Hu, C., He, L., Zhang, J., Bao, Y., Wang, J., Li, M., and Zhou, Y. (2008a) Microstructure and properties of bulk Ta_2AlC ceramic synthesized by an in situ reaction/hot pressing method. *J. Eur. Ceram. Soc.*, **28**, 1679–1685.

Hu, C., Li, F., He, L., Liu, M., Zhang, J., Wang, J., Bao, Y., Wang, J., and Zhou, Y. (2008b) In situ reaction synthesis, electrical and thermal, and mechanical properties of Nb_4AlC_3. *J. Am. Ceram. Soc.*, **91**, 2258–2263.

Hu, C., Li, F., Zhang, J., Wang, J., Wang, J., and Zhou, Y. (2007) Nb_4AlC_3: a new compound belonging to the MAX phases. *Scr. Mater.*, **57**, 893–896.

Hu, C., Sakka, Y., Nishimura, T., Guo, S., Grasso, S., and Tanaka, H. (2011) Physical and mechanical properties of highly textured polycrystalline Nb_4AlC_3 ceramic. *Sci. Technol. Adv. Mater.*, **12**, 044603.

Hu, C., Sakka, Y., Tanaka, H., Nishimura, T., and Grasso, S. (2009) Low temperature thermal expansion, high temperature electrical conductivity, and mechanical properties of Nb_4AlC_3 ceramic synthesized by spark plasma sintering. *J. Alloys Compd.*, **487**, 675–681.

Hug, G., Eklund, P., and Orchowski, A. (2010) Orientation dependence of electron energy loss spectra and dielectric functions of Ti_3SiC_2 and Ti_3AlC_2. *Ultramicroscopy*, **110**, 1054–1058.

Hug, G., Jaouen, M., and Barsoum, M.W. (2005) XAS, EELS and full-potential augmented plane wave study of the electronic structures of Ti_2AlC, Ti_2AlN, Nb_2AlC and $(Ti_{0.5},Nb_{0.5})_2AlC$. *Phys. Rev. B*, **71**, 24105.

Kanoun, M.B., Goumri-Said, S., and Reshak, A.H. (2009) Theoretical study of mechanical, electronic, chemical bonding and optical properties of Ti_2SnC, Zr_2SnC, Hf_2SnC and Nb_2SnC. *Comput. Mater. Sci.*, **47**, 491–500.

Kyrienko, O. and Shelykh, I.A. (2011) Angle-resolved reflectance and surface plasmonics of the MAX phases. *Opt. Lett.*, **36**, 3966.

Li, C., Wang, B., Li, Y., and Wang, R. (2009) First-principles study of electronic structure, mechanical and optical properties of V_4AlC_3. *J. Phys. D: Appl. Phys.*, **42**, 065407.

Li, S., Ahuja, R., Barsoum, M.W., Jena, P., and Johansson, B. (2008) Optical properties of Ti_3SiC_2 and Ti_4AlN_3. *Appl. Phys. Lett.*, **92**, 221907.

Lofland, S.E., Hettinger, J.D., Harrell, K., Finkel, P., Gupta, S., Barsoum, M.W., and Hug, G. (2004) Elastic and electronic properties of select M_2AX phases. *Appl. Phys. Lett.*, **84**, 508–510.

Lofland, S.E., Hettinger, J.D., Meehan, T., Bryan, A., Finkel, P., Hug, G., and Barsoum, M.W. (2006) Electron–phonon coupling in MAX phase carbides. *Phys. Rev. B*, **74**, 174501.

Magnuson, M., Mattesini, M., Van Nong, N., Eklund, P., and Hultman, L. (2012) The electronic-structure origin of the anisotropic thermopower of nanolaminated Ti_3SiC_2 determined by polarized x-ray spectroscopy. *Phys. Rev. B*, **85**, 195134.

Mauchamp, V., Bugnet, M., Chartier, P., Cabioch, T., Jaouen, M., Vinson, J., Jorissen, K., and Rehr, J.J. (2012) Interplay between many-body effects and charge transfers in Cr_2AlC bulk plasmon excitation. *Phys. Rev. B*, **86**, 125109.

McMillan, W.L. (1968) Transition temperature of strong-coupled superconductors. *Phys. Rev.*, **167**, 331.

Medvedeva, N., Novikov, D., Ivanovsky, A., Kuznetsov, M., and Freeman, A. (1998) Electronic properties of Ti_3SiC_2-based solid solutions. *Phys. Rev. B*, **58**, 16042–16050.

Mendoza-Galván, A., Rybka, M., Järrendahl, K., Arwin, H., Magnuson, M., Hultman, L., and Barsoum, M.W. (2011) Spectroscopic ellipsometry study on the dielectric function of bulk Ti_2AlN, Ti_2AlC, Nb_2AlC, $(Ti_{0.5},Nb_{0.5})_2AlC$ and Ti_3GeC_2 MAX phases. *J. Appl. Phys.*, **109**, 013530.

Mott, N.F. and Jones, H. (1936) *The Theory of the Properties of Metals and Alloys*, Dover Publications, New York.

Sakamaki, K., Wada, H., Nozaki, H., Onuki, Y., and Kawai, M. (1999) Carbosulfide superconductor. *Solid State Commun.*, **112**, 323–327.

Sakamaki, K., Wada, H., Nozaki, H., Onuki, Y., and Kawai, M. (2000) Evidence for superconductivity in $Nb_2SC_{0.90}$ carbosulfide. *Mol. Cryst. Liq. Cryst.*, **341**, 903–908.

Sanborn, B.A., Allen, P.B., and Papaconstantopoulos, D.A. (1989) Empirical electron–phonon coupling constants and anisotropic electrical resistivity in hcp metals. *Phys. Rev. B*, **40**, 6037.

Scabarozi, T.H., Amini, S., Finkel, P., Barsoum, M.W., Tambussi, W.M., Hettinger, J.D., and Lofland, S.E. (2008a) Electrical, thermal, and elastic properties of the MAX phase Ti_2SC. *J. Appl. Phys.*, **104**, 033502.

Scabarozi, T.H., Eklund, P., Emmerlich, J., Hogberg, H., Meehan, T., Finkel, P., Barsoum, M.W., Hettinger, J.D., Hultman, L., and Lofland, S.E. (2008b) Weak electronic anisotropy in the layered nanolaminate Ti_2GeC. *Solid State Commun.*, **146**, 498–501.

Scabarozi, T.A., Ganguly, A., Hettinger, J.D., Lofland, S.E., Amini, S., Finkel, P., El-Raghy, T., and Barsoum, M.W. (2008c) Electronic and thermal properties of $Ti_3Al(C_{0.5},N_{0.5})_2$, $Ti_2Al(C_{0.5},N_{0.5})$ and Ti_2AlN. *J. Appl. Phys.*, **104**, 073713.

Scabarozi, T.H., Benjamin, S., Adamson, B., Applegate, J., Roche, J., Pfeiffer, E., Steinmetz, C., Lunk, C., Barsoum, M.W., Hettinger, J.D. *et al.* (2012) Combinatorial investigation of the stoichiometry, electronic transport and elastic properties of $(Cr_{1-x}V_x)_2GeC$ thin films. *Scr. Mater.*, **66**, 85–88.

Stanely, J.K. (1963) Electrical and magnetic properties of metals. *Am. Soc. Metal.*, p. 51.

Tian, W., Wang, P., Zhang, G., Kan, Y., Li, Y., and Yan, D. (2006) Synthesis and thermal and electrical properties of bulk Cr_2AlC. *Scr. Mater.*, **54**, 841–846.

Toth, L. (1967) High superconducting transition temperatures in the molybdenum carbide family of compounds. *J. Less-Common Met.*, **13**, 129–131.

Wang, P., Hong, X.-L., and Zhou, W. (2007) Synthesis of Ti_2AlC by hot pressing and its mechanical and electrical properties. *Trans. Nonferrous Met. Soc. China*, **17**, 1001–1004.

Wang, X.H. and Zhou, Y.C. (2002) Microstructure and properties of Ti_3AlC_2 prepared by the solid liquid reaction synthesis and simultaneous in-situ hot pressing process. *Acta Mater.*, **50**, 3141–3149.

Ying, G., He, X., Li, M., Du, S., Han, W., and He, F. (2011) Effect of Cr_7C_3 on the mechanical, thermal, and electrical properties of Cr_2AlC. *J. Alloys Compd.*, **509**, 8022–8027.

Yoo, H.I., Barsoum, M.W., and El-Raghy, T. (2000) Ti_3SiC_2: a material with negligible thermopower over an extended temperature range. *Nature*, **407**, 581–582.

Zhou, W.B., Mei, B.C., and Zhu, J.Q. (2009) On the synthesis and properties of bulk ternary Cr_2AlC ceramics. *Mater. Sci. Poland*, **27**, 973–980.

Zhou, Y.C., Dong, H.Y., Wang, X.H., and Chen, S.Q. (2000) Electronic structure of the layered ternary carbides Ti_2SnC and Ti_2GeC. *J. Phys.: Condens. Matter*, **12**, 9617.

Zhou, Y.C., Wan, D.T., Bao, Y.W., and Wang, J.Y. (2006) In situ processing and high-temperature properties $Ti_3Si(Al)C_2/SiC$ composites. *Int. J. Appl. Ceram. Soc.*, **3**, 47–54.

6
Oxidation and Reactivity with Other Gases

6.1
Introduction

As discussed in Chapter 10, because some MAX phases, most notably Ti_3AlC_2 and Ti_3SiC_2, possess excellent high-temperature (HT) mechanical properties, they are being considered for a number of structural and nonstructural HT applications. To be used in air, however, their oxidation resistance is of paramount importance because only if a protective layer forms can they be used at elevated temperatures. A perusal of the materials currently being used for HT applications quickly establishes that only three oxides, namely, silica (SiO_2), chromia (Cr_2O_3), and alumina (Al_2O_3), form layers protective enough for use at temperatures >900 °C for extended periods. As discussed in subsequent text, for the MAX phases, the only oxide that is protective enough for extended high temperature use is alumina.

In general, the oxidation of the MAX phases occurs according to the following reaction:

$$M_{n+1}AX_n + bO_2 = M_{n+1}O_x + AO_y + X_nO_{2b-x-y} \tag{6.1}$$

Thus, the oxidation of Ti_3SiC_2, for example, results in the formation of an outer pure rutile[1)] (TiO_2) layer and an inner layer comprised of TiO_2 and SiO_2 (Barsoum, El-Raghy, and Ogbuji, 1997). In the case of $Ti_{n+1}AlC_n$ phases, Al_2O_3 and TiO_2 layers form. Oxidation of Ti_2InC results in TiO_2 and In_2O_3 (Chakraborty, El-Raghy, and Barsoum, 2003); the Sn-containing ternaries oxidize to SnO_2 and TiO_2 (Zhou, Dong, and Wang, 2004); Ti_2SC forms TiO_2 and SO_2 gas (Amini *et al.*, 2009), Ta_2AlC forms Ta_2O_5, $TaAlO_4$, and an X-ray diffraction (XRD) amorphous phase (Gupta, Filimonov, and Barsoum, 2006), and so on.

In this chapter, the gas/solid reactions/interactions of a number of MAX phases, starting with Ti_3SiC_2, are reviewed. Most of this chapter deals with the oxidation of the MAX phases in air.

As discussed throughout this chapter, in many cases the oxidation kinetics, especially at shorter times, are parabolic. This implies that either the weight gain

1) Henceforth, unless otherwise noted, TiO_2 will refer to rutile.

Δw or the oxide thickness, x, scale as

$$\left(\frac{\Delta w}{A}\right)^2 = k_w t \tag{6.2}$$

and/or

$$x^2 = 2\,k_x t \tag{6.3}$$

where t is the oxidation time and A is the surface area exposed to the atmosphere. The constants k_w and k_x are the parabolic rate constants obtained from weight gain measurements and oxide layer thickness measurements, respectively. They figure prominently in this chapter.

In what follows, the oxidation behavior of select MAX phases in air is reviewed.

6.2 Ti_3SiC_2

Before 1996, the oxidation of Ti_3SiC_2 was believed to be quite poor (Tong *et al.*, 1995). Early oxidation results on Ti_3SiC_3 (Barsoum, El-Raghy, and Ogbuji, 1997) showed that, at least for relatively short times (10 h), the oxidation kinetics were parabolic and resulted in the formation of a duplex oxide scale: an inner rutile and silica layer and an outer rutile layer (Figure 6.1a). At lower temperatures, the silica is amorphous, but at temperatures greater than 1240 °C cristobalite peaks appear in the XRD patterns and become stronger at higher temperatures. Radhakrishnan, Williams, and Akinc (1999), extended the oxidation time to 100 h, and confirmed that the oxidation kinetics in air at 1000 °C were initially parabolic, but showed that at longer times the kinetics became linear. In other words, the overall kinetics were paralinear.

In 2003, we reported on the long-term – up to 1500 h – oxidation in air, in the 875–1200 °C temperature range, of fine-grained (FG) and coarse-grained (CG) samples of Ti_3SiC_2. (Henceforth and throughout this book the abbreviations FG and CG will be used) (Barsoum *et al.*, 2003). We also studied the oxidation of Ti_3SiC_2 with 30 vol% TiC and Ti_3SiC with 30 vol% SiC, henceforth referred to as *TiC/312* and *SiC/312*, respectively. Typical scanning electron microscopy (SEM) images of SiC/312 and TiC/312 samples – both oxidized at 1125 °C for 9 h – are shown, respectively, in Figure 6.2a,b. From these micrographs it is obvious that the oxidation resistance of SiC/312 is better than that of TiC/312, with Ti_3SiC_2 (not shown) lying in between. In all cases, however, the oxidation resulted in a duplex scale: an outer TiO_2 and an inner rutile/silica layer (Figures 6.1a and 6.2). These conclusions were recently confirmed by Zhang *et al.*, who showed that the oxidation resistance of SiC/312 composites was enhanced with increasing SiC content (Zhang *et al.*, 2008).

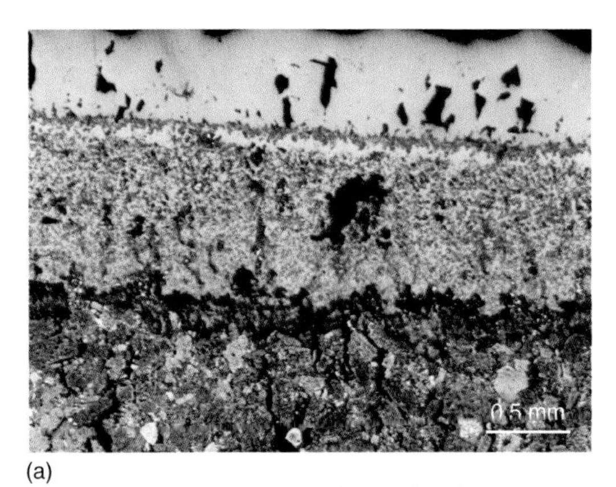

(a)

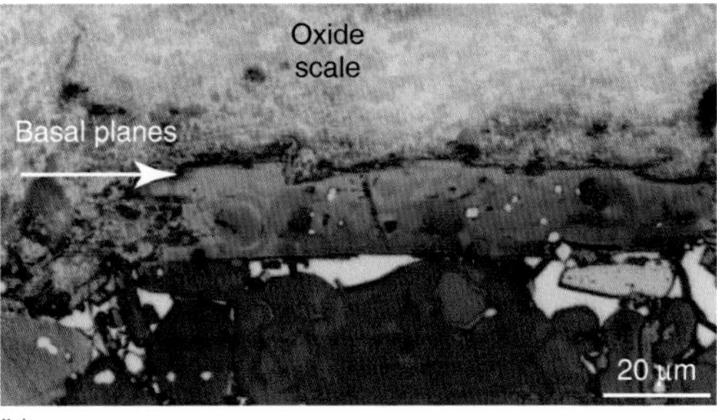

(b)

(c)

Figure 6.1 Typical etched OM images of Ti_3SiC_2 samples oxidized in air at (a) 1100 °C for 360 h (Barsoum *et al.*, 2003). The outer layer is pure TiO_2; the inner layer is comprised of TiO_2 and SiO_2. Note large pores – absent at shorter times – that form after prolonged time. (b) 1100 °C for 78 h and (c) 1150 °C for 62 h. (b, c) Show the anisotropic nature of the oxidation process: the oxidation proceeds faster along [100] than along [001].

The overall oxidation reactions for the Ti_3SiC_2, TiC/312, and SiC/312 compositions are:

$$Ti_3SiC_2 + 6O_2 = 3TiO_2 + SiO_2 + 2CO_2 \tag{6.4}$$

$$Ti_3SiC_2 + 1.5TiC + 9O_2 = 4.5TiO_2 + SiO_2 + 3.5CO_2 \tag{6.5}$$

$$Ti_3SiC_2 + SiC + 8\,O_2 = 3TiO_2 + 2SiO_2 + 3CO_2 \tag{6.6}$$

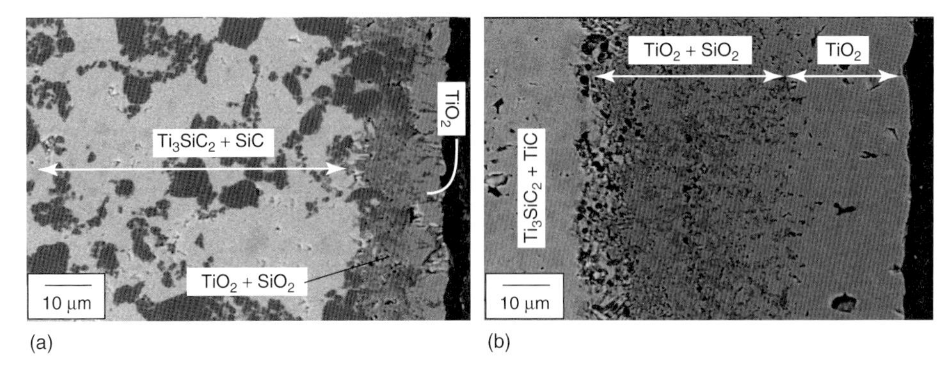

Figure 6.2 Backscattered SEM images of (a) SiC/312 and (b) TiC/312 samples oxidized in air at 1125 °C for 9 h. The dark islands in the matrix in (a) are SiC particles. The oxide scale that forms on the TiC/312 samples is thicker than the one that forms on the SiC/312 samples. (Barsoum *et al.*, 2003).

What occurs to the C is unclear at this time, but because it does not accumulate in the substrate it is assumed to diffuse through the oxide layers and oxidize at the oxide/air interface. This view is bolstered by the results of Bellucci *et al.* (2003), who made the case that at the TiC/TiO_2 interface the following reaction:

$$TiC + \gamma O = TiO_\gamma C_{1-\gamma} + \gamma C_{inter}$$

occurs. Here C_{inter} refers to a C interstitial in TiO_2. Closer to the air interface, the $TiO_\gamma C_{1-\gamma}$ phase transforms into rutile according to

$$TiO_\gamma C_{1-\gamma} + \eta O = \beta TiC_{(2-x)} + \delta\ TiC_{(1-\gamma')}O_\gamma \quad + \gamma C_{inter}$$

where $\gamma' > \gamma$, $\beta + \delta = 1$, $\gamma = [(1-\gamma) - \delta\ (\tilde{1}]/\beta$ and $\varepsilon = \gamma(\delta - 1) + \beta(2 - x)$.

More specifically, they presented evidence that the oxidation starts with O ions diffusing through the TiO_2 layer, substituting for the C in the TiC, and releasing a C_{inter}, which then diffuses through the TiO_2 and ultimately gets oxidized at the TiO_2/air interface. As important, Bellucci *et al.* (2002) showed – using an approach that is quite similar to the one outlined in subsequent text and Appendix 6.A – that the rate-limiting step in the oxidation of TiC is the diffusion of oxygen through TiO_2.

6.2.1 Oxidation Kinetics and the Nature of the Rate-Limiting Step

As discussed in subsequent text, there is little doubt that the oxide layers grow by the inward diffusion of oxygen and the outward diffusion of Ti; the Si atoms are oxidized *in situ*. At short times, the oxide layers are dense, crack free, and resistant to thermal cycling. The kinetics are parabolic and the parabolic rate constants are thermally activated (Figure 6.3).

Table 6.1 summarizes the parabolic rate constants for the oxidation of Ti_3SiC_2 and its composites. Since the activation energies Q for the oxidation of Ti_3SiC_2, TiC/312, and SiC/312 (Table 6.1 and Figure 6.3) all fall within a narrow range

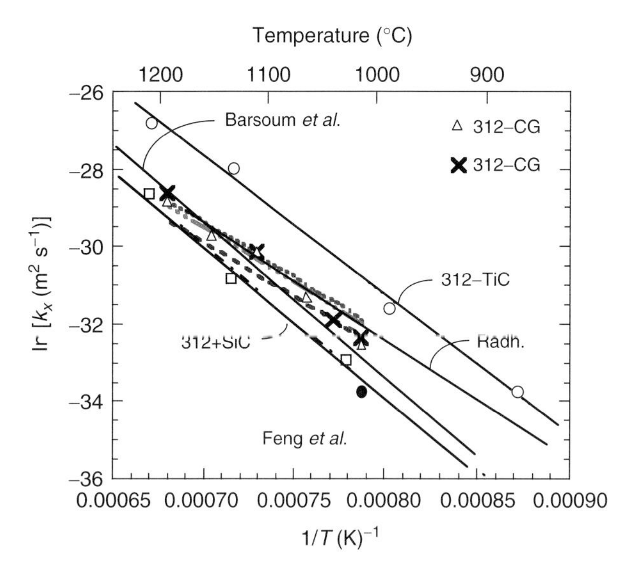

Figure 6.3 Arrhenius plots of k_x values determined at short times <30 h (Barsoum *et al.*, 2003). Also included are the results of previous work (Barsoum, El-Raghy, and Ogbuji, 1997; Feng, Orling, and Munir, 1999; Radhakrishnan, Williams, and Akinc, 1999). The dashed green, red, and blue lines are those calculated starting with the diffusivities of O and Ti for the TiC/312, Ti_3SiC_2, and SiC/312 compositions, respectively.

Table 6.1 Summary of least squares fit of the Arrhenius plots such as those shown in Figure 6.3. These results are valid in the 900–1200 °C range and for times less than 30 h. k_0 and Q are defined in Eq. (6.11).

Material	**k_x in 900–1200 °C range**		**References**
	k_o ($m^2\ s^{-1}$)	**Q ($kJ\ mol^{-1}$)**	
Ti_3SiC_2 (FG)	1.6×10^{-2}	−297.5	Barsoum *et al.* (2003)
Ti_3SiC_2 (CG)	3.3×10^{-3}	−281.8	Barsoum *et al.* (2003)
TiC/Ti_3SiC_2	6.7×10^{-2}	−297.3	Barsoum *et al.* (2003)
SiC/Ti_3SiC_2	6.0×10^{-2}	−322.5	Barsoum *et al.* (2003)
Ti_3SiC_2 (M2)[a]	1.2×10^{-2}	−296.3	Barsoum *et al.* (1997)
Ti_3SiC_2 (M1)[a]	3×10^{-2}	−305.5	Barsoum *et al.* (1997)
Ti_3SiC_2 − 7 wt% TiC	1.7×10^{-1}	−325.9[b]	Sun, Zhou, and Li (2001b)
Ti_3SiC_2 − 12 wt% TiC	1.9×10^{-4}	−252.5[b]	Sun, Zhou, and Li (2001a)
Ti_3SiC_2 (Eq. (6.11))	8.2×10^{-5}	−239	Barsoum *et al.* (2003)
SiC/312 (Eq. (6.11))	5.4×10^{-5}	−239	Barsoum *et al.* (2003)
TiC/312 (Eq. (6.11))	1.2×10^{-4}	−239	Barsoum *et al.* (2003)

[a]The fits reported here were restricted to the 900–1200 °C temperature range and consequently are slightly different from the results reported in (Barsoum, El-Raghy, and Ogbuji, 1997).
[b]These values were obtained from the data cited in Sun, Zhou, and Li (2001a).

(300 ± 15 kJ mol^{-1}), it is fair to conclude that the rate-limiting step is not only identical in all cases but, to a first and good approximation, also insensitive to the substrate chemistry. This statement is valid up to at least 30 vol% of SiC or TiC in Ti_3SiC_2 and is only true at shorter times ($t < 30$ h). Given that in all cases the ions are diffusing through TiO_2 (Figures 6.1 and 6.2), this conclusion is not very surprising.

To determine the nature of the rate-limiting step, the diffusion coefficients of O and Ti in TiO_2 are used – together with Wagner's theory for parabolic rate growth (Wagner, 1933) – to calculate the theoretical parabolic rate constants $k_{x,\text{theo}}$. The latter are then compared with the experimental results listed in Table 6.1. To calculate $k_{x,\text{theo}}$ the following implicit assumptions are made:

1) The oxide scale grows by the diffusion of O and Ti through the TiO_2 layer.
2) The TiO_2 that forms is an n-type electronic conductor (Hoshino, Peterson, and Wiley, 1985; Yahia, 1963).
3) The reaction layer is dense, coherent, and crack free, which is a good assumption at shorter times.
4) The primary defects are Ti interstitial $Ti_i^{\bullet\bullet\bullet\bullet}$,$Ti_i^{\bullet\bullet\bullet}$, oxygen vacancies $V_O^{\bullet\bullet}$, and free electrons e^{-1} (Hoshino, Peterson, and Wiley, 1985). In other words, neither oxygen interstitials $O_i^{''}$ nor Ti vacancies $V_{Ti}^{''''}$ are considered. For the sake of simplicity, it is also assumed that the titanium interstitials are fully oxidized, that is, the concentration of $Ti_i^{\bullet\bullet\bullet}$ is assumed to be nil. This assumption does not alter the final conclusions. Note that the Kroger–Vink defect notation is used throughout this chapter and book.

On the basis of these assumptions, the most likely set of defect reactions – the rationale for which is discussed in subsequent text – occurring at the oxide scale/air interface are (Barsoum *et al.*, 2003):

$$2O_2 + 8e^{-1} + 2V_O^{\bullet\bullet} = 4O_O^x + V_{Ti}^{''''} \tag{6.7}$$

Conversely, at the substrate/oxide scale interface

$$Ti = Ti_{Ti}^x + 2V_O^{\bullet\bullet} + 4e^{-1} \tag{6.8}$$

$$Ti = Ti_i^{\bullet\bullet\bullet\bullet} + 4e^{-1} \tag{6.9}$$

$$Ti_i^{\bullet\bullet\bullet\bullet} + V_{Ti}^{''''} = Ti_{Ti}^x \tag{6.10}$$

Equation (6.7) accounts for the formation of the oxygen ions that react with the $Ti_i^{\bullet\bullet\bullet\bullet}$ generated by Eq. (6.9) at the substrate/oxide scale interface (see subsequent text) and increase the thickness of the outer layer. In other words, it accounts for the formation of the lattice sites that, in turn, accommodate the $Ti_i^{\bullet\bullet\bullet\bullet}$ diffusing out. Equation (6.8) accounts for the creation of the $V_O^{\bullet\bullet}$ vacancies that diffuse outward and are ultimately responsible for the inward diffusion of oxygen ions. Equation (6.9) creates the $Ti_i^{\bullet\bullet\bullet\bullet}$ that diffuse outward and result in the growth of the outer layer. These two reactions are written separately to emphasize that the Ti plays two roles: its *in situ* oxidation at the substrate/oxide interface (Eq. (6.8)), which results in the increase in thickness of the *inner* layer, *and* its outward diffusion (Eq. (6.9)). Equation (6.10) places the Ti interstitials on regular sites in TiO_2.

Note that adding Eqs. (6.7–6.10) results in the formation of 2 moles of TiO_2, one at the air/oxide scale interface and the other at the substrate/oxide scale interface. In other words, the TiO_2 layer grows at both ends. The inner layer grows by the inward diffusion of oxygen ions, and the outer layer by the outward diffusion of Ti interstitials.[2] The Si atoms oxidize *in situ* by the inward diffusing oxygen ions.

Following Wagner's approach (Wagner, 1933) and assuming that the ionic fluxes are independent, it can be shown (Barsoum *et al.*, 2003) that $k_{x,\text{theo}}$ for the overall growth of the oxide layer is given by

$$k_{x,\text{theo}} = k_0 \exp - \frac{Q}{RT} = \beta \left(D_{\text{Ti}} + D_{\text{o}}\right) \left|\frac{\Delta G_{rxn}}{RT}\right| \tag{6.11}$$

where D_{Ti} and D_{O} are, respectively, the diffusion coefficients of Ti and O in TiO_2; R and T have their usual meanings. ΔG_{rxn} is the Gibbs free energy change – per mole of oxygen – associated with the relevant oxidation reactions, that is, reactions 6.4–6.6. The factor β is the fraction of the cross-sectional area through which the ions can flow (i.e., the fraction of the cross-sectional area that is composed of TiO_2 in the inner layers). This assumes that the diffusion of oxygen through SiC and/or SiO_2 can be safely neglected in comparison to its diffusion through TiO_2, which is an excellent assumption. Microstructurally, β ranges from $\approx$0.4 for SiC/312, to $\approx$0.5 for Ti_3SiC_2, to $\approx$0.7 for the TiC/312 samples.

In this reaction scheme, the ionic *defect* fluxes are *not* coupled to each other directly but through the local electric field that builds up with the electrons (see Appendix A in Barsoum *et al.* 2003).

The self-diffusion of Ti in TiO_2 single crystals is anisotropic and has been measured several times (Akse and Whitehurst, 1978; Arita *et al.*, 1979; Lundy and Coghlan, 1973; Marucco, Gautron, and Lemasson, 1981; Venkatu and Poteat, 1970). The agreement between the different investigations is good (Hoshino, Peterson, and Wiley, 1985). Lundy and Coghlan (1973) reported the diffusion coefficients of Ti in air along the c- and a-axes of TiO_2 as

$$D_c\left(\text{m}^2\ \text{s}^{-1}\right) = 4.6 \times 10^{-6} \exp\left[\frac{-250.6\left(\text{kJ mol}^{-1}\right)}{RT}\right] \tag{6.12}$$

$$D_a\left(\text{m}^2\ \text{s}^{-1}\right) = 2.4 \times 10^{-7} \exp\left[\frac{-202.8\left(\text{kJ mol}^{-1}\right)}{RT}\right] \tag{6.13}$$

Here, the lower of the two values, that is, Eq. (6.12), is used on the assumption that diffusion along the c-axis is rate limiting. It is important to note, however, that the other measurements (Hoshino, Peterson, and Wiley, 1985) fall within a factor of 2 of Eq. (6.12), and could have been used instead. Furthermore, the results shown in subsequent text further confirm that Eq. (6.12) – with an activation energy of 250 kJ mol^{-1} – is more consistent with the oxidation results. As far as we are aware, the

2) Note that if indeed the TiO_2 that forms upon oxidation is n-type, then the *only* ionic defects that can diffuse across the layer *have* to be positively charged. If not, a fundamental tenet of Wagner's model – local electroneutrality – would be violated.

diffusion of Ti in polycrystalline TiO_2 has not been measured. On the basis of the conclusions discussed in subsequent text, until such measurements are available, Eq. (6.12) can be used with some confidence.

The diffusion of oxygen in TiO_2 was measured twice (Arita *et al.*, 1979; Haul and Dumbgen, 1965): the two sets of results are almost identical. The values of Haul and Dumbgen, that is

$$D_O\left(m^2\ s^{-1}\right) = 2 \times 10^{-7} \exp\left[\frac{-251\left(kJ\ mol^{-1}\right)}{RT}\right] \quad (6.14)$$

are used here.

Arita *et al.* estimated their oxygen vacancy mole fraction $[V_O]$ in TiO_2 to be $\approx 2 \times 10^{-4}$. Haul and Dumbgen's TiO_2 contained on average $\approx$150 ppm Al_2O_3, which translates to a $[V_O] \approx 7.8 \times 10^{-5}$. The corresponding oxygen vacancy diffusion coefficients D_{V_O} are thus $1.7 \times 10^{-3}\ \exp[-251/RT]$ and $2.56 \times 10^{-3}\ \exp[-251/RT]$. Averaging the two sets of data, one obtains

$$D_{V_O} \approx 2 \times 10^{-3} \exp\left[\frac{-251\left(kJ\ mol^{-1}\right)}{RT}\right] \quad (6.15)$$

which is used herein.

Combining Eqs. (6.12) and (6.14) with the various parameters listed in Table 6.2 for Eqs. (6.4–6.6), together with Eq. (6.11), yields the colored dashed lines shown in Figure 6.3. Least squares fit of these lines result in the values listed in the bottom three rows in Table 6.1. The model predicts the SiC/312 composition to be the most oxidation resistant, followed by Ti_3SiC_2, with the least resistant being the TiC/312 composition, as observed. The agreement between experiment and theory for Ti_3SiC_2 and SiC/312 is excellent over the entire temperature range and times of 30 h or less. The agreement with the TiC/312 results is less satisfactory for reasons that are not fully understood. Note, however, that even at worst the difference is less than an order of magnitude. Thus the agreement between theory and experiment has to be considered good especially given that (i) the oxygen vacancy diffusivity in TiO_2 is, presumably, extrinsically controlled and thus dependent on the impurities in the starting powders (see Appendix 6.A), and (ii) some of the assumptions made in deriving Eq. (6.11), are, at best, approximations (see Appendix A in Barsoum *et al.* 2003).

Table 6.2 Summary of parameters used to calculate $k_{x,theo}$ from Eq. (6.11).

Substrate	Reaction	β^a	$\Delta G_{rxn}/RT$ (per mole of O)		
			1173 K	1273 K	1473 K
Ti_3SiC_2	6.4	0.5	65	60	51.7
TiC/Ti_3SiC_2	6.5	0.67	52	48	42
SiC/Ti_3SiC_2	6.6	0.41	53	49	42

[a] These values are assumed to be equal to the volume fraction of TiO_2 in the inner layers as estimated from micrograph such as those shown in Figs. 6.1 and 6.2.

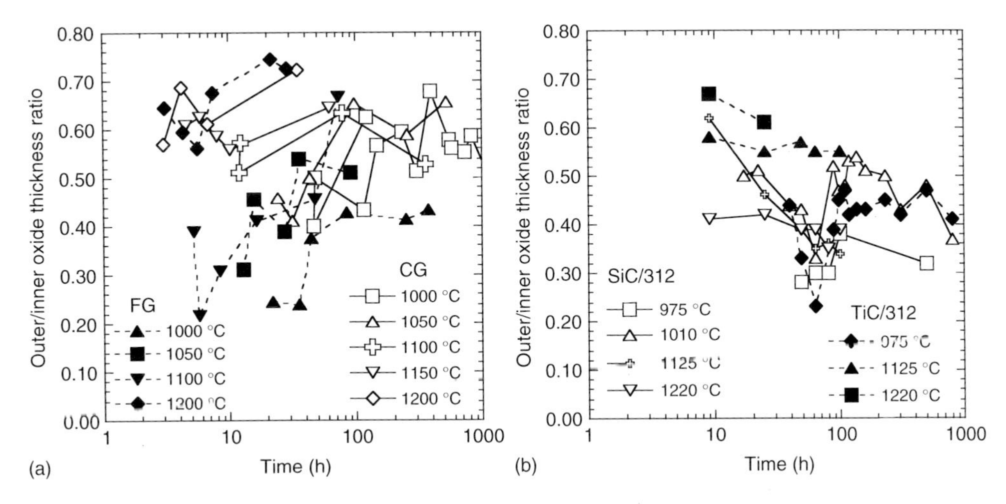

Figure 6.4 Summary of outer/inner thickness ratios of oxide scale for the, (a) CG and FG samples and (b) SiC/312 and TiC/312 samples (Barsoum *et al.*, 2003). Lines are guides to the eyes.

The oxidation rates for the FG samples were slightly higher than those of the CG ones. This difference was attributed to the anisotropy of oxidation. On the basis of the OM images shown in Figure 6.1b,c, it is fair to conclude that oxygen diffuses faster along, rather than normal to, the basal planes. This is not very surprising given the high reactivity of the Si atoms in the structure (El-Raghy and Barsoum, 1998; El-Raghy, Barsoum, and Sika, 2001).

When the outer-to-inner layer thickness ratios of the oxide scales are plotted versus time (Figure 6.4), surprisingly, and despite the scatter, the ratio falls in the relatively (for thermally activated processes) narrow range of 0.2–0.75 over the *entire* time, temperature, and composition regimes explored, a remarkable result. Since, as discussed in preceding text, the inner and outer layers grow as a result, respectively, of the diffusion of the O and Ti ions, this implies that not only the diffusion activation energies but also the pre-exponential terms must be almost identical – as observed (compare Eqs. (6.12) and (6.14)). Dividing Eq. (6.14) by Eq. (6.12), and neglecting the slight difference in activation energies, yields a ratio of 23, which is too high; that is, one of the two expressions – Eq. (6.12) or (6.14) – is incorrect. If the oxygen diffusion results (Eq. (6.12)) are assumed to be more accurate, then the pre-exponential term in Eq. (6.13) should be closer to $\approx 2 \times 10^{-7}$ $m^2\ s^{-1}$. Using this value and recalculating $k_{x,\text{theo}}$ results in values that are too low. If, on the other hand, one assumes the pre-exponential term in Eq. (6.12) to be of the order of $4.6 \times 10^{-6}\ m^2\ s^{-1}$, the ratios of the O and Ti ion fluxes are of the order of unity, as must be the case.[3]

3) Note that as long as either D_{O} or D_{Ti} is of the correct order of magnitude, Eq. (6.11) becomes quite insensitive to the value of the other. The *ratio*, however, is quite *sensitive* to the absolute values of the diffusivities; this is especially true of the activation energies.

Physically, this implies that the oxygen vacancy concentration present in the TiO_2 scales formed during oxidation of Ti_3SiC_2 and its composites is higher than those reported in the literature (Arita *et al.*, 1979; Haul and Dumbgen, 1965). This is not surprising because the compositions used to determine D_O were quite pure. See subsequent text, more specifically Appendix 6.A, for a more thorough discussion on what affects $[V_O]$ in the oxide scales.

6.2.2
Transition from Parabolic to Linear Kinetics

The reasons for the transition from parabolic to linear kinetics at longer times are currently not fully understood. The following observation, however, points to a possible culprit, namely the development of stresses in the oxide layers. The deviation from parabolic behavior is not a function of oxide thickness, as much as it is a function of temperature. For example, the kinetics of most samples oxidized at 1200 °C and higher remain parabolic despite the fact that the oxide layer thicknesses are in the 300 μm range or higher. Conversely, at lower temperatures, the deviation from parabolic behavior occurs at x values that are less than ≈100 μm. On the basis of these observations, it is fair to assume that the stresses that build up during oxidation are relieved at higher temperatures, allowing the oxide layers to remain protective despite their considerable thickness. At lower temperatures, however, the deviation occurs at oxide scale thickness values of the order of 100–200 μm.

The results for the FG and CG samples were obtained after prolonged tensile creep tests (Barsoum *et al.*, 2003). In other words, the CG and FG Ti_3SiC_2 samples were oxidized under a tensile stress. This brings up the important question as to what role, if any, externally applied stress plays. The excellent agreement between the k_x values obtained with and without stress (Figure 6.3) is good, which is indirect evidence that if tensile stresses do play a role, it is small. This is true, at least initially, when the system is in the parabolic regime. This is further confirmed by the excellent agreement between the k_x values obtained for the SiC/312 samples measured – with no external load – and those of monolithic Ti_3SiC_2 that were subjected to a tensile stress.

The situation at longer times, however, is less clear. Large pores – some almost millimeters in diameter – were observed in the oxide layers of the Ti_3SiC_2 samples (Figure 6.1a). Direct evidence (not shown) exists for the rupture of the oxide scale at a location adjacent to where a large crack was initiated in the substrate that was under a tensile load. Once ruptured, oxygen penetrated into the bulk of the sample (Barsoum *et al.*, 2003).

The fact that the oxidation kinetics become linear at $T > 1000$ °C or greater and for times greater than 30 h is problematic and will certainly limit the application of Ti_3SiC_2-based components for use in ambient air. A potential solution is to react the Ti_3SiC_2 surfaces with Si to form SiC and $TiSi_2$. Oxidation of the latter results in the formation of a protective SiO_2 layer, which enhances the oxidation resistance by roughly four orders of magnitude (El-Raghy and Barsoum, 1998). Another solution is to use the $Ti_{n+1}AlX_n$ phases instead.

6.3 $Ti_{n+1}AlX_n$

The most promising MAX phase to date with superb oxidation resistance is Ti_2AlC. After 8000 cycles from 1350 °C to room temperature, a thin, adherent, protective 15 μm α-alumina (Al_2O_3) layer was found (Figure 6.5a) (Sundberg *et al.*, 2004). The formation of this dense Al_2O_3 layer is key to HT oxidation protection. The formation of Al_2O_3 is another example of the reactivity of the A-group element *vis-à-vis* the $M_{n+1}X_n$ blocks. Here, Al_2O_3 forms despite the fact that the Al concentration is half that of Ti, which is another reactive metal. Wang and Zhou (2003b) also reported the formation of Al_2O_3 layers in Ti_3AlC_2, where the Al concentration is one-third that of Ti.

The oxidation resistance, however, is somewhat temperamental and subject to some factors that are not entirely understood. In some cases, a protective Al_2O_3 layer forms and in others – with nominally the same compositions – the majority phase is rutile (Barsoum, 2001; Barsoum *et al.*, 2001; Lee and Park, 2006). The following deals with each case separately, with the more important case first.

6.3.1 When Alumina Forms a Protective Layer

As shown in subsequent text, when Al_2O_3 forms, the oxidation kinetics are closer to cubic than parabolic. In other words, they are best described by

$$\left(\frac{\Delta w}{A}\right)^3 = k_{c,w}t \tag{6.16}$$

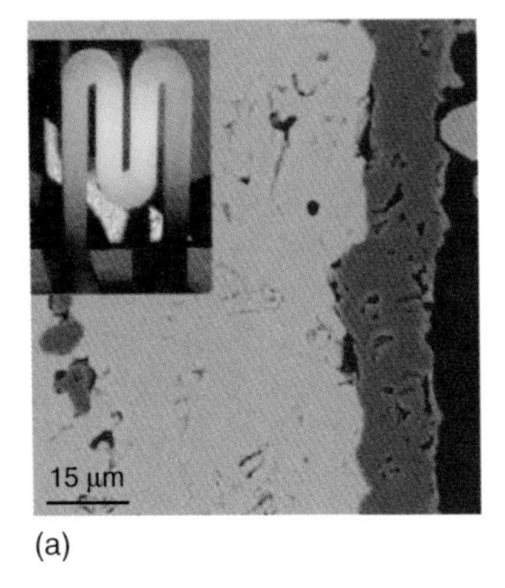

Figure 6.5 SEM image of a Ti_2AlC sample after (a) 8000 thermal cycles to 1350 °C in air, showing the presence of a thin 15-μm Al_2O_3 surface layer. The inset shows a Ti_2AlC heating element glowing at 1350 °C (Sundberg *et al.*, 2004). (b) Oxidation in air for 2800 h. Here again a protective Al_2O_3 layer forms. Not only was the oxide layer adherent to the substrate but, as importantly, it conformed to the shape of the underlying substrate, even at sharp corners, without cracking or spalling (Tallman, Anasori and Barsoum, 2013).

or

$$x^3 = k_{c,x} t \tag{6.17}$$

where $k_{c,w}$ is the cubic reaction rate constant (in kg^3 m^{-6} s^{-1}), and $k_{c,x}$ is the cubic reaction rate constant (in m^3 s^{-1}).

Furthermore, if the oxidation reaction is assumed to simply be

$$Ti_2AlC + (3x/4 + y)O_2 = x/2\ Al_2O_3 + Ti_2Al_{1-x}C_{1-y} + y\ CO_2 \tag{6.18}$$

it can be shown that to convert $k_{c,x}$ to $k_{x,w}$, the latter is to be multiplied by 9.4 $\times$ 10^{-10} m^9 kg^{-3}. This conversion factor, is useful when comparing the various studies, some in which the mass gain was measured versus some in which the oxide thickness was reported.

6.3.2 Cubic or Parabolic Kinetics

Before discussing and understanding what happens when Ti_2AlC and Ti_3AlC_2 are oxidized in air, it is imperative to establish the oxidation kinetics. To claim that the situation is muddled is an understatement at this time for the following reasons. In 2003, Wang and Zhou reported that the oxidation kinetics for Ti_2AlC were cubic (Wang and Zhou, 2003a) but those of Ti_3AlC_2 were parabolic. Others followed the lead of Wang and Zhou, and also reported that the oxidation kinetics were parabolic (Qian *et al.*, 2011; Sundberg *et al.*, 2004).

In 2007, Byeon *et al.* showed that the oxidation kinetics of Ti_2AlC were clearly cubic (Byeon *et al.*, 2007). More problematic, in a review article in 2010, Wang and Zhou (2010) – and despite the results of Byeon *et al.* (2007) – not only claimed that the oxidation kinetics of Ti_3AlC_2 were parabolic, but mysteriously, and without comment, revised their own conclusion concerning the oxidation kinetics of Ti_2AlC from cubic to parabolic.

Whether the kinetics are parabolic or cubic is important, and more than academic, because if the kinetics are indeed parabolic, then the long-term prognosis is not good. However, if the kinetics are slower than parabolic – for example, cubic – then the oxidation resistance would be good enough for practical applications. This is a crucial point that needs to be established beyond reasonable doubt.

As noted in preceding text, in 2003 Wang and Zhou quantified the oxidation kinetics of Ti_2AlC by heating it in air for 20 h in the 1000–1300 °C temperature range (Wang and Zhou, 2003a). They also studied the oxidation of Ti_3AlC_2 in the 1000–1400 °C range (Wang and Zhou, 2003b). A comparison of the results of the two papers is illuminating.

When the same authors plotted their Δw results for Ti_3AlC_2 (Wang and Zhou, 2003b), for reasons that are unclear, they claimed them to be parabolic, despite the fact that their $(\Delta w/A)^2$ versus t plots were *not* straight lines. When their results are replotted as $(\Delta w/A)^3$ versus t (not shown), the least squares fit of the results at 1300 °C resulted in an $R^2 > 0.998$. The R^2 value for the parabolic plot given by the authors (Figure 2 in Wang and Zhou, 2003b) is around 0.98. In other words, their

Table 6.3 Summary of k_c ($kg^3\ m^{-6}\ s^{-1}$) values for the oxidation of Ti_2AlC, Ti_3AlC_2, and Cr_2AlC.

Phase	1000 °C	1100 °C	1200 °C	1300 °C	1400 °C	Comments and references
Ti_3AlC_2	4.2×10^{-14}	1.7×10^{-12}	3.2×10^{-12}	2.0×10^{-11}	6.3×10^{-10}	Wang and Zhou (2003b)
Ti_2AlC[a]	3.2×10^{-13}	1.1×10^{-12}	3.0×10^{-12}	1.5×10^{-11}	—	Wang and Zhou (2003a)
Ti_2AlC	3.3×10^{-13}	1.9×10^{-12}	1.0×10^{-11}	5.1×10^{-11}	—	Air (Basu *et al.*, 2012)
Ti_2AlC	5.6×10^{-13}	2.0×10^{-12}	1.2×10^{-11}	6.0×10^{-11}	—	100% H_2O (Basu *et al.*, 2012)
Cr_2AlC	—	—	3.8×10^{-12}	3.2×10^{-11}	—	Lin *et al.* (2007a)

[a] In Wang and Zhou (2003a), the results listed in their Table 1 are wrong. The correct values based on the results they show in their Figure 6.1c, are listed here.

own results fit a cubic law better than a parabolic one. It thus makes more sense to assume the kinetics to be cubic. More importantly, if, in *both* cases, a dense Al_2O_3 layer forms, it is highly unlikely that the kinetics would be parabolic in one case and cubic in the other.

Table 6.3 lists the $k_{c,w}$ values as a function of temperature for the oxidation of Ti_3AlC_2, Ti_2AlC, and Cr_2AlC obtained from a number of papers. We note in passing that the results listed in Table 1 in Wang and Zhou's (2003a) paper are wrong; the correct values are listed in Table 6.3. A perusal of these results confirms that the oxidation kinetics of the former two related compounds are similar. This conclusion is further confirmed when the post-oxidation microstructures are compared. After 20 h oxidation at 1300 °C, the Al_2O_3 layer thickness is ≈25 μm in Ti_2AlC (Wang and Zhou, 2003a) and ≈14 μm in Ti_3AlC_2 (Wang and Zhou, 2003b). After 20 h at 1200 °C, the oxide layer thicknesses – at ≈5 μm – are almost identical for both compounds.

More importantly, Byeon *et al.* showed that when commercially available Ti_2AlC polycrystalline samples were heated in air, a continuous, adherent α-Al_2O_3 formed (Byeon *et al.*, 2007). In this work, the oxidation kinetics were found to be *cubic*. The thickness of the layer was ≈15 μm after 25 h of isothermal oxidation at 1400 °C. Roughly the same thickness was observed after 1000 one-hour cycles from ambient temperature to 1200 °C. In both cases, the layer remained adherent and protective.

More recently, Basu *et al.* (2012) showed that when commercially available Ti_2AlC samples were oxidized, (i) the oxidation kinetics up to 120 h were cubic, (ii) the activation energy was ≈270 kJ mol^{-1}, and (iii) the oxidation resulted in a continuous α-Al_2O_3 layer along with a thin TiO_2 surface layer.

To further help settle this important question, Tallman, Anasori and Barsoum (2013) measured the oxide layer thickness as a function of time for a Ti_2AlC

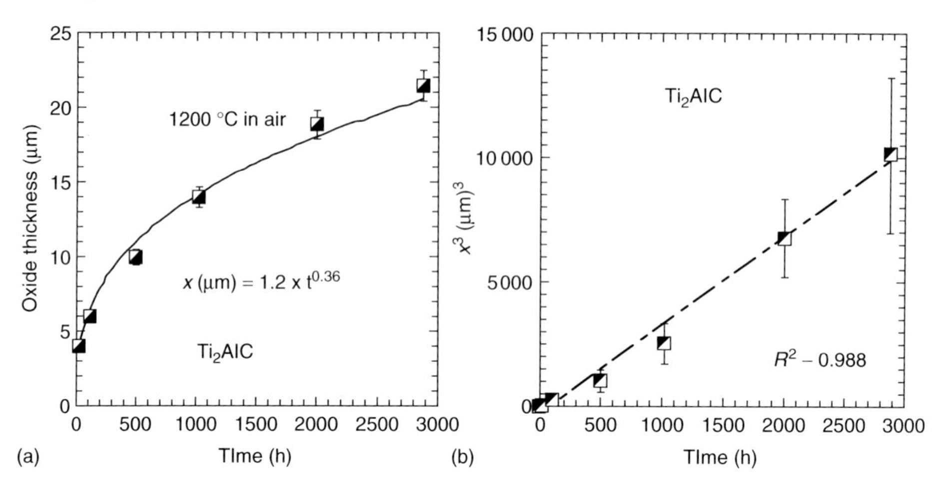

Figure 6.6 Alumina layer thickness x versus t for Ti_2AlC oxidized at 1200 °C in air plotted as (a) x versus t and (b) x^3 versus t. In both cases, it is clear that the kinetics are cubic at least up to ≈3000 h (Tallman, Anasori and Barsoum 2013).

sample oxidized at 1200 °C for almost 3000 h. The results, shown in Figure 6.6, unequivocally show that the oxidation kinetics, up to at least 3000 h, are indeed cubic. Least squares fit of the results show that the relationship

$$x\,(\mu\text{m}) = 1.2\left(\frac{t}{t_o}\right)^{0.36} \tag{6.19}$$

where $t_o = 1$ h describes the time dependence of the oxide layer thickness quite well. The R^2 value is 0.988. When the same results are plotted as x^3 versus t (Figure 6.6b), not surprisingly an excellent correlation is found. More importantly, even after 2800 h oxidation, the oxide layer was not only adherent but also conformed to the shape of the underlying substrate, even at the sharp corners without cracking or spalling (Figure 6.5b).

If one assumes the oxidation kinetics to be controlled by grain boundary diffusion then the grain coarsening kinetics of the alumina layer have to factor in the former. Furthermore, if the grain-coarsening kinetics can be described as:

$$d^m = d_o^m + Kt \tag{6.20}$$

where K, m and d_o are respectively, a constant, the grain growth exponent and the initial grain size, it can be shown that at longer times (Liu, Gao and He, 2000):

$$x^2 \approx K'\left(\frac{t}{t_o}\right)^{(m-1)/m} \tag{6.21}$$

where K′ is a constant. It follows that if the assumptions made above are correct, then:

$$n \approx (m-1)/2m \tag{6.22}$$

To test this idea Tallman, Anasori and Barsoum (2013) carefully measured the time dependencies of the grain sizes of the alumina films that formed on the Ti_2AlC sample that was oxidized for almost 3000 h. Based on least squares fit of the results, a value of 3.23 for m was obtained. This in turn implies that n obtained from Eq. 6.22 is ≈ 0.345, which coincidentally or not, is in excellent agreement with the value of n of 0.36 in Eq. 6.19, obtained from the oxidation results. This analysis is strong evidence that indeed the rate-limiting step in the oxidation of the MAX phases that form an alumina layer is grain boundary diffusion of oxygen and/or aluminum ions.

When the cubic rate constants of the various studies on Ti_2AlC, Ti_3AlC_2, and even Cr_2AlC (see subsequent text) are plotted on the same Arrhenian plot (Figure 6.7a), good agreement is obtained. This is further evidence that the kinetics are cubic and relatively independent of the underlying MAX phase composition as long as they all form Al_2O_3.

Figure 6.8a compares the mass per unit area changes obtained when Ti_2AlC made from commercial powders with those of other Al_2O_3-forming alloys during thermal cyclic oxidation at 1200 °C. Although the rate of mass gain is the highest for Ti_2AlC, the observed rate under these aggressive test conditions was still low and the oxide remained protective. More importantly, mass loss due to scale spallation was not observed, indicating a high degree of spallation resistance confirming the results of Sundberg *et al.* (2004).

One of the main reasons why the oxidation resistance of Ti_2AlC is as good as it is, and so resistant to thermal cycling, is the excellent match in thermal expansions between it and the α-Al_2O_3 that forms. Photoluminescence of the α-Al_2O_3 scale indicated that the residual stresses formed in that layer were compressive – of the order of 500 MPa – and a function of time and temperature (Figure 6.8b).

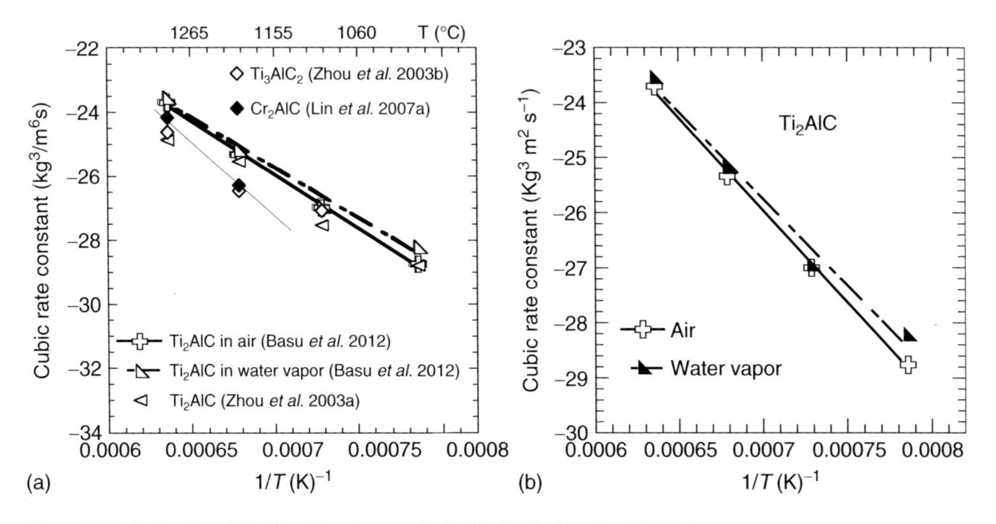

Figure 6.7 Arrhenian plot of cubic rate constants for Al_2O_3-forming MAX phases. (a) Ti_2AlC, Ti_3AlC_2 and Cr_2AlC (see Table 6.3). (b) Ti_2AlC oxidized in air and in water vapor (Basu *et al.*, 2012).

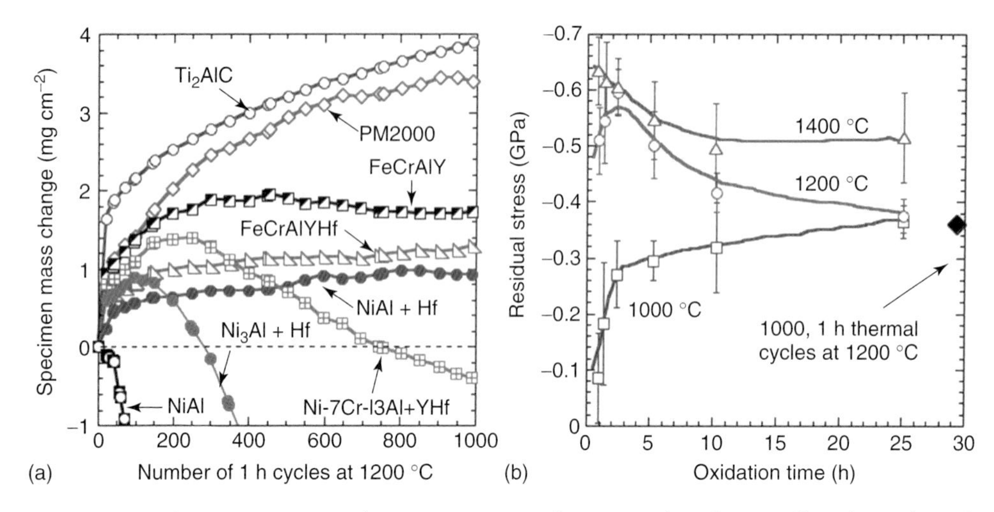

Figure 6.8 (a) Mass changes per unit area for various Al_2O_3-forming alloys during thermal cyclic oxidation at 1200 °C. (b) Magnitude of compressive residual stress within the Al_2O_3 scale determined from luminescence shifts as a function of time of isothermal oxidation at 1000, 1200, and 1400 °C (Byeon *et al.*, 2007).

Such residual stresses are considered low, and partially explain the high spallation resistance of the α-Al_2O_3 scale. We note in passing that commercially available Fe- and Ni-based Al_2O_3-forming alloys have relatively high coefficients of thermal expansion (CTEs) and typically require reactive element additions to improve their spallation resistance (Pint, 1996).

Byeon *et al.* also showed that the oxide scale developed after isothermal oxidation at 1000 and 1200 °C consisted of an inner Al-rich continuous scale and an outer Ti-rich discontinuous scale. At 1400 °C, they reported the formation of Al_2TiO_5 (Byeon *et al.*, 2007).

These results were confirmed in a recent paper that explored the oxidation of Ti_2AlC in air in the 900–1400 °C temperature range (Cui, Jayaseelan, and Lee, 2011). They proposed that at $T < 940$ °C, the oxidation resulted in titania. Yang et al. (Yang, 2011a) however, made a compelling case that it is much more much likely that the reaction occurring is reaction 6.18. At 1400 °C, titania and alumina react to form Al_2TiO_5. The formation of Al_2TiO_5 was correlated with the formation of cracks on cooling. These cracks were ascribed to thermal expansion mismatches, and, as importantly, to the high anisotropy of thermal expansion of Al_2TiO_5 (Cui, Jayaseelan and Lee, 2011). This compound expands at a rate of 10.9×10^{-6} K^{-1} along the *a*-axis, 20.5×10^{-6} K^{-1} along *b*, and shrinks at a rate of -2.7×10^{-6} K^{-1} along *c*.

Using TEM, Lin *et al.* (2006) explored the microstructures of Ti_3AlC_2 and Ti_2AlC samples after oxidation in air for 10 h at 1200 °C. An enrichment of Ti in the Al_2O_3 grain boundaries and Ti-rich precipitates in the Al_2O_3 scales were identified. They also showed little depletion of Al at the oxide/substrate interface indirectly confirming that Al diffusivity of Al in the MAX phases is quite rapid at these temperatures.

6.3.3
Effect of Water Vapor on Oxidation Kinetics

Of the three oxides that are known to be protective – SiO_2, Al_2O_3, and Cr_2O_3 – only two are not susceptible to water vapor. Silica will evaporate at higher temperatures if the ambient gases contain significant amounts of water vapor.

Basu *et al.* (2012) compared the oxidation kinetics of Ti_2AlC – made from commercially available powders – as a function of time and temperature in air and water vapor. When the cubic oxidation rate constants measured in air and water vapor were compared (Figure 6.7b), it was clear that water vapor had little effect on the kinetics. At $\approx$280 kJ mol^{-1}, the activation energy for oxidation in air was slightly higher than the 262 kJ mol^{-1} obtained in water vapor. On the basis of these results, it is reasonable to conclude that water vapor has little, or no, effect on the oxidation kinetics. This is a very important conclusion indeed, as it implies that Ti_2AlC can be used in both dry and wet environments at elevated temperatures.

6.3.4
What Determines Whether Alumina Forms

As noted in preceding text, under what conditions the oxidation of the $Ti_{n+1}AlX_n$ compounds results in a protective α-Al_2O_3 layer is not entirely clear. In retrospect, one possible reason why an α-Al_2O_3 layer did not form in our early work (Barsoum *et al.*, 2001) can be traced to the fact that the Ti used was not of high purity. More specifically, the Ti used contained a relatively large concentration of Cl.

Another factor that interferes with the formation of an Al_2O_3 layer is TiC. Wang and Zhou (2003c) showed that 5 vol% TiC in Ti_3AlC_2 was sufficient to interfere with the formation of a protective Al_2O_3 scale at 900 °C. We note in passing that the XRD results presented in that paper suggest the TiC fraction is more than 5%. This comment notwithstanding, to ensure that a continuous Al_2O_3 layer forms, it is important to minimize TiC from the microstructure.

Another less known fact has to do with electrodischarge machining (EDM). If contamination from the EDM wire is not thoroughly cleaned off – by, for example, rinsing in dilute HCl – more often than not a continuous protective Al_2O_3 layer will *not* form.

6.3.5
Intermediate Temperature Oxidation Response of Ti_2AlC

Wang and Zhou (2003c) also reported on the anomalous oxidation of Ti_3AlC_2 in air in the 500–900 °C temperature range. They showed that the oxidation was more severe at 600 °C than at 700 or 800 °C. The scales formed at 500, 600, and 700 °C consisted of anatase and rutile; those at higher temperatures were rutile. The Al_2O_3 layers at 500 and 600 °C were amorphous. Microcracks associated with the presence of anatase were postulated to be responsible for the anomalous oxidation. This is not a serious problem because a preoxidation treatment at a higher temperature, during which a continuous Al_2O_3 layer forms, should, in principle, solve the problem.

Pang *et al.* (2009) investigated the oxidation behavior of Ti_3AlC_2 over the 500–900 °C temperature range by synchrotron radiation XRD and secondary ion mass spectroscopy (SIMS), essentially confirming the conclusions of Wang and Zhou (2003c). The oxidation of Ti_3AlC_2 in the 500–900 °C was thus modified to read (Pang *et al.*, 2009)

$$2Ti_3AlC_2 + 11.5O_2 \xrightarrow[500^\circ C]{} 6TiO_2\,(\text{anatase}) + Al_2O_3\,(\text{amorphous}) + 4CO_2 \quad (6.23)$$

$$TiO_2\,(\text{anatase}) \xrightarrow[600-900^\circ C]{} TiO_2\,(\text{rutile}) \quad (6.24)$$

Pang *et al.* concluded from the SIMS results that Al was out-diffusing during oxidation. They also showed that at $T > 900\,^\circ C$, the amorphous Al_2O_3 started to crystallize to α-Al_2O_3.

6.3.6
Crack Healing

One of the most intriguing and potentially important phenomenon associated with the oxidation of Ti_2AlC and Ti_3AlC_2 at higher temperatures is their crack-healing ability (Song *et al.*, 2008; Yang *et al.*, 2011b). Crack healing in Ti_3AlC_2 was investigated by oxidizing partially precracked samples (Song *et al.* 2008). A crack near a notch was introduced into the sample by tensile deformation. After oxidation at 1100 °C in air for 2 h, the crack was completely healed, with oxidation products consisting primarily of α-Al_2O_3 as well as some rutile TiO_2 (Figure 6.9a–d). The indentation modulus and hardness of the crack-healed zone were slightly higher compared with those of the Ti_3AlC_2 base material. The preferential oxidation of Al atoms in Ti_3AlC_2 grains on the crack surface resulted in the predominance of α-Al_2O_3 particles forming in a crack less than 1 μm wide.

In 2011, the same group (Yang *et al.*, 2011b) revisited the oxidation of Ti_2AlC and carefully examined the morphology of the various oxide layers that formed both on flat and curved surfaces or cracks. They found that after oxidation at 1200 °C for 16–100 h, the α-Al_2O_3 particles that formed on flat surfaces were small (≈1 μm), densely packed, and columnar (Figure 6.9e). Those that formed in the cracks or cavities, on the other hand, were more equiaxed and less densely packed. The rutile grains, however, exhibited a broad size distribution, ranging from less than a micrometer to 10 μm. The authors also confirmed the presence of small TiO_2 particles at the α-Al_2O_3 grain boundaries.

Even more recently, the same group (Li *et al.*, 2012) showed that Ti_2AlC was capable of repeatedly repairing damage events. When the authors introduced Knoop indentations on the tensile side of Ti_2AlC flexural bars, the flexural strength dropped from 211 ± 15 to 152 ± 20 MPa. Heating the indented bars in air for 2 h at 1100 °C resulted in an increase in the flexural strengths to 224 ± 50 MPa, a value that was slightly *higher* on average than the virgin samples, albeit with larger scatter. Furthermore, after successively extending the same crack seven times and healing it between each fracture event, the fracture toughness dropped from ≈6.5 to about 3 MPa $m^{1/2}$. It is important to note here that by the end of the seventh

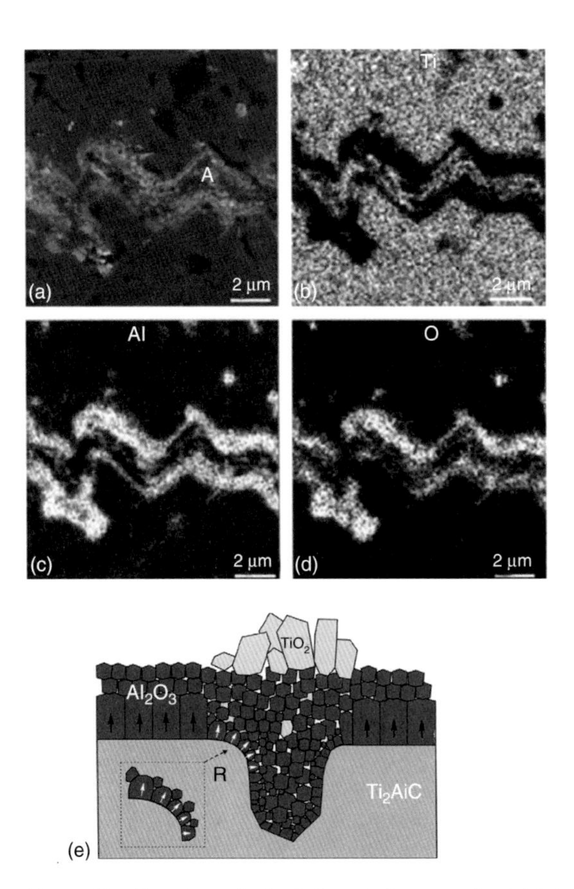

Figure 6.9 Top view of a healed crack, labelled A, in Ti_3AlC_2. (a) SEM image, (b) Ti, (c) Al, (d) O maps (Song *et al.*, 2008). (e) Schematic showing the influence of surface cavity on the formation of oxide scales during high-temperature oxidation of Ti_2AlC in air. The arrows indicate the fastest growth directions of the α-Al_2O_3 grains in the first layer. Dark gray grains represent α-Al_2O_3 (Yang *et al.*, 2011b).

cracking iteration, the filled crack was of the order of 1 mm (Figure 6.10a). As in their previous work, Li *et al.* showed that the main healing mechanism at high temperatures is the filling of the cracks by the formation of well-adhering Al_2O_3 layers and some TiO_2 (Figures 6.10b–e). The authors write in their abstract: "Self-healing ceramics have been studied for over 40 years to obtain some performance recovery and to prevent material failure during service, but so far only materials with the capability of a single healing event per damage site have been realized." They then proceeded to show how Ti_2AlC was capable of multiple healing events.

6.3.7
Ablation Resistance

Given the excellent oxidation resistance of Ti_2AlC, it was postulated (Song *et al.*, 2011) that it may be able to withstand ultrahigh temperatures for relatively short

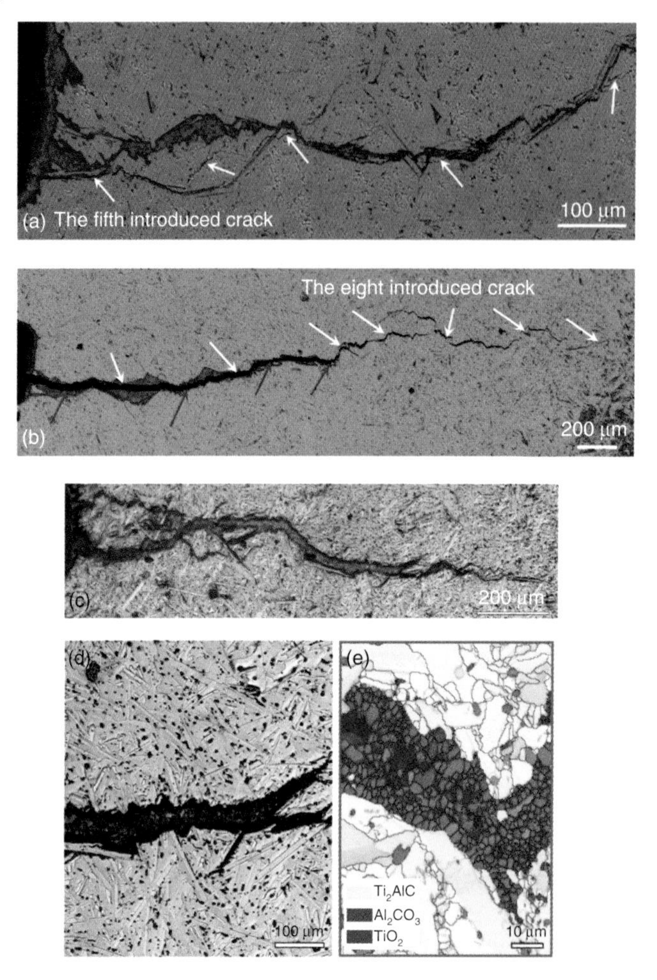

Figure 6.10 Images of fracture and crack-healing in Ti_2AlC; (a) Backscattered SEM image of crack path after four cycles of healing, and subsequent fracture. (b) Backscattered SEM crack path after seven cycles of healing, and subsequent fracture. The red arrows indicate the location of remnant crack parts. (c) OM image of a crack fractured eight times before annealing in air at 1200 °C for 100 h showing the complete filling of the crack. (d) Enlarged OM image taken from (c). Two opposite fracture surfaces were covered by the same Al_2O_3 layer (black matrix) and the gap between the two surfaces was fully filled by a mixture of Al_2O_3 and TiO_2 (white particles). (e) SEM image of the healed-damage zone obtained using electron backscatter diffraction (Li *et al.*, 2012).

periods because it should decompose into Al and a substoichiometric Ti_2Al_xC – or even into a twinned $TiC_{0.5}$ superstructure – rather than turning into a liquid as long as the temperature did not exceed the melting point of TiC_x. This conjecture was recently confirmed when the linear and mass ablation rates of Ti_2AlC under an oxyacetylene flame – at a temperature up to 3000 °C – were measured. The linear ablation rate was found to decrease from 0.14 μm s^{-1} for the first 30 s of ablation to

Table 6.4 Linear and mass ablation rates of several high-temperature materials including Ti_2AlC (Song *et al.*, 2011).

Material	Linear ablation rate ($\mu m\ s^{-1}$)	Mass ablation rate ($\mu g\ s^{-1}$)
Ti_2AlC	0.1	−240
Pure W	13.6	47
Cu-infiltrated W	11.7	85
40 vol% TiC/W composite	4.2	19
40 vol% ZrC/W composite	2.4	10
C/C (CVI) composite	3.0	2700
4 wt% ZrC-doped C/C composite	0.5	620

$0.08\ \mu m\ s^{-1}$ after 180 s. The compound even gained a small amount of weight on ablation, which was attributed to the formation of oxidation products. The ablation surface exhibited a two-layered structure: an oxide outer layer, consisting mainly of α-Al_2O_3 and TiO_2 and some Al_2TiO_5; and a porous subsurface layer containing $Ti_2Al_{1-x}C$ and TiC_xO_y. It was concluded that the major ablation mechanisms were the thermal oxidation and scouring of the viscous oxidation products by the gas torch's high-speed flow.

What is most noteworthy, however, is how low the ablation rate is for Ti_2AlC compared to other materials also tested using the same standard test (Table 6.4). Such low ablation rates suggest that Ti_2AlC could be used in ultrahigh temperature ablative environments.

6.3.8
When Alumina Does Not Form a Protective Layer

When Al_2O_3 does *not* form a protective layer, and the oxidation proceeds by the diffusion of O and Ti through a TiO_2 layer, the rate-limiting step and reaction kinetics are similar to those outlined in preceding text for Ti_3SiC_2 (Barsoum, 2001; Barsoum *et al.*, 2001). However, in contrast to Ti_3SiC_2, where the silica plays an inert role, in the presence of Al^{3+} ions a solid solution of Al_2O_3 in TiO_2 – that is, $(Ti_{1-y}Al_y)O_{2-y/2}$, where $y < 0.01$ (region D in Figure 6.11a) – in addition to Al_2O_3 forms instead. Assuming, for the sake of simplicity, that $y = 0$, the overall oxidation reaction is

$$Ti_{n+1}AlX_n + \left(2n + \frac{7}{2}\right) O_2 \rightarrow (n+1)\,TiO_2 + 0.5Al_2O_3 + nXO_2 \qquad (6.25)$$

In this reaction, the C and N atoms are assumed to diffuse through the oxide layers and ultimately oxidize to CO_2 and NO_2, respectively. Appendix 6.A deals with the oxidation of $Ti_{n+1}AlX_n$ in much more detail. In the remainder of this section, the main conclusions are highlighted.

For the most part, and especially at shorter times, the oxidation kinetics of the $Ti_{n+1}AlX_n$ phases are parabolic, with parabolic rate constants k_x defined by Eq. (6.3).

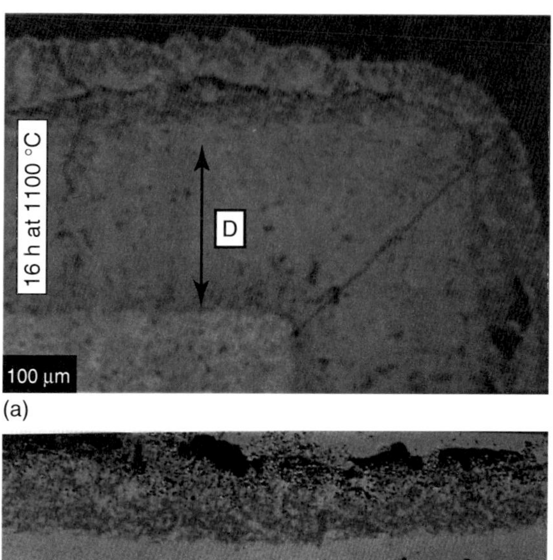

(a)

(b)

Figure 6.11 Cross-sectional backscattered SEM image of a Ti_2AlC sample heated in air at 1100 °C for (a) 16 h and (b) 64 h. Note how the more or less uniform layer labeled D in (a) is replaced by striated layers in (b) that are Al_2O_3-rich (dark layers) and TiO_2-rich. Layers of pores also appear at longer times. Note alumina diffuses towards the high oxygen partial pressure side located at the top of the micrograph. (Barsoum, 2001; Barsoum *et al.*, 2001).

A summary of the measured k_x values for a number of $Ti_{n+1}AlX_n$ compounds are listed in Table 6.A.1. In the discussion of the oxidation of Ti_3SiC_2, k_x was calculated from the values of the diffusivities of oxygen and titanium in rutile. In analyzing the oxidation kinetics of the $Ti_{n+1}AlX_n$ phases, Barsoum (2001) took a slightly different tack by converting the k_x values to diffusivities. The relationship between the two is given by Barsoum (2003) as

$$D_{RL} = k_x / \left(\ln \frac{0.21}{P_{O_2}} \right) \tag{6.26}$$

where D_{RL} is the diffusivity of the rate limiting ionic species, Ti and/or oxygen. The procedure is outlined in Appendix 6.A.

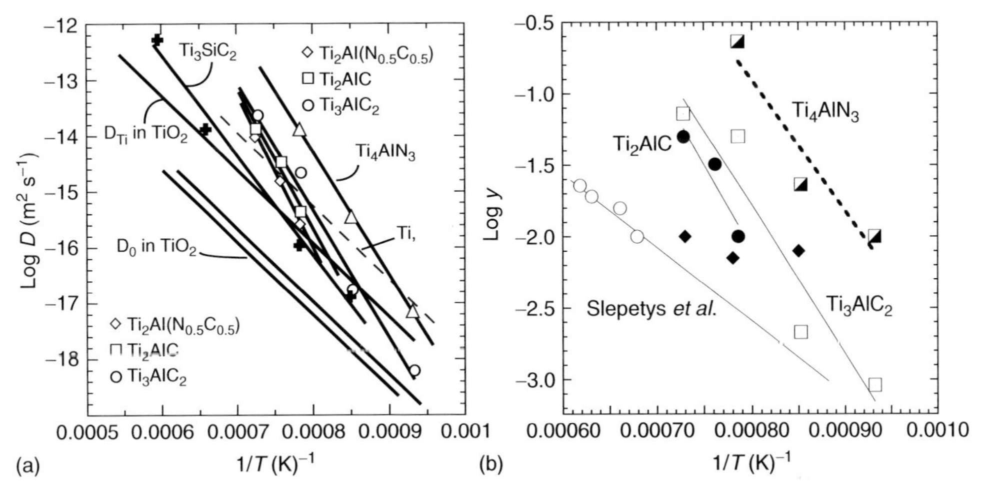

Figure 6.12 (a) Summary of diffusion coefficients determined from the k_x results listed in Table 6.A.1 (Appendix 6.A). Also included are diffusivity results of O and Ti in TiO_2 taken from the literature. The dashed line represents the diffusivity calculated from the scale that forms on Ti (b) Arrhenius plots of y in $(Ti_{1-y},Al_y)O_{2-y/2}$ deduced from the oxidation results (Table 6.A.1). Also included are literature results (Slepetys and Vaughan, 1969). The solid diamonds are EDS results taken from the oxide layers that formed on Ti_3AlC_2. (Barsoum, 2001; Barsoum *et al.*, 2001).

An Arrhenius plot of D_{RL} is shown in Figure 6.12a. Also included in Figure 6.12a are the results of D_O and D_{Ti} in TiO_2 (i.e., Eqs. (6.12) and (6.14)) and those determined from oxide layers growing on commercial Ti (Welsch and Kahveci, 1989), together with results from the oxidation of Ti_3SiC_2, subject to the same analysis. The values for the oxidation of commercial Ti (inclined dashed line labeled Ti in Figure 6.12a) intersect most of the results.

From these results it is also obvious that the presence of Al enhances the diffusivity of the rate-limiting step, especially at higher temperatures. This is best seen by comparing the *D*s deduced from the oxidation of Ti_3SiC_2 with those of the $Ti_{n+1}AlX_n$ phases; the former are lower.

The oxidation of the $Ti_{n+1}AlX_n$ compounds, at least at short times, is controlled by the inward diffusion of O and the simultaneous outward diffusion of Ti. Note that, at longer times, the oxidation kinetics revert to linear. The role that Al plays is quite intriguing. As noted above, initially, it dissolves uniformly in the rutile (Figure 6.11a). With time, a kinetic demixing occurs and the originally uniform distribution of Al devolves into repeating layers of alumina, rutile, and pores (Figure 6.11b).

Assuming $D_{RL} = D_O$, then $[V_O]$ and thus y in $(Ti_{1-y}Al_y)O_{2-y/2}$, can be calculated assuming that D_{V_O} is given by Eq. (6.15) (see Appendix 6.A). Figure 6.12b is an Arrhenius plot of y, listed in column 8 in Table 6.A.1. The last column of Table 6.A.1 lists the extrapolated temperature dependences of y calculated from the results of

Slepetys and Vaughan (1969). The agreement is quite good, and consistent with the assumption that Al is dissolved in the rutile layer.

Note that in the presence of sufficient concentrations of aliovalent ions (Al^{3+} in this case) that increase the anion vacancy concentration, such as is the case here, it is quite possible for the nature of the rate-limiting species to change from oxygen to titanium diffusion.

In our 2001 study, we did not report on the oxidation of Ti_2AlN. Recently, Cui, Jayaseelan, and Lee (2012) reported on the oxidation of Ti_2AlN fabricated by spark plasma sintering. The oxidation was found to be quite complex. However, the authors presented evidence for kinetic demixing of not a $(Ti_{1-y}Al_y)O_{2-y/2}$ layer as we observed (Barsoum, 2001; Barsoum *et al.*, 2001), but, rather, a two-phase mixture of Al_2O_3 and TiO_2. The authors, however, invoked the idea that the Al^{3+} cations diffused through the oxide layer, without specifying what they meant by the latter. If this layer is presumed to be rutile, then this mechanism is quite similar to the one we proposed. In most cases, and at most temperatures, dense outer rutile/anatase/alumina layers formed, below which thin layers of porosity were observed. The latter was ascribed to the Kirkendall effect.

Interestingly, Al_2TiO_5 formed on Ti_2AlN after 1 h oxidation at 1200 °C, while it formed after 1 h at 1400 °C on Ti_2AlC. Obviously the nucleation of Al_2TiO_5 is somehow catalyzed by the presence of nitrogen. As noted above, the presence of this phase and its large and anisotropic thermal expansion coefficients, however, render any oxide layers formed very susceptible to microcracking during thermal cycling.

6.4
Solid Solutions between Ti_3AlC_2 and Ti_3SiC_2

Lee, Nguyen, and Park (2009) and Nguyen, Park and Lee (2009) studied the oxidation of $Ti_3Al_{0.5}Si_{0.5}C_2$ and $Ti_3Al_{0.7}Si_{0.3}C_2$ solid solutions in air in the 900–1200 °C temperature range. The oxidation of the former resulted in TiO_2, Al_2O_3, and amorphous SiO_2, together with the evolution of CO or CO_2. The oxide scale consisted primarily of an outer TiO_2-rich layer, an intermediate Al_2O_3 layer, and an inner TiO_2-rich layer. Relatively thick scales formed because of the formation of a semiprotective TiO_2 and the escape of carbon. The overall oxidation reaction was thus found to be

$$Ti_3Al_{0.5}Si_{0.5}C_2 + 5.87O_2 = 3TiO_2 + 0.25Al_2O_3 + 0.5SiO_2 + 2CO_2 \quad (6.27)$$

The cyclic oxidation of $Ti_3Al_{0.7}Si_{0.3}C_2$ was also studied and found to be worse than Cr_2AlC, Ti_3AlC_2, or Ti_3SiC_2. For example, the thicknesses of the oxide scales formed during isothermal oxidation for 100 h in air were ≈15 μm at 900 °C and ≈67 μm at 1000 °C; those formed during cyclic oxidation, for the same cumulative time, were about 50 μm at 900 °C and 260 μm at 1000 °C.

Zhang *et al.* (2004) attempted to enhance the oxidation resistance of Ti_3SiC_2 by doping it with Al to make a $Ti_3Si_{0.9}Al_{0.1}C_2$ solid solution. Somewhat surprisingly, and despite the fact that only 10% of the Si was replaced with Al, at 1000 and 1100 °C a continuous Al_2O_3 inner layer and a discontinuous TiO_2 outer layer formed. At 1200 and 1300 °C, the continuous inner layer was still Al_2O_3, but the outer layer was a mixture of TiO_2 and Al_2TiO_5. The oxide layers were dense, adherent, and resistant to thermal cycling. At 1350 °C, however, the oxidation resistance was compromised by the formation of Al_2TiO_5.

6.5 Cr_2AlC

The oxidation resistance of Cr_2AlC is also quite good (Lee and Nguyen, 2008; Lee *et al.*, 2007; Lin *et al.*, 2007a; Lin, Zhou, and Li, 2007b; Tian *et al.*, 2008a). At 1000 and 1100 °C, the oxidation resistance is excellent because of the formation of a thin Al_2O_3 oxide layer with a narrow Cr_7C_3 underlayer (Figure 6.13a). At 1200 and 1300 °C, an outer $(Al_{1-x},Cr_x)_2O_3$, where $x < 0.01$, mixed oxide layer, an intermediate Cr_2O_3 oxide layer, an inner Al_2O_3 oxide layer, and a Cr_7C_3 underlayer form. Note that the latter is unique to Cr_2AlC and indirectly implies that for this compound C is *not* diffusing out as fast as Al, which results in its accumulation at the substrate/oxide interface.

The overall oxidation reaction can thus be assumed to be (Tian *et al.*, 2008b)

$$28Cr_2AlC + 25O_2 \rightarrow 14Al_2O_3 + 8Cr_7C_3 + 4CO_2 \quad (6.28)$$

$$20Cr_2AlC + 15O_2 \rightarrow 10Al_2O_3 + 4Cr_7C_3 + 4Cr_3C_2 \quad (6.29)$$

Interestingly, the formation of a Cr_3C_2 underlayer was not observed by others. The oxidation occurs by the inward diffusion of oxygen and the outward diffusion of Al. In this case, clearly Cr and C are rejected at the oxide/substrate interface. The implications of this result are discussed in subsequent text.

At 1200 °C, scale cracking and spalling is observed; at 1300 °C, the oxidation resistance deteriorates quickly as a function of cycling owing to the formation of voids and scale spallation (Lee and Nguyen, 2008). It follows that, despite its excellent oxidation resistance, it is unlikely that Cr_2AlC can be used at temperatures much higher than 1100 °C, or even 1000 °C, because of this propensity for spallation, which can be traced to the relatively high thermal expansion of this compound (Table 4.7).

6.5.1 Parabolic versus Cubic Kinetics

Here again, most authors insist that the oxidation kinetics are parabolic, when in fact they are not. The best evidence that the kinetics are not parabolic is to replot the 1200 °C weight gain results of Lin *et al.* (2007a) as $(\Delta w/A)^3$ versus t, (Figure 6.13b). Least squares fit of the resulting straight lines yield an R^2 of 0.988. When the results of Lin *et al.* are plotted as $(\Delta w/A)^2$ (not shown), the R^2 is only 0.936. In other words, $(\Delta w/A)^3$ versus t gives the better fit. The situation at 1300 °C – shown as an inset in Figure 6.13b – is even more compelling. Even the cubic fit is good only up to 60 h. As important, the k_c values calculated from Figure 4 in the paper by Lin *et al.* (2007a) are in good agreement with those obtained for Ti_2AlC (compare the

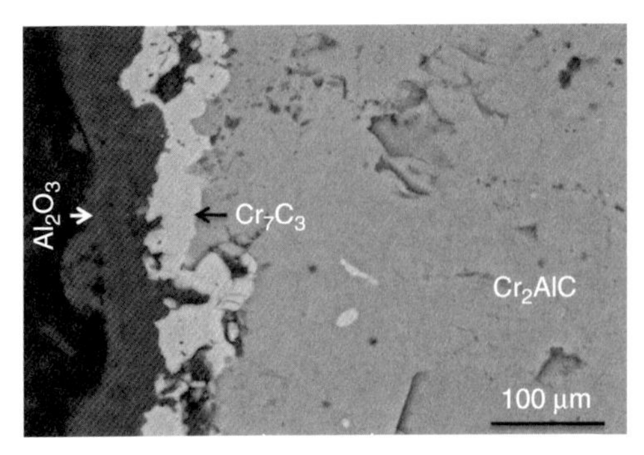

(a)

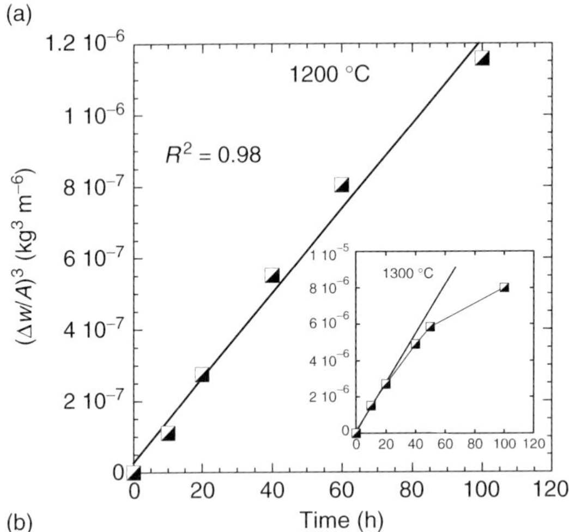

(b)

Figure 6.13 (a) SEM image of Cr_2AlC sample oxidized at 1200 °C. The outer layer is a Cr-containing Al_2O_3 and the inner layer is Cr_7C_3 (S. Gupta, unpublished results). (b) Replot of weight gain results shown in Figure 4 in Lin *et al.* (2007a) for the oxidation of Cr_2AlC in air at 1200 °C as $(\Delta w/A)^3$. Least squares fit of the results yield an R^2 of 0.988. The inset in (b) is the plot of $(\Delta w/A)^3$ versus t for oxidation at 1300 °C. In all cases, the cubic fit is better than the parabolic one (Tallman, Anasori and Barsoum, 2013).

results for Ti_2AlC and Cr_2AlC in Figure 6.7a), again not a very surprising result considering that in all cases an Al_2O_3 protective layer forms. This point needs to be emphasized: parabolic kinetics rarely result in oxidation-resistant alloys. For extended use at high temperatures the kinetics have to be slower than parabolic.

6.5.2
What Renders Cr_2AlC Different?

At this point, it is instructive to ask the question: why does Cr_2AlC form a Cr_7C_3 underlayer (Figure 6.13a), while most other MAX phases, most notably the Ti-containing ones, do not? The simplest answer is that in the latter, the alumina layer that forms is not totally impervious to the outward diffusion of C. In the case of Cr_2AlC, on the other hand, one has to conclude that the opposite is true. This is especially true at lower temperatures. For example, after 20 h at 1000 °C, the oxide layer that forms on Cr_2AlC is 400 nm thick. Under the same conditions, at 1.5 μm, the oxide layer that forms on Ti_2AlC is about three times as thick (Wang and Zhou, 2003a). This difference in kinetics can possibly be related to the presence of dissolved Cr in the Al_2O_3 layers that form and/or lower grain boundary diffusion of the C interstitials through the oxide layers.

Note that the behavior of Cr_2AlC cannot be traced to the stability of Cr_7C_3 *vis-à-vis* say TiC_x because the latter is significantly more stable. It is also unlikely that the anomalous behavior is associated with a low diffusivity of Al in Cr_2AlC because there is no reason to believe that the latter is much different from that in Ti_2AlC, which, as discussed in preceding text, is fast enough to prevent any depletion of Al at the oxide/Ti_2AlC interface.

Further evidence for the conclusion that the Al_2O_3 oxide layer that forms on Cr_2AlC is different from the one that forms on Ti_2AlC can be found in the over two orders of magnitude better corrosion resistance of Cr_2AlC to oxidation in the presence of Na_2SO_4 than Ti_2AlC or Ti_3AlC_2 (Lin *et al.*, 2007a).

In summary, the oxidation kinetics of Cr_2AlC are cubic and not parabolic, especially at longer times. The scale that forms on Cr_2AlC in the 1000 and 1100 °C temperature range is different from the one that forms on Ti_2AlC over the same temperature regime and appears to be more impervious to the diffusion of C and/or O. An important question that remains open at this time is what role the presence of an inner Cr_7C_3 layer will have on the long-term oxidation of this compound, especially if microcracks breach the Al_2O_3 layer.

6.6 Nb_2AlC and $(Ti_{0.5},Nb_{0.5})_2AlC$

The oxidation of Nb_2AlC in air was studied in the 650–800 °C temperature range (Salama, El-Raghy, and Barsoum, 2003). The oxidation reaction occurs in two steps. The first is

$$2Nb_2AlC + \frac{(7+4x)}{2}O_2 = 4NbO_x + Al_2O_3 + 2CO_2 \quad (6.30)$$

At longer times and/or higher temperatures, NbO_x and Al_2O_3 react to form $NbAlO_4$ resulting in the overall reaction

$$2Nb_2AlC + 8.5O_2 = Nb_2O_5 + 2NbAlO_4 + 2CO_2 \quad (6.31)$$

At 100%, the volume change of this reaction is relatively large.

Figure 6.14a shows a typical backscattered SEM image of the morphology of the oxide layer that forms on Nb_2AlC after an isothermal anneal in air at 700 °C for 9 h. Inset in Figure 6.14a, shows the corner of a sample oxidized for 16 h at 750 °C. The large crack at the corner (inset in Figure 6.14a) is a good indication that the residual stresses on cooling must have been large.

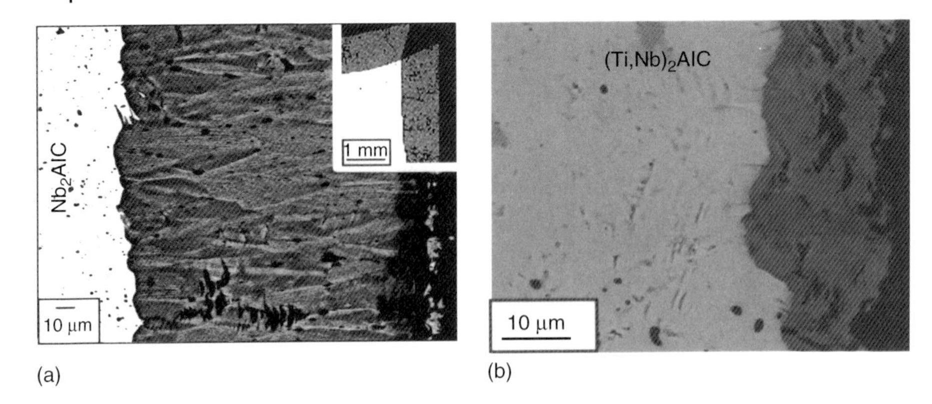

Figure 6.14 Cross-sectional backscattered SEM image of (a) Nb_2AlC sample oxidized at 700 °C for 9 h. The inset shows the corner of a sample oxidized for 16 h at 750 °C. (b) TiNbAlC sample oxidized for 64 h at 900 °C. (Salama, El-Raghy, and Barsoum, 2003).

The oxidation kinetics of $(Ti,Nb)_2AlC$, henceforth referred to as *TiNbAlC*, were also studied in the 900–1200 °C temperature range. In that range, the oxidation occurs in two steps. The first one is

$$TiNbAlC + 4O_2 = 0.5Nb_2O_5 + 0.5Al_2O_3 + TiO_2 + CO_2 \tag{6.32}$$

At longer times and/or higher temperatures, Nb_2O_5 and Al_2O_3 react to form $NbAlO_4$, resulting in the overall oxidation reaction

$$TiNbAlC + 4O_2 = NbAlO_4 + TiO_2 + CO_2 \tag{6.33}$$

At 82%, the volume change of this reaction is also relatively large.

However, given that the evidence indicates the presence of Nb_2O_5 and Al_2O_3 at all times and temperatures, a more realistic reaction is

$$\begin{aligned} TiNbAlC + 4O_2 = {} & (0.5 - x)\, Nb_2O_5 + (0.5-x)\, Al_2O_3 \\ & + 2xNbAlO_4 + TiO_2 + CO_2 \end{aligned} \tag{6.34}$$

for $x < 0.5$. In all cases, the oxidation occurs by the inward diffusion of oxygen. The C is presumed to diffuse out through the oxide layer and oxidize.

Figure 6.14b shows a typical backscattered SEM image of the morphology of the oxide layer that forms on TiNbAlC after an isothermal anneal in air for 64 h at 900 °C.

At 650 °C temperature, the oxidation kinetics of Nb_2AlC were subparabolic (Figure 6.15a), and the oxide thickness x could be described by

$$x\ (\mu m) = A\left(\frac{t}{t_o}\right)^{\nu} \tag{6.35}$$

For Nb_2AlC at 650 °C, $\nu \approx 0.4 \pm 0.02$ and $A = 16.1\ \mu m$. At 700 and 750 °C, the oxidation kinetics of Nb_2AlC are linear (Figure 6.15a), with rates that are comparable to those for the oxidation of pure Nb (Salama, El-Raghy, and Barsoum, 2003).

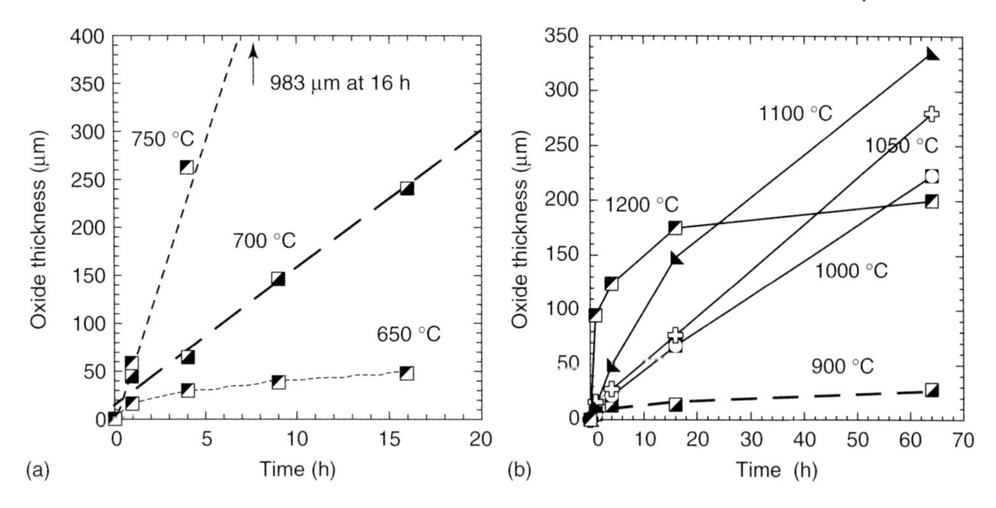

Figure 6.15 Time and temperature dependencies of total oxide thickness formed on (a) Nb_2AlC. The dashed line labeled 650 °C is a plot of x (μm) = 16 $(t/t_0)^{0.4}$, where t_0 is 1 h. The other dashed lines are linear least squares fit of the data and (b) TiNbAlC. The dashed line labeled 900 °C is a plot of x (μm) = 6.5 $(t/t_0)^{0.34}$ (Salama, El-Raghy, and Barsoum, 2003). The other lines are guides to the eyes.

A similar behavior was obtained when TiNbAlC was oxidized at 900 °C in air (Figure 6.15b), except in this case $\nu = 0.34 \pm 0.02$ and $A = 6.6$ μm (see dashed line labeled 900 °C in Figure 6.15b). This implies that this compound can be used in air at 900 °C.

At temperatures between 1000 and 1100 °C (Figure 6.15b), the oxidation kinetics for TiNbAlC start parabolic, but revert to linear at longer times (Salama, El-Raghy, and Barsoum, 2003). At 1200 °C, the oxidation is parabolic up to about 16 h; beyond that time, the oxide layers tend to spall off. At that temperature a niobium aluminide phase is observed at the oxide/MAX interface after 4 h of oxidation. Further oxidation leads to the formation of an Al_2O_3-rich oxide scale at the oxide/substrate interface. The effect of the formation of this Al_2O_3 on the kinetics can be seen in the 1200 °C results shown in Figure 6.15b. This result is tantalizing and suggests that it may be possible to engineer a TiNbAlC phase that forms a protective Al_2O_3 layer upon oxidation.

A comparison of the result shown in Figures 6.15a,b establishes that the oxidation kinetics of Nb_2AlC are significantly faster than those of TiNbAlC. Oxide spallation limited the study of the oxidation of Nb_2AlC to temperatures below 800 °C. In contrast, measurable ($x > 5$ μm) oxide layers did not form on TiNbAlC at temperatures much below 900 °C. Comparing Eqs. (6.30) and (6.32), it is reasonable to conclude that this enhancement in oxidation resistance must be due to the formation of TiO_2, a conclusion consistent with the fact that TiO_2 is a much more protective oxide than Nb_2O_5 (Kofstad, 1966). Said otherwise, the formation of TiO_2 substantially enhances the oxidation resistance of TiNbAlC relative to that of Nb_2AlC.

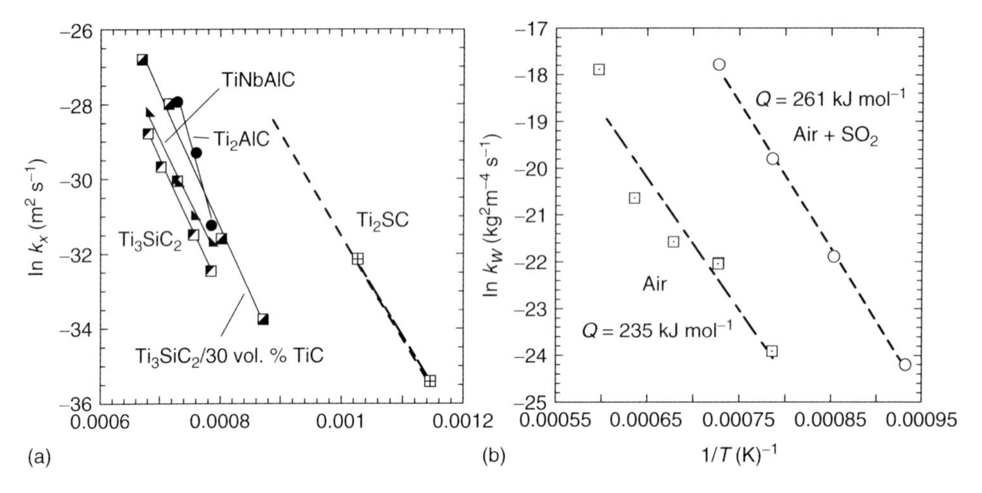

Figure 6.16 Arrhenius plots of (a) k_x for a number of rutile-forming MAX phases and (b) k_w for Ti_3AlC_2 in the absence and presence of SO_2 gas. The former are from Wang and Zhou (2003b) and the latter from Lee and Park (2011).

Furthermore, it is well established in the titanium aluminide literature that the addition of Nb enhances their oxidation resistance (Chen, Sun, and Zhou, 1992; Chen and Rosa, 1980; Choudhury, Graham, and Hinze, 1976; Mungole, Balasubramaniam, and Gosh, 2000; Roy, Balasubramaniam, and Gosh, 1996; Yoshihara and Miura, 1995). The mechanism is believed to be a reduction in oxygen vacancy concentration due to the following defect incorporation reaction:

$$Nb_2O_5 + V_O^{\bullet\bullet} \Rightarrow 2Nb_{Ti}^{\bullet} + 5O_O^x \tag{6.36}$$

It is, therefore, not surprising that the oxidation kinetics of TiNbAlC are even better than those of Ti_2AlC. This is most clearly seen in Figure 6.16a, where the k_x values for a number of MAX phases are plotted. It is obvious that the k_x values for TiNbAlC are closer to those of Ti_3SiC_2 than Ti_2AlC (Salama, El-Raghy, and Barsoum, 2003). This result suggests that the inward diffusion of oxygen is the rate-limiting step.

6.7 Ti_2SC

The oxidation behavior of fully dense, predominantly single-phase Ti_2SC sample was studied thermogravimetrically in air in the 500–800 °C temperature range (Amini, McGhie, and Barsoum, 2009; Kulkarni *et al.*, 2009). In both cases, the oxidation product was a single TiO_2 layer. At 800 °C, the oxide layer was not protective and oxidation was quite rapid. At 600 and 700 °C – and up to ~50 h – the kinetics were parabolic before they reverted to linear. It was only at 500 °C that the weight gain reached a plateau after a 50 h initial parabolic regime. Mass spectrometry of the gases evolved during oxidation confirmed that both CO_2 and

SO_2 were oxidation products (Amini, McGhie, and Barsoum, 2009). The overall oxidation reaction is thus

$$Ti_2SC + 4O_2 = 2TiO_2 + SO_2 + CO_2 \tag{6.37}$$

The oxidation occurs by the outward diffusion of Ti, S, and C and, most probably, the inward diffusion of oxygen.

Mesopores and microcracks were found in all TiO_2 layers formed except those formed at 500 °C. The presence of these defects is believed to lead to significantly higher oxidation rates as compared to other rutile-forming ternary carbides, such as Ti_3SiC_2. Figure 6.16a compares the k_x values for the oxidation of Ti_2SC with those of other rutile-forming MAX phases. Strikingly, and while the absolute k_x values for Ti_2SC are orders of magnitude higher than those of the other compounds, the activation energy for oxidation is apparently unchanged. In other words, the pores and cracks did not affect the activation energy; only the pre-exponential term. How or why this occurs is unclear at this time.

6.8 V_2AlC and $(Ti_{0.5},V_{0.5})_2AlC$

The oxidation behavior of V_2AlC and $(Ti_{0.5},V_{0.5})_2AlC$ – henceforth referred to as *TiVAlC* – was studied thermogravimetrically in air in the 500–700 °C temperature range (Gupta and Barsoum, 2004).

The oxidation kinetics for V_2AlC are shown in Figure 6.17a. At 500 and 600 °C, the oxidation rate is initially linear, but after ≈2 h it slows down appreciably. At 700 °C, the oxidation rate is significantly faster.

At 500 and 600 °C, the scales formed are layered and protective up to at least 24 h. The outermost layers are VO_2 at 500 °C and V_2O_5 at 600 °C; the inner layer

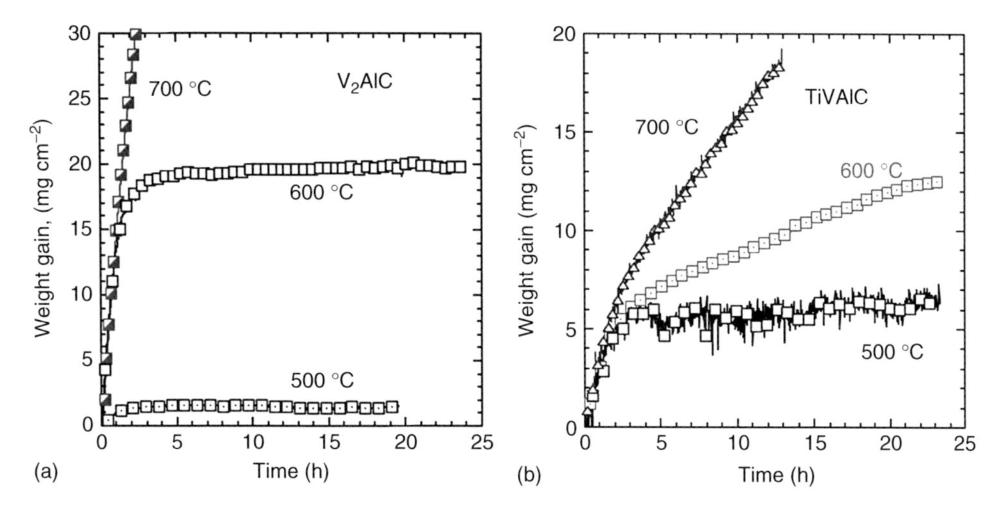

Figure 6.17 Normalized weight gain as a function of time and temperature during the oxidation of (a) V_2AlC and (b) $(Ti_{0.5},V_{0.5})_2AlC$ in air (Gupta and Barsoum, 2004).

compositions are myriad solid solutions of V, Al, and O in varying ratios. The average V oxidation state decreases from +4 or +5 at the air/oxide interface to +3 at the substrate/oxide interface.

At 500 °C the most probable oxidation reaction operative when V_2AlC is oxidized in air is

$$V_2AlC + 3.75O_2 = 2VO_2 + 0.5Al_2O_3 + CO_2 \tag{6.38}$$

The oxidation is governed by the inward diffusion of O, and, possibly, the outward migration of V; the Al atoms are essentially immobile (Gupta and Barsoum, 2004). Similar to most MAX phases, C is assumed to diffuse through the reaction layer and oxidize.

If one neglects the myriad V–Al mixed oxides that form at 600 °C, the most likely oxidation reaction is

$$V_2AlC + 4.25O_2 = 0.5V_2O_5 + VAlO_4 + CO_2 \tag{6.39}$$

A more generalized reaction would read

$$\begin{aligned} V_2AlC + (4.25 - 1.25x - 0.75y + 0.5z)\,O_2 &= 0.5\,(1-x+y)\,V_2O_5 \\ &+ (1-y)\,VAlO_4 + V_xAl_yO_z + CO_2 \end{aligned} \tag{6.40}$$

In this reaction, the values of x, y, and z depend on the distance from the oxide/air interface. At the air/oxide interface x, y, and z can be neglected, and Eq. (6.39) is operative.

At 700 °C, the oxidation reaction results in V_2O_5 and other V-Al mixed oxides with varying V oxidation states. Some of the oxides formed at this temperature are liquid and have a tendency to bleed away, which results in poor oxidation resistance.

When the time dependencies of the weight gains for TiVAlC, heated in air at various temperatures, are plotted (Figure 6.17b) it is clear that at 600 °C and above the oxides formed are not protective. The most likely oxidation reactions for TiVAlC at 500 and 600 °C are, respectively

$$(Ti_{0.5}, V_{0.5})_2AlC + 4O_2 = 3\,(Ti_{1/3}, V_{1/3}, Al_{1/3})\,O_2 + CO_2 \tag{6.41}$$

$$(Ti_{0.5}, V_{0.5})_2AlC + 3.25O_2 = \frac{3}{2}(Ti_{1/3}, V_{1/3}, Al_{1/3})_2O_3 + CO_2 \tag{6.42}$$

The oxide that forms at 500 °C is an MO_2 oxide; that at 600 °C is M_2O_3. Note that the oxidation state of all cations in Eq. (6.42) is +3.

In contrast to the oxide films that form on V_2AlC, the ones that form on the solid solution are amorphous or nanocrystalline. They are also not protective, which implies that the maximum use temperature in air of TiVAlC will have to be <500 °C. Similar to the V_2AlC, however, the oxidation occurs by the inward diffusion of oxygen. Cation diffusion appears to occur only at the local scale. (Gupta and Barsoum, 2004).

6.9 Ti_3GeC_2 and $Ti_3(Si,Ge)C_2$

The oxidation of Ti_3GeC_2 and $Ti_3(Ge_{0.5},Si_{0.5})C_2$ was studied thermogravimetrically in air in the 700–1000 °C temperature range (Gupta *et al.*, 2006). The oxidation layers formed on Ti_3GeC_2 at 700 °C were protective; at 800 °C and higher, they were not. The addition of Si to Ti_3GeC_2 slightly enhanced the latter's oxidation resistance, but at 800 °C and above, the layers formed were again not protective. In both cases, the oxidation occurred mostly by the inward diffusion of oxygen through a rutile-based $(Ti_{1-y},Ge_y)O_2$ solid solution with $y < 0.1$.

The overall reaction for the oxidation of Ti_3GeC_2 depends somewhat on temperature. At 700 °C, the operative reaction is most probably

$$Ti_3GeC_2 + 6O_2 \rightarrow 3\left(Ti_{0.9},Ge_{0.1}\right)O_2 + \left(Ti_{0.3},Ge_{0.7}\right)O_2 + 2CO_2 \tag{6.43}$$

With increasing temperatures and/or times, the rutile-based Ti-rich solid solution contains less Ge. At the highest temperatures, the concentration of Ge in the rutile was negligible. For both ternary compounds, the *a* and *c* lattice parameters of the rutile that forms were in between those of pure TiO_2 and GeO_2. This is strong evidence for the formation of a solid solution and is in agreement with the phase diagram for this system (Levin, Robbins, and McMurdie, 1964).

At higher temperatures, near the oxide/Ti_3GeC_2 interface, an oxygen-induced decomposition of the ternary according to the following simplified reaction

$$Ti_3GeC_2 + \left(2 + 1.5x - 3y\right)O_2 \rightarrow 3TiO_xC_y + Ge + \left(2 - 3y\right)CO_2 \tag{6.44}$$

is believed to occur. The reaction is simplified in that the oxycarbide that forms is assumed to be Ge-free, when in reality that may not be the case. With time, the oxycarbide is further oxidized to rutile according to the reaction

$$3TiO_xC_y + \left(3 - 1.5x + 3y\right)O_2 \rightarrow 3TiO_2 + 3yCO_2 \tag{6.45}$$

This is followed, ultimately, by

$$Ge + O_2 \rightarrow GeO_2 \tag{6.46}$$

On the basis of the XRD patterns and EDS analysis, it was concluded that at least initially, the oxidation reaction of $Ti_3(Ge_{0.5},Si_{0.5})C_2$ bulk samples is

$$Ti_3\left(Ge_{0.5}, Si_{0.5}\right)C_2 + 6O_2 \rightarrow 4\left(Ti_{0.75}, Ge_{0.125}, Si_{0.125}\right)O_2 + 2CO_2 \tag{6.47}$$

or, more likely

$$Ti_3\left(Ge_{0.5}, Si_{0.5}\right)C_2 + 6.5O_2 \rightarrow 4\left(Ti_{0.75}, Ge_{0.125}\right)O_2 + 0.5SiO_2 + 2CO_2 \tag{6.48}$$

The uncertainty here lies in not knowing whether the Si signal detected in the EDS originated from a distinct SiO_2 phase or from a quaternary – Ti, Si, Ge, O – solid solution.

At higher temperatures and/or at longer times, if the quaternary solid solution forms, it most probably dissociates according to the following simplified reaction:

$$4\left(Ti_{0.75}, Ge_{0.125}, Si_{0.125}\right)O_2 \rightarrow 3TiO_2 + 0.5SiO_2 + 0.5GeO_2 \tag{6.49}$$

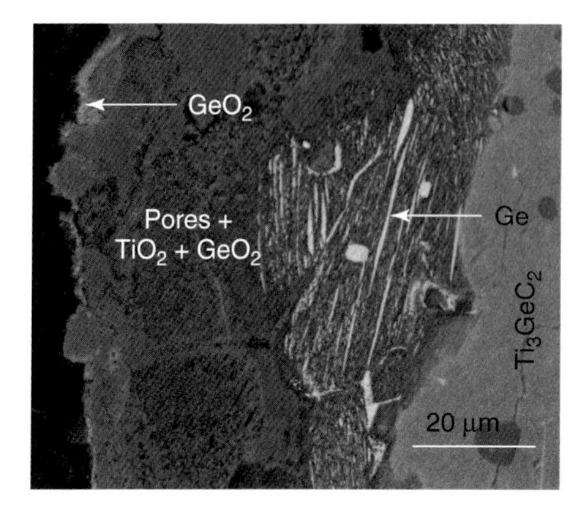

Figure 6.18 Backscattered SEM image of a Ti_3GeC_2 sample oxidized at 900 °C for 4 h showing dissociated grains at the oxide/carbide interface. The bright regions were identified to be elemental Ge (Gupta *et al.*, 2006).

This reaction is simplified in that it assumes that the TiO_2 that forms is pure, when in fact it only becomes purer with longer oxidation times and/or higher oxidation temperatures. If only a Ti–Ge–O solid solution forms (i.e., Eq. (6.48)), then the likely dissociation reaction is the same as Eq. (6.49), but without the SiO_2.

The best evidence for Eq. (6.44) can be seen in Figure 6.18. Just below the Ti_3GeC_2/oxide interface, a bright phase, identified as Ge by EDS, is clearly seen. Furthermore, the Ge formed is clearly striated and parallel to the grain's basal planes, which is strong evidence for an oxygen induced dissociation. At 1000 °C, a fraction of the Ge escapes to the surface, presumably through micropores and or fissures, and forms copious amounts of GeO_2 whiskers – with hexagonal symmetry. In the temperature range explored, there is no oxygen dissolution in the MAX matrices. This observation is generally true of most MAX phases oxidized to date.

6.10
Ta_2AlC

The oxidation behavior in air of bulk polycrystalline Ta_2AlC samples was studied in the 600–900 °C temperature range (Gupta, Filimonov, and Barsoum, 2006). In the 700–900 °C temperature range, the oxidation kinetics were linear. The weight gain at 600 °C was lower than the sensitivity level of the thermobalance used. It follows that a protective, essentially X-ray-amorphous, oxide layer comprised of Ta, Al, and O forms at 600 °C. However, this layer is not resistant to thermal cycling. At 700 °C, and above, the reaction layers are compositionally quite uniform, but porous and highly cracked. The oxide layers consisted of the crystalline phases Ta_2O_5 and $TaAlO_4$, and an XRD-amorphous phase.

It follows that the oxidation of Ta_2AlC occurs according to the following reactions:

$$Ta_2AlC + O_2 \rightarrow {}^*2Ta_2O_5/Al_2O_3{}^* + 2CO_2 \quad (6.50)$$

$${}^*2Ta_2O_5/Al_2O_3{}^* \rightarrow 2TaAlO_4 + Ta_2O_5 \quad (6.51)$$

where $*2Ta_2O_5/Al_2O_3*$ designates a microconstituent, which is not necessarily a single phase but could also be multiphasic at the nanoscale or XRD-amorphous. At higher temperatures and/or longer times, Eq. (6.51) is favored, although weak peaks for $TaAlO_4$ were observed after oxidation for 24 h at 600 °C. No Al_2O_3 peaks were observed in the XRD patterns.

6.11 Ti_2SnC, Nb_2SnC, and Hf_2SnC

Zhou *et al.* were the first to report on the oxidation of Ti_2SnC (Zhou, Dong, and Wang, 2004) in the 500–800 °C temperature range. They showed that the oxidation process was somewhat complex. Initially, the Sn is preferentially oxidized according to the following reaction:

$$Ti_2SnC + O_2 = SnO_2 + 2TiC_{0.5} \quad (6.52)$$

$TiC_{0.5}$ is then oxidized to form rutile according to

$$Ti_2C + 3O_2 = 2TiO_2 + CO_2 \quad (6.53)$$

Further oxidation occurs by the outward diffusion of Ti and C and the inward diffusion of O. As in Ti_3SiC_2, the Sn is oxidized *in situ*. As the TiO_2 layer gets thicker, however, an oxygen-induced decomposition of the ternary, according to the following simplified reaction, is presumed to occur:

$$Ti_2SnC + (x + 1 - 2y)\,O_2 \rightarrow 2TiO_xC_y + Sn + (1 - y)\,CO_2, \quad y < 1 \quad (6.54)$$

We note in passing that, while Zhou, Dong, and Wang (2004) mention and show evidence for the presence of Sn, none of their reactions accounted for its presence. Equation (6.55) is thus needed to explain that aspect of their results.

Chakraborty, El-Raghy, and Barsoum (2003) reported on the oxidation behavior in air of fully dense, predominantly (92 vol%) single phase samples of Nb_2SnC and Hf_2SnC in the 400–600 °C temperature range. The oxidation products in both cases were SnO, SnO_2, and either Nb_2O_5 or HfO_2. The oxidation was characterized by an initial incubation period, the duration of which decreased with increasing temperatures, followed by a period of near-linear oxidation kinetics. Oxidation at temperatures as low as 400 °C for 72 h resulted in the disintegration of bulk samples. Comparison with published results indicates that the oxidation kinetics of Hf_2SnC are poorer than those of Hf or HfC. Similarly, Nb_2SnC is more prone to oxidation than pure Nb metal.

6.12
Ti_2InC, Zr_2InC, $(Ti_{0.5},Hf_{0.5})_2InC$, and $(Ti_{0.5},Zr_{0.5})_2InC$

Gupta, Hoffman, and Barsoum (2006) studied the oxidation behavior of Ti_2InC, Zr_2InC, $(Ti_{0.5},Hf_{0.5})_2InC$, and $(Ti_{0.5},Zr_{0.5})_2InC$ in air in the 400–700 °C temperature range. With the exception of Ti_2InC, the oxide layers formed failed to offer any oxidation resistance at 400 °C or higher temperatures. The reaction products were In_2O_3, in which some transition metal ions were probably dissolved, and the respective transition-metal oxides. The latter were either amorphous or nanocrystalline.

In the case of Ti_2InC, a protective oxide was formed at 500 °C. However, at higher temperatures, after an incubation period, catastrophic oxidation ensued. In all cases, the oxidation occurred by the inward diffusion of oxygen.

In the remainder of this chapter, the reactions of the MAX phases with gases other than oxygen are reviewed.

6.13
Sulfur Dioxide, SO_2

In gas turbines, heat exchangers, petrochemical plants, and coal gasification, structural components are often exposed to gases containing SO_2, which can induce serious corrosion by disrupting the formation of protective scales.

Lee and Choi (2007) examined the corrosion behavior of Ti_3AlC_2 in the 800–1000 °C temperature range, in an Ar-1% SO_2 gas atmosphere. The scale formed consisted primarily of TiO_2 and α-Al_2O_3. As the corrosion progressed, the scale became thicker and developed into an outer TiO_2 layer and an inner mixed TiO_2 and α-Al_2O_3 layer. The fact that a continuous alumina layer did not form resulted in enhanced oxidation kinetics.

Lee and Park (2011) exposed both Ti_3AlC_2 and an intermetallic with a TiAl overall chemistry to an Ar-0.2 SO_2 gas mixture in the 800–1100 °C temperature range for up to 180 h. The scale thicknesses, after 40 and 60 h on Ti_3AlC_2 at 800 and 900 °C, were too thin to measure accurately. In contrast, at 800 °C, after 40 and 60 h, the scales formed on TiAl were 18 and 22 μm thick, respectively. In other words, the corrosion resistance of Ti_3AlC_2 was significantly better than an intermetallic with a significantly higher Al concentration. This is another example of the high reactivity of the A elements in the MAX phases in general, and Al in particular.

The overall reaction in the presence of SO_2 was identical to that in air, that is, Eq. (6.18). However, and even though the activation energies for oxidation were not much different from those for Ti_3AlC_2 oxidized in air, the kinetics were roughly two orders of magnitude *higher* (see line labeled SO_2 in Figure 6.16b). Since in both cases an Al_2O_3 layer is formed, here again one needs to invoke the presence of parallel oxygen paths. Note the close similarity in behavior of TiO_2-forming (Figure 6.16a) and Al_2O_3-forming (Figure 6.16b) MAX phases in the absence and presence of S. At this time, the role of S remains a mystery.

The scales that formed on Ti_3AlC_2 were thin and rich in α-Al_2O_3, and their growth rate was slow. The TiO_2 was present either as an outermost surface scale, or a mixture inside an α-Al_2O_3-rich scale. In Ti_3AlC_2, the activity and diffusivity of Ti were low, whereas those of Al were high. In the TiAl case, under identical corrosion conditions, the scales that formed consisted of thick triple layers: an outer TiO_2 layer containing some Al_2O_3 particles, an intermediate (Al_2O_3, TiO_2) mixed layer, and an inner narrow (TiS, Ti_2S) mixed layer. In other words, Al_2O_3 did not form, which is why the oxidation kinetics were greatly enhanced in the TiAl case.

This comment notwithstanding, the 2-order increase in the oxidation kinetics (Figure 6.16b) clearly show that the Al_2O_3 layer that forms in the presence of SO_2 is not as protective as the one that forms in the absence of S. More experiments, for much longer times and higher temperatures, are indicated and should be carried out.

6.14
Anhydrous Hydrofluoric, HF, Gas

When Ti_2AlC powders were exposed to anhydrous HF gas at 55 °C for 2 h, a new nanocrystalline phase of titanium aluminum fluoride, with a nominal stoichiometry of Ti_2AlF_9 – exact stoichiometry measured was $Ti_{2.1}Al_{0.9}F_9$ – formed (Naguib *et al.*, 2011). The overall reaction is presumed to be

$$Ti_2AlC + 9HF = Ti_2AlF_9 + CH_4 + 2.5\ H_2 \tag{6.55}$$

The reaction is partially topotactic in that it results in small cuboids of Ti_2AlF_9 arranged in layers that are parallel to each other and parallel to the original basal planes of the Ti_2AlC grains (Figure 6.19).

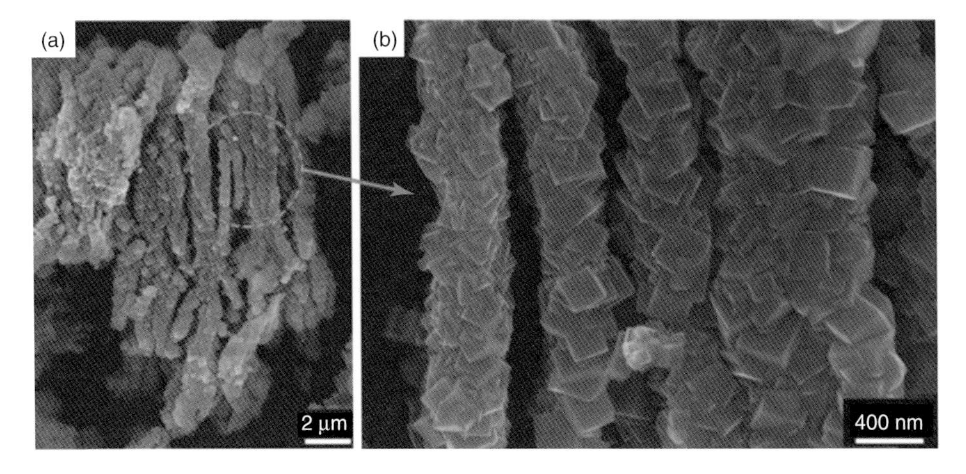

Figure 6.19 SEM images of Ti_2AlC powders after exposure to anhydrous HF at 55 °C for 2 h at (a) low magnification showing partially layered morphology. (b) The same as (a) but at higher magnification showing Ti_2AlF_9 cuboids (Naguib *et al.*, 2011).

6.15 Chlorine Gas

Reacting the MAX phases with Cl_2 gas in the 300–1100 °C temperature range results in the extraction of both the M and A layers leaving behind a microporous carbon, referred to as *carbide-derived carbons* or *CDCs* (Gogotsi *et al.*, 2003; Hoffman *et al.*, 2005; Yushin *et al.*, 2005). The overall reactions are as follows:

$$Ti_3SiC_2 + 8Cl_2 = 2C + 3TiCl_4 + SiCl_4 \quad (6.56)$$

$$Ti_3AlC_2 + 7.5Cl_2 = 2C + 3TiCl_4 + AlCl_3 \quad (6.57)$$

Here again, the reaction is topotactic in the sense that the morphology of the grains does not change as a result of the chlorination. This is best seen by comparing the fractured surfaces of a Ti_3SiC_2 sample before (Figure 6.20a) and after chlorination (Figure 6.20b). The lamellar structure of Ti_3SiC_2 is clearly seen after chlorination. The transformation to CDC generates cracks between the grains, which are presumed to occur when the grains shrink along [0001] as the Ti and Si atoms are removed (Figure 6.20c). This is best seen in Figure 6.20b, where a crack is visible between two grains that were oriented perpendicular to each other before chlorination.

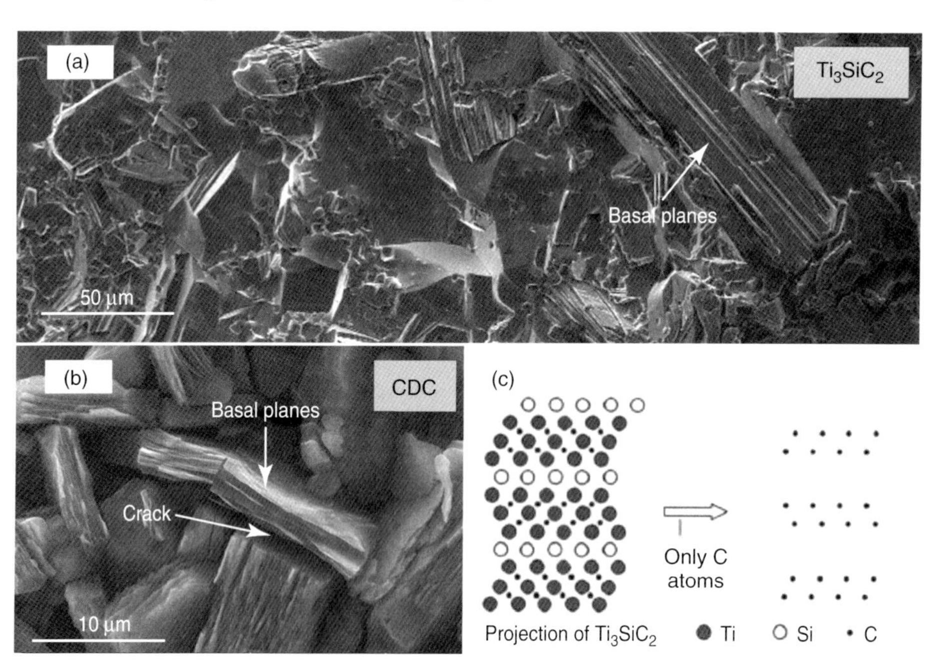

Figure 6.20 SEM images of (a) fracture surface of an as-received Ti_3SiC_2 sample, (b) CDC sample chlorinated for 3 h at 600 °C. The lamellar structure of Ti_3SiC_2 is clearly seen after chlorination. Transformation to CDC generates cracks between the grains. The latter are believed to be due to the shrinkage along [0001] when the Ti and Si atoms are removed from the structure as shown schematically in (c) (Yushin *et al.*, 2005).

6.16 Summary and Conclusions

For the most part the overall oxidation reaction for the MAX phases is given by

$$M_{n+1}AX_n + bO_2 = M_{n+1}O_x + AO_y + X_nO_{2b-x-y}$$

The nature of the oxides that form and their protectiveness, however, are a strong function of chemistry. The maximum allowable oxidation temperatures T_{max} for select MAX phases are listed in Table 6.5. T_{max} is the temperature below which the oxidation kinetics are less severe than parabolic. It is important to note that, with a few exceptions, most of the values of T_{max} listed in Table 6.5 are based on the relatively short oxidation times listed in column 3. Longer oxidation times may very well show that even these temperatures are too high.

From a technological point of view, the most promising MAX phase by far is Ti_2AlC as it forms a dense protective and cohesive α-Al_2O_3 layer. The oxidation kinetics of this phase are cubic up to $\approx$3000 h at 1200 °C and do not appear to be susceptible to severe thermal cycling or moisture. When cracks form, they self-heal repeatedly. This MAX phase is also the lightest and it can be pressureless sintered to full density from commercially available powders (Zhou *et al.*, 2006). This compound is also most readily machinable in its final, fully dense state.

The ternary Cr_2AlC also possesses excellent oxidation resistance, with cubic oxidation kinetics. It is, however, susceptible to thermal cycling because of its relatively high thermal expansion coefficient. The α-Al_2O_3 scale that forms in this case, however, appears to be even more impervious to C and O diffusion than the one that forms on Ti_2AlC. If the thermal cycling problem can be solved, this MAX phase could also become an important HT solid.

Table 6.5 Maximum temperatures T_{max} at which select MAX phases can be heated in air. In all cases, T_{max} is the temperature for which the kinetics are slower than parabolic.

MAX phase	T_{max} (°C)	Time (h)	Susceptibility to thermal cycling	References
Ti_3SiC_2	900	10	No	Barsoum, El-Raghy and Ogbuji (1997)
Ti_3SiC_2 – 30 vol% SiC	925	500	?	Barsoum *et al.* (2003)
Ti_3GeC_2	700	36	?	Gupta *et al.* (2006)
Ti_2AlC	1350	1000	—	Byeon *et al.*, (2007)
Ti_2AlC	1400	25	No	Byeon *et al.* (2007)
Ti_2AlC	1300	100	No	Basu *et al.* (2012)
Ti_2AlC	1200	3000	No	Tallman, Anasori and Barsoum (2013)
V_2AlC	600	24	?	Gupta and Barsoum (2004)
Cr_2AlC	1300	24	Yes	Lin *et al.* (2007a)
Nb_2AlC	650	15	?	Salama, El-Raghy, and Barsoum (2003)
TiNbAlC	900	65	?	Salama, El-Raghy, and Barsoum (2003)
Ti_2SC	500	300	?	Amini, McGhie, and Barsoum (2009)
Ta_2AlC	600	100	Yes	Gupta, Filimonov, and Barsoum (2006)

6.A Appendix

Oxidation of $Ti_{n+1}AlX_n$ When Alumina Does Not Form a Protective Layer

To model this system, it is important to appreciate the role Al^{3+} ions play in rutile. Given that the Schottky and Frenkel defects' enthalpies of formation in TiO_2 are quite large (Ikeda, Chiang, and Fabes, 1993; Kofstad, 1967), it is unlikely that the intrinsic diffusivity of O^{2-} ions in TiO_2 can ever be measured: a fact that is well recognized in the literature (Arita *et al.*, 1979; Haul and Dumbgen, 1965). For example, Arita *et al.* assumed $[V_O]$ to be equal to the mole fraction of MgO impurities in their crystals. Similarly, Haul and Dumbgen measured the O^{2-} diffusion D_O in TiO_2 as a function of temperature (710–1300 °C) and oxygen partial pressure P_{O_2}. They took the nondependence of D_O on P_{O_2} as evidence that D_O was dominated by the concentration of extrinsic vacancies formed as a result of the 100–200 ppm Al_2O_3 impurity level present in their crystals. As discussed subsequently, the dissolution of the Al^{3+} ions in TiO_2, and the concomitant increase in $[V_O]$, is believed to play an important role during the oxidation of the $Ti_{n+1}AlX_n$ phases.

In many of the samples oxidized, an initial uniform distribution of Al in the oxide layer was replaced, at longer times, by some areas that were Al_2O_3-rich and some that were almost totally denuded of it. This is best seen in Figure 6.11a, where after 16 h oxidation at 1100 °C, the concentration of Al in the region marked "D" was uniform. This, in turn, implies that the oxidation, initially at least, forms a solid solution of Al_2O_3 in TiO_2 or $(Ti_{1-y}Al_y)O_{2-y/2}$, where $y < 0.05$. At longer times, demixing was observed (Figure 6.11b). This observation implies that the Al^{3+} ions must first dissolve in, and then diffuse through, TiO_2.

The dissolution and outward diffusion of Al^{3+} ions from the low oxygen partial pressure $P^{II}_{O_2}$ side to the high $P^{I}_{O_2}$ side is a form of kinetic demixing, described and analyzed by Schmalzried and coworkers (Schmalzreid, 1986). On the basis of their work, the condition for kinetic demixing of a $(Ti_{1-y}Al_y)O_{2-y/2}$ solid solution is

$$\ln \frac{P^{II}_{O_2}}{P^{I}_{O_2}} > \left[\frac{D_{Al} + D_{Ti}\left({}^{y}\!/\!{}_{1-y} \right)}{6D_{Al} - 2D_{Ti}} \right] \frac{\Delta G_{TiO_2}}{RT} \qquad (6.A.1)$$

where ΔG_{TiO_2} is the free energy of formation of TiO_2. Since in this case $D_{Al} \gg D_{Ti}$, the term in brackets reduces to a value of +1/6, and the right hand side of Eq. (6.A.1) is negative. It can be shown that at all temperatures, the left-hand side is always $> \Delta G_{TiO_2}/6RT$, implying that the Al_2O_3 should ultimately precipitate at the high P_{O_2} side, as observed (Barsoum, 2001).

The most likely set of defect reactions, the rationale for which is discussed subsequently, occurring at the substrate/oxide interface during the oxidation of the $Ti_{n+1}AlX_n$ are (Barsoum, 2001)

$$y\text{Al} \xrightarrow[\text{TiO}_2]{} y\text{Al}'_{\text{Ti}} + 2y\text{V}_{\text{O}}^{\bullet\bullet} + 3ye' \tag{6.A.2}$$

$$(n+1)\,\text{Ti} \xrightarrow[\text{TiO}_2]{} (n+1)\text{Ti}_{\text{Ti}}^{x} + 2(n+1)\text{V}_{\text{O}}^{\bullet\bullet} + 4\,(n+1)\,\text{e}' \tag{6.A.3}$$

$$z\,\text{Al} \xrightarrow[\text{TiO}_2]{} z\text{Al}_{\text{i}}^{\bullet\bullet\bullet} + 3ze' \tag{6.A.4}$$

Conversely, at the oxide/air or the high P_{O_2} side

$$\left[(n+1) + \frac{3y}{4}\right]\text{O}_2(\text{g}) + \left[2\,(n+1) + \frac{3y}{2}\right]\text{V}_{\text{O}}^{\bullet\bullet} + [4\,(n+1) + 3y]\,\text{e}' \xrightarrow[TiO_2]{} \left(\frac{4\,(n+1)}{2} + \frac{3y}{2}\right)\text{O}_{\text{O}}^{x} \tag{6.A.5}$$

$$z\text{Al}_{\text{i}}^{\bullet\bullet\bullet} + 3ze' + \frac{3z}{4}\text{O}_2\,(\text{g}) \rightarrow z\text{Al}_{\text{Al}}^{x} + 1.5z\text{O}_{\text{O}}^{x} \tag{6.A.6}$$

For a net overall reaction

$$\text{Al} + (n+1)\,\text{Ti} + \left[(n+1) - \frac{y-3z}{4}\right]\text{O}_2 \Rightarrow (n+2)\left[\text{Al}_{y/n+2}\text{Ti}_{1-(1/n+2)}\right]\text{O}_{[2(n+1)/(n+2)]-(y/2(n+2))} + \frac{z}{2}\text{Al}_2\text{O}_3 \tag{6.A.7}$$

subject to the constraint that $y + z = 1$, and (see subsequent text) $y \ll z$. As noted earlier, the C and N atoms are assumed to diffuse through the layer and react with oxygen independently. They are not included in Eq. (6.A.7) for the sake of simplicity.

The rationale for the choice of defect reactions is the following:

- Equation (6.A.2) is consistent with the fact that Al-doped TiO_2 is an n-type semiconductor at low to moderate P_{O_2}s (Blumenthal *et al.*, 1966; Yahia, 1963). It is also consistent with the general conclusion that the dissolution of Al_2O_3 in TiO_2 results in the creation of V_O (Arita *et al.*, 1979; Haul and Dumbgen, 1965).
- Equation (6.A.3) assumes that the Ti^{4+} ions do not diffuse and are oxidized *in situ*. This assumption is not correct, but is made for the sake of simplicity, and at most underestimates the oxide thickness by a factor of 2.
- Equation (6.A.4) is consistent with literature results that have shown that Al_i diffuse rapidly in TiO_2 (Ikeda, Chiang, and Fabes, 1990; Yan and Rhodes, 1982).
- Equation (6.A.5) is written separately from Eq. (6.A.2) to emphasize that the Al_i diffuse outward, in parallel with the oxygen vacancies, and are presumably the ones responsible for the formation of the Al_2O_3 islands observed. The concentrations of the Al'_{Ti} and $Al_i^{\bullet\bullet\bullet}$ ions are related by (Ikeda, Chiang, and Fabes, 1990) by

$$2\text{Al}'_{\text{Ti}} + \text{V}_{\text{O}}^{\bullet\bullet} + 3\,\text{O}_{\text{O}}^{x} \xrightarrow[TiO_2]{} 2\text{Al}_{\text{i}}^{\bullet\bullet\bullet} + \frac{3}{2}\text{O}_2\,(\text{g}) + 6e' \tag{6.A.8}$$

In other words, the concentration of $Al_i^{\bullet\bullet\bullet}$ ions is high at the lower P_{O_2} side and vice versa. It is this concentration gradient that is ultimately responsible for the outward diffusion of Al.

- Equation (6.A.5) is a standard oxygen incorporation reaction and is consistent with the fact that oxygen diffusion in TiO_2 is believed to occur via oxygen vacancies.
- The precipitation of Al_2O_3 is given by Eq. (6.A.6).

At this juncture, it is possible to apply Wagner's oxidation model (Wagner, 1933) to the problem at hand to try and determine the nature of the rate-limiting step. Not surprisingly, the conclusion is the same as the one reached earlier, viz. that the oxidation rate is limited by the diffusion of the Ti and/or O ions. In this case, however, instead of calculating $k_{x,\text{theo}}$ from Eq. (6.11), the reverse is carried out; the k_x values are converted to diffusivities and compared with literature results.

Table 6.A.1 summarizes the k_x values for a number $Ti_{n+1}AlX_n$ phases. As noted in preceding text, the kinetics are typically parabolic, especially initially.

In these systems, there are five possible rate-limiting entities: electronic defects, Al^{3+}, Ti^{4+}, O^{2-}, and C or N. On the basis of the preceding discussion, the first two can be safely eliminated. The last two can also be eliminated as they have never been identified to be rate limiting in the oxidation of some well-studied systems such as SiC, Si_3N_4, TiN, or TiC. Thus, not surprisingly, the diffusion of O and/or Ti ions is rate- limiting. For the sake of simplicity, it is assumed that the diffusion of O^{2-} is rate-limiting. Wagner showed that the parabolic rate constant for the formation of a protective oxide layer – TiO_2 in this case – in air is given by (Barsoum 2003; Wagner 1933):

$$k_x = -\left(\frac{RT\Omega_{\text{TiO}_2}}{(4F)^2}\right)\frac{\sigma_{\text{def}}\sigma_e}{\sigma_{\text{def}}+\sigma_{\text{e}}}\ln\frac{P_{\text{O}_2}}{0.21} \tag{6.A.9}$$

where σ_{def} and σ_{e} are, respectively, the average values of the ionic and electronic conductivities across the growing oxide layer; P_{O_2} is the partial pressure of oxygen at the substrate/oxide interface in atm. Ω_{TiO_2} is the molar volume of TiO_2; F is Faraday constant.

Making use of the Nernst–Einstein relationship, and the fact that at $T > 800\,^\circ\text{C}$ $\sigma_{\text{e}} \gg \sigma_{\text{ion}}$, Eq. (6.A.9) can be recast to read (Barsoum, 2003)

$$D_{\text{RL}} = \frac{k_x}{\ln\left(0.21/P_{\text{O}_2}\right)} \tag{6.A.10}$$

where D_{RL} is the diffusion coefficient of the rate-limiting ion.

In deriving this relationship, the following implicit assumptions were made:

1) Al_2O_3 acts as an inert phase through which oxygen does not diffuse. This is a good assumption because Al_2O_3 is a much more protective oxide than TiO_2.
2) Al_2O_3 does not affect the cross-sectional area through which the oxygen is diffusing. This assumption is reasonable because the D_{RL} values are calculated from short-time results where the volume fraction of precipitated Al_2O_3 is low. It is also a better approximation as the ratio of Ti/Al in the MAX phases

Table 6.A.1 Summary of k_x values for the oxidation of a number of $Ti_{n+1}AlX_n$ phases and Ti_3SiC_2. The denominator in Eq. 6.26 is listed in column 4; column 5 lists D_{RL} calculated from Eq. 6.26; Column 6 lists D_{vo} calculated from Eq. (6.15), and the corresponding y values (assuming $D_{RL} = D_o$ in Eq. 6.A.14.).

Composite	Temperature (°C)	k_x ($m^2\ s^{-1}$)	$-\ln P_{O_2}/0.21$	D_{RL} ($m^2\ s^{-1}$)	D_{vo} ($m^2\ s^{-1}$)	$[V_O]$	$y = 2[V_O]$	y^a
Ti_2AlC	1000	2.7×10^{-14}	58.3	4.6×10^{-16}	1.0×10^{-13}	0.005	0.01	0.0062
	1040	1.9×10^{-13}	55.9	3.4×10^{-15}	2.1×10^{-13}	0.016	0.032	0.0082
	1100	7.4×10^{-13}	52.5	1.4×10^{-14}	5.6×10^{-13}	0.025	0.05	0.12
Ti_3AlC_2	800	4.2×10^{-17}	76.2	5.5×10^{-19}	1.2×10^{-15}	0.00045	0.0009	0.0011
	900	1.1×10^{-15}	67.8	1.7×10^{-17}	1.3×10^{-14}	0.001	0.002	0.0028
	1000	1.5×10^{-13}	60.7	2.5×10^{-15}	1.0×10^{-13}	0.025	0.05	0.0062
	1100	1.1×10^{-12}	54.7	2.0×10^{-14}	5.6×10^{-13}	0.036	0.072	0.0082
$Ti_4AlN_{2.9}$	800	4.8×10^{-16}	75.9	6.3×10^{-18}	1.2×10^{-15}	0.0053	0.01	0.0011
	900	2.1×10^{-14}	67.7	3.1×10^{-16}	1.3×10^{-14}	0.023	0.046	0.0028
	1000	7.0×10^{-13}	61.3	1.1×10^{-14}	1.0×10^{-13}	0.114	0.23	0.0062
	1100	2.9×10^{-12}	55.5	5.2×10^{-14}	5.6×10^{-13}	—	—	0.0082
Ti_3SiC_2	900	9.0×10^{-16}	67.8	1.3×10^{-17}	1.3×10^{-13}	0.0009	—	—
	1000	6.9×10^{-15}	60.7	1.1×10^{-16}	1.0×10^{-13}	0.001	—	—
	1240	7.8×10^{-13}	47.6	1.6×10^{-14}	4.3×10^{-12}	0.004	—	—
	1400	3.0×10^{-11}	40.9	7.1×10^{-13}	2.9×10^{-11}	0.025	—	—

[a] Slepetys and Vaughan (1969) and Florke (1961).

increases. Furthermore, the Al_2O_3 that forms is not continuous but forms as islands. The highest volume fraction of Al_2O_3 is ≈50 vol%.

3) The entire growth in the thickness of the layers is due to the diffusion of either O or Ti ions. This assumption is incorrect inasmuch as the outermost layers grow as a result of the diffusion of both. This assumption is, at most, off by a factor of 2.[4)]

Assuming local equilibrium, the oxygen partial pressure P_{O_2} at the substrate/oxide interface is fixed by the following reaction:

$$\mathrm{Ti}_{n+1}\mathrm{AlX}_n + (n+1)\mathrm{O}_2 \Rightarrow (n+1)\mathrm{TiO}_2 + \mathrm{Al} + n\mathrm{X} \quad (6.A.11)$$

for which at equilibrium

$$P_{O_2}^{n+1} = a_{\mathrm{Al}} a_x^n \exp \frac{\Delta G_{rxn}}{RT} \quad (6.A.12)$$

Here, ΔG_{rxn}, a_{Al}, and a_x are, respectively, the standard free energy change of reaction (6.A.11), the activities of Al and X, that is, C and/or N at the substrate/oxide interface. To calculate P_{O_2}, both the free energy of formation of the appropriate $Ti_{n+1}AlX_n$ phase, ΔG_{MAX}, a_{Al}, and a_x are required. Since that information is currently lacking, some approximations and assumptions, described in more detail in Appendix A in Barsoum *et al.* (2001), are required. Using these assumptions, it is possible to estimate the denominator term in Eq. (6.A.10). Fortunately, this term (listed in column 4 in Table 6.A.1) is quite a weak function of the assumptions made. Using these values in Eq. (6.A.10), D_{RL} can be calculated from k_x. The results are listed in column 5 in Table 6.A.1 and plotted in Figure 6.12a, together with results from the oxidation of Ti_3SiC_2, subject to the same analysis.

The results shown in Table 6.A.1 can be taken one step further. In obtaining this relationship, the implicit assumption made is that two Al^{3+} ions create one oxygen vacancy according to the following scheme:

$$\mathrm{Al}_2\mathrm{O}_3 \Rightarrow 2Al'_{\mathrm{TI}} + 3\mathrm{O}_{\mathrm{O}}^{x} + \mathrm{V}_{\mathrm{O}}^{\bullet\bullet} \quad (6.A.13)$$

It follows that $[V_O]$ in $(Ti_{1-y}Al_y)O_{2-y/2}$ is given by

$$[\mathrm{V_O}] = \frac{D_{\mathrm{O}}}{D_{\mathrm{V_O}}} = \frac{y}{2} \quad (6.A.14)$$

Assuming $D_{\mathrm{RL}} = D_{\mathrm{O}}$, then $[V_O]$ and y can be calculated assuming $D_{\mathrm{V_O}}$ is given by Eq. (6.15). Figure 6.12b is an Arrhenius plot of y, also listed in column 8 in Table 6.A.1. The last column of Table 6.A.1 lists the extrapolated temperature dependencies of y calculated from the results of Slepetys and Vaughan (1969).

Referring to Figure 6.12b, the following salient points can be made. First, at lower temperatures, and with the notable exception of Ti_4AlN_3, the agreement between the literature and our results is excellent. This is especially true considering i) the

4) Interestingly enough, the error in ignoring the outward diffusion of Ti must partially compensate for the fact that the cross-sectional area for diffusion is reduced by the presence of Al_2O_3.

very different techniques used to generate each set, ii) the numerous assumptions made to obtain our results, and iii) the fact that the measurements were carried out in two quite different temperature regimes. Excluding $Ti_4AlN_{2.9}$, the two sets of results at the highest temperatures vary by less than an order of magnitude. These results are thus consistent with the k_x values that are compatible with a rutile layer in which Al is dissolved. They also support the conclusion that dissolution of Al^{3+} cations in the TiO_2 lattice increases the oxygen vacancy concentration and thus the diffusivity of oxygen.

It is important to note that there is little doubt as to the crucial and deleterious role Al plays in the overall oxidation process: a role that is well established in the Ti-aluminide literature (see e.g. Unnam, Shenoy, and Clark, 1986). The oxidation of both Ti_3SiC_2 and Ti_3AlC_2 results in a TiO_2 matrix in which an inert phase (SiO_2 and Al_2O_3) precipitates, and yet the oxidation resistance of the former is at least an order of magnitude better than the latter (Table 6.A.1). Given the identical structures and similar chemistries, this fact alone implicates the Al. This fact also supports the assumption that the most likely rate-limiting step is the diffusion of oxygen through the scales.

As discussed in preceding text, the *initial* oxidation kinetics in air of $(Nb,Ti)_2AlC$ were shown to be parabolic, with oxidation rates lower than those of Ti_2AlC, and comparable to those of Ti_3SiC_2 (see Figure 6.16a). This was explained by postulating that the presence of Nb^{5+} cations in the TiO_2 lattice resulted in an overall reduction in the oxygen vacancy concentrations of the rutile-based scale.

In deriving Eq. (6.A.10), Wagner assumed the scale to be fully dense and diffusion to be through the bulk. In some cases, the oxide scales that formed were *not* fully dense, especially those formed at higher temperatures. Hence, the likely possibility for the inward penetration of oxygen through pores, microfissures, or by surface diffusion and/or grain boundary diffusion, especially between the Al_2O_3 and TiO_2 particles, has to be considered.

And while the aforementioned analysis may be intellectually satisfying, it is of little practical import because TiO_2 is *not* – in the long run – an oxide that is protective enough. The fact that under certain circumstances at least, Ti_2AlC and Ti_3AlC_2 form a protective Al_2O_3 layer is what is important. When the latter occurs the oxidation kinetics are greatly reduced, which is not very surprising because the permeation of oxygen through Al_2O_3 is orders of magnitude slower than that through rutile.

References

Akse, J.R. and Whitehurst, H.B. (1978) *J. Phys. Chem. Solids*, **39**, 457.

Amini, S., McGhie, A.R., and Barsoum, M.W. (2009) On the isothermal oxidation of Ti_2SC in air. *J. Electrochem. Soc.*, **156**, P101–106.

Arita, M., Hosoya, M., Kobayashi, M., and Someno, M. (1979) Depth profile measurements by secondary ion mass spectrometry for the determining the tracer diffusivity of oxygen in rutile. *J. Am. Ceram. Soc.*, **62**, 443.

Barsoum, M.W. (2001) Oxidation of $Ti_{n+1}AlX_n$ where n = 1–3 and X is C, N, part I: model'. *J. Electrochem. Soc.*, **148**, C544–C550.

Barsoum, M.W. (2003) *Fundamentals of Ceramics*, Institute of Physics, Bristol.

Barsoum, M.W., El-Raghy, T., and Ogbuji, L. (1997) Oxidation of Ti_3SiC_2 in air. *J. Electrochem. Soc.*, **144**, 2508–2516.

Barsoum, M.W., Ho-Duc, L.H., Radovic, M., and El-Raghy, T. (2003) Long time oxidation study of Ti_3SiC_2, Ti_3SiC_2/SiC and Ti_3SiC_2/TiC composites in air. *J. Electrochem. Soc.*, **150**, B166–B175.

Barsoum, M.W., Tzenov, N., Procopio, A., El-Raghy, T., and Ali, M. (2001) Oxidation of $Ti_{n+1}AlX_n$ where n = 1–3 and X is C, N, part II: experimental results. *J. Electrochem. Soc.*, **148**, C551–C562.

Basu, S., Obando, N., Gowdy, A., Karaman, I., and Radovic, M. (2012) Long-term oxidation of Ti_2AlC in air and water vapor at 1000–1300 °C temperature range. *J. Electrochem. Soc.*, **159**, C90–C96.

Bellucci, A., Di Pascasio, F., Gozzi, D., Loreti, S., and Minarini, C. (2002) Structural characterization of TiO_2 films obtained by high temperature oxidation of TiC single crystals. *Thin Solid Films*, **405**, 1–10.

Bellucci, A., Gozzi, D., Nardone, M., and Sodo, A. (2003) Rutile growth mechanism on TiC monocrystals by oxidation. *Chem. Mater.*, **15**, 1217–1224.

Blumenthal, R.N., Coburn, J., Baukus, J., and Hirthe, W.M. (1966) Electrical conductivity of non-stoichiometric rutile single crystal from 1000–1500 °C. *J. Phys. Chem. Solids*, **27**, 643–654.

Byeon, J.W., Liu, J., Hopkins, M., Fischer, W., Garimella, N., Park, K.B., Brady, M.P., Radovic, M., El-Raghy, T., and Sohn, Y.H. (2007) Microstructure and residual stress of alumina scale formed on Ti_2AlC at high temperature in air. *Oxid. Met.*, **68**, 97–111.

Chakraborty, S., El-Raghy, T., and Barsoum, M.W. (2003) Oxidation of Hf_2SnC and Nb_2SnC in air in the 400–600 °C temperature range. *Oxid. Met.*, **59**, 83–96.

Chen, Y.S. and Rosa, C.J. (1980) *Oxid. Met.*, **144**, 147.

Chen, G., Sun, Z., and Zhou, X. (1992) *Mater. Sci. Eng.*, **A152**, 597.

Choudhury, N.S., Graham, H.C., and Hinze, J.W. (1976) in *In Proc Symp on Properties of High Temperature Alloys* (eds Z.A. Foroulis and F.S. Petit), Electrochemical Society, p. 668.

Cui, B., Jayaseelan, D.D., and Lee, W.E. (2011) Microstructural evolution during high-temperature oxidation of Ti_2AlC ceramics. *Acta Mater.*, **59**, 4116–4125.

Cui, B., Sa, R., Jayaseelan, D.D., Inam, F., Reece, R., and Lee, W.E. (2012) Microstructural evolution during high-temperature oxidation of spark plasma sintered Ti_2AlN ceramics. *Acta Mater.*, **60**, 1079–1092.

El-Raghy, T. and Barsoum, M.W. (1998) Diffusion kinetics of the carburization and silicidation of Ti_3SiC_2. *J. Appl. Phys.*, **83**, 112–119.

El-Raghy, T., Barsoum, M.W., and Sika, M. (2001) Reaction of Al with Ti_3SiC_2 in the 800–1000 °C temperature range. *Mater. Sci. Eng., A*, **298**, 174.

Feng, A., Orling, T., and Munir, Z.A. (1999) Field-activated pressure-assisted combustion synthesis of polycrystalline Ti_3SiC_2. *J. Mater. Res.*, **14**, 925.

Gogotsi, Y., Nikitin, A., Ye, H., Zhou, W., Fischer, J.E., Yi, B., Foley, H.C., and Barsoum, M.W. (2003) Nanoporous carbide-derived carbon with tunable pore size. *Nat. Mater.*, **2**, 591–594.

Gupta, S. and Barsoum, M.W. (2004) Synthesis and oxidation of V_2AlC and $(Ti_{0.5},V_{0.5})_2AlC$ in air. *J. Electrochem. Soc.*, **151**, D24–D29.

Gupta, S., Filimonov, D., and Barsoum, M.W. (2006) Isothermal oxidation of Ta_2AlC in air. *J. Am. Ceram. Soc.*, **89**, 2974–2976.

Gupta, S., Ganguly, A., Filimonov, D., and Barsoum, M.W. (2006) Oxidation of Ti_3GeC_2 and $Ti_3Ge_{0.5}Si_{0.5}C_2$ in air. *J. Electrochem. Soc.*, **153**, J61–J68.

Gupta, S., Hoffman, E.N., and Barsoum, M.W. (2006) Synthesis and oxidation of Ti_2InC, Zr_2InC, $(Ti_{0.5},Zr_{0.5})_2InC$ and $(Ti_{0.5},Hf_{0.5})_2InC$ in air. *J. Alloys Compd.*, **426**, 168–175.

Haul, R. and Dumbgen, G. (1965) Sauerstoffselbs diffusion in rutilkristallen. *J. Phys. Chem. Solids*, **26**, 1.

Hoffman, E.N., Yushin, G., Barsoum, M.W., and Gogotsi, Y. (2005) Synthesis of carbide-derived carbon by chlorination of Ti_2AlC. *Chem. Mater.*, **17**, 2317–2322.

Hoshino, K., Peterson, N.L., and Wiley, C.L. (1985) *J. Phys. Chem. Solids*, **46**, 1397.

Ikeda, J.A., Chiang, Y.-M., and Fabes, B. (1990) Non-equilibrium surface segregation in Al-doped TiO_2 under an oxidizing potential: effect of redox color-boundary migration. *J. Am. Ceram. Soc.*, **73**, 1633–1640.

Ikeda, J.A., Chiang, Y.-M., Garratt-Reed, A., and Vander Sande, J. (1993) Space charge segregation at grain boundaries in TiO_2, part II. *J. Am. Ceram. Soc.*, **76**, 2447–2459.

Kofstad, P. (1966) *High Temperature Oxidation of Metals*, John Wiley & Sons, Inc., New York.

Kofstad, P. (1967) Note on the defect structure of rutile. *J. Less Common Met.*, **13**, 635–638.

Kulkarni, S., Merlini, M., Phatak, N., Saxena, S.K., Artioli, G., Amini, S., and Barsoum, M.W. (2009) Thermal expansion and stability of Ti_2SC in air and inert atmospheres. *J. Alloys Compd.*, **469**, 395–400.

Lee, D.B. and Choi, J.H. (2007) High temperature corrosion of Ti_3AlC_2 in Ar-1%SO_2 atmosphere. *Mater. Sci. Forum*, **544–545**, 343–346.

Lee, D.B. and Nguyen, T.D. (2008) Cyclic oxidation of Cr_2AlC between 1000 and 1300 °C in air. *J. Alloys Compds.*, **464**, 434–439.

Lee, D.B., Nguyen, T.D., Han, J.H., and Park, S.W. (2007) Oxidation of Cr_2AlC at 1300 °C in air. *Corros. Sci.*, **49**, 3926–3934.

Lee, D.B., Nguyen, T.D., and Park, S.W. (2009) High-temperature oxidation of $Ti_3Al_{0.5}Si_{0.5}C_2$ compounds between 900 and 1200 °C in air. *J. Alloys Compd.*, **469**, 374–379.

Lee, D.B. and Park, S.W. (2006) High-temperature oxidation of Ti_3AlC_2 between 1173 and 1473K in air. *Mater. Sci. Eng., A*, **434**, 147–154.

Lee, D.B. and Park, S.W. (2011) Corrosion of Ti_3AlC_2 at 800–1100 °C in Ar-0.2%SO_2 gas atmospheres. *Corros. Sci.*, **53**, 2645–2650.

Levin, E.M., Robbins, C.R., and McMurdie, H.F. (1964) *Phase Diagrams for Ceramists*, American Ceramic Society, Columbus, OH.

Li, S., Song, G., Kwakernaak, K., van der Zwaag, S., and Sloof, W.G. (2012) Multiple crack healing of a Ti_2AlC ceramic. *J. Eur. Ceram. Soc.*, **32**, 1813–1820.

Lin, Z., Zhuo, M., Zhou, Y., Li, M., and Wang, J. (2006) Microstructures and adhesion of the oxide scale formed on titanium aluminum carbide substrates. *J. Am. Ceram. Soc.*, **89**, 2964–2966.

Lin, Z.J., Li, M.S., Wang, J.Y., and Zhou, Y.C. (2007a) High-temperature oxidation and hot corrosion of Cr_2AlC. *Acta Mater.*, **55**, 6182–6191.

Lin, Z.J., Zhou, Y.C., and Li, M.S. (2007b) Synthesis, microstructure, and property of Cr_2AlC. *J. Mater. Sci. Technol.*, **23**, 721–746.

Liu, Z., Gao, W., and He, Y. (2000) Modeling of Oxidation Kinetics of Y-Doped Fe–Cr–Al Alloys. *Oxid Metals*, **53**, 341.

Lundy, T.S. and Coghlan, W.A. (1973) *J. Phys. Paris Colloq.*, **9**, 299.

Marucco, J.F., Gautron, J., and Lemasson, P. (1981) *J. Phys. Chem. Solids*, **42**, 363.

Mungole, M.N., Balasubramaniam, R., and Gosh, A. (2000) *Intermetallics*, **8**, 717.

Naguib, M., Presser, V., Lane, N.J., Tallman, D., Gogotsi, Y., Lu, J., Hultman, L., and Barsoum, M.W. (2011) Synthesis of a new nanocrystalline titanium aluminum fluoride phase by reaction of Ti_2AlC with hydrofluoric acid. *J. Amer. Cer. Soc.*, **94**, 4566–4561.

Nguyen, T.D., Park, S.W., and Lee, D.B. (2009) Cyclic-oxidation behavior of $Ti_3Al_{0.7}Si_{0.3}C_2$ compounds between 900 and 1100 °C in air. *Oxid. Met.*, **72**, 299–309.

Pang, W.K., Low, I.M., O'Connor, B.H., Sun, Z.-M., and Prince, K.E. (2009) Oxidation characteristics of Ti_3AlC_2 over the temperature range 500–900 °C. *Mater. Chem. Phys.*, **117**, 384.

Pint, B.A. (1996) *Oxid. Met.*, **45**, 1.

Qian, X., He, X.D., Li, Y.B., Sun, Y., Li, H., and Xu, D.L. (2011) Cyclic oxidation of Ti_3AlC_2 at 1000–1300 °C in air. *Corros. Sci.*, **53**, 290–295.

Radhakrishnan, R., Williams, J.J., and Akinc, M. (1999) Synthesis and high-temperature stability of Ti_3SiC_2. *J. Alloys Compd.*, **285**, 85–88.

Roy, T.K., Balasubramaniam, B., and Gosh, A. (1996) *Metall. Mater. Trans. A*, **27A**, 3993.

Salama, I., El-Raghy, T., and Barsoum, M.W. (2003) Oxidation of Nb_2AlC and $(Ti,Nb)_2AlC$ in air. *J. Electrochem. Soc.*, **150**, C152–158.

Schmalzreid, H. (1986) Behavior of (semi-conducting) oxide crystals in oxygen potential gradients. *React. Solids*, **1**, 117–137.

Slepetys, R. and Vaughan, P. (1969) Solid Solution of Aluminum Oxide in Rutile Titanium Dioxide. *J. Phys. Chem.*, **73**, 2157.

Song, G.M., Li, S.B., Zhao, C.X., Sloof, W.G., van der Zwaag, S., Pei, Y.T., and De Hosson, J.T.M. (2011) Ultra-high temperature ablation behavior of Ti_2AlC ceramics under an oxyacetylene flame. *J. Eur. Ceram. Soc.*, **31**, 855–862.

Song, G.M., Pei, Y.T., Sloof, W.G., Li, S.B., De Hosson, J.T.M., and van der Zwaag, S. (2008) Oxidation-induced crack healing in Ti_3AlC_2 ceramics. *Scr. Mater.*, **58**, 13–16.

Sun, Z., Zhou, Y., and Li, M. (2001a) High temperature oxidation behavior of Ti_3SiC_2-based material in air. *Acta Mater.*, **49**, 4347–4353.

Sun, Z., Zhou, Y., and Li, M. (2001b) Oxidation behavior of Ti_3SiC_2-based ceramic at 900–1300 °C in air. *Corros. Sci.*, **43**, 1095–1109.

Sundberg, M., Malmqvist, G., Magnusson, A., and El-Raghy, T. (2004) Alumina forming high temperature silicides and carbides. *Ceram. Int.*, **30**, 1899–1904.

Tallman, D., Anasori, B., and Barsoum, M.W. (2013) *Mater. Res. Lett.*, in press.

Tian, W., Wang, P., Kan, Y., and Zhang, G. (2008a) Oxidation behavior of Cr_2AlC ceramics at 1,100 and 1,250 °C. *J. Mater. Sci.*, **43**, 2785–2791.

Tian, W., Wang, P., Kan, Y., and Zhang, G. (2008b) Oxidation behavior of Cr_2AlC ceramics at 1,100 and 1,250 °C. *J. Mater. Sci.*, **43**, 2785–2791.

Tong, X., Okano, T., Iseki, T., and Yano, T. (1995) Synthesis and high temperature mechanical properties of Ti_3SiC_2/SiC composites. *J. Mater. Sci.*, **30**, 3087.

Unnam, J., Shenoy, R.N., and Clark, R.K. (1986) *Oxid. Met.*, **26**, 231.

Venkatu, D.A. and Poteat, L.E. (1970) *Mater. Sci. Eng.*, **5**, 258.

Wagner, C. (1933) *Z. Phys. Chem. B*, **21**, 25.

Wang, X.H. and Zhou, Y.C. (2003a) High temperature oxidation behavior of Ti_2AlC in air. *Oxid. Met.*, **59**, 303–320.

Wang, X.H. and Zhou, Y.C. (2003b) Oxidation behavior of Ti_3AlC_2 at 1000–1400 °C in air. *Corros. Sci.*, **45**, 891–907.

Wang, X.H. and Zhou, Y.C. (2003c) Oxidation behavior of TiC-containing Ti_3AlC_2 based material at 500–900 °C in air. *Mater. Res. Innovations*, **7**, 381–390.

Wang, X.H. and Zhou, Y.C. (2010) Layered machinable and electrically conductive Ti_2AlC and Ti_3AlC_2 ceramics: a review. *J. Mater. Sci. Technol.*, **26**, 385–416.

Welsch, G. and Kahveci, A.I. (1989) Paper presented at Oxidation of High Temperature Intermetallics, TMS, Warrendale, PA.

Yahia, J. (1963) Dependence of the electrical conductivity and thermoelectric power of pure and Al-doped rutile on equilibrium oxygen pressure and temperature. *Phys. Rev.*, **130**, 1711.

Yan, M.F. and Rhodes, W.W. (1982) Effects of contaminants in conductive TiO_2 ceramics. *J. Appl. Phys.*, **53**, 8809.

Yang, H.J., Pei, Y.T., Song, G.M. and De Hosson, J.T.M. (2011a) Comments on microstructural evolution during high-temperature oxidation of Ti_2AlC ceramics. *Scripta Mater*, **65**, 930–932.

Yang, H.J., Pei, Y.T., Rao, J.C., De Hosson, J.T.M., Li, S.B., and Song, G.M. (2011b) High temperature healing of Ti_2AlC: On the origin of inhomogeneous oxide scale. *Scr. Mater.*, **65**, 135–138.

Yoshihara, M. and Miura, K. (1995) Effects of Nb addition on oxidation behavior of TiAl. *Intermetallics*, **3**, 357.

Yushin, G., Hoffman, E., Nikitin, A., Ye, H., Barsoum, M.W., and Gogotsi, Y. (2005) Synthesis of nanoporous carbide-derived carbon by chlorination of Ti_3SiC_2. *Carbon*, **43**, 2075–2082.

Zhang, J., Wang, L., Jiang, W., and Chen, L. (2008) High temperature oxidation behavior and mechanism of Ti_3SiC_2–SiC nanocomposites in air. *Compos. Sci. Technol.*, **68**, 1531–1538.

Zhang, H.B., Zhou, Y.C., Bao, Y.W., and Li, M.S. (2004) Improving the oxidation resistance of Ti_3SiC_2 by forming a $Ti_3Si_{0.9}Al_{0.1}C_2$ solid solution. *Acta Mater.*, **52**, 3631–3637.

Zhou, A.G., Barsoum, M.W., Basu, S., Kalidindi, S.R., and El-Raghy, T. (2006) Incipient and regular kink bands in dense and 10 vol.% porous Ti_2AlC. *Acta Mater.*, **54**, 1631.

Zhou, Y.C., Dong, H.Y., and Wang, X.H. (2004) High-temperature oxidation behavior of a polycrystalline Ti_2SnC ceramic. *Oxid. Met.*, **61**, 365–377.

7 Chemical Reactivity

7.1 Introduction

There are two components to the reactivity between phases: thermodynamic and kinetic. For a reaction to occur, its free energy change, ΔG_{rxn}, has to be negative. The latter is a necessary, but not sufficient, condition; the kinetics need to be fast enough as well. A good example of the importance of the latter was discussed in Chapter 4, where the point was made that in most gaseous environments – including vacuum – the thermal stability of the MAX phases had to be kinetic in origin.

Before summarizing the reactivity of the MAX phases, it is vital to document an important fact about their reactivities – which was already displayed in the previous chapters – namely, that the $M_{n+1}X_n$ layers are chemically quite stable. By contrast, because the A layers are relatively weakly bound, they are the most reactive species. The leitmotiv of this chapter is, therefore, the various ways that the A-layers react *selectively* with their environment by diffusing out of the basal planes.

Another important consideration is whether a MAX is in equilibrium with its own A-group element, a fact that can be quickly ascertained from a perusal of the appropriate ternary phase diagram. For example, based on the Ti–Si–C, Ti–Al–N, and the Ti–Al–C ternary phase diagrams, none of the MAX phases in these systems is stable *vis-à-vis* reaction with Si or Al. It is for this reason that Ti_3SiC_2 reacts with Si to yield $TiSi_2$ and SiC. Similarly, it is why Ti_2AlC reacts with Al to form Al_4C_3 and $TiAl_3$.

For liquid metals that are not A elements, the key consideration is the solubility of the A-group element. If the solubility is high, then the propensity for reaction with the MAX phase is high, and vice versa. For example, the solubilities of Al and Si in Cu are high, which is why Ti_2AlC and Ti_3SiC_2 readily react with molten Cu. The solubility of Al in Mg, on the other hand, is quite low, which is why Ti_2AlC does not react with molten Mg to temperatures as high as 850 °C.

This chapter is divided into four parts. In the first, evidence is presented to show that the diffusivity of the A-group atoms, at least in the 312 phases, is at least two orders of magnitude higher than the M atoms. In the second part, the reactivity of select MAX phases with molten metals is outlined. In the third, their reactivity with molten salts is discussed. The penultimate section reviews

MAX Phases: Properties of Machinable Ternary Carbides and Nitrides, First Edition. Michel W. Barsoum.

their reactivity with common bases and acids such as nitric, hydrochloric, sulfuric, and hydrofluoric acids. A summary and general conclusions can be found in the last section. Oxidation and other MAX/gas reactions were discussed separately in Chapter 6.

7.2
Diffusivity of the M and A Atoms

7.2.1
Interdiffusion of A-Group Elements

To measure the diffusivities of the Si and Ge in the 312 phases, Ti_3SiC_2/Ti_3GeC_2 diffusion couples were assembled and heated in a graphite vacuum hot press under a nominal load to enhance the diffusion bond (Ganguly, Barsoum, and Doherty, 2007). Figure 7.1a plots the diffusion profiles of Si and Ge across the interface. The Matano interface is shown by a vertical dotted line. Needless to add, the Ti and C sublattices remain intact. What one obtains, therefore, is an interdiffusion coefficient D_{int}.

From the profile asymmetry, it is evident that D_{int} is higher when the Si concentration is low, and vice versa. This observation is confirmed in Figure 7.2a, which shows a semilog plot of D_{int} as a function of the ratio $x_{Si}/(x_{Ge} + x_{Si})$, where x_{Si} is the mole fraction of Si, and x_{Ge} that of Ge. For the analysis, the plausible assumption that $x_{Si} + x_{Ge} = 1$ is made. This assumption is justified because in this system the only atoms diffusing are Si and Ge. Analysis of the results at the Matano interface composition, namely, $\approx Ti_3Ge_{0.5}Si_{0.5}C_2$, yielded the results shown in Figure 7.2b. Least squares fit of these results yield:

$$D_{int}\left(m^2/s\right) = (0.3 \pm 0.2)\exp\left(\frac{-350 \text{ kJ mol}^{-1}}{RT}\right) \tag{7.1}$$

Also plotted in Figure 7.2b are the diffusivities of Si in $TiSi_2$ reported in the literature. The good agreement between the various results indirectly validates our analysis, and suggests that the Ti–Si bond strengths in $TiSi_2$ and Ti_3SiC_2 are not very different. Note that D_{int} increased with increasing Ge composition (Figure 7.2a), suggesting that the Ti–Ge bond is slightly weaker than the Ti–Si bond. This conclusion was indirectly confirmed when it was observed that under identical conditions, Ti_3GeC_2 was more prone to decomposition than Ti_3SiC_2, as shown in Figure 7.1b.

It is important to note here that when annealed at the highest temperatures, the diffusion profiles stop evolving after some time. This is best seen in Figure 7.1c, where the concentration profiles of Si and Ge are plotted as a function of annealing time. Sometime between 8 and 48 h, the concentration profile stopped evolving. While the exact reason for this state of affairs is not clear, the simplest explanation is that the outward diffusion of the Si and Ge from individual grains results in the formation of a TiC diffusion barrier around each grain, sealing the A-group element

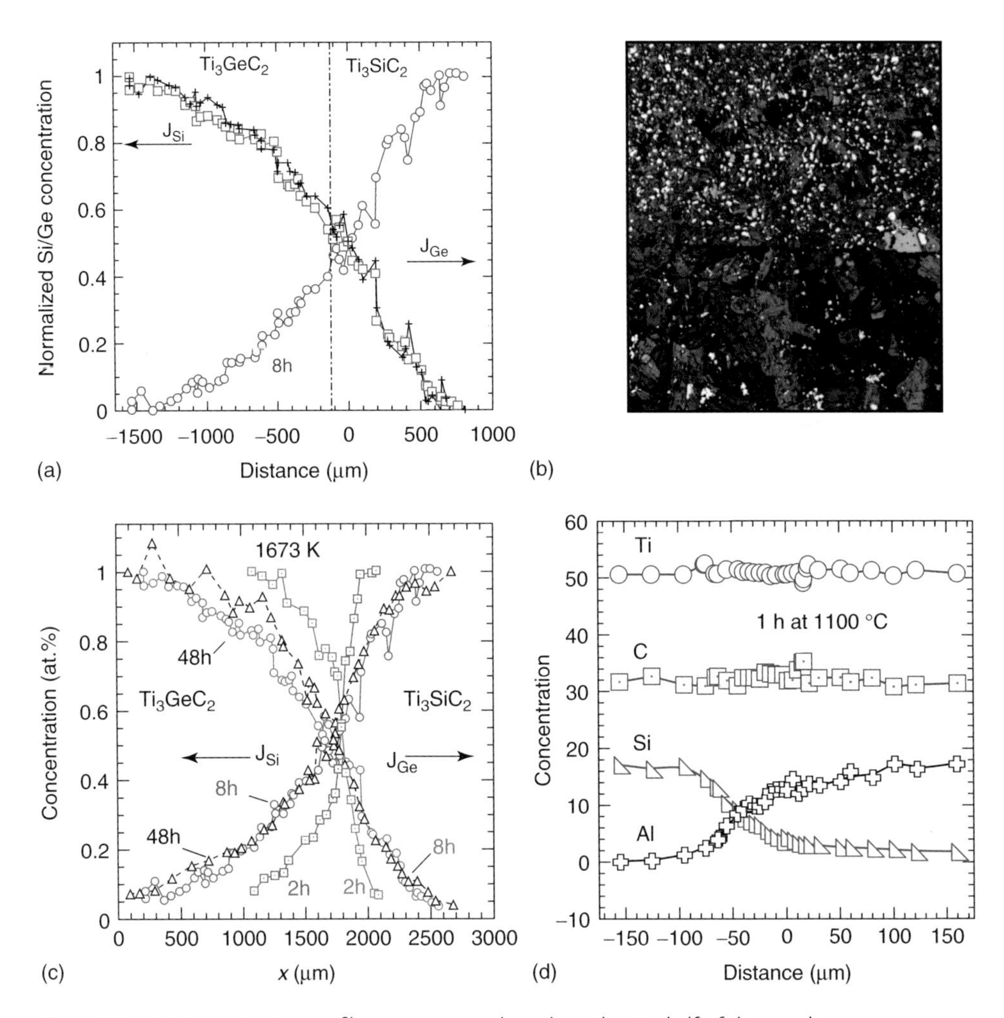

Figure 7.1 (a) Concentration profiles obtained after annealing (a) Ti_3SiC_2/Ti_3GeC_2 diffusion couples for 8 h at 1400 °C. The Si concentration is denoted by blue circles and the Ge by red squares. The location of the Matano interface is indicated by a vertical dotted line. The black line marked by small black crosses represent $x_{Ge} = 1 - x_{Si}$. The squares represent the actual normalized concentration measured. (b) Etched OM image of Ti_3SiC_2–Ti_3GeC_2 diffusion couple – where the top half of the couple is Ti_3GeC_2 – annealed at 1673 K for 48 h. The white phase in the top half is TiC_x that forms as a result of Ge loss from Ti_3GeC_2. (c) Concentration profiles obtained after annealing Ti_3SiC_2/Ti_3GeC_2 diffusion couples at 1400 °C for 2, 8, and 48 h. (Ganguly, Barsoum, and Doherty, 2007). (d) Concentration profiles obtained after annealing Ti_3SiC_2/Ti_3AlC_2 diffusion couples for 1 h at 1100 °C (Yin *et al.*, 2009).

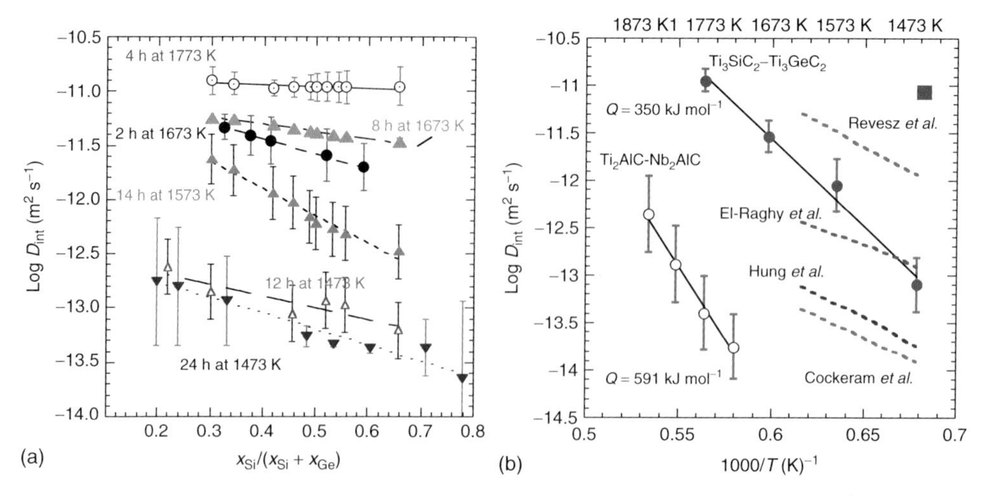

Figure 7.2 (a) Dependence of D_{int} on $x_{Si}/(x_{Si} + x_{Ge})$ in the 1473–1773 K temperature range. As defined here, x_i is not the molar concentration of i in the ternary, but rather the molar concentration of Si or Ge in the basal planes, that is, $x_{Si} + x_{Ge} = 1$. (b) Arrhenius plot of D_{int} at the Matano composition, that is, $Ti_3Ge_{0.5}Si_{0.5}C_2$ for the 312 and $(Ti_{0.5},Nb_{0.5})_2AlC$ for the 211 system. (Ganguly, Barsoum, and Doherty, 2007). Also shown are the Si diffusion coefficients in $TiSi_2$ reported in the literature by El-Raghy and Barsoum (1998), Cockeram and Rapp (1995), Hung *et al.* (1980), and Revesz *et al.* (1983). The results of Yin *et al.* (2009) at 1200 °C are shown by solid blue square at top right.

and preventing it from escaping. These comments notwithstanding, more work is needed to understand this intriguing and possibly technologically important phenomenon.

Yin *et al.* (2009) diffusion bonded Ti_3SiC_2 to Ti_3AlC_2 in the 1100–1300 °C temperature range for 1 h. The concentration profiles of the various elements after annealing at 1100 °C for 1 h are shown in Figure 7.1d. The two solids formed excellent bonds that were quite strong. Apart from the presence of Ti_5Si_3 on the Ti_3AlC_2 side adjacent to the joint interface, no other phases were detected.

For reasons that are not clear, the authors did not attempt to quantify the interdiffusion coefficients. A back-of-the-envelope estimate of the latter yields a value of 3×10^{-12} m^2 s^{-1}. This was estimated from Figure 7.1d, assuming $D = x^2/2t$, where x, the diffusion distance, is roughly 150 μm after a diffusion time t of 1 h. This value is about two orders of magnitude *higher* than reported for the Ti_3SiC_2/Ti_3GeC_2 system (see solid blue square in top right of Figure 7.2b). Such a large discrepancy in D is unlikely given how closely related – both chemically and structurally – Ti_3AlC_2, Ti_3SiC_2, and Ti_3GeC_2 are. It is thus reasonable to conclude that in the Ti_3AlC_2–Ti_3SiC_2 system, a liquid phase most probably formed that enhanced D_{int}. Said otherwise, the diffusion bonding that occurred between Ti_3AlC_2 and Ti_3SiC_2 was probably mediated by a liquid phase.

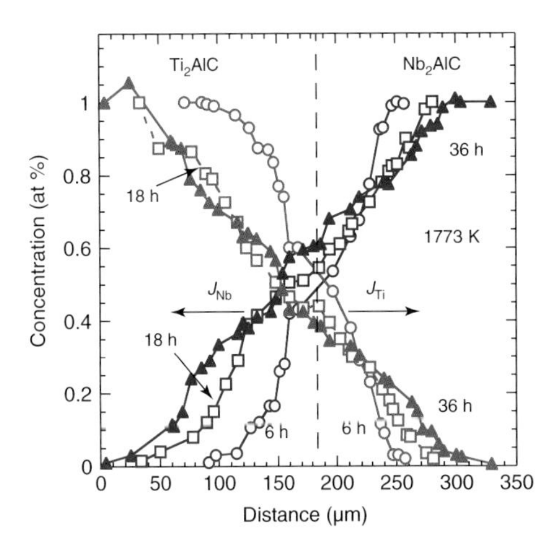

Figure 7.3 Concentration profiles of Nb (blue) and Ti (red) obtained after annealing Ti_2AlC–Nb_2AlC diffusion couples at 1773 K for 6, 18, and 36 h. In all figures, the concentrations are normalized to unity in the bulk. The Matano interface after 6 h is shown as a dotted vertical line. (Ganguly, Barsoum, and Doherty, 2007).

7.2.2
Interdiffusion of the M Atoms

Similarly, the interdiffusion of Ti and Nb in Ti_2AlC–Nb_2AlC couples was measured in the 1723–1873 K temperature range. (Ganguly, Barsoum, and Doherty, 2007). From the diffusion profiles (Figure 7.3), D_{int} for the Matano interface composition, namely, $\approx(Ti_{0.5},Nb_{0.5})_2AlC$ was measured, and found to be (Figure 7.2b)

$$D_{int}\left(m^2/s\right) = (12 \pm 3) \times 10^3 \exp\left(\frac{-591 \pm 5 \text{ kJ mol}^{-1}}{RT}\right) \tag{7.2}$$

It follows that at 1773 K the diffusivity of the M atoms is approximately 300 times lower than that of the A atoms (compare the two solid lines in Figure 7.2b), confirming that the former are better bound in the structure than the latter.

Here again, at 1500 °C, the diffusion profiles stopped evolving sometime between 18 and 36 h (Figure 7.3).

7.3
Reactions with Si, C, Metals, and Intermetallics

To understand or predict the various solid reactions that can occur with a MAX phase, it is useful to know the ternary, or higher, phase diagrams. Figure 7.4a,b plot the Ti–Si–C and Ti–Al–C ternary phase diagrams, respectively. Other MAX-related ternary phase diagrams are reproduced in Appendix 7.A.

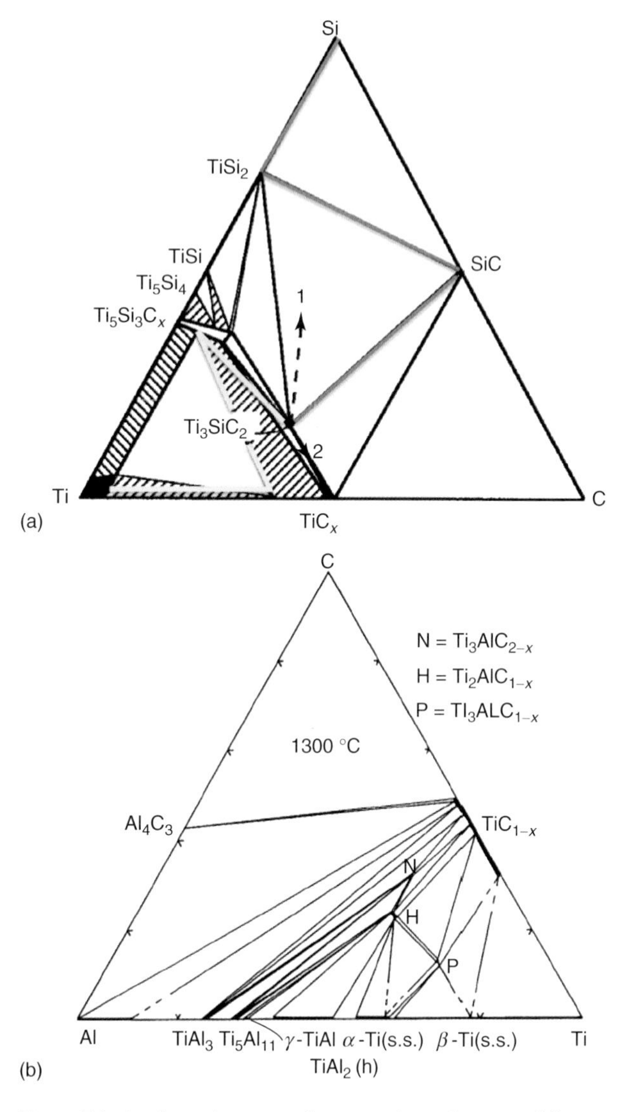

Figure 7.4 Isothermal section of ternary phase diagram of the (a) Ti–Si–C system (Wakelkamp *et al.*, 1991). The reaction path for the Ti/Ti_3SiC_2 diffusion couples is depicted by the yellow lines (Gao and Miyamoto, 2002) and for Si/Ti_3SiC_2 by the green ones. (b) Ti–Al–C ternary phase diagram at 1300 °C (Pietzka and Schuster 1994).

7.3.1
Reaction with Si

According to the Ti–Si–C phase diagram (Figure 7.4a), the reaction of Si with Ti_3SiC_2 – that is, moving along line 1 in Figure 7.4a – should shift the overall composition into the Ti_3SiC_2–$TiSi_2$–SiC compatibility triangle. Consequently, the following reaction was proposed, and confirmed (El-Raghy and Barsoum, 1998), for the silicidation process:

$$Ti_3SiC_2 + 7\ Si = 3\ TiSi_2 + 2\ SiC \tag{7.3}$$

It follows that the reaction path for this diffusion couple – depicted by the green lines in Figure 7.4a – is $Ti_3SiC_2/TiSi_2 + SiC/Si$.

The reaction layers formed were dense, in part because the volume change, ΔV, upon reaction is large and positive (120%), and in part because the silicidation temperatures are greater than 0.8 of the melting point ($\approx$1800 K) of $TiSi_2$, that is, in a regime where creep and densification can occur readily. The reaction layer was composed of a two phase mixture of $TiSi_2$ and SiC. This layer grows in two distinct morphologies: an outer layer with fine (1–5 μm) SiC particles and an inner, coarser (10–15 μm) one (Figure 7.5a). The thickness of the inner layer increases monotonically with time, and the outer layer grows to a thickness of $\approx$50 μm in 5 h and then appears to stop growing and may even shrink slightly. The continual growth of the inner layer, after the arrest of the outer one, provides strong evidence that the silicidation, after an initial transient stage, occurs by the inward diffusion of Si.

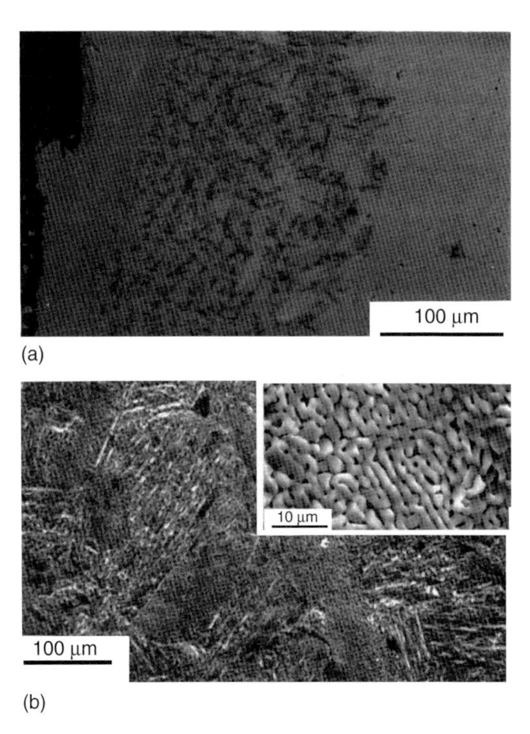

Figure 7.5 Cross-sectional SEM images of (a) silicided Ti_3SiC_2 layer formed after 16 h at 1350 °C. The Ti_3SiC_2 matrix is on the right hand side. (b) carburized layer showing topotactical nature of the reaction. Inset shows a higher magnification SEM image showing the fine nature of the porosity that forms (El-Raghy and Barsoum, 1998).

The reason for the bimodal distribution of the SiC particles is not entirely clear, but one possibility is that the interface between the coarse and fine areas – which is quite sharp – is the original interface. Initially, and as long as the reaction is not diffusion controlled, the Ti and C atoms could react with the Si to form the fine microstructure. As the reaction layer thickens, the kinetics become dominated by the inward diffusion of Si, which could, in turn, have resulted in the coarse microstructure. The argument is consistent with the fact that the interface between the two layers, despite their chemical similarity, is quite sharp (Figure 7.5a). That would also explain why the outer layer grows initially and then stops growing.

Radhakrishnan *et al.* (1996), studying $TiSi_2/SiC$ composites fabricated by a displacive reaction between TiC and Si, also showed that the SiC formed in two different morphologies: a fine needle-like shape and a more equiaxed one, a result consistent with that of El-Raghy and Barsoum (1998).

The overall reaction kinetics were parabolic, that is, followed Eq. (6.3), with a parabolic rate constant given by

$$k_x\,(\mathrm{m^2/s}) = 1 \times 10^{-8} \exp\left(\frac{-133\ \mathrm{kJ\ mol^{-1}}}{RT}\right) \tag{7.4}$$

Figure 7.6 is an Arrhenius plot of Eq. (7.4). Converting these values to a diffusivity using Wagner's approach (Chapter 6), and by comparison with published results on Si diffusion in $TiSi_2$, it was concluded that the rate-limiting step was the diffusion of Si through $TiSi_2$. The results – shown in Figure 7.2b as dashed lines on the right-hand side and labeled El-Raghy, *et al.* – clearly showed that the values obtained were consistent with those in the literature for Si diffusion in $TiSi_2$ shown by the other dashed lines on the right.

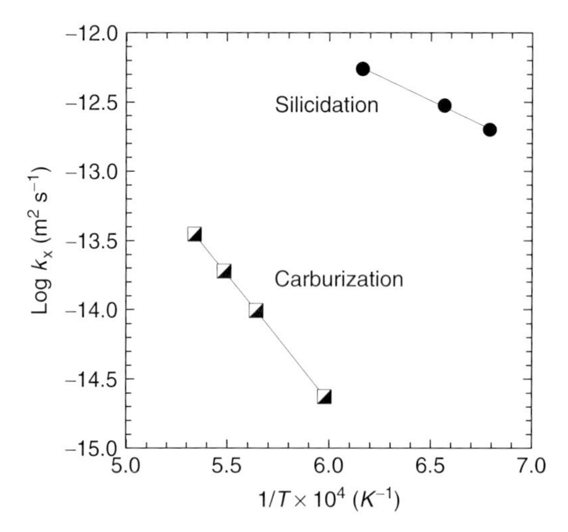

Figure 7.6 Arrhenius plot of parabolic rate constants k_x for silicidation (top line) and carburization (bottom line). (El-Raghy and Barsoum, 1998).

Both carburization (see subsequent text) and silicidation increased surface hardness; the latter also enhanced the oxidation resistance of Ti_3SiC_2 by about three orders of magnitude (El-Raghy and Barsoum, 1998).

Lastly, in this section, it is important to point out the work of Du *et al.* (2000) who carefully mapped the liquidus surfaces in the Ti–Si–C system. They also estimated the free energy of formation of Ti_3SiC_2 (see Table 4.6).

7.3.2
Reactivity in C-Rich Environments

In the 1400–1600 °C temperature range, the reaction of Ti_3SiC_2 with graphite foils resulted in the formation of a 15 vol% porous surface layer of TiC_x (with $x > 0.8$) (El-Raghy and Barsoum, 1998). The operative reaction was surmised to be

$$Ti_3SiC_2 + x\,C = Si\ (\text{vapor}) + 3\ TiC_{0.66+x}\ \text{with}\ x > 0.2 \tag{7.5}$$

The reaction was topotactic. This is best seen in Figure 7.5b, where a SEM image of a carburized surface clearly shows the shapes and contours of the original grains. On the basis of these images and many since, it is reasonable to assume that the Si atoms diffuse out of the basal planes, through the grain boundaries and into the vacuum of the hot press. Concomitant with the loss of Si is a transformation of the remaining TiC into a cubic rock-salt structure (see Figure 2.7). This transformation is accompanied by a 15% volume shrinkage, which manifests itself as very finely dispersed pores in the structure (inset in Figure 7.5b).

The carburization kinetics were parabolic and given by

$$k_x\,(\text{m}^2/\text{s}) = 2.5 \times 10^{-4} \exp\left(\frac{-351.6\ \text{kJ mol}^{-1}}{RT}\right) \tag{7.6}$$

The Arrhenius plot of Eq. (7.6) is shown in Figure 7.6. Comparing these results with the silicidation results, it is obvious that the carburization kinetics are more than two orders of magnitude slower than those of silicidation. When the k_x values were converted to diffusivities using the Wagner's formalism (Chapter 6), the resulting values suggested that the carburization kinetics were rate limited by the diffusion of C through dense TiC_x with $x > 0.8$. Given the porosity of the reaction layer formed, this result was somewhat surprising (El-Raghy and Barsoum, 1998).

This early work was the first in which the topotactical nature of the reactivity of the MAX phases was established. As discussed subsequently, it is a leitmotiv of this chapter.

7.3.3
Reactions with Ti and Ti–6Al–4V

In 2002, Gao and Miyamoto studied the Ti/Ti_3SiC_2 and $Ti{-}6Al{-}4V/Ti_3SiC_2$ diffusion couples in the 1200–1400 °C temperature range in a vacuum of 10^{-3} Pa (Gao and Miyamoto, 2002). The reaction path (outlined in yellow in Figure 7.4a) was determined to be $Ti_3SiC_2/Ti_5Si_3C_x/Ti_5Si_3C{+}TiC_x/TiC_x/Ti$. This result is in total agreement with the work of Naka, Feng, and Schuster (1997), who investigated SiC/Ti diffusion couples and reported the diffusion path to be $SiC/Ti_3SiC_2/Ti_5Si_3C_x/$ $Ti_5Si_3C+TiC_x/TiC_x/Ti$. Said otherwise, the operative simplified reaction is

$$Ti + 3\,Ti_3SiC_2 = Ti_5Si_3C + 5\,TiC \quad \Delta V = -3.2\% \tag{7.7}$$

The reaction is simplified because the Ti_5Si_3C chemistry varied over the range $(Ti_{0.64-0.8},Si_{0.2-0.36})C_{0.45-0.55}$. Furthermore the TiC_x that formed at shorter times had a significant Si content, which reduced with time. Below 1350 °C, the reaction rate was controlled by a solid-state diffusion process. Above 1350 °C, a liquid phase formed that dramatically increased k_x. At 1300 °C, k_x was 3.1×10^{-14} m² s⁻¹; at 1350 it was 1.0×10^{-13} $m^2\ s^{-1}$. A Ti-rich liquid phase appeared as a result of a peritectic eutectic reaction in the Ti–Si–C system at 1341 °C. When the k_x values were converted to diffusion coefficients, the results suggested that Si diffusion through Ti_5Si_3C was rate limiting.

At higher temperature and longer holding time, the TiC_x tended to grow at the boundary with the Ti–6Al–4V alloy.

7.3.4 Reactions with Ni and NiTi

If the MAX phases are to be used extensively in high-temperature environments, it would be beneficial if their bonding with Ni-based superalloys – the current materials of choice for high-temperature applications – be understood. Yin, Li, and Zhou (2006) studied the reactions between Ti_3SiC_2 and Ni in the 800–1100 °C temperature range in vacuum. The overall reaction rate was parabolic, with a rate constant given by [1)]

$$k\left(\mathrm{ms}^{-1/2}\right) = 0.002 \exp\left(\frac{-76.54\ \mathrm{kJ\ mol^{-1}}}{RT}\right) \tag{7.8}$$

or a parabolic rate constant given by

$$k_x\left(\mathrm{m^2s^{-1}}\right) = 4.0 \times 10^{-6} \exp\left(\frac{-153\ \mathrm{kJ\ mol^{-1}}}{RT}\right) \tag{7.9}$$

The reaction resulted in a duplex layer. The diffusion path was determined to be $Ni/Ni_{31}Si_{12} + Ni_{16}Ti_6Si_7 + TiC_x/Ti_3SiC_2 + Ti_2Ni + TiC_x/Ti_3SiC_2$. The diffusion of Ni through the reaction zone toward Ti_3SiC_2 and the counter diffusion of Si were assumed to be the main controlling steps in the bonding process. From thermodynamic considerations and other similar work (see subsequent text), the more likely diffusion path is $Ni/Ni_{31}Si_{12} + Ni_{16}Ti_6Si_7 + TiC_x/Ti_5Si_3C_x + Ti_2Ni +$ TiC_x/Ti_3SiC_2.

The reactivity of Ti_3SiC_2 with NiTi was studied using diffusion couples held for 1 h in the 1000–1300 °C temperature range (Basu *et al.*, 2011). No bonding was evident at 1000 °C and the samples completely reacted at 1300 °C. Homogeneous

1) Equation (1) in the paper by Yin, Li, and Zhou (2006) is wrong and does not reproduce the results shown in their Figure 8. The correct expression is given in Eq. (7.8).

diffusion bonding was observed at 1100 and 1200 °C, however. The predominant diffusing species was Si. Its diffusion from Ti_3SiC_2 into the interfacial layer, and its reaction with NiTi, were presumed to be the major event during the joining process. Evidence for Ti diffusion into the interface from the NiTi was also found. At 1100 °C, the reaction path between Ti_3SiC_2 and NiTi was surmised to be $Ti_3SiC_2/Ti_5Si_3C_x + TiC_{1-y}$/NiTiSi + Ni_3Ti_2Si/Ni_2Ti_3Si/NiTi. At 1200 °C, the reaction path was Ti_3SiC_2/NiTiSi + TiC_{1-y}/Ni_2Ti_3Si/NiTi. In essence, the Si diffuses out from Ti_3SiC_2 and reacts with the NiTi to form a host of ternary intermetallics.

Interestingly, the reaction layer formed at 1200 °C was thinner than the one formed at 1100 °C for the same reaction time. This anomaly was attributed to the faster diffusion of Si at the higher temperature and its reaction with NiTi to form a thin Ni_2Ti_3Si layer that *prevented* further reaction. The reaction occurs in the presence of a liquid phase. These results are promising because they suggest that it is possible to bond Ti_3SiC_2 to NiTi.

7.3.5 Reactions with NiAl

Hajas *et al.* (2010) studied the reactivity of Cr_2AlC with NiAl and concluded that the operative reaction was

$$Cr_2AlC + NiAl_x \Rightarrow NiAl_y + Cr_7C_3 \tag{7.10}$$

where $y > x$. The possibility that the reaction products containing Cr_3C_2 could not be ruled out, however. Here again is an example of the A-group element, Al in this case, out-diffusing into the surrounding NiAl and the concomitant conversion of the Cr_2AlC matrix to chromium carbides. This reaction was found to initiate at temperatures as low as 1000 °C, which would limit the application temperature of Cr_2AlC–NiAl composites to $T < 1000$ °C.

The same authors reported the reaction onset temperature to be 1280 °C for a powder mixture containing 50 at% Ti_2AlC and 50 at% NiAl.

7.3.6 Aluminum

El-Raghy, Barsoum, and Sika (2001) were the first to study the reactivity of Ti_3SiC_2 in molten Al. They found the reaction kinetics to be linear and followed the relation

$$k_x\,(\mathrm{m\,s^{-1}}) = 7.2\exp\left(\frac{-207\pm5\ \mathrm{kJ\,mol^{-1}}}{RT}\right) \tag{7.11}$$

The reaction layer consisted of two interconnected, or interpenetrating, networks of $TiC_{0.67}$ and molten Al (Figure 7.7a). The samples preserved their original shape and dimensions after the reaction. Some TiC particles, as well as $TiAl_3$ platelets, were sometimes observed in the Al bath. The rate-limiting step and driving force for this reaction was presumed to be Si dissolution into the molten Al, with the most likely reaction being

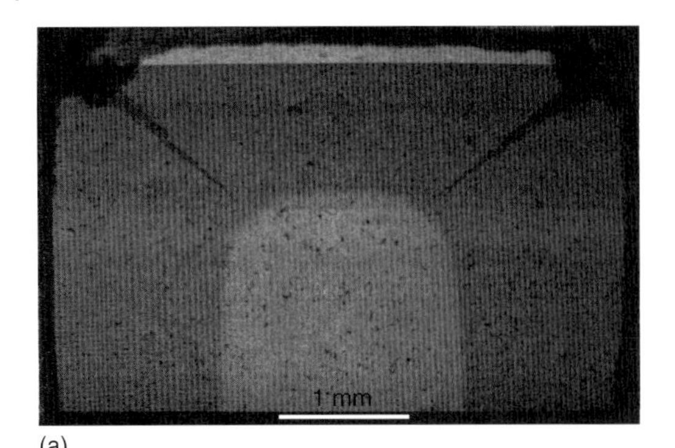

(a)

(b)

Figure 7.7 SEM images of (a) the cross section of a sample held in molten Al at 900 °C for 96 h. Note unreacted core and Al-filled cracks at the corners. (b) Reaction layer after etching in 1:1:1 vol% solution of $HF/HNO_3/H_2O$ for 10 s. The etchant dissolves the Al, leaving behind a porous TiC_x network. The porosity level is ≈40 vol%. (El-Raghy, Barsoum, and Sika 2001).

$$Ti_3SiC_2 + Al = 3\,TiC_{0.67} + Si\,(Al) \tag{7.12}$$

The volume change ΔV of this reaction is −15 vol%. This shrinkage was manifested most prominently at the corners of samples immersed in the molten Al; the corners, presumably as a result of the volume shrinkage, crack during the reaction or more likely during cooling, and are penetrated by molten Al (Figure 7.7a). As noted earlier, the reaction occurs essentially with no overall change in the shape of the samples. It follows that as the Si diffuses into the melt, the remaining TiC_x layers shrink, with the volume so created being filled with molten Al. The interpenetrating nature of the porous network and the fine scale of the pores that form are shown in Figure 7.7b.

In addition to Eq. (7.13) El-Raghy, Barsoum, and Sika (2001) also found evidence for the following reaction:

$$TiC_x + 3y\,Al = y\,TiAl_3 + Ti_{(1-y)}C_x \tag{7.13}$$

In 2003, Gu, Yan, and Zhou (2003) investigated the same system, but in the 600–650 °C temperature range and essentially reached the same conclusions as El-Raghy, Barsoum, and Sika (2001). In Gu *et al.*'s work, the Al to Ti_3SiC_2 volume fractions were varied. When the Al content was ≈50 vol%, the reaction was sluggish. However, when the Al content was raised to 90 vol%, the reaction was much more vigorous because the sink for Si expanded, which in turn drove Eq. (7.12) to the right.

In the process of fabricating Al-matrix composites reinforced with Ti_3AlC_2, Wang *et al.* (2011) studied their reactivity. After 1 h at 600 °C, they observed no reaction. At 700 °C, X-ray diffraction (XRD) peaks belonging to TiC and $TiAl_3$ were observed. At higher temperatures, peaks belonging to Al_4C_3 were observed. At even higher temperatures (900 and 1000 °C), the XRD peaks for Al_4C_3 and Ti_3AlC_2 disappeared. Combining these results with the Ti–Al–C ternary diagram (Figure 7.4b), the most likely, simplified, reactions occurring sequentially, from lower to higher temperatures, are

$$32\ Al + 3\ Ti_3AlC_2 = 2\ Al_4C_3 + 9\ TiAl_3 \tag{7.14a}$$

$$2\ Al_4C_3 + 6\ TiAl_3 = 6\ TiC + 26\ Al \tag{7.14b}$$

For an overall reaction at temperatures >900 °C

$$2\ Al + Ti_3AlC_2 = 2\ TiC + TiAl_3 \tag{7.15}$$

These reactions are simplified in that the Ti/C atomic ratios in the TiC are assumed to be unity at all times. Note that these results imply that Al_4C_3 is unstable above about 900 °C, in agreement with the results of Nukami and Flemings (1995) who reported that Eq. (7.14b) is operative above 887 °C.

It is important to note that Ti_3SiC_2 is thermodynamically stable in Al–Si alloys with small Si concentrations (El-Raghy, Barsoum, and Sika, 2001). This is an important result because it confirms that the driving force for the reaction is the dissolution of the Si in the Al. This result also implies that many of the aforementioned reactions would be suppressed when the activity of the A-group element in the surroundings is elevated to the point that ΔG for Eq. (7.12) is positive.

Vincent *et al.*, (1998) studied the reactivity of Ti_2SnC and Al at 1000 °C for 4 h and concluded that the following reaction was operative:

$$Ti_2SnC + 3\ Al = TiAl_3 + TiC + Sn\ (liq.) \tag{7.16}$$

At 720 °C, Al_4C_3 was present.

7.3.7
Reactions with Cu

When Ti_3AlC_2 and Cu powders (50 vol%) were mixed and hot pressed in the 800–1050 °C temperature range, little reaction was observed when the temperature was 800 °C (Zhang, Wang, and Zhou, 2007). At 950 °C and higher temperatures, the reaction

$$Ti_3AlC_2 + Cu = 3\ TiC_{0.67} + Al\ (Cu) \tag{7.17}$$

was proposed. Here again, the Al deintercalated from the basal planes – by diffusing into and forming a solid solution with the Cu – leaving behind $TiC_{0.67}$, in which the vacancies were partially ordered. The transformation leads to a ≈21 vol% shrinkage, which in turn caused the TiC_x to crack and allowed the Cu to penetrate between the former. Within the defective Ti_3AlC_2 layer, Cu was found with the following crystallographic relationships: $(1\bar{2}10)$ $Ti_3AlC_2||1\bar{1}0$ Cu and (0001) $Ti_3AlC_2||$ (111) Cu. In some regions, both the $TiC_{0.67}$ and the Cu were heavily twinned. An ordered, hexagonal TiC_x was identified as a transition phase linking Ti_3AlC_2 and cubic TiC_x.

If the temperature is kept below the melting point of Cu, no noticeable reactions between Cu and Ti_3AlC_2 (Peng, 2007), Ti_3SiC_2 (Zhang, Zhimei, and Zhou, 1999), or Ti_2SnC (Wu, Zhou, and Yan, 2005) are observed.

7.3.8 Lead and Lead Bismuth Alloys

Barnes, *et al.* (2008) tested the corrosion resistance of Ti_3SiC_2 and Ti_2AlC in circulating molten lead (Pb) at 650 and 800 °C for possible application as cladding or structural materials in Pb-cooled fast nuclear reactors. The extent of reaction was minimal in both melts. The only observed interaction with Pb was a result of surface cracks in the Ti_2AlC produced by machining before exposure to the Pb.

Similarly, Heinzel, Müller, and Weisenburger (2009) studied the compatibility of Ti_3SiC_2 in stagnant Pb and Pb–Bi melts containing oxygen at temperatures between 550 and 750 °C for up to 4000 h. Two different oxygen concentrations were chosen. On exposure to the melts, a thin TiO_x layer was formed at the Pb/Ti_3SiC_2 interface, which presumably prevented further reaction/dissolution. Rivai and Takahashi (2008) exposed Ti_3SiC_2 to stagnant Pb–Bi alloys containing 5×10^{-6} wt% oxygen at 700 °C for 1000 h and found no trace of corrosion.

More recently, the corrosion behavior of Ti_3SiC_2 was investigated in molten Pb with an oxygen concentration of $\approx 1 \times 10^{-6}$ wt%, flowing at ≈ 1 m s^{-1} at 500 °C for 2000 h, and, again, the corrosion resistance was deemed to be excellent (Utili *et al.*, 2011).

The corrosion characteristics of Ti_3SiC_2 in Pb–Bi under transient temperature conditions in the 550–800 °C temperature range were also investigated (Rivai and Takahashi, 2010). Two cases of transient temperature conditions were explored. In one, the specimens were immersed in Pb–Bi at 550 °C for 12 h, and then the temperature was increased to 800 °C and kept there for 12 h. In the second case, the specimens were immersed in Pb–Bi at 550 °C for ≈500 h before the temperature was increased up to 800 °C and held for 15 h. In neither case was there any trace of corrosion.

On the basis of these results, it is reasonable to conclude that Ti_3SiC_2 has the potential to be an excellent material for containing molten Pb or Pb–Bi alloys in nuclear reactors. In a recent paper, Sienicki *et al.* (2011) suggested Ti_3SiC_2 could be used for an improved natural circulation Pb-cooled, small modular fast reactor.

7.3.9
Reactions with Mg

In one of our early papers on the reactivity of Mg with Ti_2AlC at 750 °C, we reported that when Ti_2AlC porous preforms were infiltrated with molten Mg, an interdiffusion of Ti into the Mg and Mg into the MAX phase occurred (Amini, Ni, and Barsoum, 2009). More recent results, however, suggest that if a reaction occurs at all, it is mild. It follows that neither Ti_3SiC_2 nor Ti_2AlC appears to react with molten Mg up to temperatures as high as 850 °C.

7.4
Reactions with Molten Salts

7.4.1
Cryolite and LiF

Immersion of polycrystalline Ti_3SiC_2 samples in molten cryolite – with an approximate composition of 47 wt% NaF, 43 wt% AlF_3, 5 wt% CaF_2, and 5 wt % Al_2O_3 – at 960 °C resulted in the preferential out-diffusion of Si atoms to form a partially ordered cubic phase with approximate chemistry $Ti(C_{0.67},Si_{0.06})$ (Barsoum *et al.*, 1999). The overall reaction is

$$Ti_3SiC_2 = 3Ti\left(C_{0.67}, Si_{0.06}\right) + 0.94Si \;\left(\text{in cryolite}\right) \tag{7.18}$$

Etched optical OM images of the reaction layers (Figure 7.8) confirmed that the Si diffused out of the basal planes and through the grain boundaries to the exterior of the sample (Figure 7.8b). In Figure 7.8b, a grain that was partially denuded of Si is apparent. The central region of the grain maintains the Ti_3SiC_2 chemistry and retains its color. In other words, the reaction occurs topotactially, with minimal microstructural disruption.

This work was also the first to show evidence for a detwinning of the TiC_x with the egress of Si. The process is shown schematically in Figure 2.7. The TiC forms in domains where the (111) planes are related to each other by mirror planes; that is, the loss of Si results in the *detwinning* of the Ti_3C_2 layers. Raman spectroscopy, XRD, OM, SEM, and TEM (transmission electron microscopy) all indicated that the Si exits the structure topotactically, in such a way that the vacancies remained partially ordered in the cubic phase.

More recently, Naguib *et al.*, (2011b) showed that when Ti_2AlC powders were immersed in molten LiF, at 900 °C in air for 2 h, a Ti–C–O–F phase formed via a topotactic transformation not unlike the one that occurred when Ti_3SiC_2 was immersed in cryolite. In this case, the Al diffuses out of the structure and reacts with LiF to form Li_3AlF_6. The simplified overall reaction is conjectured to be

$$0.5\,Ti_2AlC + 2.9\,LiF + 0.625\,O_2 = TiC_{0.5}O_{0.4}F_{0.5}Al_{0.1} + 0.4\,Li_3AlF_6 + 0.85\,Li_2O \tag{7.19}$$

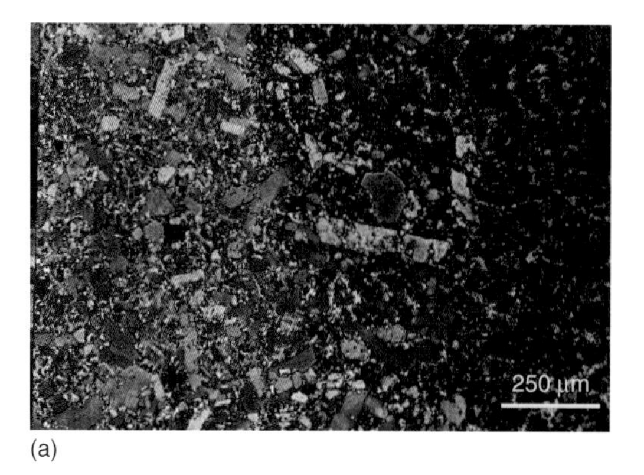

(a)

(b)

Figure 7.8 Polished and etched OM images of a Ti_3SiC_2 sample after immersion in cryolite at 960 °C for 100 h. (a) Lower magnification image showing reaction layer (dark) and bulk Ti_3SiC_2 (colored region). (b) Higher magnification image of reaction layer showing grain boundary nature of attack. This image also shows the preferential removal of Si surrounding pre-existing TiC_x particles that appear white (Barsoum *et al.*, 1999).

It is important to note that the reaction kinetics are quite sensitive to the O content in the melt. At ≈20 µm, the reaction layer that formed when a bulk sample was heated in a vacuum hot press for 8 h was significantly thinner than the 200–400 µm thick layers that formed in air after just 2 h at the same temperature. These results are not surprising because it is well established that oxygen plays a key role when metals react with molten fluoride salts (Delpech *et al.*, 2010). It is quite possible that in the absence of oxygen, no reaction would have occurred.

At ≈2%, ΔV in Eq. (7.19) is small, which in part explains, and is consistent with, the absence of cracks at the corners of a sample that was immersed in the molten salt for 2 h (Figure 7.9a). Compare this image to the one shown in Figure 7.7a, where the volume change was higher and resulted in corner cracks.

The transformation is topotaxial, and again involves detwinning of the hexagonal Ti_2AlC structure (Figure 7.9b, c and d). At 10 nm, the domains that form are quite small.

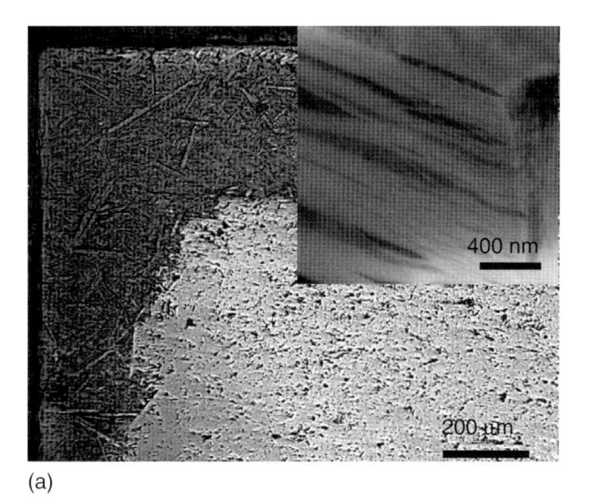

(a)

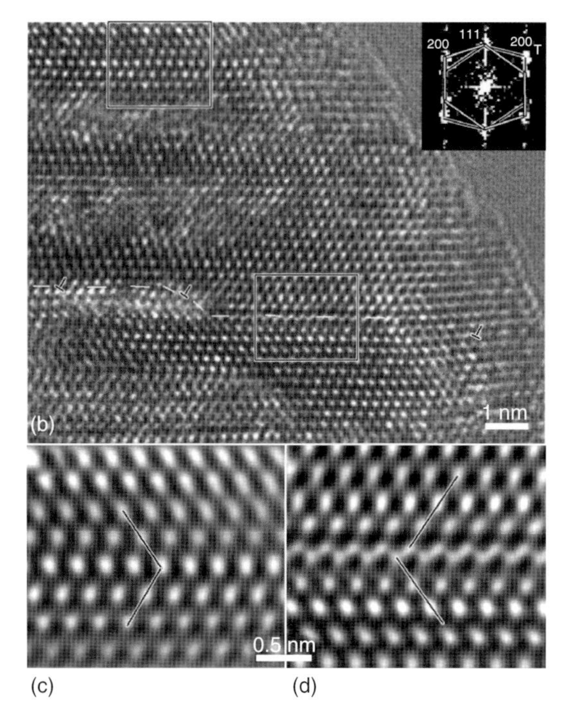

(c) (d)

Figure 7.9 (a) Cross-sectional OM images of polished bulk Ti_2AlC sample after LiF treatment for 2 h at 900 °C in air showing the reaction layer and sharp corner after reaction. Inset shows the very fine scale of the delaminations in the reaction layer. (b) High-resolution TEM image showing two twins: a coherent twin on top and an incoherent twin in the middle with some dislocations shown as inverted Ts. The inset shows the corresponding fast Fourier transform pattern, again showing evidence for the presence of twins. (c) and (d) are enlarged images of the coherent and incoherent twins outlined by top and bottom rectangles in (b), respectively (Naguib *et al.*, 2011b).

7.4.1.1 Eutectic K_2CO_3 and Li_2CO_3 Mixtures

The separator plates in the molten carbonate fuel cell are subject to a corrosive carbonate environment. The possibility of using Ti_3SiC_2 for this application was investigated by Liu *et al.* (2003a), who immersed polycrystalline Ti_3SiC_2 samples in a eutectic Li_2CO_3 (68 at%) and K_2CO_3 (32 at%) mixture in the 650–850 °C temperature range in air. In all cases, the immersed samples lost weight, which implies that whatever surface oxides formed – TiO_2 and SiO_2 in this case – were rapidly dissolved in the eutectic melt. The most likely reactions are

$$Li_2CO_3 + SiO_2 = Li_2SiO_3 + CO_2 \tag{7.20}$$

$$Li_2CO_3 + TiO_2 = Li_2TiO_3 + CO_2 \tag{7.21}$$

Direct evidence for the formation of Li_2TiO_3 was found in XRD diffractograms of the reacted sample. No evidence was found for Li_2SiO_3, however, suggesting that its solubility in the melt was significantly higher. It is worth noting that Eq. (7.20) is the one that occurs when SiC, with a native SiO_2 oxide, is immersed in molten K_2CO_3.

7.4.2 Other Alkali Metal Chlorides and Fluorides

In Chapter 6, the reaction of Ti_2AlC powders with anhydrous HF was reviewed. In this section, the reactivity of the MAX phases with other molten alkali metal salts is discussed.

7.4.2.1 Sodium Sulfate

In some high-temperature applications such as heat engines or heat exchangers, in addition to O, other reactants such as sulfur, nitrogen, carbon, and chlorine may be present. Moreover, deposits of NaCl and Na_2SO_4 may also accumulate on exposed surfaces. These deposited salts can lead to severe hot corrosion attack and accelerate a material's degradation.

When polycrystalline Ti_3SiC_2 samples were covered with thin Na_2SO_4 films and oxidized in the 900 and 1000 °C temperature range in air, their oxidation kinetics were enhanced by a factor of ≈ 2 relative to those that were not covered (Liu, Li, and Zhou, 2003b). In an attempt to enhance the corrosion resistance of Ti_3SiC_2 at 850 °C in 75 wt% Na_2SO_4 –25 wt% NaCl mixtures, Liu, Li, and Zhou (2006) preoxidized Ti_3SiC_2 at 1200 °C. The preoxidation resulted in the formation of a duplex oxide scale: an outer TiO_2 and inner SiO_2/TiO_2 layer (e.g., Figure 6.1a). The outer rutile layer enhanced the corrosion resistance, but only temporarily. Once the rutile layer was dissolved or penetrated by the molten salt, the corrosion became as severe as before preoxidation.

Wang and Zhou (2004) showed that the oxidation/corrosion of Ti_3AlC_2 covered by a thin layer of Na_2SO_4 was also severe in the 900–1000 °C temperature range. The presence of the salt prevented the formation of a protective alumina layer (Chapter 6). Sulfur-rich layers were also detected at the substrate/scale interface. However, when the Ti_3AlC_2 samples were preoxidized in air (Lin *et al.*, 2006b) to form a thin Al_2O_3 layer before exposure to the salt, the specimens survived 20 h at 900 °C with minor weight losses. The latter was believed to be due to the reduction of the salt to Na_2O or Na_2S and subsequent expulsion of gaseous species such as SO_2 or SO_3.

The same authors also explored the corrosion resistance of Ti_2AlC below and above the melting point (884 °C) of Na_2SO_4 (Lin *et al.*, 2006a). The latter was brushed on the surface of the specimens to correspond to a concentration of $\approx 4\,mg\,cm^{-2}$. At 850 °C, that is, below the melting point of Na_2SO_4, the reaction kinetics were parabolic. At 900 and 1000 °C, the reaction was quite severe and resulted in the formation of the following sodium titanates: $Na_{0.23}TiO_2$, $Na_2Ti_9O_{19}$, and $Na_2Ti_6O_{13}$. As in the case of Ti_3AlC_2, S segregation was observed at the substrate/scale interface. Here again, the presence of the salt interfered with the formation of a protective Al_2O_3 layer that forms otherwise (see Chapter 6). When the Ti_2AlC samples were preoxidized in air at 1000 °C for 2 h before application of the salt, the corrosion rate was dramatically reduced, which is consistent with the fact that Al_2O_3 does not react with Na_2SO_4, a fact that has long been appreciated. For example, Lawson, Pettit, and Blachere (1964) found that the corrosion of Al_2O_3 by Na_2SO_4 is negligible and mainly depended on impurities, such as MgO and SiO_2, in the starting polycrystalline Al_2O_3.

Lin, Li, and Wang (2007) compared the oxidation of Cr_2AlC with, and without, a Na_2SO_4 coating at 900 and 1000 °C. They concluded that in the presence of the molten salt, the oxidation kinetics were slightly higher than those in the uncoated samples; in other words they were excellent. They attributed the corrosion resistance to the presence of a dense Al_2O_3 layer. Why in this case preoxidation is not required to impart resistance is unclear at this time, but suggests that the driving force for the formation of an Al_2O_3-rich layer in Cr_2AlC is higher than that for Ti_2AlC.

7.5 Reactions with Common Acids and Bases

The corrosion rates of the MAX phases vary from composition to composition and depend on the corrosive medium. Figure 7.10 plots the weight losses recorded during a six-month immersion of Ti_3SiC_2 samples in sodium hydroxide (NaOH) as well as in hydrochloric (HCl), sulfuric (H_2SO_4), nitric (HNO_3), and hydrofluoric (HF) acids (Travaglini *et al.*, 2003). Table 7.1 summarizes some early results on the corrosion rates of Ti_3SiC_2 in solutions of differing concentrations. The results show that with the exception of HNO_3, Ti_3SiC_2 is quite resistant to the other acids and NaOH. In the following sections, these results and others are reviewed.

7.5.1 Hydrochloric Acid

When Ti_3SiC_2 was immersed in 1M HCl for six months, the corrosion rate (Figure 7.10a and Table 7.1) was negligible (Travaglini *et al.*, 2003). Cyclic

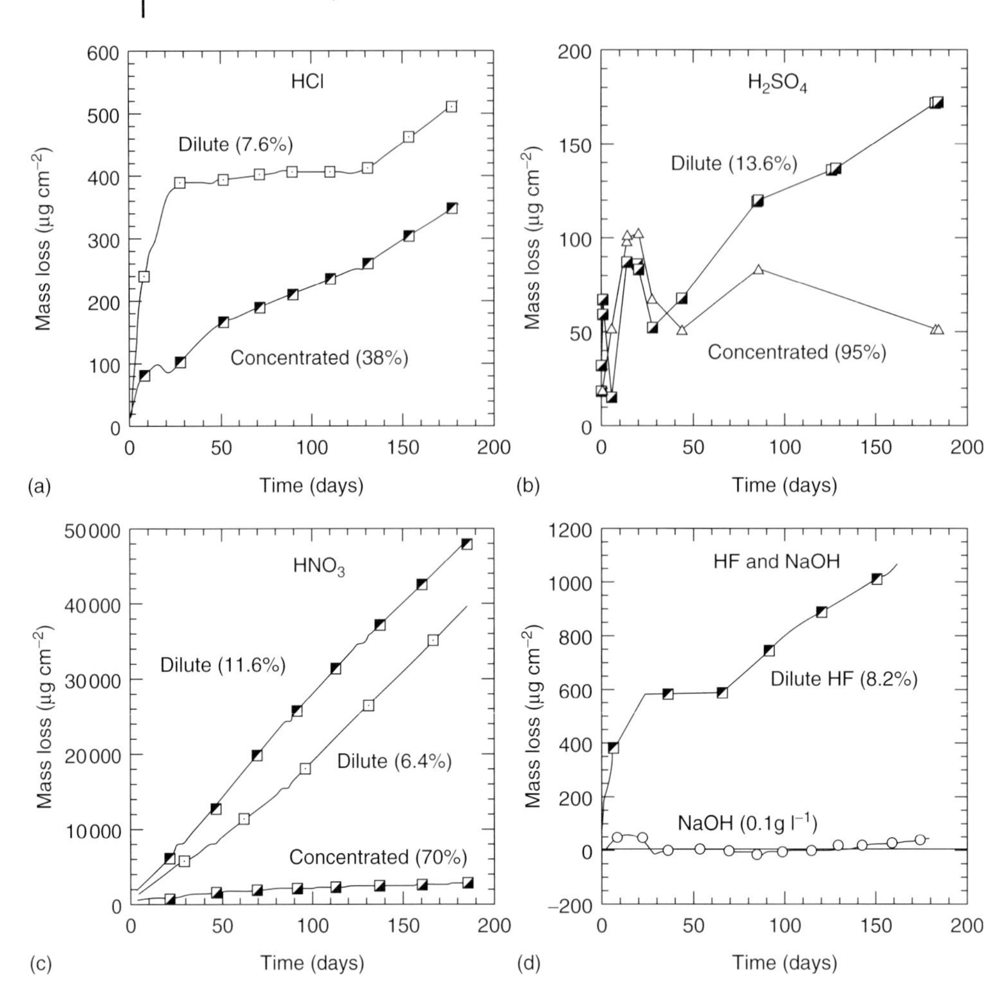

Figure 7.10 Weight loss during six-month immersion of Ti_3SiC_2 samples in (a) HCl, (b) H_2SO_4, (c) HNO_3, and (d) HF and NaOH. The error bars in (c) are less than the width of the symbols (Travaglini *et al.*, 2003).

polarization and potentiostatic i–t transients in dilute HCl strongly suggest that a thin, irreversible, electrically insulating layer forms on the surface. Exposing a sample to a constant current density of 0.6 mA cm^{-2} in a 22 vol% HCl solution for two days resulted in the formation of a 5-µm thick SiO_2-based surface layer. It follows that in dilute HCl, the Ti atoms are dissolved, while the Si atoms are oxidized *in situ* forming a passivating layer that prevents further dissolution. In other words, the operative reaction is most probably

$$Ti_3SiC_2 + 3O_2 = 3Ti\,(sol) + SiO_2 + 2CO_2 \tag{7.22}$$

Further evidence for this conclusion can be found in Figure 7.11a, in which the zeta potentials for Ti_3SiC_2 powders immersed in HNO_3, HCl, and H_2SO_4 are plotted as

Table 7.1 Summary of chemical corrosion rates of Ti_3SiC_2 in common acids and NaOH of differing concentrations (Travaglini *et al.*, 2003). Also listed for comparison purposes are the corrosion rates of unalloyed Ti.

Solution (vol%)	pH	Chemical corrosion rates (μm per year)			
		Ti_3SiC_2		Unalloyed Ti	
		Weight loss	SEM	Corrosion rate	References
70% HNO_3	<0	12.4	25 ± 5	<5	Schutz and Thomas (1987)
11.6% HNO_3	<0	210	400 ± 200	<5	McQuillan and McQuillan (1956)
6.4 % HNO_3	≈0	162	243 ± 200	<5	Kelly (1982)
38% HCl	<0	≈0	—	2500	Schutz and Thomas (1987)
25% HCl	—	—	—	550	McQuillan and McQuillan (1956)
16% HCl			—	5045 (37 °C)	Kelly (1982)
7.6% HCl	0.22	1.6	—	1270	Yu *et al.* (1999)
95% H_2SO_4	—	≈0	—	1570	Schutz and Thomas (1987)
13.6% H_2SO_4	<0	0.8	—	1270	
8.2% HF	0.85	≈0	—	Soluble	—
0.1 g l^{-1} NaOH	11.5	≈0	—	<3	—

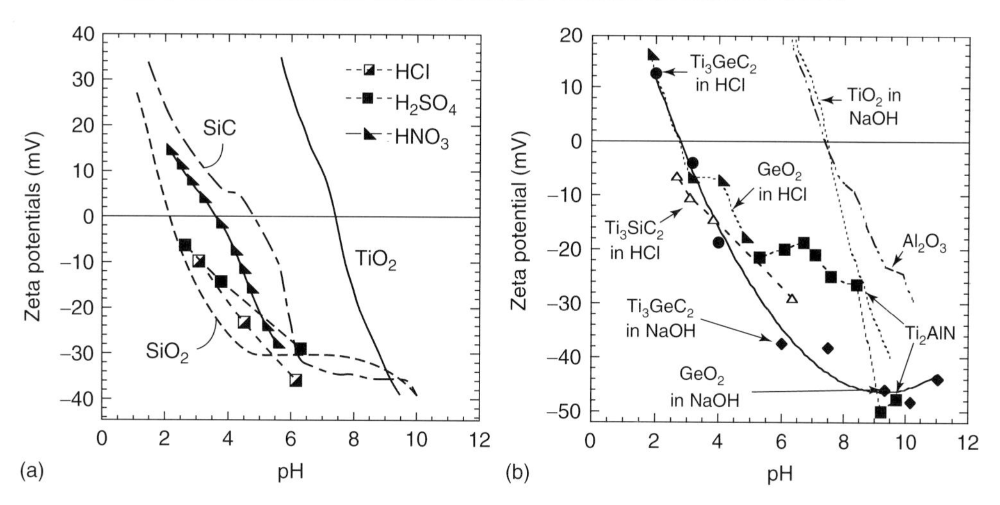

Figure 7.11 Zeta potentials as a function of pH for (a) Ti_3SiC_2 powders immersed in HNO_3, HCl, and H_2SO_4 (Travaglini *et al.*, 2003). Also included are results for SiC, SiO_2, and TiO_2 and (b) Ti_3GeC_2 and Ti_2AlN immersed in various solutions (Jovic *et al.*, 2006a). Also included are results for GeO_2, TiO_2, and Al_2O_3 powders. Lines are guides to the eyes.

a function of pH. Also included are the results for SiC, SiO_2, and TiO_2. In all cases, it is clear that the Ti_3SiC_2 surfaces behave more like those of SiO_2 than TiO_2.

Subsequent work (Jovic and Barsoum, 2004) confirmed that in both HCl and H_2SO_4, a thin SiO_2 passivating layer forms on the surface of Ti_3SiC_2. Anodic oxidation in HCl resulted in an oxide film that grew rapidly to a thickness of 10 μm in the first 100 h. Subsequent growth was slower.

The corrosion rate of Ti_3SiC_2 in HCl was significantly lower than pure Ti (Table 6.1). For example, at 65 °C, the corrosion rate of Ti_3SiC_2 in dilute HCl was ≈2000 times lower than that of pure Ti. This low corrosion rate under the aforementioned conditions bodes well for the use of Ti_3SiC_2 as a base material for dimensionally stable anodes for chlorine production (Jovic and Barsoum, 2006).

In 1 M HCl solutions, Ti_2AlC, V_2AlC, and V_2GeC actively dissolve and Ti_4AlN_3 and Ti_3GeC_2 – like Ti_3SiC_2 – passivate (Jovic *et al.*, 2006b). Depending on the potential, $(Ti,Nb)_2AlC$ and Cr_2AlC showed trans-passive behavior. In 1M HCl, Ti_2AlN showed trans-passive behavior (Jovic *et al.*, 2006b).

Recently, Xie *et al.* (2012) immersed Ti_2AlC, Ti_3AlC_2, Ti_2AlN, Ti_4AlN_3, Ti_3SiC_2, TiC, and TiN powders in a HCl solution (2.5 mol l^{-1}), and heated the mixture in an autoclave at 200 °C for 24 h. Not surprisingly, given the reactivity of the A-group element, they found that the reactivities increased in the following sequences: TiC $< Ti_3AlC_2 < Ti_2AlC$ and TiN $< Ti_4AlN_3 < Ti_2AlN$. When a reaction did occur, in all cases, rutile was formed. They also showed that Ti_3SiC_2 was more stable than Ti_3AlC_2, in total accordance with the fact that Ti_3SiC_2 forms a protective silica layer in HCl (Jovic and Barsoum, 2004). It is also in accordance with the fact that HF selectively etches Al from Ti_3AlC_2 and Ti_2AlC, but not Si from Ti_3SiC_2 (see subsequent text).

7.5.2
Sulfuric Acid

When Ti_3SiC_2 was immersed in dilute and concentrated H_2SO_4 solutions for six months, the corrosion rates were negligible (see Figure 7.10b and Table 7.1) (Travaglini *et al.*, 2003). In concentrated H_2SO_4, (Figure 7.10b), the corrosion rate after the first day corresponded to 111 μm per year. This rate, however, dropped dramatically after about a month to essentially zero, again suggesting the formation of a passivating layer. Cyclic polarization and potentiostatic i–t transients in dilute H_2SO_4 strongly suggest that a thin, irreversible, electrically insulating layer forms on the Ti_3SiC_2 surface. This conclusion is further bolstered by the results shown in Figure 7.11a, where it is reasonable to assume that the Ti atoms are dissolved while the Si atoms are oxidized *in situ* to form a protective layer.

Similar to Ti_3SiC_2, in 1 M H_2SO_4 solutions, Ti_2AlC, $(Ti,Nb)_2AlC$, Ti_4AlN_3, Ti_2AlN, and Ti_3GeC_2 – passivate. V_2AlC and V_2GeC show active dissolution, while Cr_2AlC exhibits trans-passive behavior (Jovic *et al.*, 2006b).

More recently, Li *et al.* showed that Ti_3AlC_2 had poor corrosion resistance in 1 M H_2SO_4 owing to the formation of a permeable Ti sub-oxide layer that acts as a pseudo-passivating film (Li *et al.*, 2010). The authors also show clear evidence for grain boundary corrosion after 25 days immersion or anodic polarization. However, since the purity and homogeneity of the starting Ti_3AlC_2 material was not given, the conclusion that a permeable suboxide layer forms needs to be confirmed by further work. This is especially true since not all grain boundaries were attacked. It is thus possible that what the authors observed is the dissolution of minor second phases at the grain boundaries. More work on well characterized samples is indicated.

7.5.3 Nitric Acid

When Ti_3SiC_2 was placed in concentrated HNO_3 for six months, the corrosion rate was linear (Figure 7.10c) and equivalent to $\approx$13 µm per year (Table 7.1) (Travaglini *et al.*, 2003). In contrast to Ti metal, the weight losses of Ti_3SiC_2 in dilute HNO_3 were higher (250–320 µm per year) and depended on the acid concentration. Post-immersion SEM images of samples immersed in HNO_3 indicated that an oxygen-rich, Si-based layer formed on the surface of the samples. This implies that the Ti atoms are again leached out into the HNO_3 solution, leaving behind a Si-rich layer that is ultimately oxidized. Why this layer is not protective in this case is unclear.

7.5.4 Hydrofluoric Acid

When bulk polycrystalline Ti_3SiC_2 samples were immersed in dilute HF for six months, the corrosion rate was $\approx$5 µm per year (Figure 7.10d and Table 7.1) (Travaglini *et al.*, 2003). Here again, one must conclude that a passivating layer forms. Further, since Ti dissolves in dilute HF (Table 7.1), it is tempting to ascribe the passivation to the formation of a silica layer, whose nature is unknown at this time.

Quite recently, the reactivities of the MAX phases in HF were shown to depend on the HF concentration (Naguib *et al.*, 2012). For example, Ti_2AlC powders were found to completely dissolve in concentrated (50 vol%) HF. However, when the same powders were immersed, at room temperature, in 10 vol% HF for 10 h, the Al was selectively etched resulting in exfoliated two-dimensional (2-D) sheets of Ti_2C bonded together by weak secondary bonds (see subsequent text).

In contrast, Ti_3AlC_2 powders were more resistant to dissolution: immersion in 50% HF for 2 h at room temperature resulted in the formation of 2-D nanosheets (Naguib *et al.*, 2011a). The following simplified reactions occur when Ti_3AlC_2 is immersed in HF:

$$Ti_3AlC_2 + 3HF = AlF_3 + \frac{3}{2}H_2 + Ti_3C_2 \quad (7.23)$$

$$Ti_3C_2 + 2H_2O = Ti_3C_2(OH)_2 + H_2 \quad (7.24)$$

$$Ti_3C_2 + 2HF = Ti_3C_2F_2 + H_2 \quad (7.25)$$

Equation (7.23) is essential, and is followed by Eq. (7.24) and/or (7.25). These reactions result in the exfoliation of 2-D Ti_3C_2 layers, with OH and/or F surface groups.

Table 7.2 Summary of conditions needed to exfoliate various MAX phase powders by immersion in HF. The c lattice parameters before (c_i) and after (c_f) HF treatment are listed as well as the domain size λ_d along [0001] calculated from the Scherrer formula (Naguib *et al.*, 2012).

Compound	HF (conc%)	Time (h)	c_i (Å)	c_f (Å)	λ_d (Å)
Ti_2AlC	10	10	13.6	15.04	60
Ta_4AlC_3	50	72	24.08	30.33	380
				28.43	180
$(Ti_{0.5},Nb_{0.5})_2AlC$	50	28	13.79	14.88	50
$(V_{0.5},Cr_{0.5})_3AlC_2$	50	69	17.73	24.26	28
Ti_3AlCN	30	18	18.41	22.28	70
Ti_3AlC_2	50	2	18.42	20.51	110

Equations (7.24) and (7.25) are simplified in that they assume the terminations are OH or F, respectively, when in fact they most probably are a combination of both (Naguib *et al.*, 2011a).

The ternary Ti_3AlC_2 is not the only composition that can be exfoliated in this manner. Table 7.2 summarizes the conditions needed to exfoliate various MAX phases by immersion in HF solutions of various potencies (Naguib *et al.*, 2012). Also shown are the c lattice parameters before (c_i) and after (c_f) HF treatment as well as the domain sizes λ_d along [0001]. In all cases, $c_f > c_i$. It is also obvious that the domain size along [0001] λ_d is quite small and for the most part ranges from 50 to 100 nm.

Figures 7.12 and 7.13 show, respectively, typical SEM and TEM images of various exfoliated MAX phases. In both cases, it is clear that the scale of the exfoliation is quite fine indeed. TEM has also shown that these 2D layers behave and configure in structures that are reminiscent of graphene. For example, after sonication, some of the 2D Ti_3C_2 layers roll into conical scrolls and some are a few multilayer thick. It is for this reason that this new class of solids was labeled "MXenes" (Naguib *et al.*, 2011a).

Sonication of the flakes results in their dispersion. When the latter are deposited on glass slides, the thinner ones were optically transparent despite their good electrical conductivity (Naguib *et al.*, 2012). The conductivity of pressed discs of HF-treated powders is comparable to that of multilayer graphene. The pressed MXenes surfaces also showed hydrophilic behavior.

7.5.5 Sodium Hydroxide

When Ti_3SiC_2 was immersed in 1 M NaOH for six months, the dissolution rate was low (Figure 7.10d and Table 7.1) (Travaglini *et al.*, 2003). Ti_3AlC_2 also displayed good corrosion resistance in 1 M NaOH by forming a dense and protective passivating film of titanium oxides (Li *et al.*, 2010).

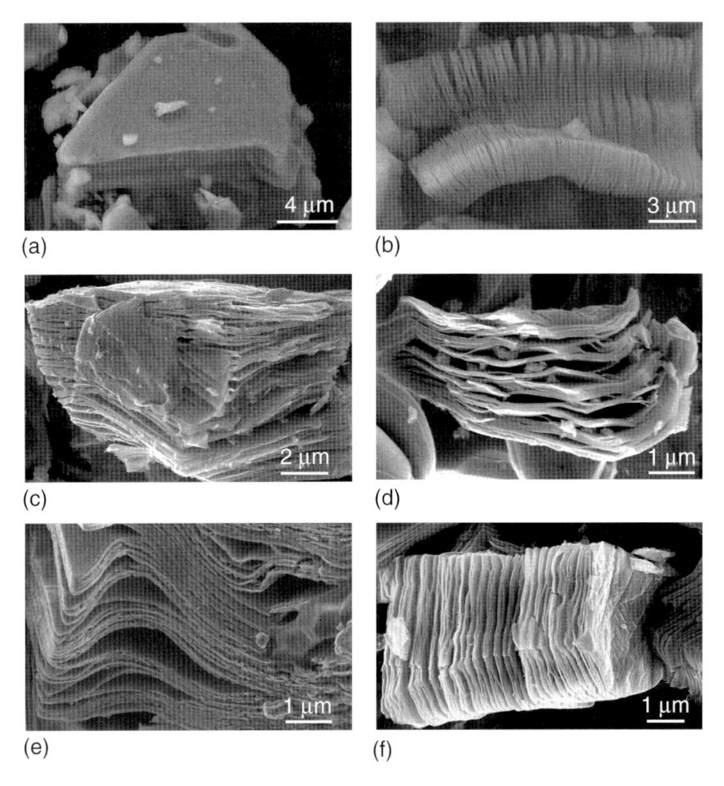

Figure 7.12 Secondary electron SEM images for (a) Ti_3AlC_2 particle before treatment, which is typical for unreacted MAX phases, (b) Ti_3AlC_2, (c) Ti_2AlC, (d) Ta_4AlC_3, (e) TiNbAlC, and (f) Ti_3AlCN after HF treatment. In (b–f) the exfoliation as a result of HF treatment is obvious (Naguib *et al.*, 2012).

When Ti_3GeC_2 was immersed in 1 M NaOH, at potentials more positive than 0.0 V with respect to a saturated silver/silver chloride reference electrode, a semiconducting GeO_2-based passivating layer – stable down to a pH of 3 – formed (Jovic *et al.*, 2006a). This is best evidenced in Figure 7.11b, where the isoelectronic point (IEP) of Ti_3GeC_2 powders is plotted as a function of pH together with those of GeO_2 and other relevant oxides. The fact that the electrophoresis results of the Ti_3GeC_2 powders are quite similar to those of GeO_2 powders – over a wide pH range (Figure 7.11b) – is important evidence for this conclusion. Had the layers formed been TiO_2-based, the IEP would have been closer to 8. Figure 7.14 further confirms the similarities between Ti_3GeC_2 and Ti_3SiC_2. On the basis of these results, it is reasonable to assume that passivating GeO_2 or SiO_2 layers form, respectively.

The IEP results for Ti_2AlN in NaOH are slightly more ambiguous. The zeta potential at low pH is closer to that of GeO_2. At pH around 9 and 10, the response is more reminiscent of Al_2O_3 or TiO_2 powders. More work is thus needed to better understand what is occurring when Ti_2AlN is immersed in 1 M NaOH.

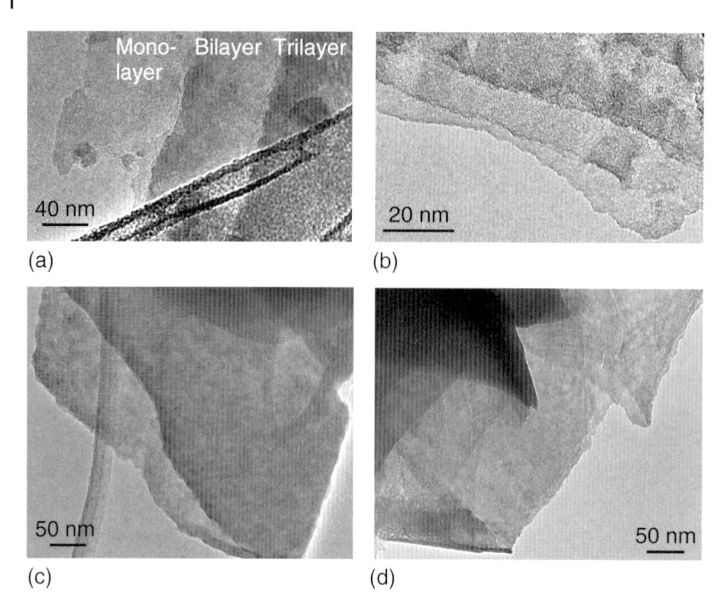

Figure 7.13 Typical TEM images of HF treated powders of (a) Ti_3AlC_2, (b) Ti_3AlCN, (c) $(Ti_{0.5}Nb_{0.5})_2AlC$, and (d) Ta_4AlC_3. In all cases, the energy dispersive spectra did not show an Al signal (Naguib *et al.*, 2012).

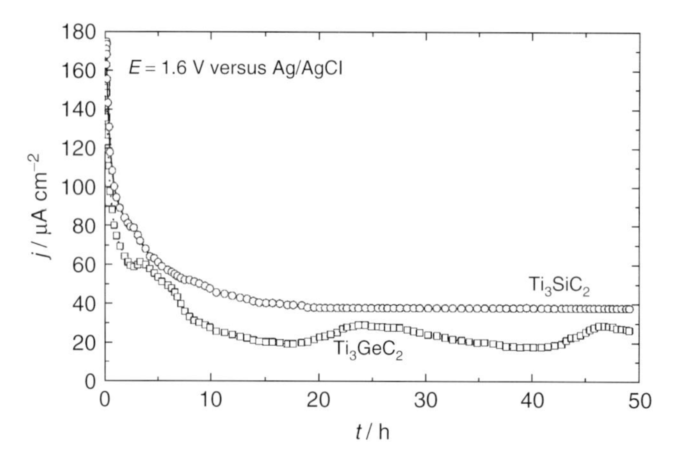

Figure 7.14 Current density as a function of time recorded at a constant potential of 1.6 V versus Ag/AgCl for Ti_3GeC_2 and Ti_3SiC_2 samples immersed in 1 M NaOH (Jovic *et al.*, 2006a).

To eliminate the possibility of dissolution, fine Ti_2AlN powders produced by a file were placed in 1 M NaOH and 1 M HCl solutions for a month at room temperature. No apparent dissolution was observed.

7.6
Summary and Conclusions

In general, the $M_{n+1}X_n$ layers in the MAX phases are quite stable *vis-à-vis* reaction. In many cases, the preferred reaction path is one where the A-group element out-diffuses from the MAX phase into the surrounding medium, leaving behind the binary carbides. The A-group element, on the other hand, tends to either form intermetallics or simply dissolve in the reacting medium. If no intermetallics form *and* the A-group element has no solubility in the reacting medium (e.g., Mg, Pb, Pb–Bi), then no reaction is observed.

The reaction with aggressive molten salts seems to be a function of oxygen content in the latter. In the absence of O, it may very well be that the MAX phases are stable. In the presence of oxygen, the MAX phases form oxides that, in turn, dissolve in the salts, causing more oxide formation, and so on. To find MAX phases that are stable in molten salts in the presence of oxygen, the oxides that form should not be soluble in the salt.

Select MAX phases are stable in various common acids and NaOH. Immersion of many Al-containing MAX phases in HF solutions at room temperature of various potency results in the selective etching of the Al layers and the conversion of the 3D MAX phases to their 2D counterpart. The latter was labeled MXene to emphasize the loss of the A element and as importantly the similarity of the resulting 2D solids to graphene. In only the most extreme conditions – such as in hot HCl, or in contact with anhydrous HF gas – will they react to form rutile, anatase, and/or new phases.

7.A Appendix

Figures 7.A.1–7.A.4 plot, respectively, the ternary phase diagrams in the Ti–Al–N (Procopio, El-Raghy, and Barsoum, 2000), V–Al–C, Cr–Al–C (Schuster, Nowotny, and Vaccaro, 1980), Ti–Ge–C (Kephart and Carim, 1998), Ti–In–C (Ganguly, Barsoum, and Schuster, 2005), Nb–Sn–C (Barsoum *et al.*, 2002), and Ti–Sn–C (Vincent *et al.*, 1998) systems.

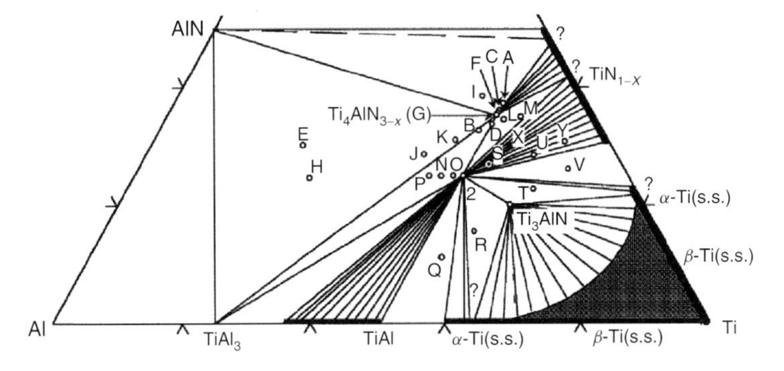

Figure 7.A.1 Isothermal section of the Ti–Al–N ternary phase diagram at 1300 °C (Procopio, El-Raghy, and Barsoum, 2000).

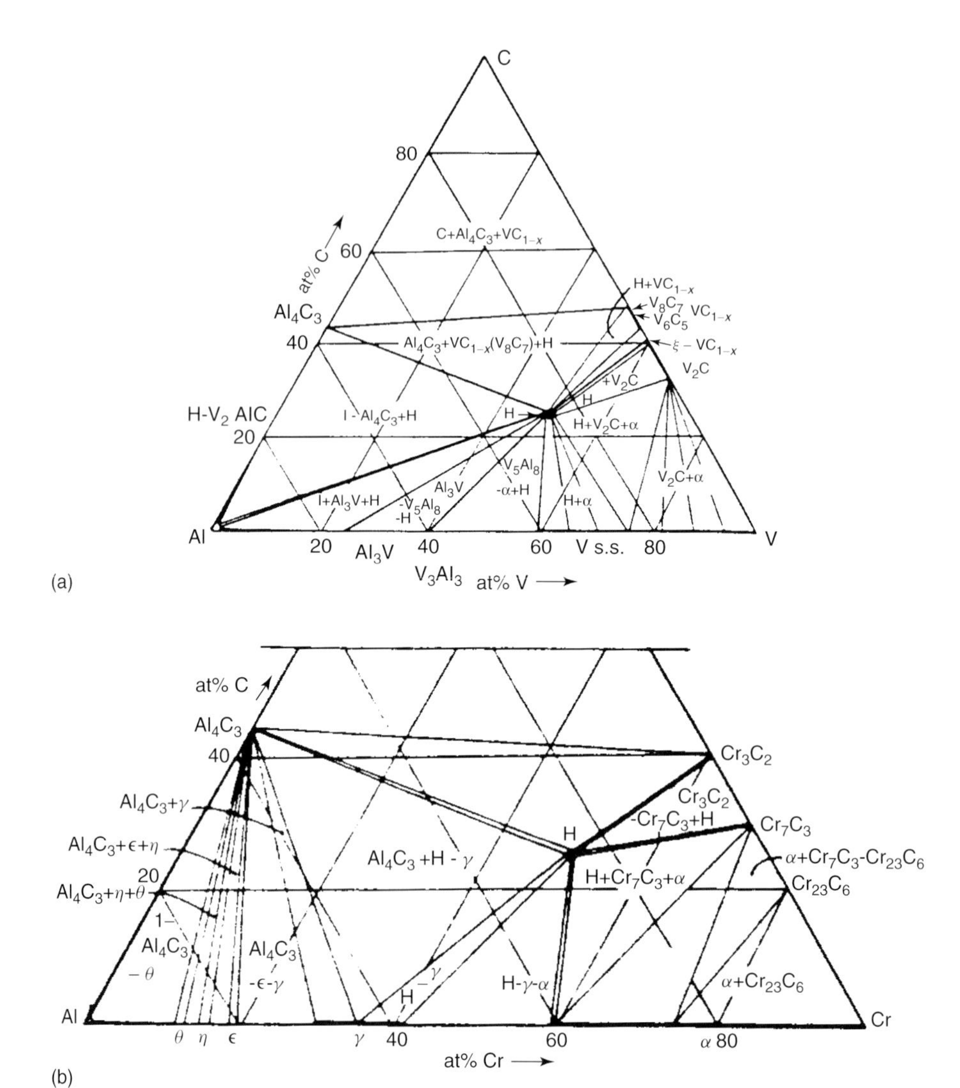

Figure 7.A.2 Ternary phase diagrams at 1000 °C in (a) V–Al–C and (b) Cr–Al–C systems (Schuster, Nowotny, and Vaccaro, 1980).

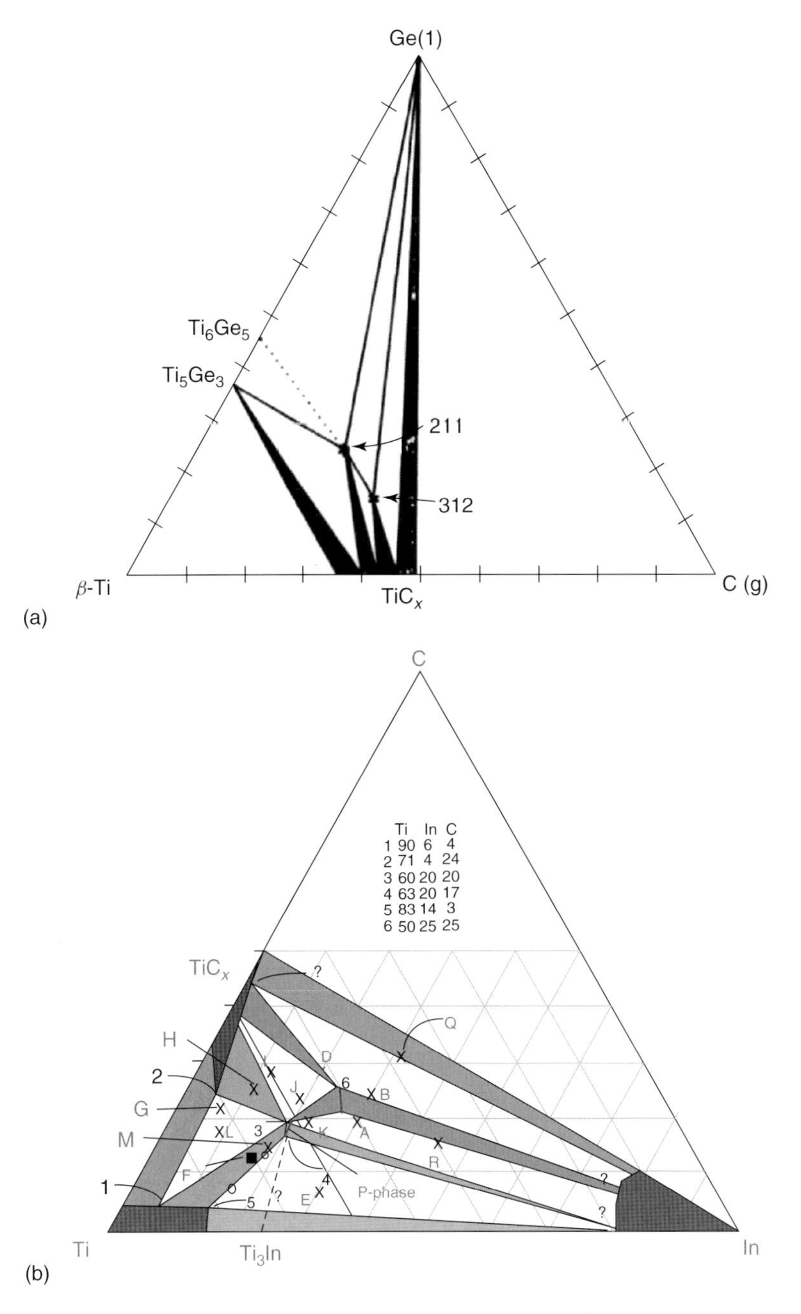

Figure 7.A.3 Ternary phase diagrams in (a) Ti–Ge–C at 1200 °C (Kephart and Carim, 1998) and (b) Ti–In–C at 1300 °C (Ganguly, Barsoum, and Schuster, 2005) systems.

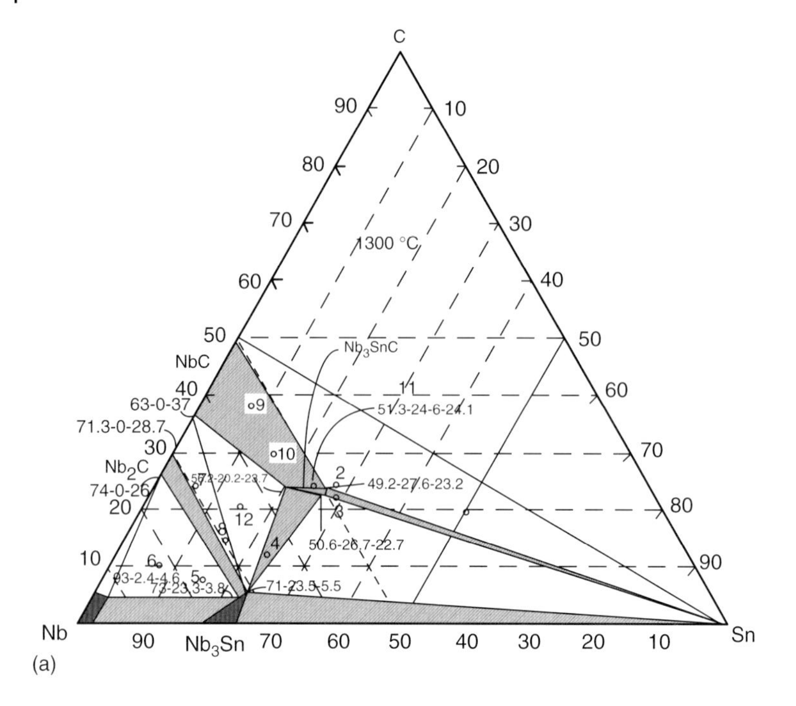

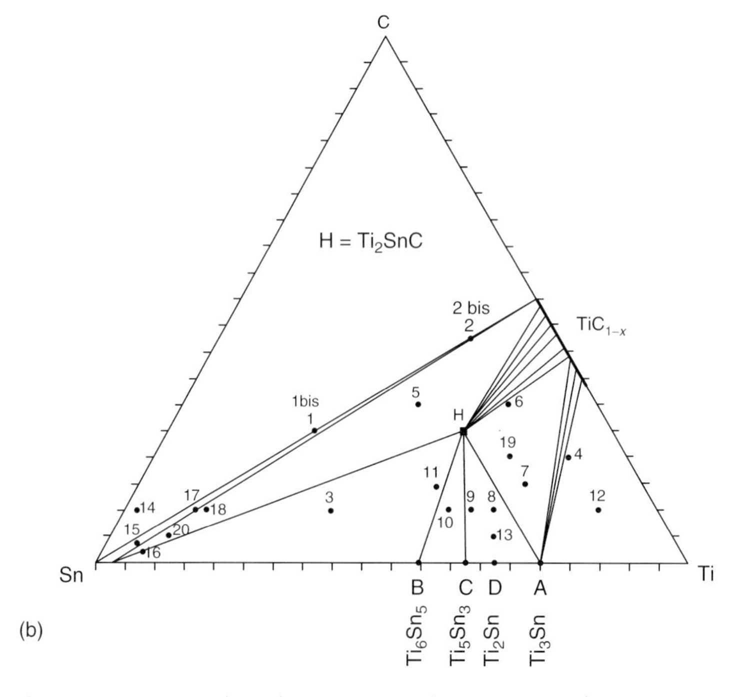

Figure 7.A.4 Ternary phase diagrams in (a) Nb–Sn–C at 1300 °C (Barsoum *et al.*, 2002) and (b) Ti–Sn–C (Vincent *et al.*, 1998) systems.

References

Amini, S., Ni, C., and Barsoum, M.W. (2009) Processing, microstructural characterization and damping of a Ti_2AlC/nanocrystalline Mg-matrix composite. *Compos. Sci. Technol.*, **69**, 414–420.

Barnes, L.A., Dietz Rago, N.L., and Leibowitz, L. (2008) Corrosion of ternary carbides by molten lead. *J. Nucl. Mater.*, **373**, 424–428.

Barsoum, M.W., El-Raghy, T., Farber, L., Amer, M., Christini, R., and Adams, A. (1999) The topotaxial transformation of Ti_3SiC_2 to form a partially ordered cubic $TiC_{0.67}$ phase by the diffusion of Si into molten cryolite. *J. Electrochem. Soc.*, **146**, 3919–3923.

Barsoum, M.W., Ganguly, A., Siefert, H.J., and Aldinger, F. (2002) The 1300 °C isothermal section in the Nb-Sn-C ternary phase diagram. *J. Alloys Compd.*, **337**, 202–207.

Basu, S., Ozaydin, M.F., Kothalkar, A., Karaman, I., and Radovic, M. (2011) Phase and morphology evolution in high-temperature Ti_3SiC_2–NiTi diffusion-bonded joints. *Scr. Mater.*, **65**, 237–240.

Cockeram, B. and Rapp, R. (1995) Kinetics of multilayered titanium-silicide coatings grown by the pack cementation method. *Metall. Mater. Trans. A*, **26**, 777–791.

Delpech, S., Cabet, C., Slim, C., and Picard, G.S. (2010) Molten fluorides for nuclear applications. *Mater. Today*, **13**, 34.

Du, Y., Schuster, J., Seifert, H., and Aldinger, F. (2000) Experimental investigation and thermodynamic calculation of the titanium-silicon-carbon system. *J. Am. Ceram. Soc.*, **83**, 197–203.

El-Raghy, T. and Barsoum, M.W. (1998) Diffusion kinetics of the carburization and silicidation of Ti_3SiC_2. *J. Appl. Phys.*, **83**, 112–119.

El-Raghy, T., Barsoum, M.W., and Sika, M. (2001) Reaction of Al with Ti_3SiC_2 in the 800–1000 °C temperature range. *Mater. Sci. Eng., A*, **298**, 174.

Ganguly, A., Barsoum, M.W., and Doherty, R.D. (2007) Interdiffusion between Ti_3SiC_2-Ti_3GeC_2 and Ti_2AlC-Nb_2AlC diffusion couples. *J. Am. Ceram. Soc.*, **90**, 2200–2204.

Ganguly, A., Barsoum, M.W., and Schuster, J. (2005) The 1300 °C isothermal section in the Ti-In-C ternary phase diagram. *J. Am. Ceram. Soc.*, **88**, 1290–1296.

Gao, N.F. and Miyamoto, Y. (2002) Joining of Ti_3SiC_2 with Ti–6Al–4V alloy. *J. Mater. Res.*, **17**, 51–59.

Gu, W.L., Yan, C.K., and Zhou, Y.C. (2003) Reactions between Al and Ti_3SiC_2 in temperature range of 600–650 °C. *Scr. Mater.*, **49**, 1075–1080.

Hajas, D.E., Scholz, M., Ershov, S., Hallstedt, B., Palmquist, J.-P., and Schneider, J. (2010) Thermal and chemical stability of Cr_2AlC in contact with α-Al_2O_3 and NiAl. *Int. J. Mater. Res.*, **101**, 12.

Heinzel, A., Müller, G., and Weisenburger, A. (2009) Compatibility of Ti_3SiC_2 with liquid Pb and PbBi containing oxygen. *J. Nucl. Mater.*, **392**, 255–258.

Hung, L.S., Gyulai, J., Mayer, J.W., Lau, S.S., and Nicolet, M.A. (1980) Kinetics of $TiSi_2$ formation by thin Ti films on Si. *J. Appl. Phys.*, **54**, 5076–5080.

Jovic, V.D. and Barsoum, M.W. (2004) Corrosion behavior and passive film characteristics formed on Ti, Ti_3SiC_2 and Ti_4AlN_3 in H_2SO_4 and HCl. *J. Electrochem. Soc.*, **151**, B71–B76.

Jovic, V.D. and Barsoum, M.W. (2006) Electrolytic cell and electrodes for use in electrochemical processes. U.P. Office, ed. (USA). 7,001,494.

Jovic, V.D., Barsoum, M.W., Jovic, B.M., and Ganguly, A. (2006a) Corrosion behavior of Ti_3GeC_2 and Ti_2AlN in 1M NaOH. *J. Electrochem. Soc.*, **153**, B238–B243.

Jovic, V.D., Barsoum, M.W., Jovic, B.M., Gupta, S., and El-Raghy, T. (2006b) Corrosion behavior of select MAX phases in NaOH, HCl and H_2SO_4. *Corros. Sci.*, **48**, 4274–4282.

Kelly, E.J. (1982) Electrochemical-behavior of titanium. *Mod. Aspects Electrochem.*, **14**, 319–424.

Kephart, J.S. and Carim, A.H. (1998) Ternary compounds and phase equilibria in Ti-Ge-C and Ti-Ge-B. *J. Electrochem. Soc.*, **145**, 3253.

Lawson, M.G., Pettit, F.S., and Blachere, J.R. (1993) Hot corrosion of alumina. *J. Mater. Res.*, **8**, 1964.

Li, D., Liang, Y., Liu, X., and Zhou, Y.C. (2010) Corrosion behavior of Ti_3AlC_2 in NaOH and H_2SO_4. *J. Eur. Ceram. Soc.*, **30**, 3227–3234.

Lin, Z.J., Li, M.S., Wang, J.Y., and Zhou, Y.C. (2007) High-temperature oxidation and hot corrosion of Cr_2AlC. *Acta Mater.*, **55**, 6182–6191.

Lin, Z., Zhou, Y., Li, M., and Wang, J. (2006a) Hot corrosion and protection of Ti_2AlC against Na_2SO_4 salt in air. *J. Eur. Ceram. Soc.*, **26**, 3871–3879.

Lin, Z., Zhou, Y., Li, M., and Wang, J. (2006b) Improving the Na_2SO_4-induced corrosion resistance of Ti_3AlC_2 by pre-oxidation in air. *Corros. Sci.*, **48**, 3271–3289.

Liu, G., Li, L., Zhou, Y., and Zhang, Y. (2003a) Corrosion behavior and strength degradation of Ti_3SiC_2 exposed to a eutectic K_2CO_3 and Li_2CO_3 mixture. *J. Eur. Ceram. Soc.*, **23**, 1957–1962.

Liu, G., Li, M., and Zhou, Y. (2003b) Hot corrosion of Ti_3SiC_2-based ceramics coated with Na_2SO_4 at 900 and 1000 °C in air. *Corros. Sci.*, **45**, 1217–1226.

Liu, G., Li, M., Zhou, Y., and Zhang, Y. (2006) Influence of pre-oxidation on the hot corrosion of Ti_3SiC_2 in the mixture of Na_2SO_4–NaCl melts. *Corros. Sci.*, **48**, 650–661.

McQuillan, A.D. and McQuillan, M.K. (1956) *Titanium*, Butterworths, London.

Naguib, M., Kurtoglu, M., Presser, V., Lu, J., Niu, J., Heon, M., Hultman, L., Gogotsi, Y., and Barsoum, M.W. (2011a) Two dimensional nanocrystals produced by exfoliation of Ti_3AlC_2. *Adv. Mater.*, **23**, 4248–4253.

Naguib, M., Presser, V., Tallman, D., Lu, J., Hultman, L., Gogotsi, P., and Barsoum, M.W. (2011b) On the topotactic transformation of Ti_2AlC into a Ti-C-O-F cubic phase by heating in molten lithium fluoride in air. *J. Am. Ceram. Soc.*, **94**, 4566–4561.

Naguib, M., Mashtalir, O., Carle, J., Lu, J., Hultman, L., Gogotsi, Y., and Barsoum, M.W. (2012) Two-dimensional transition metal carbides. *ACS Nano*, **6**, 1322–1331.

Naka, M., Feng, J.C., and Schuster, J.C. (1997) Phase reaction and diffusion path of the SiC/Ti system. *Metall. Mater. Trans. A*, **28**, 1385.

Nukami, T. and Flemings, M.C. (1995) In situ synthesis of TiC particulate reinforced aluminum matrix composites. *Metall. Mater. Trans. A*, **26A**, 1877–1884.

Peng, L.M. (2007) Fabrication and properties of Ti_3AlC_2 particulates reinforced copper composites. *Scr. Mater.*, **56**, 729–732.

Pietzka, M.A., and Schuster, J. (1994) Summary of Constitution Data of the System Al-C-Ti. *J. Phase. Equilibria*. **15**, 392.

Procopio, A.T., El-Raghy, T., and Barsoum, M.W. (2000) Synthesis of Ti_4AlN_3 and phase equilibria in the Ti-Al-N system. *Metall. Mater. Trans. A*, **31A**, 373.

Radhakrishnan, R., Henager, C.H., Brimhall, J.L., and Bhaduri, B.B. (1996) Synthesis of Ti_3SiC_2/SiC and $TiSi_2$/SiC composites using displacement reactions in the Ti-Si-C system. *Scr. Mater.*, **34**, 1809.

Revesz, P., Gyimesi, J., Pogany, L., and Peto, G. (1983) Lateral growth of titanium silicide over a silicon dioxide layer. *J. Appl. Phys.*, **54**, 2114–2115.

Rivai, A.K. and Takahashi, M. (2008) Compatibility of surface-coated steels, refractory metals and ceramics to high temperature lead-bismuth eutectic. *Prog. Nucl. Energy*, **50**, 560–566.

Rivai, A.K. and Takahashi, M. (2010) Corrosion characteristics of materials in Pb–Bi under transient temperature conditions. *J. Nucl. Mater.*, **398**, 139–145.

Schuster, J.C., Nowotny, H., and Vaccaro, C. (1980) The ternary systems: Cr-Al-C, V-Al-C and Ti-Al-C and the behavior of the H-phases (M_2AlC). *J. Solid State Chem.*, **32**, 213.

Schutz, R. and Thomas, D.E. (1987) Corrosion of titanium and titanium alloys, in *Metals Handbook*, 9th edn (ed J.R. Davis), ASM, Metals Park, OH.

Sienicki, J.J., Moisseytsev, A., Bortot, S., Lu, Q., and Aliberti, G. (2011) SUPERSTAR: an improved natural circulation, lead-cooled, small modular fast reactor for international deployment. Paper presented at Proceedings of ICAPP, Nice, France, 2011.

Travaglini, J., Barsoum, M.W., Jovic, V., and El-Raghy, T. (2003) The corrosion behavior of Ti_3SiC_2 in common acids and dilute NaOH. *Corros. Sci.*, **45**, 1313–1327.

Utili, M., Agostini, M., Coccoluto, G., and Lorenzini, E. (2011) Ti_3SiC_2 as a candidate material for lead cooled fast reactor. *Nucl. Eng. Des.*, **241**, 1295–1300.

Vincent, H., Vincent, C., Mentzen, B.F., Partor, S., and Bouix, J. (1998) Chemical interaction between C and Ti dissolved in liquid Sn: crystal structure and reactivity of Ti_2SnC with Al. *Mater. Sci. Eng.*, **A256**, 83–91.

Wakelkamp, W.J.J., Loo, F.J.V., and Metselaar, R. (1991) Phase relations in the Ti-Si-C system. *J. Eur. Ceram. Soc.*, **8**, 135.

Wang, W.J., Gauthier-Brunet, V., Bei, G.P., Bonneville, J., Joulain, A., and Dubios, S. (2011) Powder metallurgy processing and compressive properties of Ti_3AlC_2/Al composites. *Mater. Sci. Eng. A*, **530**, 168.

Wang, X.H. and Zhou, Y.C. (2004) Hot corrosion of Na_2SO_4-coated Ti_3AlC_2 in air at 700–1000 °C. *J. Electrochem. Soc.*, **151**, B505–B511.

Wu, J., Zhou, Y., and Yan, C. (2005) Mechanical and electrical properties of Ti_2SnC dispersion-strengthened copper. *Z. Metallk.*, **96**, 847–852.

Xie, J., Wang, X., Li, A., Li, F., and Zhou, Y. (2012) Corrosion behavior of selected $M_{n+1}AX_n$ phases in hot concentrated HCl solution: effect of a element and MX layer. *Corros. Sci.*, **60**, 129–135.

Yin, X., Li, M., Xu, J., Zhang, J., and Zhou, Y. (2009) Direct diffusion bonding of Ti_3SiC_2 and Ti_3AlC_2. *Mater. Res. Bull.*, **44**, 1379–1384.

Yin, X.H., Li, M.S., and Zhou, Y.C. (2006) Microstructure and mechanical strength of diffusion-bonded Ti_3SiC_2/Ni joints. *J. Mater. Res.*, **21**, 2415–2421.

Yu, S., Brodrick, C.W., Ryan, M.P., and Scully, J.R. (1999) Effects of Nb and Zr alloying additions on the activation behavior of Ti in hydrochloric acid. *J. Electrochem. Soc.*, **146**, 4429–4438.

Zhang, J., Wang, J.Y., and Zhou, Y.C. (2007) Structure stability of Ti_3AlC_2 in Cu and microstructure evolution of Cu–Ti_3AlC_2 composites. *Acta Mater.*, **55**, 4381–4390.

Zhang, Y., Zhimei, S., and Zhou, Y. (1999) Cu/Ti_3SiC_2 composite: a new electrofriction material. *Mater. Res. Innovations*, **3**, 80–84.

8
Dislocations, Kinking Nonlinear Elasticity, and Damping

8.1
Introduction

As discussed in this and the next chapter, one of the main deformation mechanisms of the MAX phases is the formation of kink bands (KBs). KB formation is quite ubiquitous in nature and has been invoked to explain the deformation of numerous materials and structures such as organic crystals (Mugge, 1898), card decks (Gay and Weiss, 1974), rubber laminates (Robertson, 1969), oriented polymer fibers (Bassett and Attenburrow, 1979; Rider and Keller, 1966; Robertson, 1969; DeTeresa *et al.*, 1988; Zaukelies, 1962), wood (Keith and Cote, 1968), graphite fibers (Hathorne and Teghtsoonian, 1975; Jones and Johnson, 1971), laminated C–C, and C–epoxy composites (Argon, 1972; Weaver and Williams, 1975; Gupta, Anand, and Kryska, 1994), amongst others. This chapter does not deal with such KBs but rather with dislocation-based ones. It is thus imperative, at the outset, to understand dislocations and their arrangements in the MAX phases.

8.2
Dislocations and Their Arrangements

Morgiel, Lis, and Pampuch (1996) were the first to note that most of the dislocations in Ti_3SiC_2 were basal dislocations. Farber *et al.* (1998) confirmed this finding and showed the Burgers vector to be $1/3 <1\bar{2}10>$.

Regular and high-resolution transmission electron microscopy (TEM) (Barsoum *et al.*, 1999b; Farber, Levin, and Barsoum, 1999; Kooi *et al.*, 2003) studies of Ti_3SiC_2 revealed the presence of only perfect dislocations lying in the basal planes with Burgers vector equal to the *a* lattice parameter. Every dislocation is of a mixed nature with an edge and screw component (Figure 8.1) (Farber, Levin, and Barsoum, 1999). Note that since nonbasal dislocations would have Burgers vectors greater than *c*, that is, >11–23 Å, their presence is unlikely and, even if present – as a result of growth, for example – would not play a major role in the deformation under normal circumstances.

MAX Phases: Properties of Machinable Ternary Carbides and Nitrides, First Edition. Michel W. Barsoum.

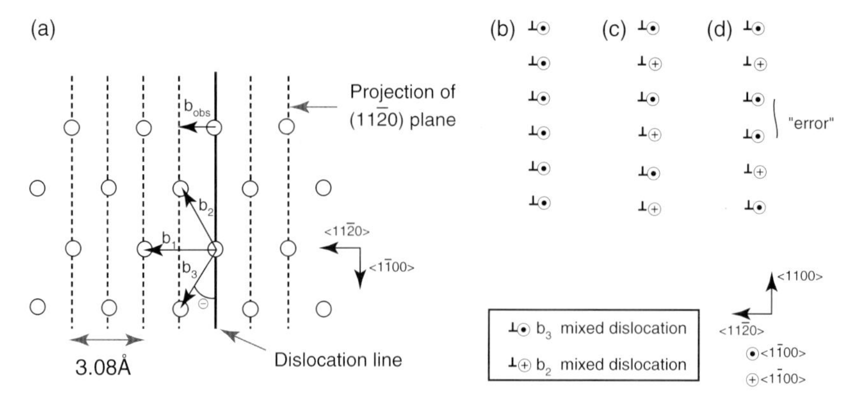

Figure 8.1 (a) Schematic representation of the 0001 projection showing the dislocation line, and projections of observed and actual Burgers vectors, **b**2 and **b**3. (b–d) Possible arrangements of mixed dislocations **b**2 and **b**3 in a dislocation wall. (b) Screw components are all in the same direction. (c) Alternating screw components. (d) Same as (c) but showing a stacking error in the screw component which gives rise to twist. Configuration (b) is unstable, whereas (c) and (d) are stable (Farber *et al.*, 1999).

The dislocations arrange themselves either in walls, that is, low- or high-angle grain boundaries normal to the basal planes (Figure 8.2a,b), or pile-ups on the same basal planes (Figure 8.2c). Given that the low angle kink boundaries are mobile, they are henceforth referred to as mobile dislocation walls, MDWs. The walls have both tilt and twist components (Farber, Levin, and Barsoum, 1999; Kooi *et al.*, 2003). To account for both, the boundary was interpreted to be composed of parallel, *alternating*, mixed perfect dislocations, with two different Burgers vectors lying in the basal plane at an angle of 120° relative to one another (Figure 8.1a). The excess of one type of dislocation accounts for the twist (Figure 8.1c) (Farber, Levin, and Barsoum, 1999; Kooi *et al.*, 2003).

Except under extreme conditions, such as under regions adjacent to indention marks (Tromas *et al.*, 2011), dislocation interactions, other than orthogonal and those within the basal plane (see subsequent text), have not been reported to date. As discussed later, this fact has far-reaching ramifications in that deformation can now occur without work-hardening in the classic sense.

8.2.1
Dislocation–Dislocation Interactions

In the remainder of this chapter, the case will be made that dislocations can glide in a fully reversible manner in the MAX phases and other kinking nonlinear solids. Recent density functional theory (DFT) calculations have shown that the

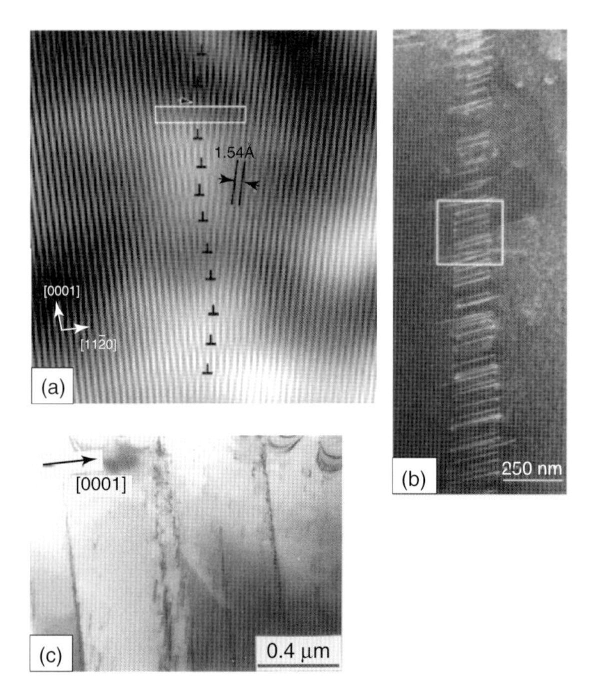

Figure 8.2 (a) Bright-field image of a low angle boundary or MDW filtered in 11$\bar{2}$1 reflection using fast Fourier transform. An example of the Burgers circuit drawn around one of the dislocations is shown by the rectangle; the termination of the extra plane is indicated by the arrow. The locations of all other dislocations are designated by the inverted "T" symbol (Farber *et al.*, 1999). (b) Dislocation wall in which dislocations are parallel and positioned in different basal planes one under another. (c) Bright-field image of an area containing dislocation arrays. The specimen is close to the orientation where basal planes are in an edge-on position (Barsoum *et al.*, 1999a).

critical resolved shear stress (CRSS) of basal dislocations in hexagonal metals (Lane *et al.*, 2011), and by extension, most other hexagonal solids including the MAX phases, is very low indeed. It is postulated that the experimental CRSS values reported (see subsequent text and Chapter 9) most probably result from dislocation–dislocation interactions. Until recently, there was little evidence for such interactions. Quite recently, however, such evidence has been obtained on Ti_2AlN samples that were compressively deformed to a strain of about 5%, with confining pressure (Guitton, *et al.*, 2012). One example of a reaction between basal dislocations is shown in Figure 8.3a. A schematic of the reaction is shown in Figure 8.3b. An example of a dipole formed by two dislocation segments on different basal planes is shown in Figure 8.3c. A schematic of the dipole is shown in Figure 8.3d. It is worth noting that such interactions and dipoles are common in other layered solids, such as graphite and mica (Kelly, 1981; Meike, 1989).

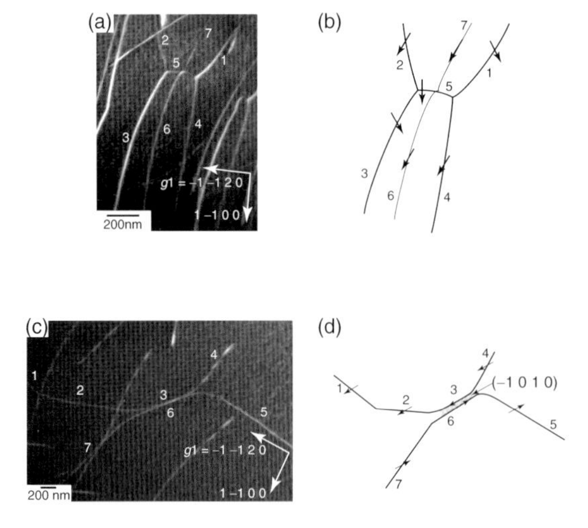

Figure 8.3 (a) TEM micrographs of reactions between dislocations observed with $g_1 = [\bar{1}\bar{1}20]$. (b) Schematic of reaction. (c) TEM micrograph observed with $g_1 = [0\bar{1}13]$ showing dipole. (d) Schematic of dipole. In (b) and (d), the Burgers vector directions are depicted by arrows (Guitton *et al.*, 2012).

8.3
Kink Band Formation in Crystalline Solids

Outside geology (Christoffersen and Kronenberg, 1993; Kronenberg, Kirby, and Pinkston, 1990; Patterson and Weiss, 1996), the formation of KBs in crystalline solids has been more of an afterthought, for the past 50 years since first reported on by Orowan in metals (Orowan, 1942). Orowan induced KBs in Cd single crystals by compressing them with their basal planes almost parallel to the compression axis. He concluded that kink boundaries consisted of planes that bisect the angle between the glide planes on either side of them, and along which dislocations were concentrated (see inset in Figure 9.4a). The boundaries are, in turn, composed of excesses of one kind of dislocations.

In 1949, Hess and Barrett (Hess and Barrett, 1949) proposed a qualitative model to explain KB formation by the regular glide of dislocations. The major elements of their model are summarized schematically in Figure 8.4a–c. Initially on loading, elastic bending (Figure 8.4a) creates maximum shear stress in the center of the column or grain (Figure 8.4b). Above a critical value, this shear stress is sufficient to create, within the volume that is to become the KB, pairs of dislocations of opposite signs that move in opposite directions (Figure 8.4c). The end result is two regions of severe lattice curvature, separated from each other and from the unkinked crystal, by well-defined kink boundaries BC and DE (Figure 8.4d). These kink planes, or boundaries, have an excess of edge dislocations of one sign, which, in turn, are responsible for the lattice rotations observed. The combination of the two kink boundaries and the region between them defines a KB. Examples of KBs

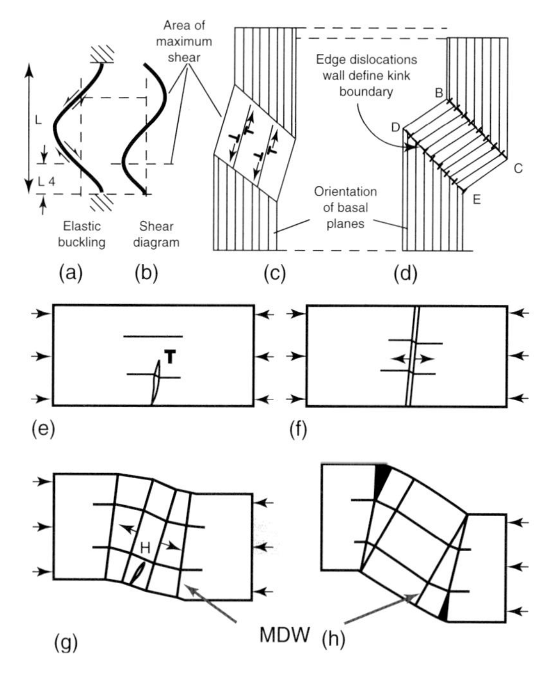

Figure 8.4 Schematic of kink band formation: (a–d) After Hess and Barrett (1949) and (e–h) after F&S (Frank and Stroh, 1952). (a) Elastic buckling. (b) Corresponding shear diagram. (c) Initiation of pairs of dislocations in areas of maximum shear. (d) Kink band and kink boundaries comprised of edge dislocations of one sign giving rise to the signature stove-pipe configuration. (e) Initiation of kink band at tip of narrow kink T. (f) Intersection of T with free surface removes the attractive energy between the walls and allows them to separate and move in opposite directions. (g, h) Repetition of same process to creates more mobile dislocation walls, which ultimately become kink boundaries.

at various lengths scales are shown in Figure 8.5 and 8.6. Most are characterized by the stove-pipe configuration shown schematically in Figure 8.4d.

Hess and Barrett's model was qualitative. In 1952, Frank and Stroh (F&S hereafter) (Frank and Stroh, 1952) used a Griffith-like approach to derive an expression for the *critical* remote shear stress (see Eq. 8.2) above which a subcritical KB will become critical and, like a crack, rapidly and autocatalytically grow. In their model, pairs of dislocations of opposite signs nucleate and grow at the tip of a thin elliptical kink (labeled T in Figure 8.4e) when the remote applied shear stress τ exceeds some critical value. The minimum value of lattice rotation that will allow the dislocation wall to grow via this mechanism was estimated to be $\approx 3^\circ$ for metals. A continuing stress then forces the walls apart to form mobile dislocation walls (MDWs). The process of wall formation can then be repeated, at the same source, resulting in the generation of new MDWs (Figure 8.4g,h). The component dislocations in successive walls can then collapse into a thin region forming a kink boundary. The collapse does not necessarily occur simultaneously along the

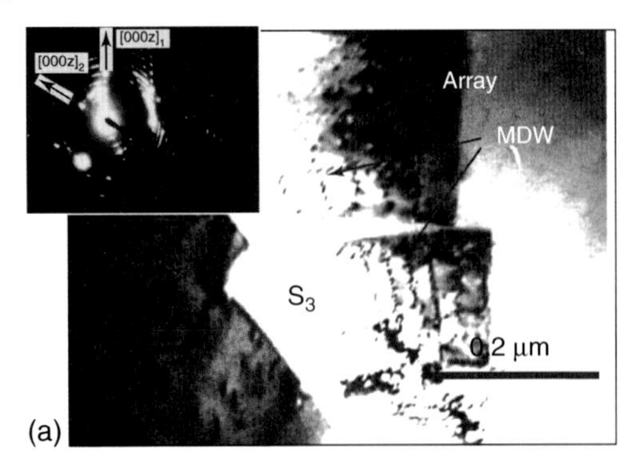

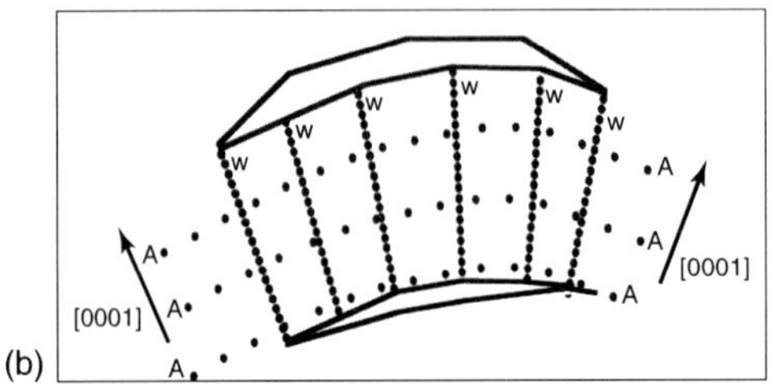

Figure 8.5 (a) Bright-field image of a bent region delaminated in three slices (only one labeled here as S3 is shown). Each slices containing separated walls and arrays. The inset shows the selected area diffraction from the same region, but tilted to an edge-on position for the basal planes demonstrating lattice rotation around [1100]. (b) Schematic of section S3 in (a) showing intersecting walls (W) and arrays (A). Herein, such separated walls, because they are mobile, are termed *mobile dislocation walls* (MDWs) (Barsoum *et al.*, 1999a).

whole wall but can occur sequentially starting at one end of the kink boundary and moving to the other end (e.g., Figure 8.4h).

A kink boundary thus results from the accumulation of several MDWs in a relatively narrow region. Hess and Barrett suggested that the first wall can be stopped by some defect, resulting in the accumulation of walls near this defect. Alternatively, F&S suggested that, since each wall makes a different angle to the external load, as a result of successive changes of lattice direction (Figure 8.4h), the increase in the shear stress for each successive MDW formed may result in their moving faster, leading to their accumulation. Clear evidence of MDWs in Ti_3SiC_2 can be seen in the TEM micrograph shown in Figure 8.5a. The nearly parallel MDWs, shown in Figure 8.5a, resulted in the lattice rotation anticipated as shown in the inset of Figure 8.5a (Barsoum *et al.*, 1999a). The typical arrangements of dislocations walls and arrays is shown schematically in Figure 8.5b. Note their orthogonal relationship. Examples of kink bands at various lengths scales are shown in Figure 8.6a to c.

One of the important ramifications of kink band formation is their hysteretic nonlinear elastic behavior upon loading. When polycrystalline Ti_3SiC_2 samples were repeatedly, and cyclically, compressed, fully and spontaneously reversible hysteretic

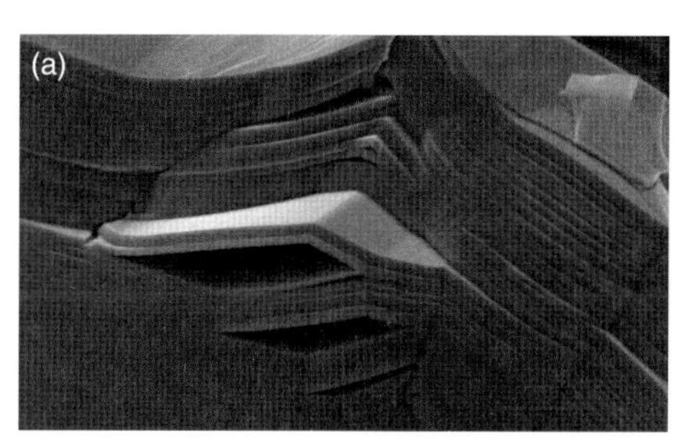

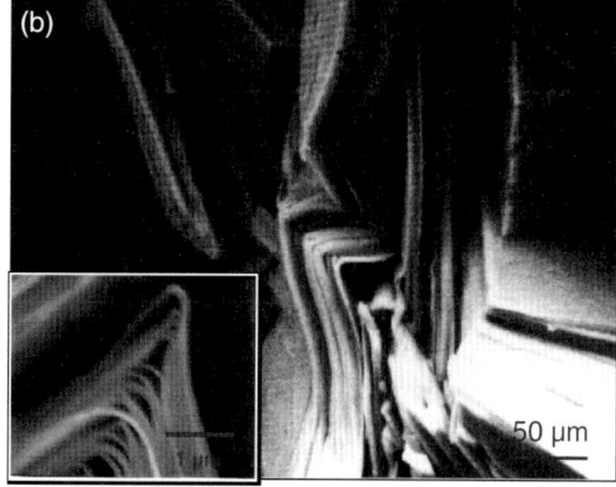

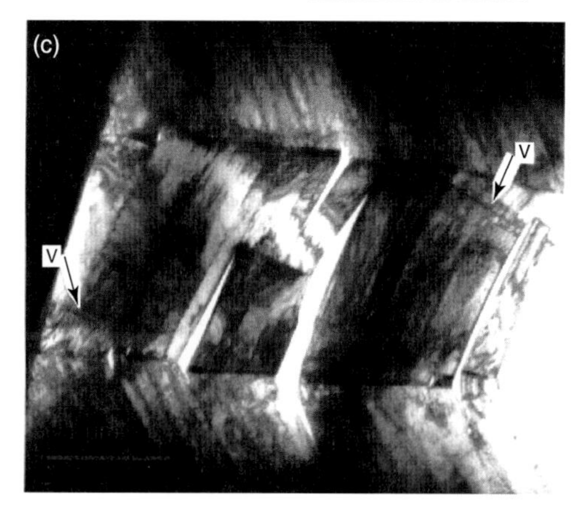

Figure 8.6 Kink bands and delaminations at different length scales as imaged in (a, b), SEM (Sun *et al.*, 2005; Barsoum, Brodkin, and El-Raghy, 1997) and (c) TEM (Barsoum *et al.*, 1999a).

stress–strain loops (Figure 8.7a) even under stresses on the order of 1 GPa were generated (Figure 8.7b) (Barsoum *et al.*, 2003). The loops generated were *not* a function of strain rate, at least in the 10^{-3}–10^{-5} s^{-1} range. For example, the three small loops in Figure 8.7a were obtained at strain rates of 10^{-3}, 10^{-4}, and 10^{-5} s^{-1}. Within the scatter of our experiment, the three loops were indistinguishable (Barsoum *et al.*, 2003; Zhen, Barsoum, and Kalidindi, 2005). When the samples were cycled to progressively larger stresses, nested loops were obtained with a single loading trajectory (Figure 8.7b,c). If the samples were first loaded to a maximum stress and then cycled from that stress to progressively lower stresses, nested loops were also obtained, but in this case a single unloading trajectory was obtained (see loops labeled 90% dense in Figure 8.7c). More important, and to eliminate microcracking as a possible mechanism, a fine-grained (FG) sample was cycled 100 times to 700 MPa. When the first and last loops were compared (Figure 8.7d), it was obvious that, if anything, the last loop was slightly stiffer, precluding microcracking as the origin of the loops. Also shown in Figure 8.7d are all 100 loops for a coarse-grained (CG) sample cycled 100 times. Here again, it is difficult to argue – as

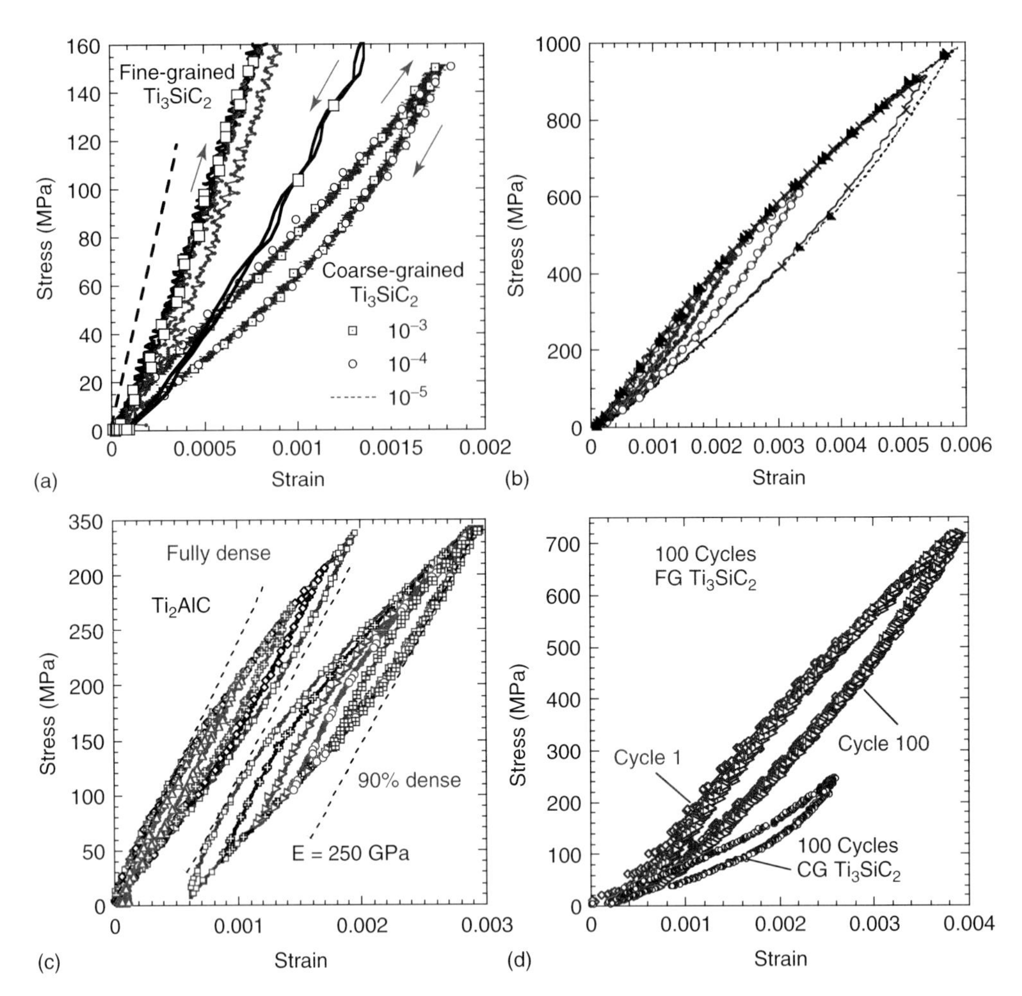

Figure 8.7 Typical cyclic compressive spontaneously reversible stress/strain loops in (a) Ti_3SiC_2 showing the dramatic difference in response between coarse-grained (CG) (small blue loops) and fine-grained (FG) samples (black and red loops). Note the small loops are comprised of three different loops, obtained at three strain rates, differing from each other by an order of magnitude each. Within the resolution of the measurement, the three loops are identical. The red loop – obtained on the FG sample at 250 MPa – is significantly stiffer and smaller in area than the CG loops at the same stress; (b) FG Ti_3SiC_2 loaded up to a stress of 1 GPa. Note nested loops with a single loading trajectory; (c) Fully dense (right-hand side loops) and 10 vol% porous (left-hand side loops) Ti_2AlC. The latter dissipates more energy, on an absolute scale (Zhou *et al.*, 2006); (d) CG and FG Ti_3SiC_2, both cycled 100 times; the FG sample to 700 MPa and the CG to 250 MPa (Zhen, Barsoum, and Kalidindi, 2005).

some have done recently – that the sample is microcracking. What is noteworthy, however, is the reproducibility of the loops over the entire experiment.

From these and other results, some of which are discussed in subsequent text, the dissipated energy per unit volume per cycle, W_d, represented by the loop areas (Figure 8.8c), was found to

1) increase as the square of the maximum applied stress, that is, σ^2 with a threshold stress;
2) be a strong function of grain size, with larger grains dissipating significantly more energy at the same stresses (compare red and blue FG and CG loops in Figure 8.7a); and
3) be higher for porous solids than for fully dense ones (Figure 8.7c) (Fraczkiewicz, Zhou, and Barsoum, 2006; Sun *et al.*, 2005; Zhou *et al.*, 2006).

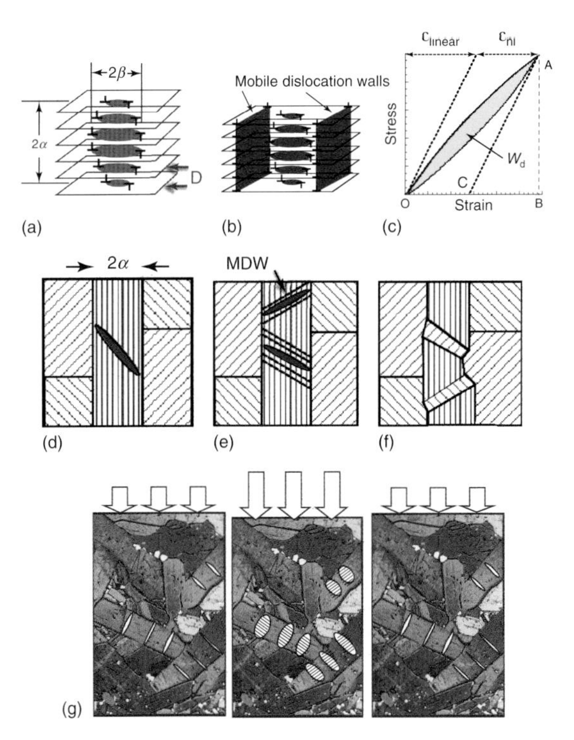

Figure 8.8 (a) Schematic of an IKB with length 2α and width 2β. D is the distance between the horizontal slip planes. (b) Same as (a) but also showing MDWs. (c) Typical stress–strain loop and definitions of nonlinear strain ε_{NL} and W_d. (d) IKB forming within a grain. The IKB extends to the grain boundaries where it is arrested. (e) at higher stresses and/or temperatures the IKBs can devolve into MDWs; (f) The formation and collapse of a number of MDWs leads to the formation of kink bands and boundaries. Note that, after deformation, the domain or grain size is effectively reduced (Barsoum *et al.*, 2003). (g) Schematic of how IKB would form in a polycrystalline sample. At a threshold stress given by Eq. (8.2), IKBs are nucleated. Increasing the stress results in their growing fatter (i.e., increase in 2β), with a constant 2α. Removal of the load results in the shrinking of the IKBs and their ultimate annihilation.

8.4
Incipient Kink Bands

To explain these results, the existence of incipient kink bands (IKBs) had to be invoked (Barsoum *et al.*, 2003). An IKB consists of multiple, coaxial, parallel dislocation loops separated from each by a distance D (Figure 8.8a). An IKB is a KB whose walls, for whatever reason, do *not* dissociate into MDWs (e.g., Figure 8.8d). A schematic of how an IKB dissociates into MDWs with the further nucleation of an IKB within the first set of MDWs is shown in Figure 8.8e. Because of their shape, IKBs only remain open when a load – over a certain threshold – is applied; when the load is removed, they shrink, and are, thus by definition, *fully and spontaneously reversible* (e.g. Figure 8.8g).

IKBs are believed to be the micromechanism responsible for the behavior shown in Figure 8.7. It is this realization that led to labeling such solids as kinking nonlinear elastic (KNE): elastic because of the absence of plastic deformation; nonlinear because the stress–strain curves are clearly nonlinear, and kinking because the energy dissipated is due to the formation of dislocation-based IKBs.

A *sufficient* condition for a solid to be KNE is *plastic anisotropy*, a good measure of which is the c/a ratio in hexagonal crystals (Barsoum *et al.*, 2005b). What is meant by plastic anisotropy here is that the dislocations are confined to two dimensions. For ratios greater than ≈ 1.4, most solids are KNE. It is thus not surprising that the MAX phases with $c/a > 3$ are KNE. We note in passing that KNE solids are quite ubiquitous in nature (see subsequent text) (Barsoum and Basu, 2010). This is particularly true when it is appreciated that a majority of ceramics and minerals, are, almost by definition, KNE because, for the most part, they are plastically anisotropic. The fact that ice, mica, and, by extension, most, if not all, layered silicates are KNE solids means that their understanding is crucial for understanding structural geology (Barsoum *et al.*, 2004b; Barsoum *et al.*, 2001; Basu and Barsoum, 2009).

The remainder of this chapter deals with the MAX phases' KNE behavior with emphasis on how to model their response to stress and quantify it. The following section deals with our microscale model, which in turn is based on the work of F&S (Frank and Stroh, 1952).

8.5
Microscale Model for Kinking Nonlinear Elasticity

Referring to Figure 8.8c, the total strain ε_{tot} at a stress σ can be additively decomposed into a linear elastic strain, that is, σ/E, and a nonlinear elastic strain ε_{NL}, that is,

$$\varepsilon_{\text{tot}} = \frac{\sigma}{E} + \varepsilon_{\text{NL}} = \frac{\sigma}{E} + \varepsilon_{\text{IKB}} + \varepsilon_{\text{DP}} \tag{8.1}$$

where E is Young's modulus. In the most general case, and in the absence of phase transitions and/or microcracking, the nonlinear, fully reversible strain ε_{NL}

comprises two components. We assume the first, ε_{IKB}, is due to IKBs; the second, ε_{DP}, is due to dislocation pile-ups (DPs). Note that because the dislocations are confined to 2D, both strains are in principle fully reversible. In what follows, we only deal with ε_{IKB}; ε_{DP} will be assumed to be small (Barsoum *et al.*, 2005b).

F&S (Frank and Stroh, 1952) considered an elliptic KB – in a single crystal – with length 2α and width 2β such that $\alpha \gg \beta$ (Figure 8.8a). They showed that the remote critical shear stress τ_c needed to render such a subcritical KB unstable is given by

$$\tau_c \approx \sqrt{\frac{4G^2 b\gamma_c}{\pi^2 2\alpha} \ln \frac{b}{w\gamma_c}}$$

and b, G, and w are, respectively, the Burgers vector, shear moduli, and dislocation core widths; γ_c is the critical kinking angle (Eq. (8.3)).

Since most of our work has been on polycrystalline materials loaded axially with a load corresponding to a remote axial stress σ_t, we modified the F&S equation to (Barsoum *et al.*, 2005b):

$$\tau_t \approx \frac{\sigma_t}{M} = \sqrt{\frac{4G^2 b\gamma_c}{\pi^2 2\alpha} \ln \frac{b}{w\gamma_c}} \tag{8.2}$$

where M is the Taylor factor relating τ_t at the single-crystal level to σ_t at the polycrystalline level. For randomly oriented microstructures $M \approx 3$.

We have repeatedly shown that, at least for the MAX phases, 2α can be equated with the grain dimension along [0001] (Fraczkiewicz, Zhou, and Barsoum, 2006; Zhou and Barsoum, 2010; Zhou *et al.*, 2006). In that case, the threshold stress – measured experimentally (see subsequent text) – can be directly used to estimate w, which is an important parameter that is quite difficult to measure otherwise. As noted in the preceding text, γ_c is the critical kinking angle calculated assuming (Frank and Stroh, 1952; Hull, 1965)

$$\gamma_c = \frac{b}{D} \approx \frac{3\sqrt{3}\,(1-\nu)\,\tau_{loc}}{2G} \approx \frac{3\sqrt{3}\,(1-\nu)}{8\pi e}\left(\frac{b}{w}\right) \tag{8.3}$$

where ν is the Poisson's ratio and τ_{loc} is the local shear stress needed to nucleate a dislocation pair and D is the distance between dislocation loops along [0001] (Figure 8.8a). If one assumes, as F&S did, that $\tau_{loc} \approx G/30$, then γ_c is of the order of 3°. Note that assuming $G/30$ implicitly assumes $w = b$ (Zhou, Basu, and Barsoum, 2008). If, as assumed here, $w = 5b$, then γ_c is on the order of 0.01 rad: that is, it is quite small.

An IKB consists of multiple parallel dislocation loops (Figure 8.8a). As a first approximation, each loop can be assumed to be made up of two edge and two screw dislocation segments with lengths, $2\beta_x$ and $2\beta_y$, respectively. The latter are related to the applied stress σ and 2α assuming (Frank and Stroh, 1952; Zhou, Basu, and Barsoum, 2008)

$$2\beta_x \approx \frac{2\alpha\,(1-\nu)}{G\gamma_c}\frac{\sigma}{M} \quad \text{and} \quad 2\beta_y \approx \frac{2\alpha}{G\gamma_c}\frac{\sigma}{M} \tag{8.4}$$

These expressions were modified from the original F&S paper to account for the polycrystalline nature of the materials tested by the introduction of M.

The formation of an IKB can be divided into two stages: nucleation and growth (Zhou *et al.*, 2010b; Zhou and Barsoum, 2010). Since the former is not well understood, our model only considers IKB growth from $2\beta_{xc}$ and $2\beta_{yc}$ to $2\beta_x$ and $2\beta_y$, respectively. The dislocation segment lengths of an IKB nucleus, β_{xc} and $2\beta_{yc}$, are presumed to either preexist, or are nucleated during prestraining. Based on a few studies on KNE solids, it appears that sometimes a prestraining is needed to observe KNE behavior. For example, when a $Ti_3AlC_{0.5}N_{0.5}$ cylinder was first loaded to a stress of 400 MPa, the response was linear elastic (Fig. 8.9a). After loading to 750 MPa, unloading, and reloading to 400 MPa, a small stress–strain loop was clearly visible (Zhou and Barsoum, 2010). The same is true of Mg single crystals; prestraining is needed to observe loops (Burke and Hibbard, 1952).

It follows that for $\sigma > \sigma_t$, the latter given by Eq. 8.2, the IKB nuclei grow. Assuming Eq. 8.4 is operative, it can be readily shown that the IKB-induced axial strain resulting from their growth is given by (Zhou *et al.* 2010b and Zhou and Barsoum 2010):

$$\begin{aligned} \varepsilon_{\text{IKB}} &= \frac{\Delta V N_k \gamma_c}{k_1} = \frac{N_k \gamma_c 4\pi\alpha\left(\beta_x\beta_y - \beta_{c,x}\beta_{c,y}\right)}{3k_1} \\ &= \frac{4\pi(1-\nu)N_k\alpha^3}{3k_1 G^2\gamma_c M^2}\left(\sigma^2-\sigma_t^2\right) = m_1\left(\sigma^2-\sigma_t^2\right) \end{aligned} \tag{8.5}$$

where m_1 is the coefficient before the term in brackets in the last term; N_k is the number of IKBs per unit volume; ΔV is change in the volume kinked as the IKBs grow from their critical size at σ_c to their size at σ. It follows that the volume fraction of the material that is kinked, ν_f, is given by the product $V \times N_k$. The factor k_1 relates the volumetric strain due to the IKBs to the axial strain along the loading direction. Note that once m_1 is determined experimentally (see subsequent text), N_k can be estimated from Eq. (8.5) because in principle all the other values are known.

The W_d resulting from the growth of the IKBs from β_{ic} to β_i and back to β_{ic} is given by Barsoum *et al.* (2005b) and Zhou and Barsoum (2010)

$$\begin{aligned} W_d &= \frac{4\Omega\pi N_k\alpha}{D}\left(\beta_x\beta_y - \beta_{xc}\beta_{yc}\right) \\ &= \frac{4\pi(1-\nu)N_k\alpha^3}{G^2\gamma_c M^2}\frac{\Omega}{b}\left(\sigma^2-\sigma_t^2\right) = m_2\left(\sigma^2-\sigma_t^2\right) \end{aligned} \tag{8.6}$$

where Ω is the energy dissipated by a dislocation line sweeping a unit area. It follows that Ω/b should be proportional, if not equal, to the CRSS of an IKB dislocation loop. Combining Eqs. (8.5) and (8.6) yields

$$W_d = 3k_1\frac{\Omega}{b}\varepsilon_{\text{IKB}} = \frac{m_2}{m_1}\varepsilon_{\text{IKB}} \tag{8.7}$$

Experimentally, one can determine m_1, m_2, σ_t, and $3k_1(\Omega/b)$. Note that to estimate Ω/b, only knowledge of k_1 in Eq. (8.7) is required.

Once nested stress–strain loops are obtained (e.g., Figure 8.7c) – and the plots shown in Figure 8.9b–d are plotted – Eq. (8.7) can be used to estimate Ω/b, assuming $k_1 = 2$. Experimentally, m_1 can be determined from the slopes of ε_{NL} versus σ^2 plots (e.g., Figure 8.9b). Similarly, m_2 can be determined from W_d versus σ^2 plots (e.g., Figure 8.9c). It follows that if our assumptions are correct, and, more importantly, if the micromechanism causing the dependence of ε_{NL} on σ (i.e., Eq. (8.5)) is the *same* as the one responsible for W_d (Eq. (8.6)), then the *ratio* m_2/m_1 should equal $3k_1\Omega/b$. In other words, if both expressions give the same values for Ω/b – which they do – that would be strong evidence that our assumptions are correct and, more importantly, that the same micromechanism that results in the parabolic dependence of σ on ε_{NL} is the one responsible for W_d as well.

If the IKBs are assumed to be cylinders with radii β_{av} – where $\beta_{av} = (\beta_x + \beta_y)/2$ – then the *reversible* dislocation density ρ_{rev} due to the IKBs is given by (Barsoum *et al.* 2005b)

$$\rho_{rev} = \frac{2\pi N_k 2\alpha\beta_{av}}{D} = \frac{4\pi N_k \alpha\beta_{av}\gamma_c}{b} \tag{8.8}$$

8.6
Experimental Verification of the IKB Model

Figures 8.7 and 8.9a show the typical response of the MAX phases to cyclic loading. Before delving into the details of the model, it is important to qualitatively understand the response observed. To do so, it is crucial to appreciate that, for the most part, 2α is nothing but the grain dimension along the *c*-axis (Fraczkiewicz, Zhou, and Barsoum, 2006; Zhou *et al.*, 2006). In other words, for the MAX phases, 2α is not the grain diameter but its thickness (Figure 8.8g). From Eq. (8.2), it follows that σ_t is a function of grain size, with the nucleation of IKBs occurring at lower stresses in larger grains than in finer ones as observed (Barsoum *et al.*, 2003, 2005b). This can best be seen in Figure 8.7a, where the loop at 260 MPa in the FG sample (narrow, red loop in Figure 8.7a) is considerably smaller than the loop at 150 MPa of a CG sample of the same material, namely, Ti_3SiC_2.

Probably the strongest evidence to date that our IKB model is valid is shown in Figure 8.7c, where it is obvious that a 10% porous Ti_2AlC sample (loops on the right) dissipates *more* energy per unit volume per cycle on an *absolute* scale than its fully dense counterpart (loops on left). This straightforward result essentially eliminates all mechanisms, such as DPs and/or twinning, that scale directly with the volume of the material tested. It is, however, in full agreement with our IKB model in that kinking is a form of plastic instability, or buckling. This in turn explains why a less rigid solid is more prone to kinking that a fully dense one (Zhou *et al.*, 2006). Similar conclusions were reached for porous Ti_3SiC_2 samples where the enhancements in W_d could be accounted for by a reduction in G (Fraczkiewicz, Zhou, and Barsoum, 2006; Sun *et al.*, 2005).

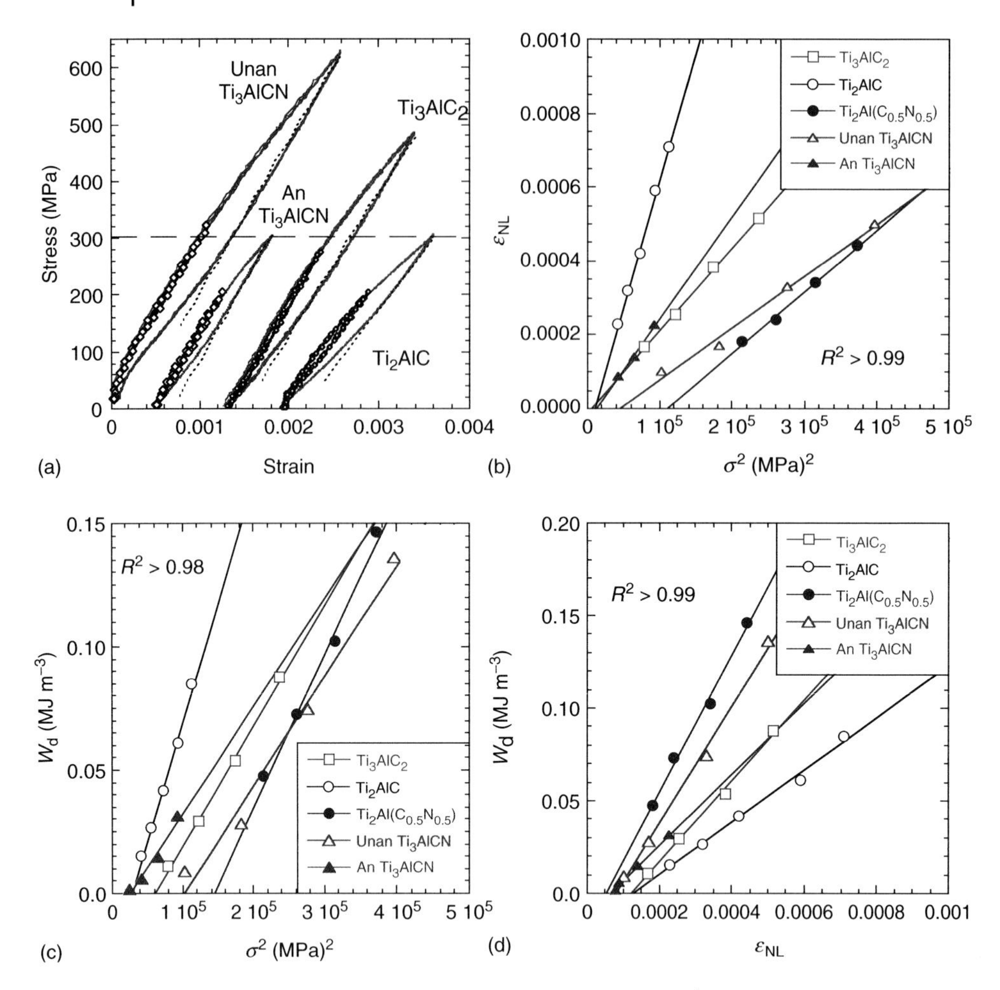

Figure 8.9 (a) Cyclic compressive stress–strain curves for a number of $Ti_{n+1}AlX_n$ phases showing two loops, one of which was loaded to 300 MPa (dashed line). "An" and "Unan" refer to a sample that was tested after and before annealing respectively. (b) ε_{NL} versus σ^2. (c) W_d versus σ^2. (d) W_d versus ε_{NL}. In all cases, the least squares R^2 values are noted on the graphs (Zhou and Barsoum, 2010).

On the basis of our model, KNE solids can be characterized by three parameters: σ, ε_{NL}, and W_d, all obtainable from the hysteretic stress–strain curves (Figure 8.8c). If $N_k\alpha^3$ is not a strong function of stress, then, according to Eqs. (8.5–8.7), plots of W_d versus ε_{IKB}, ε_{IKB} versus σ^2 and W_d versus σ^2 should all yield straight lines, the latter two with x-axes intercepts at $\sigma_t{}^2$. The assumption that $N_k\alpha^3$ is a weak function of stress is an excellent one because, as α increases, N_k decreases and vice versa (Table 8.1).

Figure 8.9a compares the hysteretic loops of Ti_2AlC, Ti_3AlC_2, $Ti_2Al(C_{0.5},N_{0.5})$, and Ti_3AlCN (Zhou and Barsoum, 2010). Figure 8.9b–d, respectively, plot ε_{NL}

Table 8.1 Summary of measured and calculated parameters of microscale model. In all cases $M = 3$, $k_1 = 2$, $b = 3$ Å, w = 5b, and $\nu = 0.2$ for all compounds, except Cr_2GeC for which $\nu = 0.29$ and $b = 2.95$ Å and $w = 4\,b$. The values for Cr_2GeC do not match the entries in the original reference (Amini *et al.*, 2008) since our methodology has since been refined. Also the values of the nonlinear strains in Figure 7a in Amini *et al.* (2008) are incorrect. The stress at which the various parameters were calculated is listed in the penultimate column.

	G^a (GPa)	Ω/b^b (MPa)	2α (μm)	$\sigma_t^{\,c}$ (MPa)	N_k (m^{-3})	$2\beta_{xc}$ (μm)	$2\beta_x$ (μm)	ρ_{rev} (m^{-2})	σ (MPa)	References
Ti_3AlC_2	124	36	10	244	1.8×10^{16}	0.48	0.96	4.3×10^{13}	486	Zhou and Barsoum (2010)
Un-annealed Ti_3AlCN	137	53	10	270	1.4×10^{16}	0.48	1.12	4×10^{13}	629	
Annealed Ti_3AlCN	30	23	180	2.2×10^{15}	0.74	1.25	3.3×10^{13}	305		
$Ti_2Al(C_{0.5},N_{0.5})$	124	62	4	376	2×10^{17}	0.3	0.48	9.7×10^{13}	610	
Ti_2AlC	118	24	19	170	7.3×10^{15}	0.67	1.3	4.6×10^{13}	336	
Ti_3SiC_2	143	30	42	136	4.7×10^{14}	1	2.7	1.2×10^{13}	370	—
Cr_2GeC	80	22	32	92	2.3×10^{15}	0.8	4.6	9.2×10^{13}	550	Amini *et al.* (2008)
Ti_3AlC_2	108	52	25	125	5.9×10^{14}	—	—	4.9×10^{12}	457	Bei *et al.* (2013)
$Ti_3(Al,Sn)C_2$	104	79	6	246	1.6×10^{16}	—	—	1.3×10^{14}	624	

[a] Measured by ultrasound (Radovic *et al.*, 2006).
[b] From W_d versus ε_{NL} plots (e.g., Figure 8.9d).
[c] From plots such as shown in Figure 8.9c.

versus σ^2, W_d versus σ^2, and W_d versus ε_{NL}. Least squares fit of the results yield correlation coefficients R^2 values >0.97 in all cases (Figure 8.9b to d).

Table 8.1 summarizes the microscale parameters that describe kinking in these solids. From these results, it is obvious that the σ_ts for the unannealed solid solutions are substantially greater than those of their corresponding carbide end members. The main reason for this state of affairs is that despite our best efforts to fabricate samples with identical 2α values, the solid solution compositions consisted of large grains embedded in a matrix of much smaller grains. The small grains – whose influence is essentially the opposite of pores – appear to prevent, or constrain, the larger grains from kinking, resulting in an increase in σ_t, through a reduction in the average 2α (column 4, Table 8.1). Consistent with this interpretation is that when the Ti_3AlCN sample was annealed to rid it of the fine grains, σ_t decreased significantly. As a check on the model, the values of 2α calculated from σ_t for the various microstructures were comparable to those measured microstructurally (Zhou and Barsoum, 2010).

Note that the 2α values calculated from the model are an "equivalent" or "effective," average grain size. These values will always be less than the largest grains in a material. It is for this reason that the 2α values calculated from W_d versus σ^2 plots are always greater than those directly obtained from where the stress–strain curves deviate from linearity. The latter, presumably, depend on the largest and most favorably oriented grains to kinking rather than the "average."

On the basis of the values for Ω/b listed in Table 8.1, it would be reasonable to conclude that what is observed is an increase in the CRSS due to solid solution strengthening. While tempting and possibly, at least partially, correct, the results for the annealed Ti_3AlCN sample suggest otherwise. However, since it is not clear why annealing would result in an almost twofold decrease in Ω/b, this explanation must be viewed as suspect.

The second, and more likely, explanation is that the effect observed is essentially a Hall–Petch-like effect in which Ω/b, is itself a function of grain size. This explanation is bolstered when Ω/b is plotted as a function of $1/\sqrt{2\alpha}$ (Figure 8.10). Least squares analysis of the data result in an $R^2 > 0.92$. Note that we have previously shown that in Ti_3SiC_2, Ω/b is also a function of the grain size (Barsoum *et al.*, 2005b). It is also important to note that the least squares fit of the data points goes through the origin as it should. By modeling the $11\bar{2}1$ twin boundary, we have recently shown that the CRSSs of basal dislocations in hexagonal metals are so low that they could not be calculated (Lane *et al.*, 2011). The same should be true for the basal slip in all hexagonal solids in general and the MAX phases in particular. One must thus conclude that the CRSS reported in Table 8.1 results from dislocation-dislocation interactions. In other words, the CRSS is a measure of the dislocation density in the basal planes prior to measuring the loops. Note that this implies a relationship between dislocation density and grain size.

Also included in Table 8.1 are the results on Cr_2GeC (Amini *et al.*, 2008), a perusal of which indicates that the parameters describing its kinking characteristics are quite comparable to the others in the table despite a significantly lower value of G. More recently, Bei *et al.* (2013) concluded that Ti_3AlC_2 and

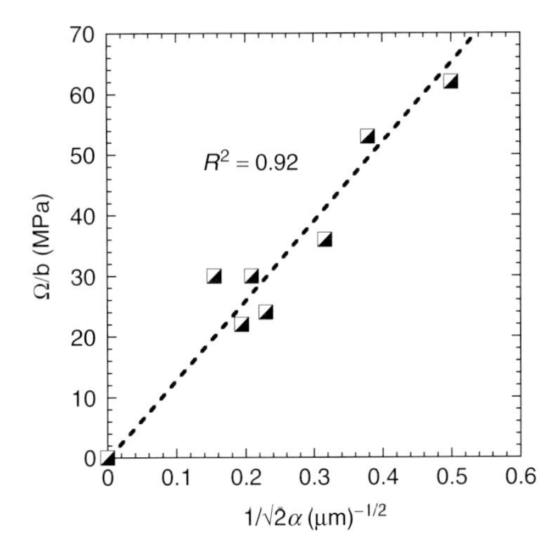

Figure 8.10 Hall–Petch plot of Ω/b or CRSS of basal planes as a function of $1/\sqrt{2\alpha}$. Note that the least squares fit line through the data points goes through the origin, as it should (Zhou and Barsoum, 2010).

$Ti_3(Al_{0.8},Sn_{0.2})C_2$ were indeed KNE solids because their response to cyclic loading was well described by Eqs. (8.2–8.8). Note that for most calculations, including Bei *et al.*'s, very similar assumptions were made, that is, $w = 3b$ to $5b$, $M = 3$, and $k_1 = 2$.

At $\approx 1-9 \times 10^{13}$ m^{-2}, the values of ρ_{rev} are reasonable given the maximum stresses applied, listed in penultimate column in Table 8.1. Recall that ρ_{rev} is not the dislocation density in the sample when the load is removed, but rather that which is due solely to the IKBs: that is, ρ_{rev} given by Eq. (8.8). Referring to Table 8.1, what is remarkable is that the values of ρ_{rev} vary by less than an order of magnitude despite the fact that (i) the maximum applied stresses vary over a factor of 2, (ii) G varies by almost a factor of 2, and (iii) the N_k values differ by over three orders of magnitude. This observation suggests that an equilibrium ρ_{rev} exists to which all systems migrate. In solids with high shear moduli and small grains, many small IKBs form; for solids with low shear moduli and large grains, many fewer, but larger, IKBs form.

In summary, our proposed model shows good agreement between the various parameters and their inter-relationships. The fact that the ratio $m_2/m_1 \approx 3k_1\Omega/b$ is strong evidence that the micromechanism responsible for the strain nonlinearities is the same as the one responsible for W_d.

8.7 Effect of Porosity

Since kinking is a form of buckling (e.g., Figure 8.6), it is reasonable to assume that porosity would enhance the effect. This important conjecture has been borne out by experiment several times. For example, as shown in Figure 8.7c, introducing 10% porosity in a Ti_2AlC sample resulted in an *absolute* increase in W_d. It is worth noting, again, that this result cannot be reconciled with any micromechanism that scales with the volume of the material, such as DPs, for example.

Figure 8.11a compares the stress–strain loops obtained when a FG, a CG, and a 43 vol% porous Ti_3SiC_2 sample were cyclically compressed. The curve labeled P assumes the stresses in the porous sample to be those applied. The curve labeled EP in Figure 8.11a, corrects the applied σ to an effective stress, σ_{eff}, by dividing the former by the effective load-bearing cross-sectional area of a specimen containing 43% porosity, that is, dividing by 0.57. The elastic response – as determined from ultrasound experiments on fully dense Ti_3SiC_2 – is also plotted as a straight line

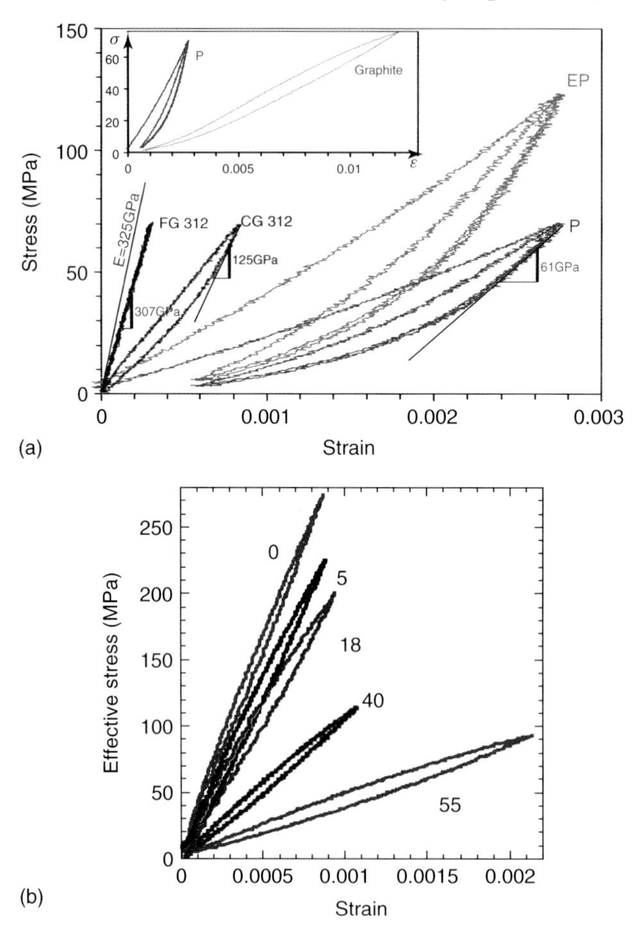

Figure 8.11 Effect of porosity on shape and size of reversible stress–strain loops in (a) a 43 vol% porous Ti_3SiC_2 sample. Also shown for comparison are the loops obtained on FG and CG fully dense samples. The inclined line on far left represents the purely elastic response (Sun *et al.*, 2005). The inset compares the response of polycrystalline graphite to that of the 43 vol% sample. (b) 0, 5, 18, 40, and 55 vol% porosity Ti_3SiC_2 samples. The first cycles are *not* shown here (Fraczkiewicz *et al.*, 2006).

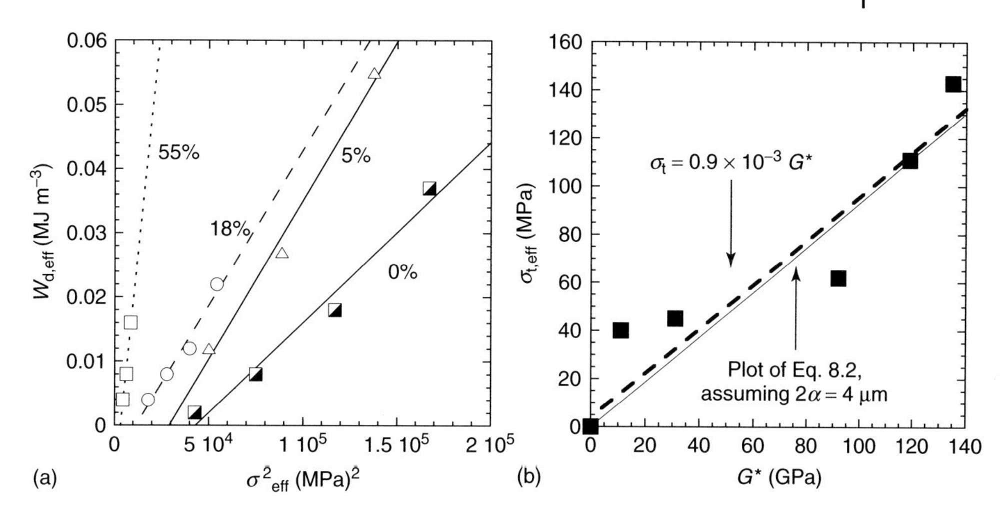

Figure 8.12 Effect of porosity on (a) $W_{d,eff}$ versus σ^2_{eff} plots and (b) $\sigma^2_{t,eff}$ versus G^* (Fraczkiewicz *et al.*, 2006), where G^* is the effective modulus.

(extreme left) with a slope that corresponds to a Young's modulus of 325 GPa. The effect of porosity on the shape and size of the reversible stress–strain loops is shown in Figure 8.11b. In these loops, the first cycles, after which some plastic deformation is observed (e.g. Fig. 8.11a) were omitted.

From these results, the following important conclusions can be reached: (i) The first cycle (Figure 8.11a) for the porous sample is open; subsequent cycles are closed and stiffer than the first. This cyclic hardening is important because it precludes microcracking as a viable mechanism. (ii) Increasing the porosity level reduces the effective moduli of the stress–strain loops (Figure 8.11a,b). (iii) For a given σ, introducing pores dramatically increases W_d. The effect of porosity on the latter can be better appreciated when the effective $W_{d,eff}$ is plotted as a function of σ^2_{eff} (Figure 8.12a).

By systematically varying the porosity level in Ti_3SiC_2, Fraczkiewicz *et al.*, (2006) showed that the major effect of porosity can be taken into account in our KNE model by simply replacing G in Eqs. (8.2–8.6) by G^* given by Gibson and Ashby (1988):

$$G* = \frac{E_s}{2\,(1+\nu)}\left(\frac{\rho*}{\rho_S}\right)^2 \tag{8.9}$$

where the asterisk refers to the porous solid and the subscript s refers to the fully dense solid with a density of ρ_s and Young's modulus E_s.

8.8
Experimental Evidence for IKBs

To date, most of the evidence for the existence of IKBs has been circumstantial. Direct evidence for the presence of IKBs is lacking, which begs the question: why?

The answer lies in the fact that IKBs disappear when the load is removed and thus have to be observed *in situ* under load, which suggests either *in situ* diffraction and/or loading in a TEM. However, since grain boundaries are believed to play an important role in confining the IKBs, it follows that unless one is lucky, any IKBs that would form in a thin TEM foil would most likely rapidly devolve into MDWs (Figure 8.5a). Another inherent disadvantage of TEM is that because at some level kinking is a buckling phenomenon, it is much more prevalent in compression than in tension. Carrying out compression experiments on very thin foils in a TEM is nontrivial.

The more promising approach is *in situ* diffraction during cyclic loading. IKBs can be considered to be twins that grow simultaneously in two dimensions rather than one. However, a fundamental difference between twins and IKBs is the shear angle. In twinning, the shear angle is crystallographic, usually quite large, and can be easily picked up in *in situ* neutron diffraction (ND) experiments of bulk samples (e.g., Clausen *et al.*, 2008). At this point it is worth noting that the $11\bar{2}1$ twin in hexagonal solids is unique in that it can be considered to be a special kink boundary, where the distance between basal dislocations, *D*, is the c lattice parameter. Said otherwise, when the $11\bar{2}1$ twin is formed, a basal dislocation must be nucleated every *c*. This twin can therefore be formed by the glide of basal plane dislocations alone, with an associated strain of equal to a/c. (Freise and Kelly, 1961).

As noted in the preceding text, the IKB angle, on the other hand, is of the order of 3° or less and, consequently, the concomitant lattice rotations are more difficult to observe. To make matters worse, the associated total strains are, even at the highest stresses, of the order of 0.5%, i.e. quite small.

These comments notwithstanding, Figure 8.13 presents some intriguing recent ND results that are still being analyzed. In this figure, the full-width at half maximum of the 0004 reflections in a highly oriented Ti_2AlC sample – where the basal planes are parallel to the loading direction – is plotted versus applied stress. Such spontaneous and reversible broadening of ND peaks has, to the best of our knowledge, never been reported to date. It is thus reasonable to ascribe it to the formation of IKBs and the mosaicity they would engender (Caspi *et al.* 2013).

As important, the ND results were unambiguous on another point: *No evidence of twinning was found.* The ND results thus confirmed an important tenet of our model: that is, *the kinking angle $\gamma_{c,}$ is small, as predicted.* From the broadening shown in Fig. 8.13 γ_c, is estimated to be 0.016, which compares very favorably with the value of 0.012 we calculate from Eq. 8.3, assuming w = 5b (Caspi *et al.* 2013).

KNE solids are ubiquitous in nature and span the gamut of bonding in solids, from metallic, to covalent, to ionic, to van der Waals and combinations thereof (Barsoum and Basu, 2010). Figure 8.14 is a plot of c_{44} versus the *c*/*a* ratio in hexagonal crystals. All solids to the right of a *c*/*a* ratio of 1.4 are KNE because they are plastically anisotropic. The asterisk denotes solids that exhibit KNE behavior at room temperature in polycrystalline form. The dagger denotes KNE behavior, but only observed under a nanoindenter.

As noted several times above, plastic anisotropy is a sufficient condition for a solid to be classified as KNE. The MAX phases, with their high *c*/*a* ratios, are

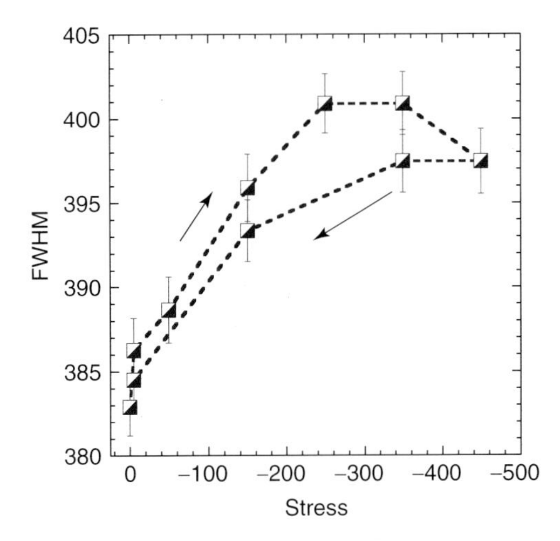

Figure 8.13 Full Width at Half Maximum (FWHM) of the {0004} ND reflection in a highly oriented Ti_2AlC sample as a function of applied stress. Such a spontaneous reversal in FWHM has never been reported on before (Caspi *et al.* 2013).

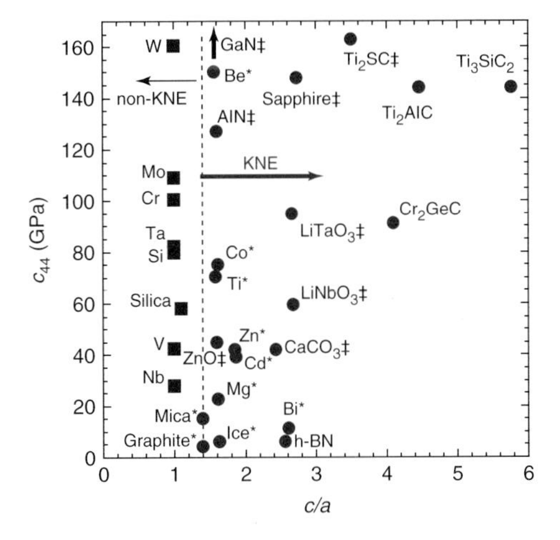

Figure 8.14 Plot of c_{44} versus c/a ratio in select materials. All solids to the right of the c/a ratio of 1.4 are KNE because they are plastically anisotropic. The asterisk denotes KNE behavior at room temperature in polycrystalline form. The dagger denotes KNE behavior but only under a nanoindenter (Barsoum and Basu, 2010).

per force extremely plastically anisotropic and thus it is not surprising that they are KNE solids. Lastly and quite crucially, in Chapter 9, the case is made that the energy stored by DPs is at least two orders of magnitude lower than that observed in the MAX phases and thus cannot explain the results obtained. Note the energy stored at any stress is given by the area of the triangle OAC shown in Figure 8.8c (Barsoum *et al.* 2005b).

8.9
Why Microcracking Cannot Explain Kinking Nonlinear Elasticity

A recent paper made the case that the KNE response in the MAX phases can be explained by microcracking (Poon *et al.*, 2011). A plethora of circumstantial evidence exists, however, that simply cannot be explained by microcracking. Amongst them are the following:

1) The observance of KNE behavior, which is governed by the same set of equations listed herein, in hexagonal metals (Zhou and Barsoum, 2009; Zhou *et al.*, 2010b; Zhou, Basu, and Barsoum, 2008) in which microcracking is very unlikely. KNE behavior has also been observed in Mg single crystals (Robertson and Hartmann, 1964) in which it is quite unlikely that microcracking is responsible for the reversible stress–strain loops observed.
2) *Cyclic hardening* – When CG samples of Ti_3SiC_2 are cyclically loaded at temperatures greater than the brittle-to-plastic transition, cyclic *hardening* is observed (Figure 10.13). Moreover, when the same sample was cooled down and reloaded, the response was significantly stiffer than the original CG material (Figure 10.14). More importantly, cyclic hardening was observed in porous samples. Referring to Figure 8.11a, it is obvious that the second cycle is harder and stiffer than the first. Such observations are totally incompatible with microcracking.
3) *Fatigue resistance* – In systems that microcrack, there is a steady loss in stiffness with cycling. This comes about as the friction between the crack surfaces wear. For the MAX phases, the loop shapes and areas do *not* appear to be a function of cycle number even when the applied stress was 700 MPa for 100 cycles (Figure 8.7d). If anything, the evidence suggests cycling hardening.
4) *Macroscopic kink boundaries* – That the MAX phases form KBs is a fact. These KBs must nucleate somehow. It is unlikely that DPs and/or twins are responsible for, or can result in, KBs. More importantly, as noted above the energy stored by DPs is al least two orders of magnitude lower than observed.
5) The CRSSs obtained from the stress–strain curves of KNE solids are functions of chemistry and grain size. If friction from microcrack faces were responsible for the loops obtained, it is not apparent why that should be the case.
6) At a given stress, FG samples, with their much higher number of intergranular contacts dissipate less energy than their CG counterparts.
7) Lastly, it was recently shown that dislocations in graphite – and by extension all layered solids – can be modeled by a cohesive crack model (Yang *et al.*, 2011). It follows that at some level, the differences between basal plane dislocations in layered solids and cracks can easily blur.

These comments notwithstanding, it is quite possible that microcracking is a requisite step for the observance of KNE because these cracks would allow the IKBs room to nucleate and grow with increasing stress.

8.10
The Preisach–Mayergoyz Model

Hysteresis is quite common in nature and includes magnetic, ferroelectric, and mechanical, among others. Herein we are interested in the latter. Mechanical hysteresis is common in some solids when they are cyclically loaded. For example, granular solids, such as rocks, exhibit hysteresis that has been related to the internal friction of cracks common in such solids (Aleshin, and Van Den Abeele, 2007; Lawn and Marshall, 1998; Walsh, 1966). One of the hallmarks of this mechanism is a reduction in modulus with cycling (Walsh, 1965). The hysteresis described here, on the other hand, is believed to be due to the reversible motion of dislocations; if intergranular friction and microcracks play a role, it is not believed to be a major one (see subsequent text).

First developed to describe ferromagnetic hysteresis, the Preisach model is based on the idea that macroscopically observed irreversible processes can be decomposed into *independent* switching events described by *independent* bistable relays (Preisach, 1935). More recently, Mayergoyz, recognizing that the Preisach model offered a general mathematical framework for the description of hysteresis of different physical origins, derived the *necessary and sufficient* conditions for representation of any given hysteresis by the Preisach model (Mayergoyz, 1986, 2003). These conditions are as follows: first, that each local stress maximum wipes out the effect of other local stress maxima below it; second, the hysteresis loops obtained via cycles with the same end points of input, but different prehistories, are congruent. Mayergoyz called these properties *wiping out* and *congruency*, respectively. Since then the Preisach model has been renamed the Preisach–Mayergoyz (PM) model.

Guyer, McCall, and Boitnott (1995) and Ortin (1992) successfully applied this model to describe the nonlinear elastic response of granular geological materials and a shape-memory alloy, respectively. And while previous work had clearly shown wiping out, as far as we are aware, until our report (Zhou *et al.*, 2010a), congruency had *never* been reported in mechanical systems.

Before the PM model can be used, it is essential to establish that wiping out and congruency are indeed valid. Figure 8.15a clearly illustrates the wiping-out property of Ti_3SiC_2. When the sample was loaded to a stress σ, all minor and intermediate loops obtained below that stress are wiped out; that is, there is complete loss of memory or load history. The same is true for the 10 vol% porous Ti_2AlC sample (Figure 8.15b); during unloading, all minor loops are closed and are wiped out by a larger unload.

Congruency is illustrated in Figure 8.15c. Here, the minor loops – obtained by loading a CG Ti_2AlC sample – are congruent when the stress is cycled between 200 and 310 MPa, and independent of whether the loops were obtained on loading *or* unloading. It follows that the two criteria needed for a hysteresis to be adequately described by the PM model are satisfied. The full implications of this statement are far-reaching and beyond the scope of this book. We briefly touch on the salient highlights; the interested reader is referred to the original paper for the details

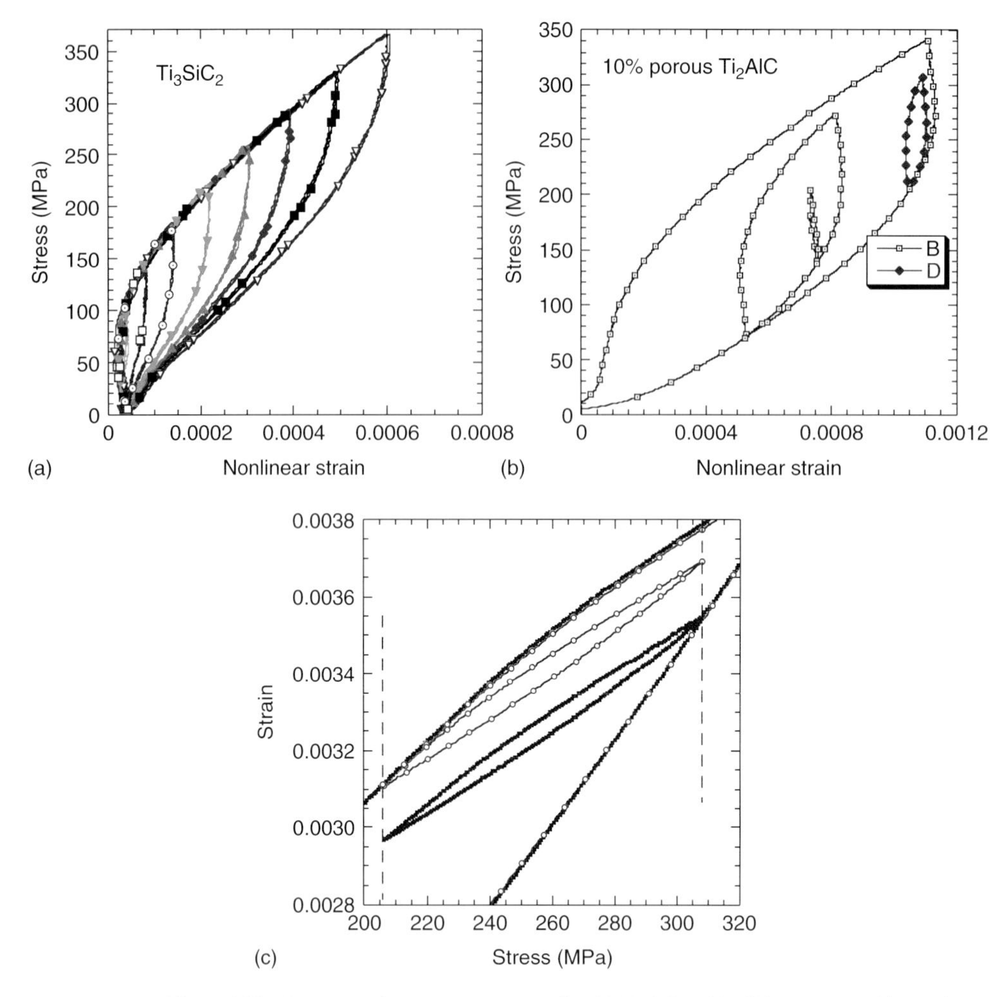

Figure 8.15 Stress–nonlinear strain curves for (a) Ti_3SiC_2, (b) 10% porous Ti_2AlC, and (c) CG Ti_2AlC. (a, b) Show the wiping-out property and (c) shows congruency (Zhou *et al.*, 2010a). Note in these figures the linear elastic strain terms were subtracted out.

(Zhou *et al.*, 2010a). Once a solid is deemed describable by the PM model, the model is capable of predicting the response of these materials to complex stress histories quite accurately. As important, the model can be used to calculate the *distributions* of the onset and friction stresses associated with the IKBs.

8.11 Damping

Typically, the price paid for high stiffness is a lack of damping and vice versa (Figure 8.16). In general, damping can be achieved by (i) physical friction between

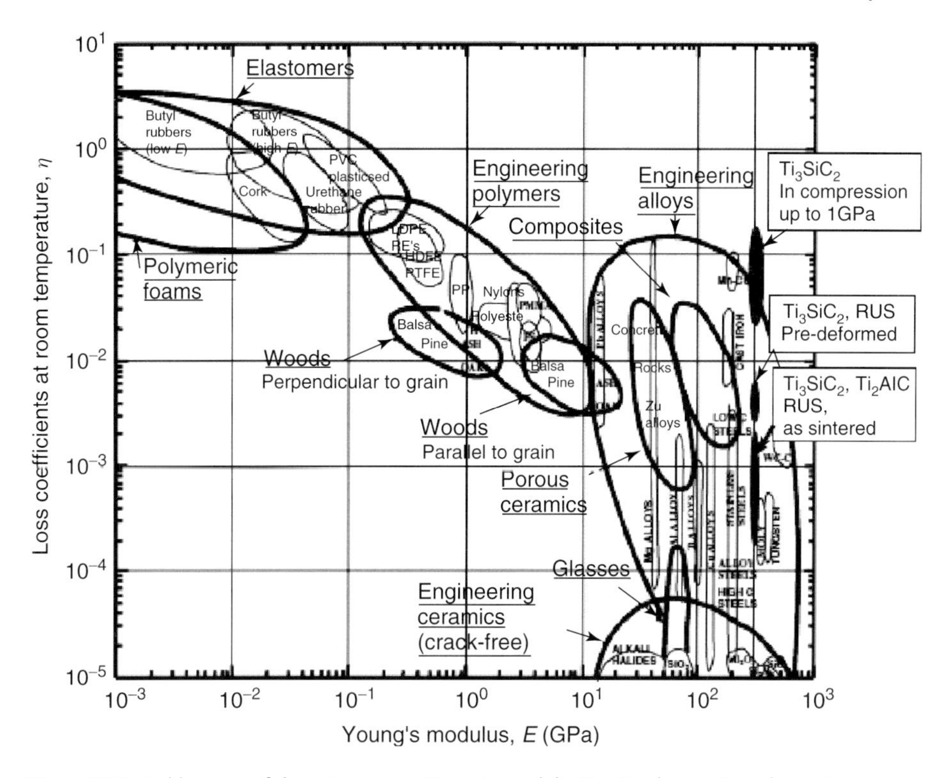

Figure 8.16 Ashby map of damping versus Young's moduli. Structural ceramics, shown in lower right, are some of the least damping solids known. The results for Ti_3SiC_2 measured in compression or resonant ultrasound spectroscopy are indicated in the plot.

constituents of a composite or microcracks, (ii) phase transitions, (iii) elastomeric materials, (iv) internal friction due to atoms jumping between various defect sites, and/or (v) dislocation motion. Physical friction and phase transitions lead to materials with poor fatigue properties and low moduli. Elastomeric materials have low moduli and cannot be used at higher temperatures. Damping by dislocation motion in metals is, typically, quite low. Damping by dislocation motion is also contingent on keeping the applied loads quite low to avoid work-hardening.

This lack of damping is especially true for structural ceramics – mapped in the lower right in Figure 8.16. It follows that the discovery of kinking nonlinear elasticity in the MAX phases is an important breakthrough in that it is now possible to engineer solids that are not only stiff but also highly damping. The damping is – with important exceptions described in subsequent text – a weak function of temperature, up to temperatures on the order of 800 °C. Furthermore, according to Eq. 8.6, the damping increases as the square of the applied stress as observed (Fig. 8.9c).

There are several techniques by which damping is quantified. There are also a number of definitions of damping, which depend on the method by which it is measured. Probably the most direct method is to compare the areas of the stress–strain hysteretic loops, to the total energy stored. If one defines the loss coefficient as the ratio of W_d to the area OAB shown in Figure 8.8c, then the results shown in Figure 8.7b fall in the top right rectangle shown in Figure 8.16: clearly outside the Ashby envelope.

8.12
Nonlinear Dynamic Effects

The interaction of elastic waves with solids is an important subject in solid-state physics and materials science (Truell, Chick, and Elbaum, 1969; Guyer, and Johnson, 1999). The localized nonlinear scattering of acoustic waves by microstructural defects, such as cracks, grain boundaries, or inhomogeneously distributed damage, has been used to explore the nature of these defects and to carry out nondestructive evaluation of solids under load (Nagy, 1999). For example, some materials display a strain dependence of the resonance frequency, which, in turn, implies a decreased modulus with increasing strains (Guyer and Johnson, 1999; McCall and Guyer, 1994) Originally, these nonlinear effects were studied in rocks and *granular* materials (Guyer and Johnson, 1999; McCall and Guyer, 1994; Guyer, McCall, and Boitnott, 1995). They were also observed in some sintered ceramics and damaged solids containing cracks and/or macrodefects. These materials also often exhibit nonlinear, and usually hysteretic, stress–strain relationships.

In the following sections, we discuss how ultrasound interacts with the MAX phases to give rise to nonlinearities. There are several techniques by which these nonlinearities have been studied. One is resonant ultrasound spectroscopy (RUS); another is the stress–bias technique. The results from each are discussed in the next two sections.

8.12.1
Resonant Ultrasound Spectroscopy

RUS is a relatively new technique developed by Migliori *et al.* for determining the complete set of elastic moduli from the resonant spectra of freely suspended single crystals (Migliori and Sarrao, 1997; Migliori *et al.*, 1993). The technique measures the resonance peak frequencies of a freely suspended single crystal and uses their location to calculate the complete set of elastic constants. In the case of polycrystals, the peak frequencies yield E and G of the solid. We used RUS to measure the Young's and shear moduli of select MAX phases as a function of temperature (e.g., Figure 3.12). The values were in excellent agreement with previous results obtained using more traditional ultrasound techniques (Radovic *et al.*, 2006).

A major advantage of RUS, however, over the more traditional ultrasound techniques is that it can concomitantly measure mechanical damping. The latter is manifested as an increase in the widths of the peaks and is defined as (Migliori and Sarrao, 1997; Migliori *et al.*, 1993)

$$Q^{-1} = \frac{\Delta\omega_\kappa}{\omega_{ko}} \tag{8.10}$$

where ω_{ko} is the frequency associated with the kth eigenmode and $\Delta\omega_k$ is the FWHM of that mode.

Before proceeding further, it is crucial to underscore the differences, apart from the obvious increases in frequency, between the results obtained with RUS and those obtained in quasistatic tests such as shown in Figures 8.7 and 8.9a. The strains generated during RUS are orders of magnitude lower than those in the quasistatic tests. Said otherwise, by their nature, RUS experiments are restricted to low acoustic wave excitations in the kilohertz frequency range and hence low (10^{-8} to 10^{-6}) strain amplitudes (Finkel *et al.*, 2009).

8.12.2 Nonlinear Effects at Ambient Temperatures

Figure 8.17a plots Q^{-1} – as determined by RUS – as a function of frequency, microstructure, and predeformation. From these results, it is clear that Q^{-1} is *not* a function of grain size. Also plotted in Figure 8.17a are the *room-temperature* Q^{-1} values measured *after* a CG Ti_3SiC_2 sample was compressively deformed 4% at 1573 K under a load corresponding to a stress of 25 MPa (Barsoum *et al.*, 2005a). This relatively modest deformation resulted in roughly an order of magnitude increase in Q^{-1} (compare solid circles with squares and open circles in Figure 8.17a) at room temperature.

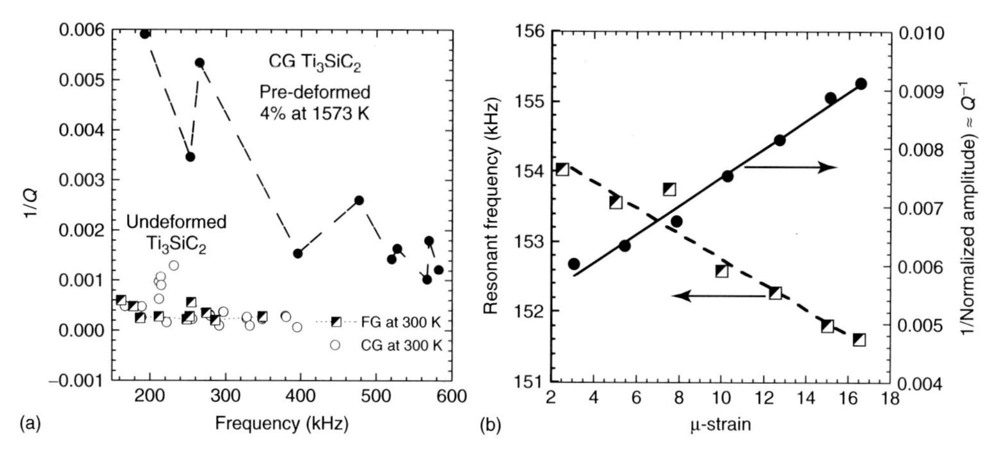

Figure 8.17 (a) Effect of grain size and frequency on Q^{-1} of Ti_3SiC_2. Also included are the room temperature results of a sample that was deformed 4% in compression at 1573 K. The large increase in Q^{-1} is presumably due to an increase in basal plane dislocation density introduced during the high-temperature deformation (Barsoum *et al.*, 2005a). (b) Effect of drive amplitudes on the ambient temperature resonant frequency (left y-axis) and Q^{-1} (right y-axis) for a polycrystalline Ti_3AlC_2 sample (Finkel *et al.*, 2009).

The strong function of damping on deformation history (Figure 8.17a) is compelling evidence that it is dislocation based. When combined with the fact that the 4% deformation at 1573 K also resulted in stiffening and hardening (Fig. 10.14) – a fact that can only be accounted for an increase in dislocation density – the evidence becomes almost irrefutable. The increase in Q^{-1} after deformation must thus be due to an increase in dislocation density.

It is important to note that the nondependence of Q^{-1} on grain size shown in Figure 8.17a also rules out grain boundaries as the damping source because there are more of them in the FG material. The ultrasound energy must thus couple with dislocation line segments, causing them to either vibrate – string model – and/or move – hysteretic model (Schaller, Fentozzi, and Gremaud, 2001).

The variations in Q^{-1} with frequency shown in Figure 8.17a are not noise. The uncertainty in the determination of Q^{-1} from the resonance spectra (i.e., Eq. (8.10)) is smaller than the width of the data points. These variations reflect the various eigenmodes of the sample and are quite typical of RUS results.

8.12.3 Modified RUS

To further bolster these conclusions, two other experiments were carried out: a modified RUS, and a stress–bias experiment (Finkel *et al.*, 2009). In the former, two transducers are attached to two parallel faces of a sample using epoxy. One transducer drives the sample, and the second measures the response. In addition, as a cross-check, the sample's z-displacement – measured using a laser Doppler vibrometer (LDV) – was transformed into a strain with an accuracy better than 10 microstrain.

As typical of nonlinear systems, the resonant frequency decreases linearly with increasing strain (left y-axis in Figure 8.17b). Concomitantly, Q^{-1} increases in a linear manner (right y-axis in Figure 8.17b) (Guyer and Johnson, 1999; Guyer, McCall, and Boitnott, 1995). It is important to note that the shifts in resonant frequencies observed – of the order of 200 Hz – are at least an order of magnitude larger than those in granular solids (Guyer and Johnson, 1999; Guyer, McCall, and Boitnott, 1995) or those measured by RUS on the same materials (Radovic *et al.*, 2006) due to the higher strain amplitudes used to obtain the results shown in Figure 8.17b. This decrease in resonant frequencies is of the order of 1.6%, which corresponds to a 2.6% decrease in modulus. These results are fully consistent with the interaction of the sound waves with preexisting dislocations.

8.12.4 Stress–Bias Technique

In this technique, an ultrasound wave is sent through a sample and its attenuation is measured as a function of applied stress (see lower inset in Figure 8.18a). Typically, in such experiments, the attenuation changes are quite small, and, unless the

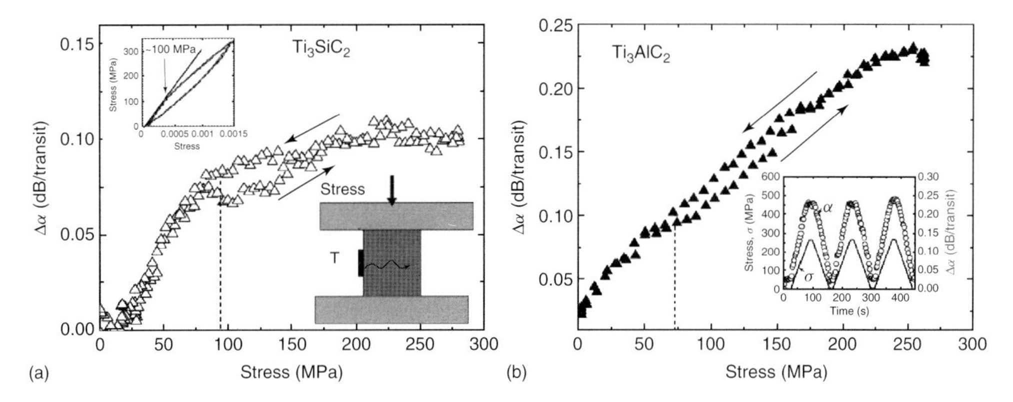

Figure 8.18 Attenuation changes as a function of stress in (a) CG Ti_3SiC_2 and (b) CG Ti_3AlC_2 samples. The bottom right inset in (a) shows schematic of the experimental bias–stress setup; left top inset shows the stress–strain curve obtained for the same sample. Note the fully reversible nature of the deformation. The inset in (b) shows attenuation and stress changes as a function of time (Finkel *et al.*, 2009).

applied strains are minuscule, irreversible (Gremaud, Bujard and Benoit 1987). This is not the case for KNE solids in general, and the MAX phases in particular.

The effects of cyclic compressive stresses on samples of Ti_3SiC_2 and Ti_3AlC_2 on changes in attenuation $\Delta\alpha$ are shown in Figure 8.18a,b, respectively. For both compounds, for stresses less than about 75 MPa, which for a Young's modulus of $\approx$300 GPa corresponds to a strain of 2×10^{-4}, $\Delta\alpha$ increases linearly with strain. Above a strain of $\approx 2 \times 10^{-4}$, hysteretic behavior is observed. Not coincidentally, the shift from one regime to the other occurs at a stress at which the mechanical hysteretic loops are observed (top inset in Figure 8.18a). At higher stresses, the responses of the two samples are slightly different; $\Delta\alpha$ for the Ti_3AlC_2 sample increases more or less linearly with stress despite the hysteresis (Figure 8.18b); for the Ti_3SiC_2 sample, $\Delta\alpha$ appears to saturate somewhat. Note that the $\Delta\alpha$ changes for an Al polycrystalline sample used as a control were below the system resolution of $\approx$0.01 dB/transit for stresses up of $\approx$80 MPa (Finkel *et al.*, 2009).

8.12.5
Effects of Temperature

Figure 8.19a plots the temperature dependence of Q^{-1} for a number of MAX phases. In all but one case, Q^{-1} remained more or less constant up to $\approx$1173 K, before it increased dramatically.

The fact that the changes in E and G with temperature are small and linear (e.g., Figure 3.12) as the imaginary component, that is, Q^{-1}, increases dramatically is unusual. For most solids, they are strongly coupled.

As noted in the preceding text, ultrasound strains applied during RUS are insufficient to nucleate IKBs in the MAX phases at least upto $\approx$1173 K. The fact that the elastic moduli do not drop dramatically at $\approx$1173 K, as they do in tension

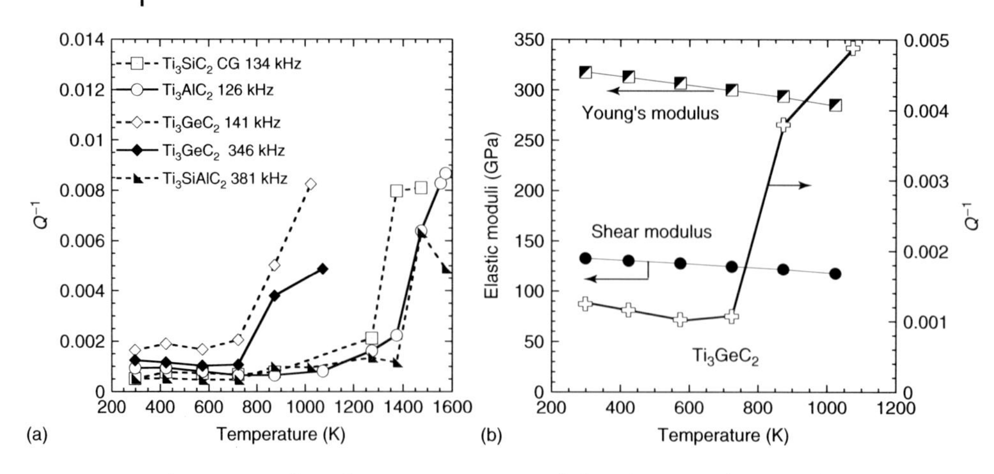

Figure 8.19 Effect of temperature on (a) Q^{-1} for select MAX phases (Radovic *et al.*, 2006). With the notable exception of Ti_3GeC_2, a dramatic increase in Q^{-1} at $T>1300$ K is observed. (b) Q^{-1} (right y-axis), E and G (left y-axis).

(Radovic *et al.*, 2002), compression (El-Raghy *et al.*, 1999), or bending (Barsoum, Ali, and El-Raghy, 2000; Boa and Zhou, 2003; Li *et al.*, 2001) – a drop attributed to the formation of KBs, MDWs, and dislocation arrays (Chapter 10) – is further evidence that IKBs are not nucleated in the RUS experiments. This is an important result because some have attributed the increases in damping to the brittle-to-plastic transition that occurs in all MAX phases (Chapter 10). And, while at first blush this seems to be a reasonable assumption because, for most MAX phases, the brittle to plastic transition temperature (BPTT) are in the vicinity of those at which the damping increases dramatically, the relationship is most probably coincidental. For example, as Figure 8.19b shows, in Ti_3GeC_2 the damping increases dramatically at temperatures ($\approx$800 K) that are significantly lower than the BPTT for this material, measured to be $\approx$1300 K in compression (Barsoum and El-Raghy, 1997). Others have attributed the increase in damping to a drop in modulus. That, again, is most probably coincidental because, at least in the case of Ti_3GeC_2, there is no such drop (Figure 8.19b).

It follows that the reason for the sharp and dramatic increase in Q^{-1} at temperatures greater than $\approx$1173 K (Figure 8.19a) for most of the MAX phases shown and $\approx$800 K for Ti_3GeC_2 is unclear at this time. However, since this increase is fully reversible (when the samples are cooled the original spectra are recovered), it is fair to conclude that the microstructure does not change in any permanent way. In other words, the increase cannot be due to the formation of KBs, DPs, microcracks, or an increase in dislocation density. The increase in damping with temperature may be due to an increase in dislocation mobility and/or other defects. Note that here mobility does *not* imply long-range glide of dislocations but rather a localized hopping in response to the ultrasound waves.

A more likely possibility, however, is that the increase in damping is associated with a dramatic increase in error bars and a discontinuity in the thermal motion parameters of the Ti_{II} atoms, shown in Fig. 4.11b, in Ti_3GeC_2 between 300 and 500 °C during both heating and cooling (Lane, Vogel, and Barsoum, 2010). Why this anomalous behavior occurs in Ti_3GeC_2 and not Ti_3SiC_2, for example, and how it so influences damping, is a subject of ongoing research.

8.13 Summary and Conclusions

The MAX phases have been classified as KNE solids that are characterized by fully and spontaneously reversible stress–strain loops. The microstructural element postulated to lead to KNE behavior is the IKB. The basic idea is that above a threshold stress, IKBs (coaxial, fully reversible dislocation loops) nucleate and grow in two dimensions. Removal of the stress results in the spontaneous shrinking/ elimination of the IKBs.

The following factors, all embodied in Eq. (8.6), have been shown to affect the size and shapes of the reversible loops:

1) **Grain size**: The larger the grain size, the larger 2α and hence the lower the threshold stresses, which, for a given σ, in turn, results in larger stress–strain loops and larger W_d values. The size to which the dislocation loops can grow is also larger.
2) **Shear moduli**: The lower the values of G, the larger the W_d. This is a strong factor because W_d is inversely proportional to G^2 (Eq. (8.6)). This factor also explains why, at a given stress, porous MAX phases dissipate more energy than their fully dense counterparts.
3) **CRSS or Ω/b**: This is an important factor and is essentially a measure of the CRSSs of the IKB dislocations. All else equal, solids with higher CRSS should dissipate more energy than those that have lower values. This comment notwithstanding, because there is a strong correlation between CRSS and G and the effect of the latter is important, the net result of having solids with high values of CRSS to enhance W_d has to be considered on a case-by-case basis.
4) **Dislocation core width w**: In principle, this factor is related to the critical kinking angle γ_c. The narrower the dislocations, the higher γ_c, and the lower the values of W_d (Eq. (8.6)). However, since γ_c, is also a function of G, here again its effect may be masked by changes in G.
5) **Texture**: While not explicitly addressed herein, texture can and will affect the Taylor factor M and k_1 in Eq. (8.6) and will thus also play a role (Caspi *et al.* 2013).

The ramifications of the existence of KNE solids are far-reaching and have already yielded many dividends in terms of our understanding of hitherto unexplained phenomena associated with the deformation of quite a number of diverse solids.

First, they elucidate why graphite responds to stress the way it does, a 50+ year old problem (Barsoum *et al.*, 2004a). Second, they explain and *quantify* the microyielding and damping in hexagonal metals Mg, Zn, Co, Ti, and Zr, which are two questions that, surprisingly, given the amount of work carried out on these metals, had remained unanswered (Zhou, Basu, and Barsoum, 2008).

Furthermore, the ability to easily measure the CRSSs of IKB dislocations and, in principle, their core width from a simple compression experiment of a polycrystalline solid is noteworthy and should help better understand not only the response of dislocations to stress in KNE solids but in general.

Given the diversity and ubiquity of KNE solids (Figure 8.14), it is clear that IKBs play a much more important role in our daily life than has hitherto been appreciated: from geology to high damping solids. Most importantly, it is reasonable to claim that IKBs are one of the last few, but crucial, missing pieces in the deformation-of-solids puzzle.

Lastly, the fact that the MAX phases kink has important ramifications on their confined deformations, such as under nanoindenters. This aspect of their deformation is discussed in the next chapter.

To recapitulate, large grains, less rigid solids, and higher stresses increase W_d, the first two by reducing the threshold stresses needed for IKB formation and the latter by increasing the diameters of the dislocation loops formed. Removal of the load results in the spontaneous collapse of the IKBs. The energy dissipated, or damping, results from the friction associated with the to-and-fro motion of IKB dislocations.

References

Aleshin, V. and Van Den Abeele, K. (2007) Friction in unconforming grain contacts as a mechanism for tensorial stress–strain hysteresis. *J. Mech. Phys. Solids*, **55**, 765.

Amini, S. and Barsoum, M.W. (2010) On the effect of texture on the mechanical properties of nano-crystalline Mg-matrix composites reinforced with MAX phases. *Mater. Sci. Eng., A*, **527**, 3707–3718.

Amini, S., Zhou, A., Gupta, S., DeVillier, A., Finkel, P., and Barsoum, M.W. (2008) Synthesis and elastic and mechanical properties of Cr_2GeC. *J. Mater. Res.*, **23**, 2157–2165.

Argon, A.A.S. (1972) Fracture of composites. *Treatise Mater. Sci. Technol.*, **1**, 79.

Barsoum, M.W., Ali, M., and El-Raghy, T. (2000) Processing and characterization of Ti_2AlC, Ti_2AlCN and $Ti_2AlC_{0.5}N_{0.5}$. *Metall. Mater. Trans. A*, **31**, 1857–1865.

Barsoum, M.W. and Basu, S. (2010) in *Encyclopedia of Materials Science and Technology* (eds R.W.C.K.H.J. Buschow, M.C. Flemings, B. Ilschner, E.J. Kramer, S. Mahajan, and P. Veyssiere), Elsevier, Oxford, pp. 1–23.

Barsoum, M.W., Brodkin, D., and El-Raghy, T. (1997) Layered machinable ceramics for high temperature applications. *Scr. Metall. Mater.*, **36**, 535–541.

Barsoum, M.W. and El-Raghy, T. (1997) A progress report on Ti_3SiC_2, Ti_3GeC_2 and the H-phases, M_2BX. *J. Mater. Synth. Process.*, **5**, 197–216.

Barsoum, M.W. and El-Raghy, T. (1999) Room temperature ductile carbides. *Metall. Mater. Trans. A*, **30**, 363–369.

Barsoum, M.W., Farber, L., El-Raghy, T., and Levin, I. (1999a) Dislocations, kink bands and room temperature plasticity of Ti_3SiC_2. *Metall. Mater. Trans. A.*, **30**, 1727–1738.

Barsoum, M.W., Farber, L., Levin, I., Procopio, A., El-Raghy, T., and Berner, A. (1999b) High-resolution transmission electron microscopy of Ti_4AlN_3, or

$Ti_3Al_2N_2$ revisited. *J. Am. Ceram. Soc.*, **82**, 2545–2547.

Barsoum, M.W., Murugaiah, A., Kalidindi, S.R., and Gogotsi, Y. (2004a) Kink bands, nonlinear elasticity and nanoindentations in graphite. *Carbon*, **42**, 1435–1445.

Barsoum, M.W., Murugaiah, A., Kalidindi, S.R., and Zhen, T. (2004b) Kinking nonlinear elastic solids, nanoindentations and geology. *Phys. Rev. Lett.*, **92**, 255508–255501.

Barsoum, M.W., Radovic, M., Finkel, P., and El-Raghy, T. (2001) Ti_3SiC_2 and ice. *Appl. Phys. Lett.*, **79**, 479–481.

Barsoum, M.W., Radovic, M., Zhen, T., and Finkel, P. (2005a) Dynamic elastic hysteretic solids and dislocations. *Phys. Rev. Lett.*, **94**, 085501.

Barsoum, M.W., Zhen, T., Zhou, A., Basu, S., and Kalidindi, S.R. (2005b) Microscale modeling of kinking nonlinear elastic solids. *Phys. Rev. B*, **71**, 134101.

Barsoum, M.W., Zhen, T., Kalidindi, S.R., Radovic, M., and Murugahiah, A. (2003) Fully reversible, dislocation-based compressive deformation of Ti_3SiC_2 to 1 GPa. *Nat. Mater.*, **2**, 107–111.

Bassett, D.C. and Attenburrow, G.E. (1979) Compliances and failure modes of oriented chain-extended polyethylene. *J. Mater. Sci.*, **14**, 2679–2687.

Basu, S. and Barsoum, M.W. (2009) On spherical nanoindentations, kinking nonlinear elasticity of mica single crystals and their geological implications. *J. Struct. Geol.*, **31**, 791–801.

Bei, G.-P., Laplanche, G., Gauthier-Brunet, V., Bonneville, J., and Dubois, S. (2013) The compressive behavior of Ti_3AlC_2 and $Ti_3Al_{0.8}Sn_{0.2}C_2$ MAX phases at room temperature. *J. Amer. Cer. Soc.*, **96**, 331–664.

Boa, Y.W. and Zhou, Y.C. (2003) Evaluating high-temperature modulus and elastic recovery of Ti_3SiC_2 and Ti_3AlC_2 ceramics. *Mater. Lett.*, **57**, 4018–4022.

Burke, E.C. and Hibbard, W.R. (1952) Plastic deformation of magnesium single crystals. *Trans. AIME*, **194**, 295–303.

Caspi, E., Shamma, M., Clausen, D., Vogel, B., Brown, S., Presser, V., Amini, S., Yeheskel, O., and Barsoum, M.W. (2013) In situ neutron diffraction evidence for incipient kink bands in highly textured polycrystalline Ti_2AlC. Sub. for pub.

Christoffersen, R. and Kronenberg, A.K. (1993) Dislocation interactions in experimentally deformed biotite. *J. Struct. Geol.*, **15**, 1077.

Clausen, B., Tome, C.N., Brown, D.W., and Agnew, S.R. (2008) Reorientation and stress relaxation due to twinning: modeling and experimental characterization for Mg. *Acta Mater.*, **56**, 2456–2468.

DeTeresa, S.J., Porter, R.S., and Farris, R.J. (1988) *J. Mater. Sci.*, **23**, 1886–1894.

El-Raghy, T., Barsoum, M.W., Zavaliangos, A., and Kalidindi, S.R. (1999) Processing and mechanical properties of Ti_3SiC_2: II, effect of grain size and deformation temperature. *J. Am. Ceram. Soc.*, **82**, 2855–2860.

Farber, L., Barsoum, M.W., Zavaliangos, A., El-Raghy, T., and Levin, I. (1998) Dislocations and stacking faults in Ti_3SiC_2. *J. Am. Ceram. Soc.*, **81**, 1677–1681.

Farber, L., Levin, I., and Barsoum, M.W. (1999) HRTEM study of a low-angle boundary in plastically deformed Ti_3SiC_2. *Philos. Mag. Lett.*, **79**, 163.

Finkel, P., Zhou, A.G., Basu, S., Yeheskel, O., and Barsoum, M.W. (2009) Direct observation of nonlinear acousto-elastic hysteresis in kinking nonlinear elastic solids. *Appl. Phys. Lett.*, **94**, 241904.

Fraczkiewicz, M., Zhou, A.G., and Barsoum, M.W. (2006) Mechanical damping in porous Ti_3SiC_2. *Acta Mater.*, **54**, 5261–5270.

Frank, F.C. and Stroh, A.N. (1952) On the theory of kinking. *Proc. Phys. Soc. London*, **65**, 811–821.

Freise, E.J. and Kelly, A. (1961) Twinning in graphite. *Proc. Phys. Soc. London, Sect. A*, **264**, 269–276.

Gay, N.C. and Weiss, L.E. (1974) The relationship between principal stress directions and the geometry of kinks in foliated rocks. *Tectonphysics*, **21**, 287.

Gibson, L.J. and Ashby, M.F. (1988) *Cellular Solids, Structure and Properties*, Pregamon Press, Oxford.

Gremaud, G., Bujard, M., and Benoit, W. (1987). The Coupling Technique: A Two-wave Acoustic Method for the Study of Dislocation Dynamics. *J. Appl. Phys.*, **61**, 1795.

Guitton, A., Joulain, A., Thilly, L., and Tromas, C. (2012) Dislocation analysis of Ti_2AlN deformed at room temperature under confining pressure. *Philos. Mag. Lett.*, **92**, 4536–4546.

Gupta, V., Anand, K., and Kryska, M. (1994) *Acta Metall. Mater.*, **42**, 781–795.

Guyer, R.A. and Johnson, P.A. (1999) Nonlinear mesoscopic elasticity: evidence for a new class of materials. *Phys. Today*, **52**, 30–36.

Guyer, R.A., McCall, K.R., and Boitnott, G.N. (1995) Hysteresis, discrete memory and nonlinear wave propagation in rock: a new paradigm. *Phys. Rev. Lett.*, **74**, 3491.

Hathorne, H.M. and Teghtsoonian, E. (1975) Axial compression fracture in carbon fibres. *J. Mater. Sci.*, **10**, 41–51.

Hess, J.B. and Barrett, C.S. (1949) Structure and nature of kink bands in zinc. *Trans. AIME*, **185**, 599–606.

Hull, D. (1965) *Introduction to Dislocation*, Pergamon Press, Oxford.

Jones, W.R. and Johnson, J.W. (1971) Intrinsic strength and non-hookean behaviour of carbon fibres. *Carbon*, 9, 645–655.

Keith, C.T. and Jr Cote, W.A. (1968) Microscopic characterization of slip lines and compression failures in wood cell walls. *Forest Prod. J.*, **18**, 67–78.

Kelly, B.T. (1981) *Physics of Graphite*, Applied Science Publishers, London.

Kooi, B.J., Poppen, R.J., Carvalho, N.J.M., De Hosson, J.T.M., and Barsoum, M.W. (2003) Ti_3SiC_2: a damage tolerant ceramic studied with nanoindentations and transmission electron microscopy. *Acta Mater.*, **51**, 2859–2872.

Kronenberg, A., Kirby, S., and Pinkston, J. (1990) Basal slip and mechanical anisotropy in biotite. *J. Geophys. Res.*, **95**, 19257.

Lane, N.J., Simak, S.I., Mikhaylushkin, A., Abrikosovv, I., Hultman, L., and Barsoum, M.W. (2011) A first principles study of dislocations in HCP metals through the investigation of the $(11\bar{2}1)$ twin boundary. *Phys. Rev. B*, **84**, 184101.

Lane, N.J., Vogel, S.C., and Barsoum, M.W. (2010) High temperature neutron diffraction and the temperature-dependent crystal structures of Ti_3SiC_2 and Ti_3GeC_2. *Phys. Rev. B*, **82**, 174109.

Lawn, B.R. and Marshall, D.B. (1998) Nonlinear stress–strain curves for solids containing closed cracks with friction. *J. Mech. Phys. Solids*, **46**, 85.

Li, J.F., Pan, W., Sato, F., and Watanabe, R. (2001) Mechanical properties of polycrystalline Ti_3SiC_2 at ambient and elevated temperatures. *Acta Mater.*, **49**, 937–945.

Mayergoyz, I.D. (1986) Mathematical models of hysteresis. *Phys. Rev. Lett.*, **56**, 1518.

Mayergoyz, I.D. (2003) *Mathematical Models of Hysteresis and their Applications*, Elsevier, New York.

McCall, K.R. and Guyer, R.A. (1994) Equation of state and wave-propagation in hysteretic nonlinear elastic-materials. *J. Geophys. Res.*, **99**(23), 887.

Meike, A. (1989) In situ deformation of micas: a high-voltage electron-microscope study. *Am. Mineral.*, **74**, 780.

Migliori, A. and Sarrao, J.L. (1997) *Resonant Ultrasound Spectroscopy: Applications to Physics, Materials Measurements and Nondestructive Evaluation*, John Wiley & Sons, Inc., New York.

Migliori, A., Sarrao, J.L., Visscher, W.M., Bell, T.M., Lei, M., Fisk, Z., and Leisure, R.G. (1993) Resonant ultrasound spectroscopic techniques for measurement of the elastic moduli of solids. *Physica B*, **183**, 1–24.

Morgiel, J., Lis, J., and Pampuch, R. (1996) Microstructure of Ti_3SiC_2,-based ceramics. *Mater. Lett.*, **27**, 85–89.

Mugge, O. (1898) *Neues Jahrb. Mineral.*, **1**, 71.

Nagy, P. (1999) *Ultrasonics*, **36**, 375.

Orowan, E. (1942) A type of plastic deformation new in metals. *Nature*, **149**, 463–464.

Ortin, J. (1992) Preisach modeling of hysteresis for a pseudoelastic Cu–Zn–Al single crystal. *J. Appl. Phys.*, **71**, 1454.

Patterson, M.S. and Weiss, L.E. (1996) Experimental deformation and folding in phyllite. *Geol. Soc. Am. Bull.*, **77**, 343–373.

Poon, B., Ponson, L., Zhao, J., and Ravichandran, G. (2011) Damage accumulation and hysteretic behavior of MAX phase materials. *J. Mech. Phys. Solids*, **59**, 2238–2257.

Preisach, F. (1935) Uber die magnetische nachwirkung. *Z. Phys.*, **94**, 277.

Radovic, M., Barsoum, M.W., El-Raghy, T., Wiederhorn, S.M., and Luecke, W.E. (2002) Effect of temperature, strain rate and grain size on the mechanical response of Ti_3SiC_2 in tension. *Acta Mater.*, **50**, 1297–1306.

Radovic, M., Ganguly, A., Barsoum, M.W., Zhen, T., Finkel, P., Kalidindi, S.R., and Lara-Curzio, E. (2006) On the elastic properties and mechanical damping of Ti_3SiC_2, Ti_3GeC_2, $Ti_3Si_{0.5}Al_{0.5}C_2$ and Ti_2AlC in the 300–1573 K temperature range. *Acta Mater.*, **54**, 2757–2767.

Rider, J.G. and Keller, A. (1966) *J. Mater. Sci.*, **1**, 389–398.

Robertson, R.E. (1969) *J. Polym. Sci.*, **7**(Part A-2), 1315–1328.

Robertson, J.M. and Hartmann, D.E. (1964) The temperature dependence of the microyield points in prestrained magnesium single crystals. *Trans. Met. Soc. AIME*, **230**, 1125–1133.

Schaller, R., Fentozzi, G., and Gremaud, G. (eds) (2001) *Mechanical Spectroscopy Q-1 2001 with Applications to Materials Science*, Trans Tech Publication, Zurich.

Sun, Z.-M., Murugaiah, A., Zhen, T., Zhou, A., and Barsoum, M.W. (2005) Microstructure and mechanical properties of porous Ti_3SiC_2. *Acta Mater.*, **53**, 4359–4366.

Tromas, C., Villechaise, P., Gauthier-Brunet, V., and Dubios, S. (2011) Slip line analysis around nanoindentation imprints in Ti_3SnC_2: a new insight into plasticity of MAX-phase material. *Philos. Mag.*, **91**, 1265–1275.

Truell, R., Chick, B.B., and Elbaum, C. (1969) *Ultrasonic Methods in Solid-State Physics*, Academic Press, New York.

Walsh, J.B. (1965) The effect of cracks in rocks on Poisson's ratio. *J. Geophys. Res.*, **70**, 399.

Walsh, J.B. (1966) Seismic wave attenuation in rock due to friction. *J. Geophys. Res.*, **71**, 2591.

Weaver, C.W. and Williams, J.G. (1975) *J. Mater. Sci.*, **10**, 1323–1333.

Yang, B., Barsoum, M.W., and Rethinam, R.M. (2011) Nanoscale continuum calculation of basal dislocation core structures in graphite. *Philos. Mag.*, **91**, 1441–1463.

Zaukelies, D.A. (1962) *J. Appl. Phys.*, **33**, 2797–2803.

Zhen, T., Barsoum, M.W., and Kalidindi, S.R. (2005) Effects of temperature, strain rate and grain size on the compressive properties of Ti_3SiC_2. *Acta Mater.*, **53**, 4163–4171.

Zhou, A. and Barsoum, M.W. (2009) Kinking nonlinear elasticity and the deformation of Mg. *Metall. Mater. Trans. A*, **40**, 1741–1756.

Zhou, A., Basu, S., Finkel, P., Friedman, G., and Barsoum, M.W. (2010a) Hysteresis in kinking nonlinear elastic solids and the Preisach-Mayergoyz model. *Phys. Rev. B*, **82**, 094105.

Zhou, A., Brown, D., Vogel, S., Yeheskel, O., and Barsoum, M.W. (2010b) On the kinking nonlinear elastic deformation of polycrystalline cobalt. *Mater. Sci. Eng., A*, **527**, 4664–4673.

Zhou, A.G. and Barsoum, M.W. (2010) Kinking nonlinear elastic deformation of Ti_3AlC_2, Ti_2AlC, $Ti_3Al(C_{0.5},N_{0.5})_2$ and $Ti_2Al(C_{0.5},N_{0.5})$. *J. Alloys Compd.*, **498**, 62–67.

Zhou, A.G., Barsoum, M.W., Basu, S., Kalidindi, S.R., and El-Raghy, T. (2006) Incipient and regular kink bands in dense and 10 vol.% porous Ti_2AlC. *Acta Mater.*, **54**, 1631.

Zhou, A.G., Basu, B., and Barsoum, M.W. (2008) Kinking nonlinear elasticity, damping and microyielding of hexagonal close-packed metals. *Acta Mater.*, **56**, 60–67.

9
Mechanical Properties: Ambient Temperature

9.1
Introduction

Given the close structural and chemical similarities between the MAX and their corresponding MX phases it is not surprising – as outlined so far – that they share some common attributes and properties. For example, they are both metallic-like conductors dominated by d–d bonding. Their phonon conductivities are both susceptible to the presence of vacancies and the thermal expansion coefficients of the ternaries track those of the binaries, to name a few.

In sharp contradistinction, the mechanical properties of the MAX phases cannot be more different than those of their binary cousins. Arguably, and by far the most striking, is the machinability of the former. The MX compounds are characterized, and prized, for their hardness and wear resistance, which is why they are used as abrasives and cutting tools in industry. The MAX phases, on the other hand, are characterized and sometimes prized for their machinability.

Section 8.2 dealt with dislocations and their arrangements – either in dislocation pileups (DPs), or mobile dislocation walls (MDWs), that are normal to the basal planes. The fact that basal, and only basal plane dislocations multiply and are mobile at all temperatures, implies that the number of independent slip systems is two. This chapter summarizes the ambient temperature mechanical properties of the MAX phases that inhabit a middle ground between metals – with their 5+ independent slip systems – and structural ceramics, including the MX phases, with effectively zero slip systems at ambient temperatures. The MAX phases thus occupy an interesting middle ground in which in *constrained* deformation modes, highly oriented microstructures, and/or at higher temperatures, discussed in Chapter 10, they are pseudo-plastic. In unconstrained deformation, thin form, and especially in tension at lower temperatures, they behave in a brittle fashion.

In the following section, the response of large, quasi-single crystals to compressive loads is described. The evidence presented highlights the importance of kink band (KB) formation within individual grains *and* that of shear band formation to the overall deformation. Sections 9.3–9.6 deal, respectively, with the response of polycrystalline MAX phases to compressive, shear, flexural, and tensile loads. Section 9.7 explores the hardness of the MAX phases at various lengths scales.

MAX Phases: Properties of Machinable Ternary Carbides and Nitrides, First Edition. Michel W. Barsoum.

Fracture toughness, as determined by K_{1c}, is discussed in Section 9.8. Sections 9.9 and 9.10 deal with fatigue resistance and damage tolerance, respectively. The micro-mechanisms responsible for the high K_{1c}'s, R-curve behavior, and fatigue resistance are described in Section 9.11. Thermal shock resistance is discussed in Section 9.12. Strain rate effects are discussed in Section 9.13. Solid solution effects are taken up in Section 9.14. The penultimate section deals very briefly with machinability. The last section summarizes the chapter.

9.2
Response of Quasi-Single Crystals to Compressive Loads

When highly oriented, Ti_3SiC_2 samples – fabricated by sinter forging of porous Ti_3SiC_2 preforms – with grains of the order of 2 mm, shown in Figures 9.1 and 9.2, were loaded in compression the response, shown in Figure 9.3b, was quite anisotropic (Barsoum and El-Raghy, 1999). When the basal planes were oriented in such a way that allows for slip (z-direction in Figure 9.3a) the samples yielded at $\approx$200 MPa (Figure 9.3b) and the deformation occurred by the formation of classic shear bands (Figure 9.1a,b). By contrast, when the slip planes were parallel to the applied load (x-direction in Figure 9.3a), and deformation by ordinary dislocation glide was suppressed. The stress–strain curves showed clear maxima at stresses between 230 and 290 MPa, followed by a strain-softening regime, before recovering at a stress of $\approx$200 MPa (Figure 9.3b). Here, the deformation occurred by a combination of KB formation at the corners of the tested cubes, delaminations within individual grains, and ultimately shear-band formation (Figure 9.2a,b). The fact that the KBs were nucleated at the corners is not a coincidence but fully consistent with kinking being a form of plastic buckling that is more prevalent when a grain is unconstrained *and* the basal planes are loaded edge on.

Note the KB visible on the bottom left hand side of the cube in Figure 9.2b did *not* result in the total delamination of the outermost grain in which it was initiated. This observation is *compelling microstructural evidence that KBs are potent suppressors* of delaminations and is one important reason the MAX phases are as damage tolerant as they are (Barsoum and El-Raghy, 1999). This topic is discussed further below.

As noted in Ch. 8, Orowan (1942) was the first to describe KB formation in metals when he loaded Cd single crystals with their basal planes edge on. The inset in Figure 9.4a reproduces a schematic of a kink boundary – in which an excess of one kind of dislocations is present – as depicted by Orowan. Figure 9.4a is a SEM micrograph of the lower left corner of the cube shown in Figure 9.2b that is uncannily similar to Orowan's depiction. An example of typical multiple KBs and nested delaminated layers is shown in Figure 9.4b.

The extraordinary extent by which a single, ultra-large, Ti_3SiC_2 grain can deform at room temperature via the formation of multiple KBs and kink boundaries when unconstrained is shown in Figure 9.5a. Figure 9.5b shows the fractured surface of a cube similar to the one shown in Figure 9.1, which again shows the flexibility of the basal planes and their propensity to kink. Inset in Figure 9.5b, is a higher

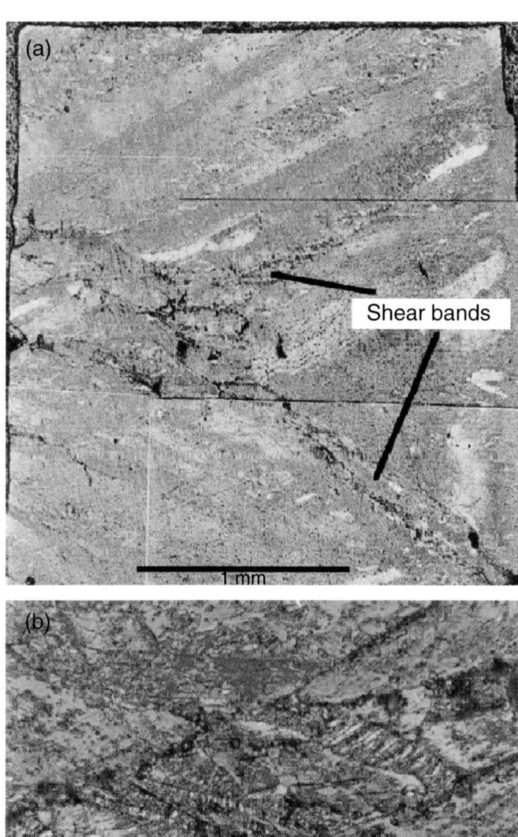

Figure 9.1 (a) OM micrograph of polished and etched Ti_3SiC_2 sample after loading along the *z*-direction in Figure 9.3a. Two shear bands are clearly visible. (b) Same as (a), but at a higher magnification focusing on one of the shear bands (Barsoum and El-Raghy, 1999).

magnification of the feature labeled a in Figure 9.5b. Note extremely sharp radii of curvatures possible. It is crucial to note that such features can *only* occur by an excess of dislocations of one kind.

We note in passing that similar features – characterized by very sharp radii of curvature – were observed in single crystals of graphite (Barsoum *et al.*, 2004a), mica (Basu and Barsoum, 2009), sapphire, $LiNbO_3$, and $LiTaO_3$ (Anasori *et al.*, 2011; Basu, Zhou, and Barsoum, 2008) after indenting them with spherical indenters. Indeed, based on the totality of the work we have carried out in this area to date, such features are sufficient to conclude that a solid is probably a kinking nonlinear elastic (KNE) solid.

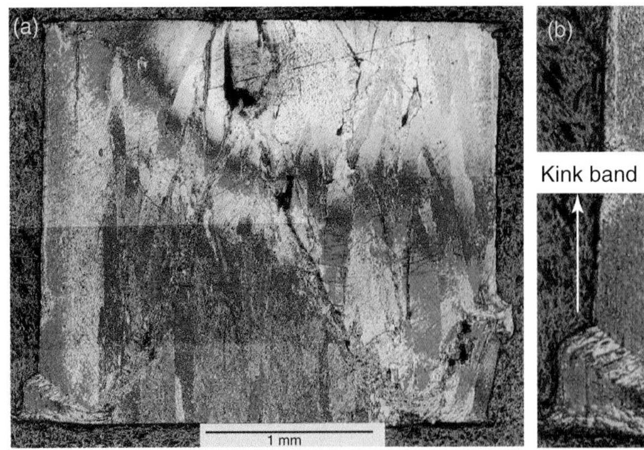

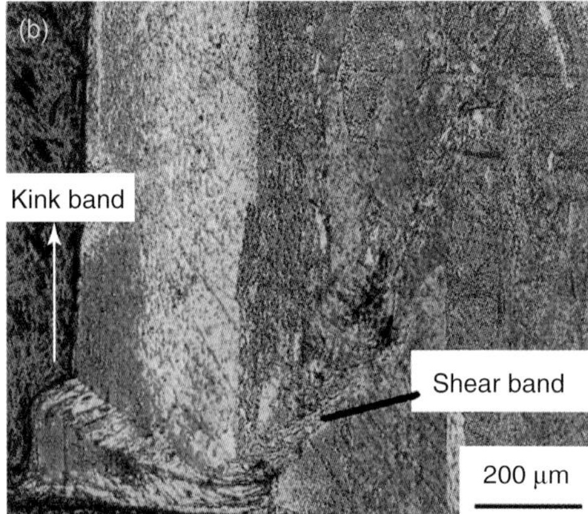

Figure 9.2 (a) OM micrograph of polished and etched sample after deformation along the *x*-direction in Figure 9.3a, that is, where the basal planes are loaded edge-on. Note kinking at corners and (b) higher magnification view of lower left corner in (a), emphasizing the kink band and, as important, the genesis of the shear band that cuts across the cube face (Barsoum and El-Raghy, 1999).

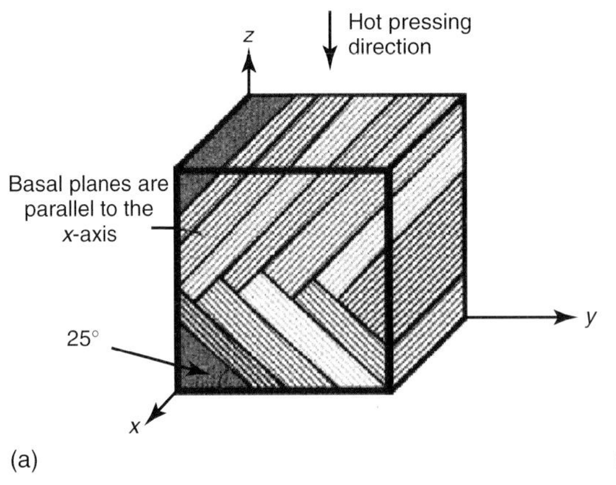

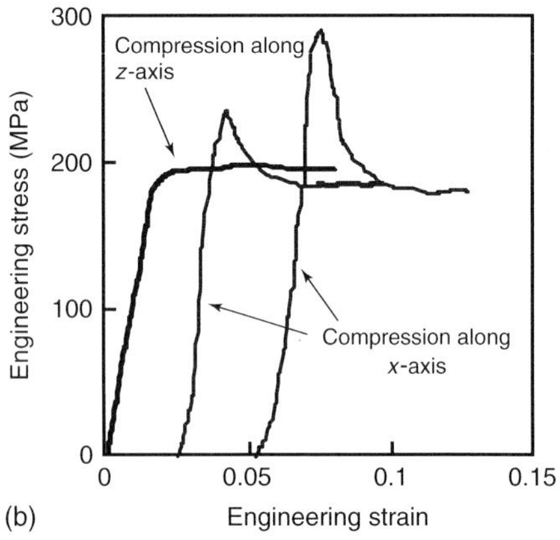

Figure 9.3 (a) Schematic of highly oriented macro-grained Ti_3SiC_2sample formed by sinter forging of porous Ti_3SiC_2 performs. The lines represent basal planes. (b) Engineering stress–strain curves of 2 mm cubes shown in Figures 9.1 and 9.2. The response was anisotropic and depended on the angle between the loading direction and the basal planes (Barsoum and El-Raghy, 1999).

As discussed in detail in the following sections, the response of polycrystalline MAX phase samples to stress is – like most solids – a function of grain size. Figure 9.6a,b, respectively show typical OM micrographs of a fine-grained (FG), and a coarse-grained (CG), Ti_3SiC_2 sample. In what follows, reference to FG and CG Ti_3SiC_2 samples will, unless otherwise noted, correspond to the microstructures shown in Figure 9.6.

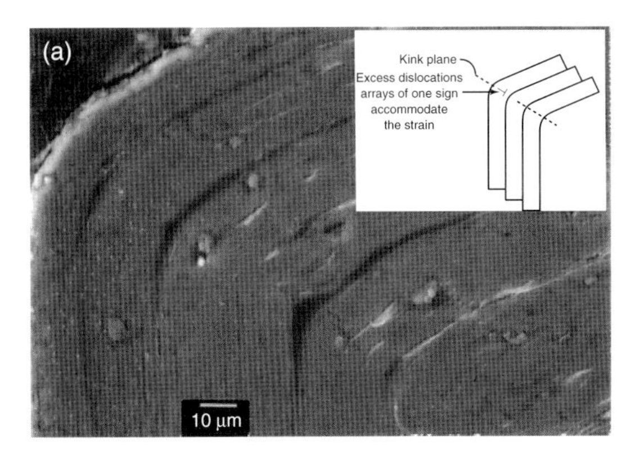

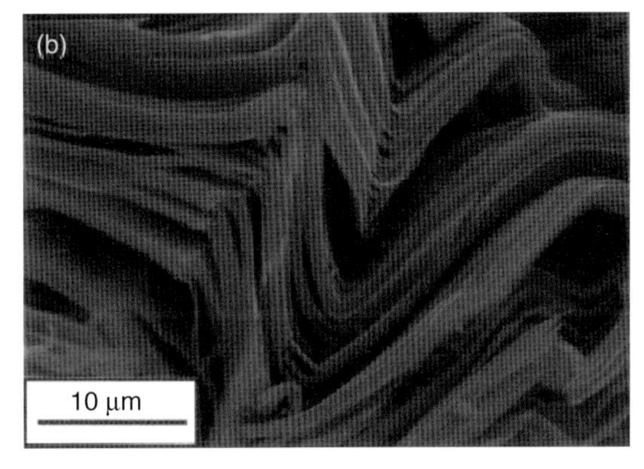

Figure 9.4 SEM micrographs of, (a) lower left corner shown in Figure 9.2a (Barsoum and El-Raghy, 1999). Inset shows schematic of a kink boundary as depicted by Orowan (1942). (b) Typical fractured MAX phase surface showing multiple, nested kink bands.

9.3 Response of Polycrystalline Samples to Compressive Stresses

The ultimate compressive stresses (UCSs), of the MAX phases are a function of grain size, with CG samples being weaker (Table 9.1). The UCSs range from $\approx$300 MPa to $\approx$2 GPa (El-Raghy *et al.*, 1999; Hu *et al.*, 2007, 2008b; Procopio, Barsoum, and El-Raghy, 2000; Wang and Zhou, 2002a). In all cases, the CG samples failed at lower stresses than their FG counterparts.

The compressive failure modes are also a function of grain size, loads at failure, and strain rates. When loaded at high strain rates, FG samples, especially when the UCSs are of the order of 1.0 GPa or higher, tend to fail suddenly in a brittle fashion (Barsoum, Brodkin, and El-Raghy, 1997). However, not all MAX phases fail suddenly; some, especially CG 211 and 312 phases, exhibit graceful-failure characteristics in that the stress–displacement curves are closer to an inverted shallow V (Figure 9.7a), than a sharp drop at the UCS (Barsoum, Ali, and El-Raghy, 2000a; Tzenov and Barsoum, 2000).

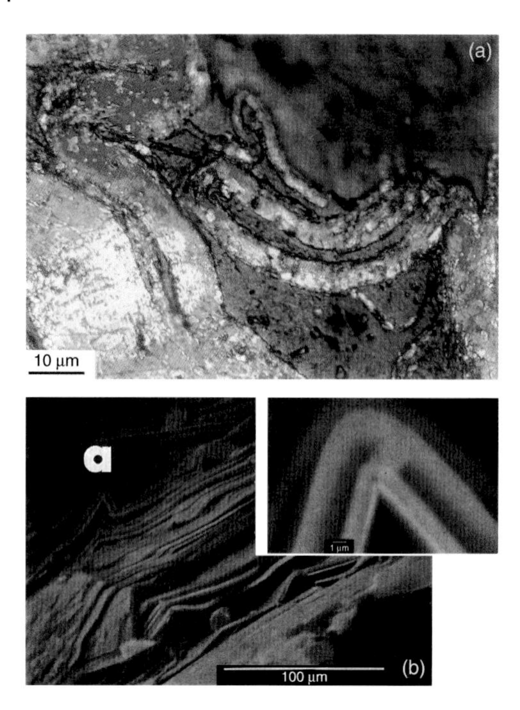

Figure 9.5 (a) Polished and etched OM micrograph of a near-surface feature of a sample such as the ones shown in Figures 9.1 and 9.2, that was deformed to a strain of more than 50% at room temperature. (b) SEM micrograph of fractured surface of cube shown in Figure 9.2. Inset shows typical very sharp radii of curvatures possible when the MAX phases are deformed at room temperatures (Barsoum and El-Raghy, 1999).

Figure 9.6 OM micrographs of, (a) fine-grained, FG and (b) coarse-grained, CG, Ti_3SiC_2 sample (El-Raghy *et al.*, 1999).

Table 9.1 Summary of ambient temperature mechanical properties of select MAX phases under various loading conditions. Flex. refers to 3 or 4-point flexure, while comp. refers to compression. D refers to the diameter of the grains and W to their thickness.

Compound	Grain size (μm)	Load	σ (MPa)	References
		413 phases		
$Ti_4AlN_{2.9}$	20*D*	4-Flex.	350 ± 15	Procopio, Barsoum, and El-Raghy (2000)
$Ti_4AlN_{2.9}$	20*D*	Comp.	475 ± 15	Procopio, Barsoum, and El-Raghy (2000)
Nb_4AlC_3	50*D*; 17*W*	3-Flex.	350 ± 40	Hu *et al.* (2008c)
Nb_4AlC_3	21*D*; 9*W*	3-Flex.	455 ± 7	Hu *et al.* (2009)
Nb_4AlC_3	50*D*; 17*W*	Comp.	500 ± 50	Hu *et al.* (2008c)
Nb_4AlC_3	50*D*; 17*W*	Shear	120 ± 40	Hu *et al.* (2008c)
Ta_4AlC_3	10*D*; 3*W*	3-Flex.	370 ± 20	Hu *et al.* (2007)
Ta_4AlC_3	10*D*; 3*W*	Comp.	800 ± 100	Hu *et al.* (2007)
Ta_4AlC_3	10*D*; 3*W*	Shear	250 ± 20	Hu *et al.* (2007)
		312 phases		
Ti_3SiC_2	≈100*D*	4-Flex.	260 ± 20	Barsoum and El-Raghy (1996)
Ti_3SiC_2	100*D* (Figure 9.6b)	4-Flex.	320	El-Raghy *et al.* (1999)
Ti_3SiC_2	5 (Figure 9.6a)	4-Flex.	600	El-Raghy *et al.* (1999)
Ti_3SiC_2	10–20*D*	3-Flex.	410 ± 25	Gao, Miyamoto, and Zhang (1999)
Ti_3SiC_2	26*D*; 12*W*	4-Flex.	300	Li *et al.* (2001)
Ti_3SiC_2	20–50*D*; 5–8*W*	3-Flex.	390 ± 10	Bao, Zhang, and Zhou (2004)
$Ti_3Si(Al)C_2$	15 ± 12*D*; 4 ± 2	3-Flex	460 ± 25	Wan *et al.* (2008)
$Ti_3Si(Al)C_2$	50 ± 40*D*; 16 ± 10	3-Flex	306 ± 15	Wan *et al.* (2008)
Ti_3SiC_2	≈100*D*	Comp.	580 ± 20	Barsoum and El-Raghy (1996)
Ti_3SiC_2	44 ± 40	Comp.	710	Barsoum, Brodkin, and El-Raghy (1997)
Ti_3SiC_2	100*D* (Figure 9.6b)	Comp.	720	El-Raghy *et al.* (1999)
Ti_3SiC_2	5 (Figure 9.6a)	Comp.	1050	El-Raghy *et al.* (1999)
Ti_3SiC_2	20–50*D*; 5–8*W*	Comp.	750 ± 80[a]	Bao, Zhang, and Zhou (2004)
Ti_3SiC_2	20–50*D*; 5–8*W*	Comp.	840 ± 80[b]	Bao, Zhang, and Zhou (2004)
Ti_3SiC_2	20–50*D*	Comp.	935	Zhang and Sun (2005)
Ti_3SiC_2	20–50*D*; 5–8*W*	Shear	138 ± 12	Bao, Zhang, and Zhou (2004)
Ti_3SiC_2	3–5 (Figure 9.6a)	Tensile	225	Radovic *et al.* (2000)
Ti_3SiC_2	100–200*D* (Figure 9.6a)	Tensile	200	Radovic *et al.* (2002)
Ti_3SiC_2	3–5 (Figure 9.6b)	Tensile	300	Radovic *et al.* (2002)
Ti_3AlC_2	≈20*D*	3-Flex.	340	Wang and Zhou (2002a)
Ti_3AlC_2	30 ± 10*D*; 6 ± 2*W*	3-Flex.	320 ± 12	Wan *et al.* (2008)
Ti_3AlC_2	75 ± 25*D*; 20 ± 10*W*	3-Flex.	170 ± 13	Wan *et al.* (2008)
Ti_3AlC_2	25*D*; 10*L*	Comp.	560 ± 20	Tzenov and Barsoum (2000)
Ti_3AlC_2	10–30*D*	Comp.	760	Wang and Zhou (2002a)

[a] Load parallel to hot pressing direction.
[b] Load normal to hot pressing direction.

Table 9.1 (*Continued*)

Compound	Grain size (μm)	Load	σ (MPa)	References
Ti_3AlC_2	60–100*D*; 10–40*W*	Comp.	545	Bei *et al.* (2013)
$Ti_3Al_{0.8}Sn_{0.2}C_2$	10–80*D*; 2–15*W*	Comp.	840	Bei *et al.* (2013)
Ti_3GeC_2	FG	Comp.	1270	Barsoum, Brodkin, and El-Raghy (1997)
Ti_3GeC_2	46 ± 25	Comp.	467 ± 12	Ganguly, Zhen, and Barsoum (2004)
$Ti_3Ge_{3/4}Si_{1/4}C_2$	70 ± 60	Comp.	670 ± 10	Ganguly, Zhen, and Barsoum (2004)
$Ti_3Ge_{1/2}Si_{1/2}C_2$	40 ± 30	Comp.	540 ± 10	Ganguly, Zhen, and Barsoum (2004)
Ti_3GeC_2	46 ± 25	Flex.	217 ± 4	Ganguly, Zhen, and Barsoum (2004)
$Ti_3Ge_{1/2}Si_{1/2}C_2$	40 ± 30	Flex.	254 ± 15	Ganguly, Zhen, and Barsoum (2004)
		211 phases		
Ti_2AlC	100–200*D*	Comp.	≈390	Barsoum, Brodkin, and El-Raghy (1997)
Ti_2AlC	25*D*	Comp.	540	Barsoum, Ali, and El-Raghy (2000a)
Ti_2AlC	75*D*; 15*W*	3-Flex.	275	Wang and Zhou (2002b)
Ti_2AlN	100–200	Comp.	≈470	Barsoum, Brodkin, and El-Raghy (1997)
Ti_2AlN	100*D*	Comp.	380 ± 30	Barsoum, Ali, and El-Raghy (2000a)
$Ti_2AlC_{0.5}N_{0.5}$	25*D*	Comp.	800 ± 80	Barsoum, Ali, and El-Raghy (2000a)
Ti_2SC	3–5	Comp.	1400	Amini, Barsoum, and El-Raghy (2007)
Ti_2GeC	FG	Comp.	≈1750	Barsoum, Brodkin, and El-Raghy (1997)
TiNbAlC	15*D*	4-Flex.	350 ± 15	Salama, El-Raghy, and Barsoum (2002)
TiNbAlC	45*D*	4-Flex.	310 ± 10	Salama, El-Raghy, and Barsoum (2002)
V_2AlC	49*D*; 19*W*	3-Flex.	263 ± 23	Hu *et al.* (2008a)
V_2AlC	108*D*; 37*W*	3-Flex.	290 ± 6	Hu *et al.* (2008a)
V_2AlC	119*D*; 47*W*	3-Flex.	270 ± 12	Hu *et al.* (2008a)
V_2AlC	405*D*; 106*W*	3-Flex.	61 ± 15	Hu *et al.* (2008a)
V_2AlC	49*D*; 19*W*	Comp.	740 ± 90	Hu *et al.* (2008a)
V_2AlC	108*D*; 37*W*	Comp.	600 ± 30	Hu *et al.* (2008a)
V_2AlC	119*D*; 47*W*	Comp.	530 ± 12	Hu *et al.* (2008a)
V_2AlC	405*D*; 106*W*	Comp.	390 ± 16	Hu *et al.* (2008a)
Cr_2AlC	≈40*D*; 10*W*	3-Flex	378	Tian *et al.* (2006)
Cr_2AlC	5	3-Flex.	480 ± 30	Tian *et al.* (2007)
Cr_2AlC	5.5 ± 3	4-Flex	555 ± 11	Tian *et al.* (2009)
Cr_2AlC	10	3-Flex.	500	Zhou, Mei, and Zhu (2009)

Table 9.1 (*Continued*)

Compound	Grain size (μm)	Load	σ (MPa)	References
Cr_2AlC	2	3-Flex.	510 ± 15	Li *et al.* (2011)
Cr_2AlC	35	3-Flex.	310 ± 10	Li *et al.* (2011)
Cr_2AlC	10–40	3-Flex.	470 ± 30	Ying *et al.* (2011b)
Cr_2AlC	5	Comp.	1160 ± 25	Tian *et al.* (2007)
Cr_2AlC	10–40	Comp.	950 ± 25	Ying *et al.* (2011b)
Cr_2AlC	10	Comp.	630	Zhou, Mei, and Zhu (2009)
Cr_2GeC	$20 \pm 10D$; $10 \pm 5W$	Comp.	770	Amini *et al.* (2008)
Nb_2AlC	$15D$	4-Flex.	415 ± 20	Salama, El-Raghy, and Barsoum (2002)
Nb_2AlC	$17D$	3-Flex.	480 ± 40	Zhang *et al.* (2009)
Nb_2AlC	$17D$	4-Flex.	440 ± 30	Zhang *et al.* (2009)
Zr_2SC	10	3-Flex.	275 ± 10	Opeka *et al.* (2011)
Ta_2AlC	$15D$; $3W$	3-Flex.	360 ± 20	Hu *et al.* (2008b)
Ta_2AlC	$15D$; $3W$	Comp.	800	Hu *et al.* (2008b)
Ta_2AlC	$15D$; $3W$	Shear	110 ± 25	Hu *et al.* (2008b)

9.3.1
Shear Band Formation

When the failure is graceful it usually occurs along a plane inclined $\approx 20-40^{\circ}$ to the loading axis. The shear fracture angle, θ_c, upon compression is typically in the range of $23-35^{\circ}$ (Table 9.2). These values clearly deviate from the maximum shear stress plane, viz. 45°. Zhang and Sun (2005) noted that such behavior was reminiscent of other brittle solids, such as ice and bulk metallic glasses. It follows that the appropriate failure criteria to use is the Mohr–Coulomb criterion. Said otherwise, failure occurs when:

$$\tau_n + \mu\, \sigma_n \geq \tau_c \tag{9.1}$$

where τ_c is the critical shear fracture stress, μ is an effective friction coefficient across the shear plane, σ_n and τ_n are the normal and shear stresses on the shear plane given, respectively, by:

$$\sigma_n = \sigma \sin^2\theta \tag{9.2}$$

$$\tau_n = \sigma \sin\theta \cos\theta \tag{9.3}$$

where θ is the angle between the stress axis and the shear plane. By combining Eqs. (9.1–9.3) and by finding the minimum with respect to θ it can be shown that (Zhang and Sun, 2005):

$$\mu = \frac{\cos 2\theta_c}{\sin 2\theta_c} \tag{9.4}$$

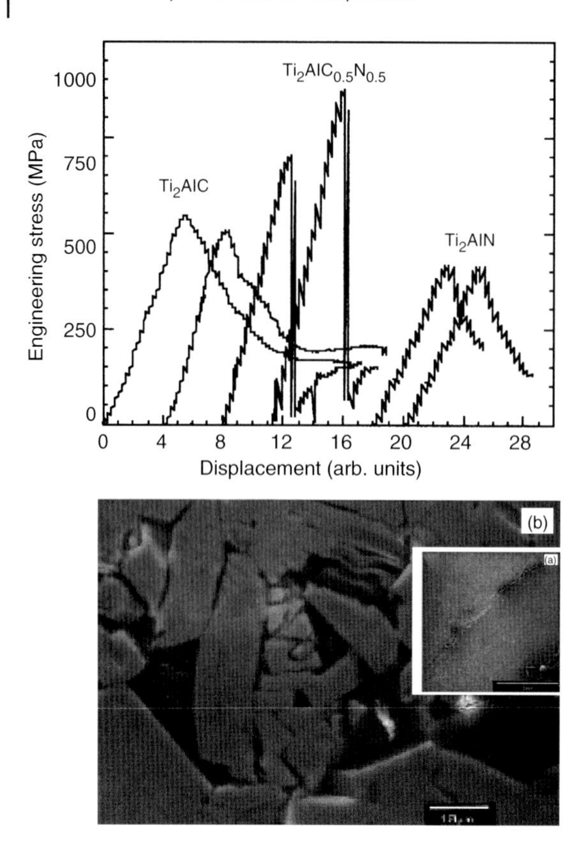

Figure 9.7 (a) Characteristic stress–displacement curves obtained when coarse-grained Ti_2AlX samples are loaded in compression. Note that the $Ti_2AlC_{0.5}N_{0.5}$ composition is significantly stronger, and more brittle than the end members. The various curves are shifted to the right for clarity's sake (Barsoum, Ali, and El-Raghy, 2000a) and (b) Higher magnification SEM micrograph of shear band formed when a CG Ti_3AlC_2 sample was loaded in compression showing the extent of damage to the grains in the vicinity of the shear band shown in inset at lower magnification (Tzenov and Barsoum, 2000).

and

$$\tau_c = \sigma_f \sin\theta_c \left(\cos\theta_c - \mu \sin\theta_c\right) \tag{9.5}$$

where σ_f is the normal stress at failure. Knowing σ_f and θ_c from experiment, one can calculate τ_c which represents the resistance to shear of a polycrystalline sample in the absence of friction.

Table 9.2 summarizes uniaxial compression results of select MAX phases subjected to the aforementioned analysis. From these results it is clear that:

1) τ_c is a function of grain size. Larger grains result in smaller values of τ_c.
2) μ falls in the range of 0.2 to $\approx$1.0.

Table 9.2 Summary of compression results subjected to Mohr–Coulomb analysis (i.e., Eqs. (9.1–9.5)). In Table D refers to the diameter of the grains and W to their thickness.

Composition	Grain size (μm)	UCS (MPa)	θ_c	μ	τ_c (MPa)	References
Ti_3SiC_2	2000	200	≈35	0.36	≈73[a]	Barsoum and El-Raghy (1999)
Ti_3AlC_2	≈25	375 ± 15	≈40	≈0.2	258	Barsoum, Ali, and El-Raghy (2000)
Ti_3SiC_2	20–50*D*	935	23	≈1	200	Zhang and Sun (2005)
Ti_3AlC_2: Set A	20–50*D* 5–8*W*	750 ± 80	38	0.25	360 ± 40	Bao *et al.* (2004)
Ti_3AlC_2: Set B	20–50*D* 5–8*W*	840 ± 80	26	0.78	330 ± 20	Bao *et al.* (2004)
Ti_3AlC_2	60–100*D* 10–40*W*	545	34	0.4	185	Bei *et al.* (2013)
$Ti_3Al_{0.8}Sn_{0.2}C_2$	10–80*D* 2–15*W*	840	30	0.6	242	Bei *et al.* (2013)

[a] In the original reference τ_c was estimated assuming θ_c was 45°.
Set A was loaded in the same direction as the hot pressing direction; set B was loaded such that the loading direction was normal to the hot pressing direction.

Working with polycrystalline Ti_3SiC_2 samples with grain diameters, D, of the order of 20–50 μm, Zhang and Sun (2005) determined that at ambient temperature, θ_c was 23°, $\mu \approx 1$, and τ_c was ≈200 MPa. The latter is significantly lower than the value of 73 MPa reported on the cubes shown in Figure 9.1a (Barsoum and El-Raghy, 1999). The difference is most probably related to the large differences in grain sizes between the two samples.

Interestingly, μ for bulk metallic glasses is of the order of 0.1 (Zhang, Eckert, and Schultz, 2003a), which is roughly an order of magnitude lower than those of the MAX phases. This conclusion is consistent with the fact that the fracture surfaces of bulk metallic glasses are significantly smoother than those of the MAX phases.

Inset in Figure 9.7b shows a typical shear band formed when a CG Ti_3AlC_2 sample was loaded in compression. A higher magnification SEM micrograph of the shear band (Figure 9.7b) shows the extent of damage to the grains in the vicinity of the shear band. In that region typical MAX phases' energy absorbing mechanisms – grain bending, grain decohesion, pull-out and push in, crack deflections at grain boundaries, and KBs – are evident. On the basis of this micrograph, it is reasonable to assume that the deformation of the grains, and the concomitant ligaments that reach across the shear plane, are what prevent a sudden loss in load bearing capability, demonstrated in Figure 9.7a. In general, the propensity for shear band formation increases with increasing grain size and reduced strain rates.

More recently (Bei *et al.*, 2013) also demonstrated that in addition to the typical energy absorbing mechanisms listed above, transgranular cracks, in which the lamella fail in an apparently brittle fashion (Figure 9.8a and b), play a role. KBs are also ubiquitous.

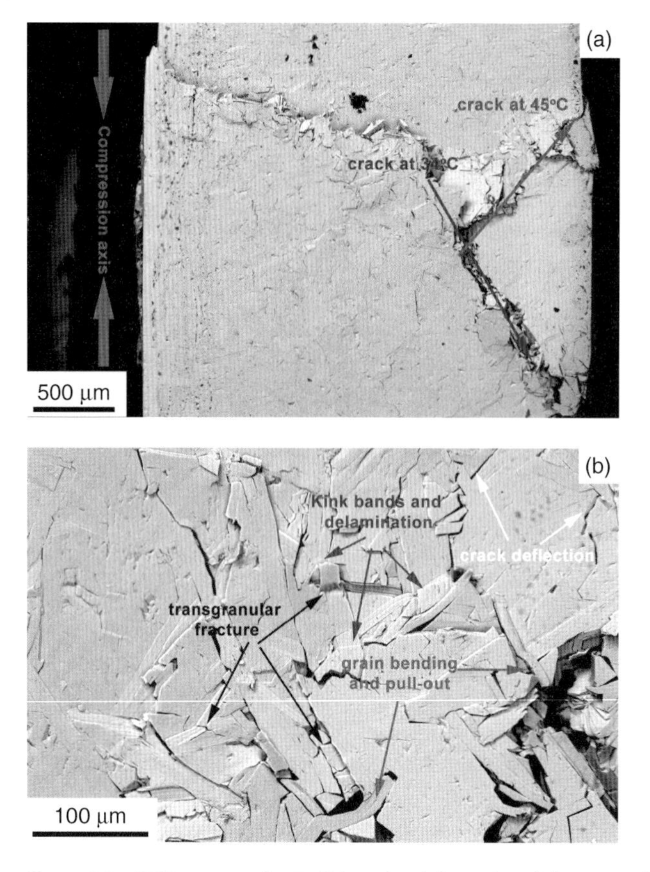

Figure 9.8 SEM images of a Ti_3AlC_2 cube deformed to failure at ambient temperatures at, (a) low magnification showing the shear bands formed and (b) higher magnification with focus on region comprising the shear band. Note multiple energy dissipating mechanisms (Bei *et al.*, 2013).

9.3.2 Effect of Grain Size and the Hall–Petch Relationship

A perusal of Table 9.1 makes it amply clear that, like in most crystalline solids, finer grained materials are stronger than their CG counterparts. As pointed out early on (El-Raghy *et al.*, 1999), this relationship can be explained in one of two possible ways. The first is to invoke the Griffith criterion; the second is to invoke Hall–Petch.

For the former, typically the failure strengths of ceramics, σ_f, are related to their fracture toughness, K_{1c}, assuming:

$$\sigma_f = \xi \frac{K_{1c}}{\sqrt{\pi a}} \tag{9.6}$$

where a is the flaw or crack size. ξ is a geometric factor that is a function of testing method and is of the order of unity for flexure and $\approx$15 for compression (Barsoum, 2003). Assuming a UCS of 1 GPa, and conservatively a K_{1c} of 5 MPa $m^{1/2}$ (Section 9.8), then according to Eq. (9.6), the resulting flaw size would be $\approx$120 μm, which is at least an order of magnitude larger than the typical grain sizes ($\approx$5 μm) of samples that fail at 1 GPa. It is thus unlikely that the Griffith criterion is applicable.

In the metallurgical literature, on the other hand, it has long been established that the yield points follow the empirical Hall–Petch relationship, namely:

$$\sigma = \sigma_o + \frac{k_o}{\sqrt{d}} \tag{9.7}$$

where σ_o and k_o are constants and d is the grain size. It is for this reason, for example, that when the values of Ω/b, or the critical resolved shear stress (CRSS), obtained from our cyclic loading experiments are plotted versus $1/\sqrt{2\alpha}$ not only is an excellent correlation obtained, but also the least squares fit of the results should, and do, go through the origin (Figure 8.10). Note that such a plot has to go through the origin as the CRSSs of basal plane dislocations in hexagonal materials tend to zero as the grain size increases and the dislocation density decreases (Lane *et al.*, 2011). In other words, for dislocation-free MAX phases, $\sigma_o \approx 0$.

As far as we are aware there have been no systematic studies that have attempted to probe whether Eq. (9.7) was valid for the MAX phases. The closest is a study by Hu *et al.* (2008a) who explored the effects of grain size on the mechanical properties of V_2AlC and simply noted that the strengths decreased with increasing grain size. Figure 9.9a replots their UCSs results as a function of $1/\sqrt{D}$ (blue squares in Figure 9.9a) and $1/\sqrt{W}$ (red solid circles in Figure 9.9a), where D and W are the average grain diameters and thicknesses, respectively. From this plot it is clear that indeed Eq. (9.7) is obeyed. It is thus reasonable to conclude that the UCSs of the MAX phases follow the Hall–Petch relationship with $\sigma_o \approx 0$. These comments notwithstanding, this conclusion has to be considered tentative until more careful results are obtained on the effects of grain size on compressive strengths. Note that the V_2AlC samples all failed by the formation of shear bands (C. Hu, Private communication).

Interestingly, when the flexural strengths (Section 9.5) are plotted on the same plot (solid triangles and open crosses in Figure 9.9a), the relationship is not as good for reasons that are not clear, but probably have to do with the complex state of stress that evolves as a bar is flexed to failure.

In summary, when polycrystalline MAX phase samples are loaded in compression they fail either by the formation of shear bands that are inclined 25–40° to the loading direction or in a brittle fashion. The latter is more prevalent when the UCSs and strain rates are high and the grains are small. The Hall–Petch relationship appears to be applicable to the UCSs despite the fact that failure occurs mostly by the formation of shear bands.

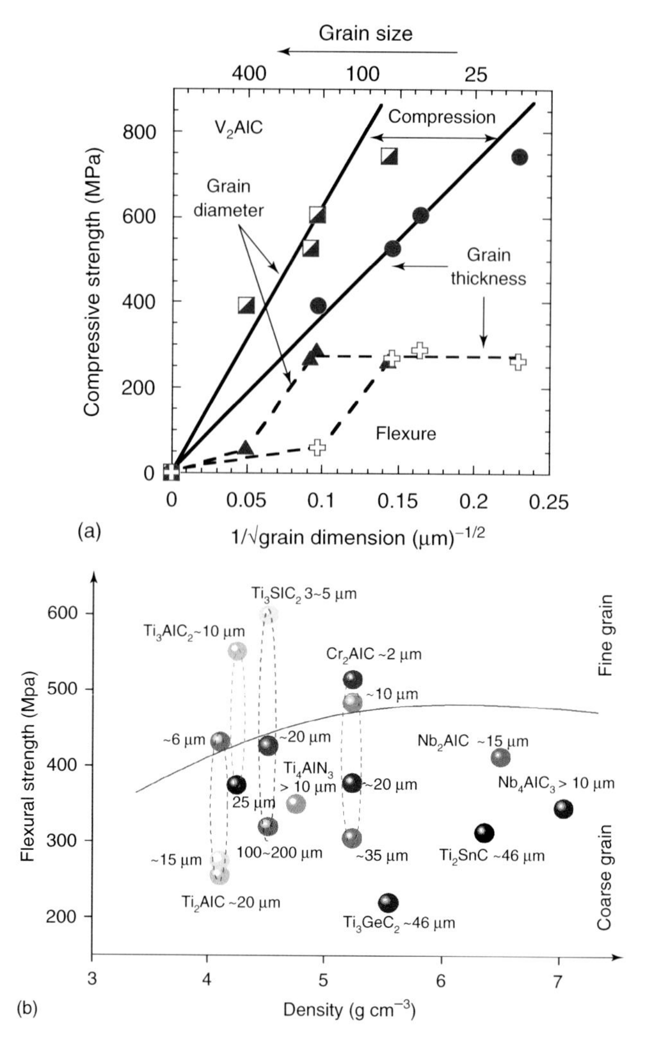

Figure 9.9 Dependencies of, (a) UCSs of polycrystalline V_2AlC samples on $1/\sqrt{D}$ (blue squares) and $1/\sqrt{W}$ (red circles). Also plotted are dependencies of three-point flexural strengths on $1/\sqrt{D}$ (solid blue triangles) and $1/\sqrt{W}$ (open crosses). Dashed lines are guides to the eye. Data taken from Hu *et al.* (2008a). (b) Flexural strengths of select MAX phases on grain size (Li *et al.*, 2011). In all cases, coarser-grained samples failed at lower stresses.

9.3.3
Anisotropic Effects

Before moving on it is worth noting the effects of anisotropy on compressive strengths. Bao *et al.* (2004) measured the compressive strengths parallel to and normal to the hot pressing direction and found that when the applied load on polycrystalline Ti_3AlC_2 samples was parallel to the hot-pressing direction the UCS, at 750 MPa, was lower than the 840 MPa obtained when the applied load was normal to the hot pressing direction (compare sets A and B in Table 9.2).

9.3.4
Confined Deformation

Like other layered solids, such as mica and graphite, when the deformation is constrained, both the yield points and the strains to failure are greatly enhanced. The effect of confinement on the compressive strengths of Ti_3AlC_2 and $Ti_3Al_{0.8}Sn_{0.2}C_2$ was recently studied by Bei (2011). By applying a confining pressure during the compression of Ti_3AlC_2 polycrystalline cylinders, the UCS increased from around 550 MPa to almost 1.2 GPa. The yield point of the $Ti_3Al_{0.8}Sn_{0.2}C_2$ composition was increased from 840 MPa to over 1.4 GPa.

As importantly the confining pressure reduced the propensity for shear banding (Figures 9.7b and 9.8) leading to a more uniform distribution of deformation and damage. The strain to failure was also increased from essentially zero to $>6\%$. The simplest explanation for these observations is that the hydrostatic pressure applied close the cracks that tend to form upon loading.

9.4
Response of Polycrystalline Samples to Shear Stresses

The fact that the MAX phases fail in compression by the formation of shear bands implies that their resistance to shear is low. There have been few studies on the failure of the MAX in shear. Bao *et al.* (2004) compared the failure of polycrystalline Ti_3AlC_2 samples (20–50 μm diameter grains, 5–8 μm thick) in three different configurations: uniaxial compression, double-notched, and punch shear tests. At $\approx 350 \pm 50$ MPa, the τ_c values for the compression samples (Table 9.2), were found to be significantly higher than those obtained for the double notched (100 ± 25 MPa) or the punch test (140 ± 12 MPa) samples. The authors concluded that the punch test was the best measure of shear resistance. Other advantages of this method are its simplicity, both in terms of sample preparation and data analysis.

Given the layered nature of the MAX phases it is not surprising that one of their weakest failure modes is by shear. Table 9.1 lists a few other shear strengths of select MAX phases measured using the punch test. The values range from a high of ≈ 250 MPa for Ta_4AlC_3 to a low of about 110 MPa for Ta_2AlC (Hu *et al.* 2007 and 2008b). At this time, there has been no systematic study of the effect of grain size on the shear strengths, but a perusal of the available results indicates that, yet one more time, the shear strengths increase with decreasing grain size.

9.5
Response of Polycrystalline Samples to Flexure Stresses

The flexural strengths of the MAX phases are typically ≈50% lower than the UCSs. Figure 9.9b and Table 9.1 summarize the range of flexural stresses reported for select MAX phases. The flexural strengths range from lows of about 200 MPa to a highs of about 600 MPa.

9.5.1
Grain Size Effects

As clearly shown in Figure 9.9b, the flexural strengths are functions of grain size; finer grains result in higher flexural strengths. However as noted above, when the flexural strengths of the V_2AlC samples were plotted versus grain diameters or grain thicknesses (Figure 9.9a) the relationship was complicated. It follows that how the grain size determines the strengths of the MAX phases in flexure is not clear at this time.

9.5.2
Anisotropic Effects

Neither Table 9.1 nor Figure 9.9b include some exciting recent results on highly textured Nb_4AlC_3 samples (Hu *et al.*, 2011a,b). Interestingly, the flexural strengths were a weak function of orientation. When the samples were tested such that the loading direction was parallel to the *c*-axis, the three-point flexural strength were ≈1200 ± 300 MPa. When the basal planes were loaded edge on, the flexural strength, at ≈1200 ± 100 MPa, was statistically indistinguishable from the other orientation, albeit with smaller scatter. When un-textured hot pressed samples were tested, at ≈350 MPa, the flexural strengths were significantly lower.

These are quite intriguing results that the authors ascribe to orienting the grains. However, there are a number of factors that need to be taken into account before this interpretation hardens into conventional wisdom. The grain sizes of the two sets of samples were quite different. The grains in the textured samples were only ≈2 μm in diameter and more or less equiaxed (see Fig. 9.10), while those in the hot pressed samples were ≈50 μm in diameter and ≈17 μm thick. The grain refinement was ascribed to the presence of ≈15 vol% Nb–Al–O particles that were distributed homogeneously in the grains and at the grain boundaries. In other words, as the authors concede, that in addition to texture, grain size plays a crucial role.

It is also important to note that the deflections of the propagating cracks occurred at the millimeter scale (Figure 9.10a) despite the fact that the grains were ≈2 μm in diameter (Figure 9.10b). Indeed the failure mode (Figure 9.10a) is more reminiscent of laminated ceramic composites than polycrystalline ceramics. It would thus

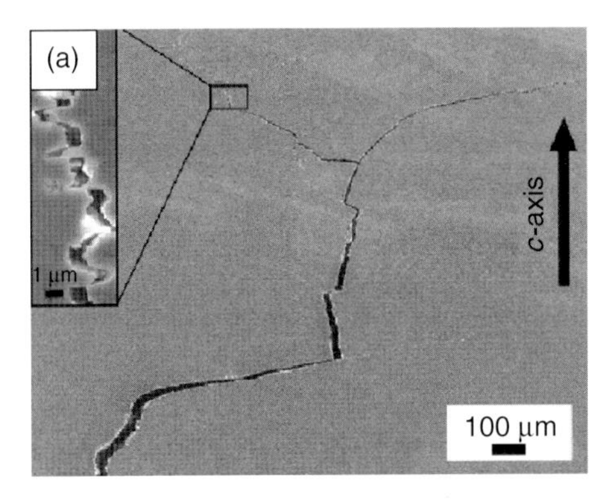

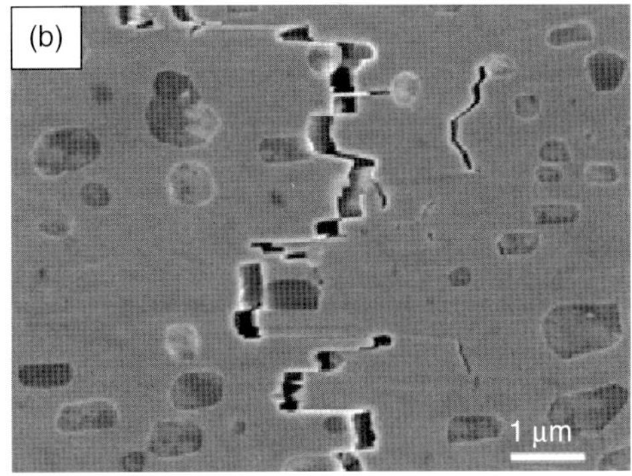

Figure 9.10 Typical SEM micrographs of the crack fracture paths in highly oriented Nb_4AlC_3 samples at, (a) low magnification and (b) higher magnification (Hu *et al.*, 2011b). Inset in (a) shows higher magnification of region outlined by the small rectangle in (a).

appear that the combination of grain refinement together with texturing that leads to laminations, at a different length scale – compare Figure 9.10a,b – resulted in these record values of flexural strengths. The effect of aligning the grains on fracture toughness is discussed below.

In a more recent paper, the same group (Hu *et al.*, 2011b) were able to grow textured samples, with only 3 vol% Nb–Al–O particles. This resulted in slightly coarser grains ($\approx$5 μm *D*, 1 μm *W*) and concomitantly lower flexural strengths (880 MPa parallel to, and 790 MPa perpendicular to the *c*-axis, i.e., edge on). In other words, in this case orientation made a difference; when the basal planes were loaded edge on, the samples failed at lower stresses. At this time, the information available is not sufficient to make any general conclusions. More systematic work is needed to explore the effects of highly orienting the grains on the mechanical properties of the MAX phases.

9.6
Response of Polycrystalline Samples to Tensile Stresses

In contrast to their response to compressive or flexural stresses, the tensile properties of the MAX phases have not been as well documented. However, akin to other structural ceramics, the ultimate tensile strengths (UTSs), are the lowest and range between 100 and $\approx$300 MPa (Table 9.1) (Radovic *et al.*, 2000, 2002). In all cases, larger grained samples result in weaker solids. The failure is brittle, with little to no ductility. It follows that two of the main challenges facing the MAX phase community that need to be addressed before these solids are used in structural applications are: (i) enhancing the UTSs and (ii) endowing them with at least some ductility at ambient temperatures.

9.7
Hardness

9.7.1
Vickers Hardness

Unlike their MX cousins, the MAX phases are relatively soft and exceedingly damage tolerant. The Vickers hardness, H_v, values of polycrystalline MAX phases fall in the range of 2–8 GPa (Table 9.3). They are thus softer than most structural ceramics, but harder than most metals.

Working with chemically vapor deposited (CVD) single crystals, Nickl, Schweitzer, and Luxenberg (1972) were the first to note that the hardness of Ti_3SiC_2 was anisotropic, relatively low, and higher when loaded along the *c*-direction. This result was later confirmed by Goto and Hirai (1987), who were the first to show that the hardness was a function of indentation load.

On the basis of this early and more recent work, the following characteristics are now established for the hardness of the MAX phases:

1. With decreasing load, H_v increases (Figure 9.11a); below a certain load it is not measurable, as no trace of the indentations is found (Murugaiah *et al.* 2004; El-Raghy *et al.*, 1997).
2. H_v is a function of grain size. Finer-grained materials are harder than their coarser-grained counterparts (Figure 9.11a,b).
3. The finer the grains, the less pronounced is the decrease in H_v with load. This is manifest in Figure 9.11a in which H_v of Ti_3SiC_2 is plotted as a function of load (Kuroda *et al.*, 2001). The same is true of Cr_2AlC as shown in Figure 9.11b (Li *et al.*, 2011; Lin *et al.*, 2005; Ying *et al.*, 2011a).

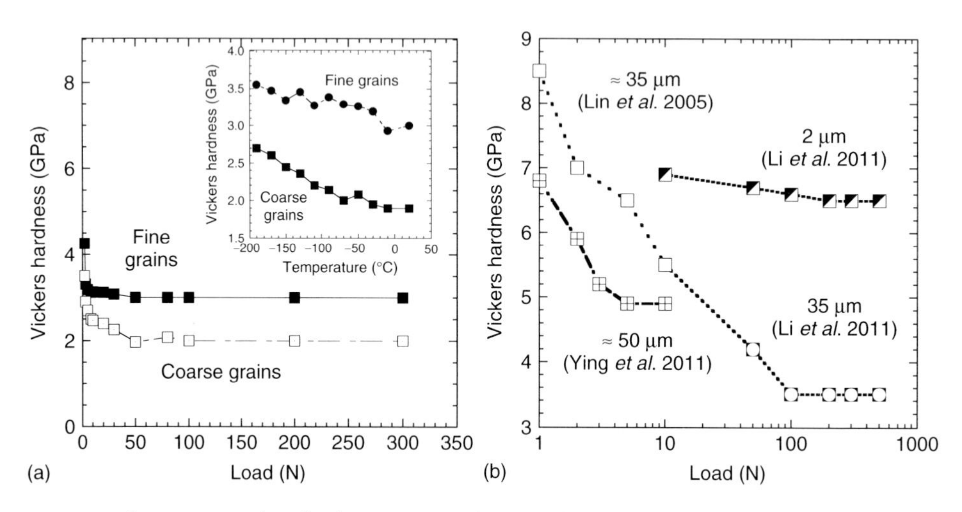

Figure 9.11 Vickers hardness versus indentation load as a function of grain size for, (a) Ti_3SiC_2. Inset plots effect of temperature on H_v (Kuroda *et al.*, 2001). (b) Cr_2AlC (Li *et al.*, 2011; Lin *et al.*, 2005; Ying *et al.*, 2011a). Note semi-log nature of plot (b).

Table 9.3 Summary of Vickers hardness, V_H, values of select MAX phases. For the grain sizes, when available, both the diameter, D and the width, W, of the hexagonal plates are listed. When not specified, the values given are those for the average diameters, D, of the grains.

Compound	Grain size (μm)	H_v	Load (N)	References
		413 phases		
$Ti_4AlN_{2.9}$	20	2.5	100	Procopio, Barsoum, and El-Raghy (2000)
Nb_4AlC_3	50*D*; 17*W*	2.6 ± 0.2	200	Hu *et al.* (2008c)
Nb_4AlC_3	21*D*; 9*W*	3.7 ± 0.5	10	Hu *et al.* (2009)
Ta_4AlC_3	10*D*; 3*W*	5.1	50	Hu *et al.* (2007)
		312 phases		
Ti_3SiC_2	150*D*; 25*W*	6	10	Goto and Hirai (1987)
Ti_3SiC_2	—	6	100	Lis *et al.* (1993)
Ti_3SiC_2	3–5	4	100	El-Raghy *et al.* (1997)
Ti_3SiC_2	3–5	2	300	Low (1998)
Ti_3SiC_2	50–200	2.5	1000	Low *et al.* (1998)
$Ti_3Si(Al)C_2$	15*D*; 4*W*	3	10	Wan *et al.* (2008)
$Ti_3Si(Al)C_2$	50*D*; 16*W*	4 ± 0.1	10	Wan *et al.* (2008)
Ti_3AlC_2	25*D*	3.5	300	Tzenov and Barsoum (2000)
Ti_3AlC_2	30*D*; 6*W*	3.5 ± 0.2	10	Wan *et al.* (2008)
Ti_3AlC_2	74*D*; 20*W*	2.2 ± 0.2	10	Wan *et al.* (2008)
Ti_3GeC_2	FG	5.0	10	Barsoum, Brodkin, and El-Raghy (1997)
		211 phases		
Ti_2AlC	100–300	5.5	10	Barsoum, Brodkin, and El-Raghy (1997)
Ti_2AlC	25	4.5	10	Barsoum, Ali, and El-Raghy (2000)
Ti_2AlC	≈50	2.8	10	Wang and Zhou (2002b)
$Ti_2AlC_{0.7}$	6	5.8	50	Bai *et al.* (2009)
Ti_2AlN	≈100	3.5	10	Barsoum, Brodkin, and El-Raghy (1997)
Ti_2AlN	100	4.0	10	Barsoum, Ali, and El-Raghy (2000)
$Ti_2AlC_{0.5}N_{0.5}$	25	5.5	10	Barsoum, Ali, and El-Raghy (2000)
Ti_2SC	3–5	7.5	300	Amini, Barsoum, and El-Raghy (2007)
Ti_2SC	10–20	6	300	Amini, Barsoum, and El-Raghy (2007)
Ti_2SnC	NA	3.5	10	El-Raghy, Chakraborty, and Barsoum (2000)
Ti_2GeC	FG	5.5	10	Barsoum, Brodkin, and El-Raghy (1997)
V_2AlC	75*D*; 30*W*	2.5 ± 0.5	100	Hu *et al.* (2008a)
Cr_2AlC	—	5.5 ± 0.5	10	Lin *et al.* (2005)
Cr_2AlC	20	3.5	50	Tian *et al.* (2006)
Cr_2AlC	5	5.2	10	Tian *et al.* (2007)
Cr_2AlC	10	5.2	1	Zhou, Mei, and Zhu (2009)

(continued overleaf)

Table 9.3 (*Continued*)

Compound	Grain size (μm)	H_v	Load (N)	References
Cr_2AlC	35	3.5	500	Li *et al.* (2011)
Cr_2AlC	2	6.4	500	Li *et al.* (2011)
Cr_2AlC	10–40	4.9	10	Ying *et al.* (2011b)
Cr_2GeC	20 ± 10	2.5	200	Amini *et al.* (2008)
Cr_2GaC	≈ 10	1.4	2	Sun (2011)
Nb_2AlC	14	6.5	10	Salama, El-Raghy, and Barsoum (2002)
Nb_2AlC	17	4.5	300	Zhang *et al.* (2009)
NbTiAlC	16	5.5	10	Salama, El-Raghy, and Barsoum (2002)
Nb_2SnC	—	3.8	10	El-Raghy, Chakraborty, and Barsoum (2000)
Hf_2SnC	NA	3.8	10	El-Raghy, Chakraborty, and Barsoum (2000)
Hf_2PbC	NA	3.8 ± 0.7	10	El-Raghy, Chakraborty, and Barsoum (2000)
Ta_2AlC	15*D*; 3*W*	4.4	50	Hu *et al.* (2008b)
Zr_2SnC	NA	3.9 ± 0.3	10	El-Raghy, Chakraborty, and Barsoum (2000)
Zr_2PbC	NA	3.2 ± 0.5	10	El-Raghy, Chakraborty, and Barsoum (2000)
Zr_2SC	10	6.5	200	Opeka *et al.* (2011)

4. The change in H_v with temperature is mild, at least in the 77–300 K temperature range (inset in Figure 9.11a) (Kuroda *et al.*, 2001). This is an important result as it implies that dislocations are mobile and multiply at temperatures as low as 77 K. It also implies the dislocations' CRSSs, are most probably not thermally activated; if they are, then the activation energy is low.

In the late 1980s to mid-1990s, Pampuch and coworkers (Lis *et al.*, 1993; Pampuch and Lis, 1995; Pampuch *et al.*, 1989) were the first to note that, in sharp contradistinction to other structural ceramics, it was quite difficult to induce cracks from the corners of Vickers indentations in Ti_3SiC_2. Instead they observed large pileups – not unlike those shown in Figure 9.12a – around indentations that suggested some ductility. Pampuch *et al.* were also the first to note that at about 0.012, the H_v/E ratio of Ti_3SiC_2 was closer to low ductility, high-strength metals than to ceramics.

Working with Ti_3SiC_2, on the other hand, we were the first to document a number of important energy absorbing mechanisms around Vickers indentations (El-Raghy *et al.*, 1997). These include diffuse microcracking, delamination, crack deflection, grain pull-out (Figure 9.12b), and the kinking of individual grains (Figure 9.12c).

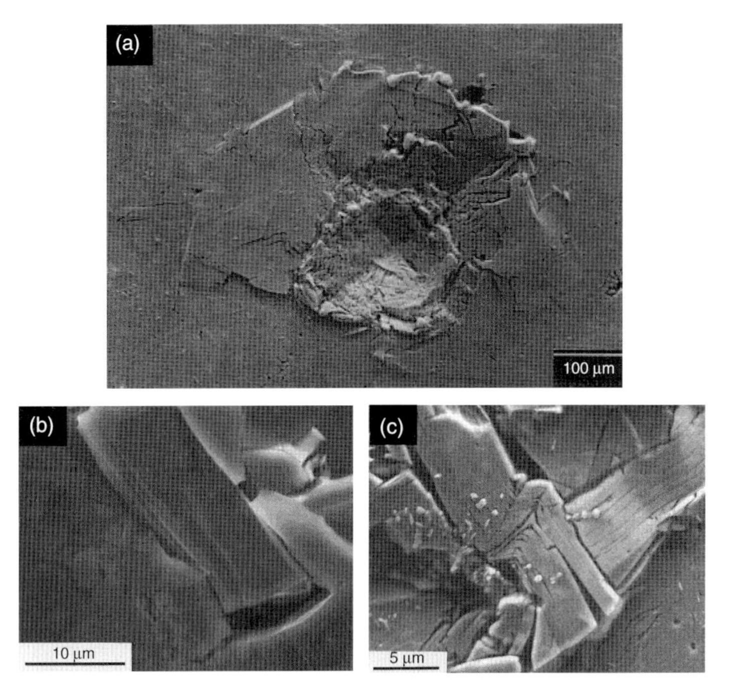

Figure 9.12 SEM micrographs of, (a) Vickers indentation mark made in $Ti_3SiC_2$with a load of 100 N at room temperature; (b) grain pull-out and deformation, and (c) multiple kinking of a grain (El-Raghy *et al.*, 1997).

The pileup of material around the indentation mark, together with the aforementioned energy absorbing mechanisms and the lack of cracks emanating from the corners of Vickers indent, can be considered three of the defining characteristics of the MAX phases. This response has been observed in most, if not all, MAX phases studied to date. This comment notwithstanding, in some cases, especially for FG MAX phases, small cracks have been observed to emanate from the corners of Vickers indentation marks (Amini, Barsoum, and El-Raghy, 2007).

Lastly in this section it is worth noting that, compared to H_v values of other MAX phases, those for Ti_2SC and Zr_2SC – when all other factors such as grain size and indentation load are taken into consideration – are significantly higher. Given the low values of the c-parameter in these S-containing MAX phases, these result are not too surprising and were anticipated (Barsoum, 2000).

9.7.2
Nanoindentation

Given their plastic anisotropy, the response of the MAX phases to nanoindentation (NI) is also anisotropic; when the basal planes are parallel to the surface, the extent of plastic deformation is higher and the hardness is lower than if the basal planes are loaded edge-on (compare Figure 9.13a,b) (Kooi *et al.*, 2003; Murugaiah *et al.*,

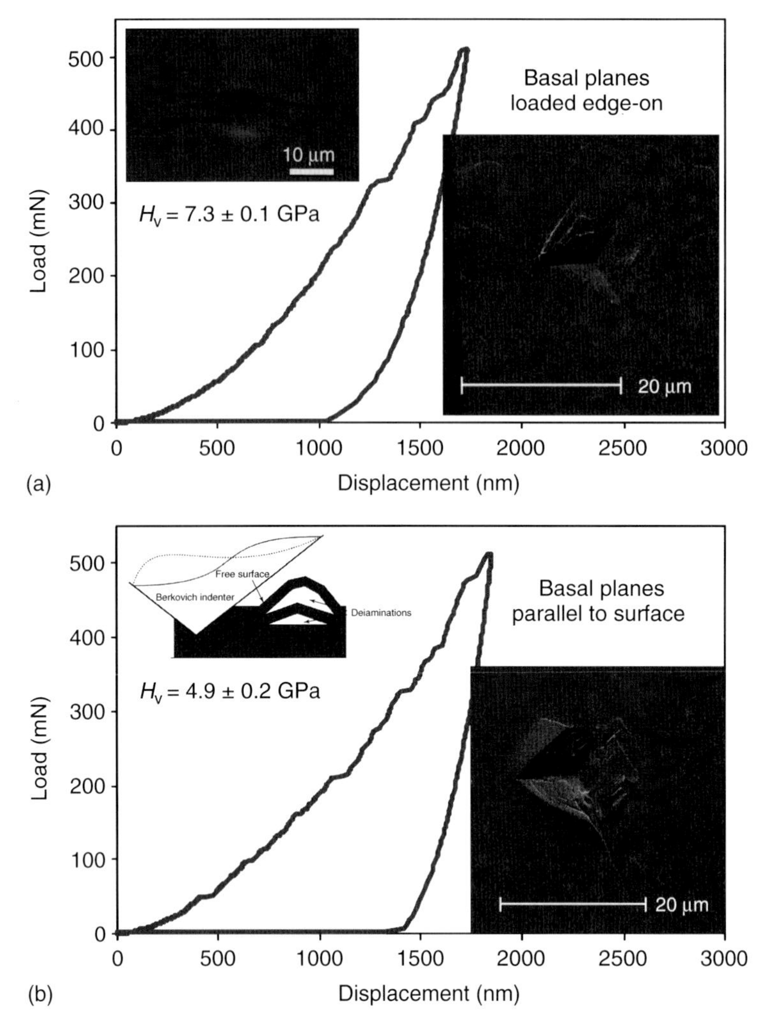

Figure 9.13 Load–displacement curves of a Berkovich indentation in a Ti_3SiC_2 grain with the basal planes, (a) perpendicular and (b) parallel to the surface (Kooi *et al.*, 2003). Right inset in (a) is SEM image of the indentation in a Ti_3SiC_2 grain with the basal plane perpendicular to the surface made with a Berkovich indenter. Left inset is the same made with a spherical nanoindenter (Murugaiah *et al.*, 2004). In both cases, there is little to no pileup. Instead larger cracks that are parallel to the basal planes form; smaller cracks normal to the larger ones also form. Right inset in (b) is SEM image of a Berkovich indentation in a Ti_3SiC_2 grain with the basal plane parallel to the surface. Left inset in (b) is a schematic of how an indentation results in delamination and kink band formation on the surface for Ti_3SiC_2 when the basal planes are parallel to the surface (Kooi *et al.*, 2003).

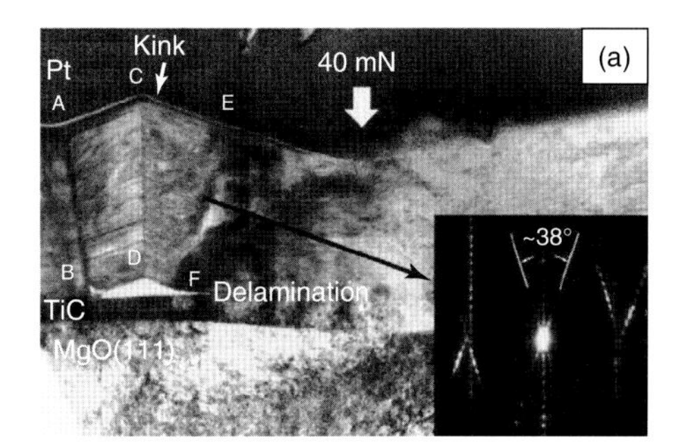

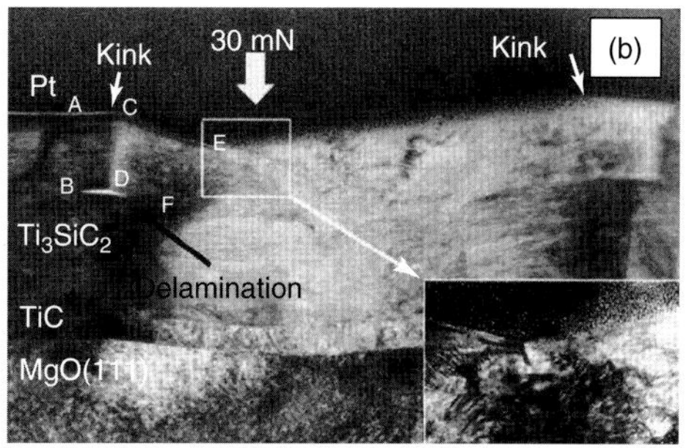

Figure 9.14 Cross-sectional TEM images of Berkovich indents made in epitaxial thin films, where the basal planes are parallel to the surface of Ti_3SiC_2 at maximum loads of, (a) 40 mN and (b) 30 mN. The SAD pattern in (a) was taken from the kink band. Note clear evidence for kink band formation, delaminations and resulting pileups (Molina-Aldareguia *et al.*, 2003).

2004). In the former case, it is easier to form KBs because the top surface is unconstrained. A schematic of how the pileup around the indentation develops is shown schematically in the left inset in Figure 9.13b. To form the pileup, both delaminations and kinking are required.

When the basal planes are loaded edge-on (Figure 9.13a), they tend to delaminate *without* a pileup and in a manner that is totally consistent with the lack of obvious kinking (insets in Figure 9.13a). It is more likely than not that the layers right under the indentation mark are heavily deformed and kinked.

The relationship between kinking and the shape of the indentations that form under a Berkovich indenter when the basal planes are parallel to the surface can most clearly be seen in the cross-sectional transmission electron microscopy (TEM) micrograph shown in Figure 9.14a,b as a function on indentation load (Molina-Aldareguia *et al.*, 2003). When the indentation load was 40 mN, the basal planes tilt by $\approx 19^\circ$ around $[1\bar{1}00]$ across the AB kink boundary, and by $\approx 38^\circ$ across the CD kink plane (Figure 9.14a). When the indentation load was 30 mN, the tilt across the AB plane was just a few degrees (Figure 9.14b); that across the CD kink plane was half that for the 40 mN indent. In other words, increasing the load increases the severity of the kinking and the concomitant pile-up of material around the indenter. The SAD shown in inset in Figure 9.14a shows direct evidence for lattice rotations. Furthermore, the area directly beneath the indented surface was also characterized by a high density of Moire fringes (not shown) consistent with the superimposition of regions of different orientations.

9.7.3
Repeated Spherical Indentations and IKBs

A few years ago we showed that repeated NIs, into a given location, was a powerful, and relatively simple, nondestructive technique to demonstrate that a solid is KNE. Typical results are shown in Figure 9.15a, where the load displacement curves for

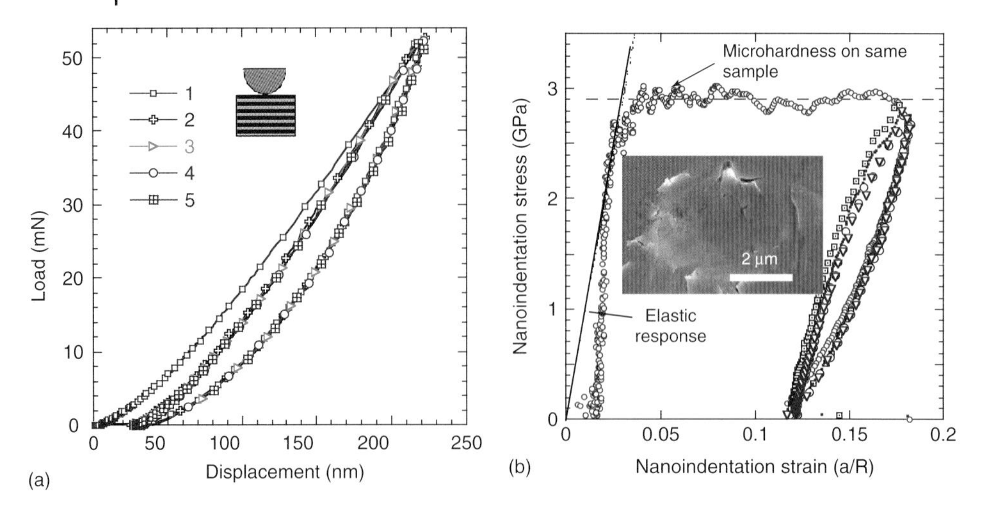

Figure 9.15 Typical (a) load–displacement curves for a polycrystalline Ti_2AlC sample loaded five times in the same location to 50 mN using a 13.5 μm hemispherical indenter. (b) Corresponding stress–strain curves. After an elastic regime, plastic deformation at ≈3 GPa results in a 10% increase in a/R. Upon unloading and reloading, fully and spontaneously reversible, stress–strain loops – characteristic of KNE solids – are obtained. Inset shows a SEM micrograph of indented region. (Zhou *et al.*, 2006a).

a Ti_2AlC sample loaded five times into the same location with a hemi-spherical indenter are shown. In such experiments, the first cycle is typically open, but all subsequent cycles are fully and spontaneously reversible.

One advantage of using spherical indenters is that it is possible to convert the load displacement curves to their corresponding NI stress–strain curves (Basu, Moseson, and Barsoum, 2006; Moseson, Basu, and Barsoum, 2008). Figure 9.15b plots the NI stress–strain curves for the results shown in Figure 9.15a, from which it is clear that:

1. Initially and during the first loading cycle the response is linear elastic.
2. At about 3 GPa, the response becomes almost perfectly plastic. Coincidentally or not the H_v of this sample (depicted by horizontal dashed line in Fig. 9.15b) was 3 GPa (Zhou *et al.*, 2006a).
3. The plastic deformation results in an a/R value of ≈10%. Here the strain is defined as a/R, where a is the contact radius and R is the indenter diameter.
4. Upon unloading and reloading fully and spontaneously reversible, stress–strain loops – characteristic of KNE solids – are obtained. Note that since the domains that form during the first cycle of plastic deformation are small, the threshold stress for incipient kink band (IKB) nucleation is now ≈1.5 GPa. The threshold stress can be equated to the point in the loading curves of cycles 2 and higher deviate from linearity.

To shed further light on what occurs during the first cycle, it is instructive to review what happens to another layered solid, viz. graphite single crystals, when loaded parallel to their basal planes. During the first cycle, the plastic deformation that occurs at a given yield point and/or during massive pop-ins, the single crystal breaks up into much smaller domains (Figure 9.16a). The domains are typically delineated by kink boundaries. Upon reloading, it is presumed that IKBs nucleate within the freshly created domains (Figure 9.16b). But since now the domain sizes are quite small, the thresholds stresses for the nucleation of IKBs can be quite high. Here graphite was used to illustrate this phenomenon because the formation of the smaller domains after the first cycle is unambiguous. There is no reason to believe that the same is not occurring in the MAX phases. Because the latter, however, are more ductile than graphite the breakup is not apparent.

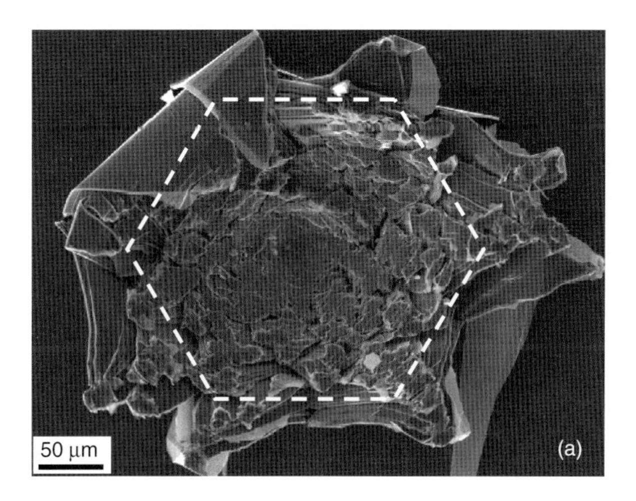

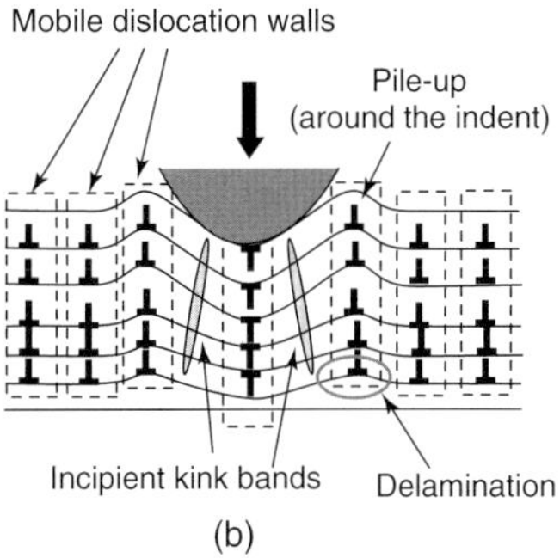

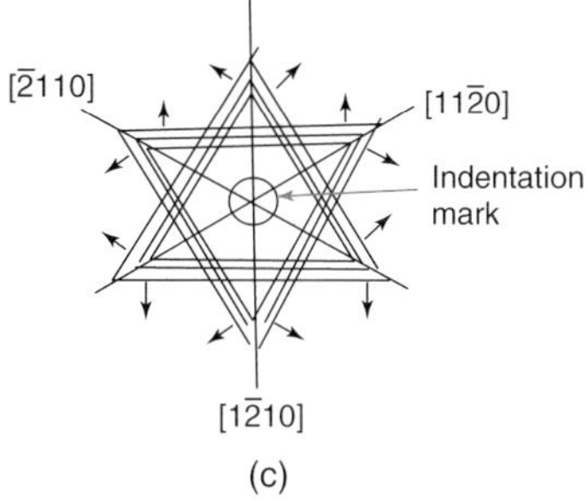

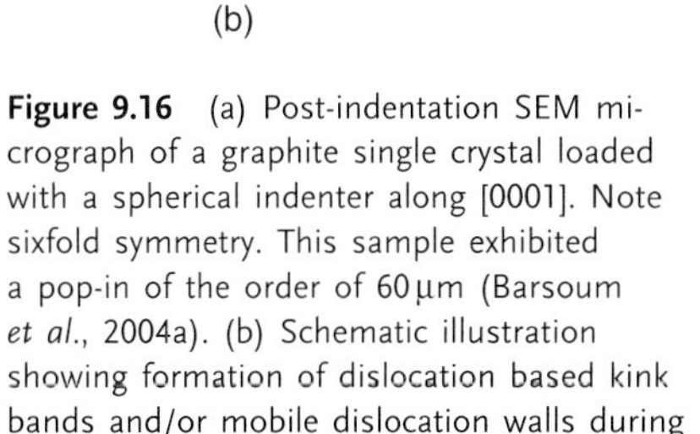

Figure 9.16 (a) Post-indentation SEM micrograph of a graphite single crystal loaded with a spherical indenter along [0001]. Note sixfold symmetry. This sample exhibited a pop-in of the order of 60 μm (Barsoum *et al.*, 2004a). (b) Schematic illustration showing formation of dislocation based kink bands and/or mobile dislocation walls during spherical indentation of a hexagonal crystal along [0001]. The horizontal lines represent basal plane dislocation arrays/pileups. (c) Schematic illustration of how mobile dislocation walls radiate away from the indentation mark – denoted by a small circle (Basu and Barsoum, 2007). These MDWs end up forming a Star of David shaped defect that was observed by Bradby *et al.* in ZnO (2002).

As important, relatively far away from the indentation mark (denoted by a small circle in Figure 9.16c) Bradby *et al.* observed a Star of David shaped defect when single crystals of ZnO were indented with a 4.2 μm hemi-spherical indenter (Bradby *et al.*, 2002). On the basis of the ideas propounded in Chapter 8, it is reasonable to assume that the defects observed by Bradby *et al.* are MDWs that were nucleated under the indenter. Repulsion between MDWs of the same sign should then push them radially outward to form the hexagonal net seen in Figure 9.16c. The hexagonal symmetry of the indentation mark shown in Figure 9.16a, presumably arises from the same physics. Note that as the size of the defect observed by Bradby *et al.* was significantly larger than the indentation mark (Figure 9.16c), it follows that whatever defects formed must have been mobile over relatively large distances, indirectly proving that the MDWs are indeed mobile. The relationship between KB formation (Figure 9.16b) and the cross-sectional TEM micrographs (Figure 9.14) should also be clear at this time.

In short, probably the simplest technique to show that a solid is KNE is to repeatedly thrust a spherical nanoindenter into the same location. If fully, and spontaneously reversible, load–displacement loops are obtained, whose areas are not a function of cycle number, one can reasonably conclude that a solid is KNE. And while the aforementioned results apply to ZnO and graphite, there is no reason to believe they would be any different for other hexagonal solids in general, and other MAX phases in particular. In most of our NI work, hexagonal single crystals were indented along [0001]), and we have repeatedly shown that – for numerous solids, including graphite (Barsoum *et al.*, 2004a), mica (Barsoum *et al.*, 2004b; Basu and Barsoum, 2009), ZnO (Basu and Barsoum, 2007), sapphire (Basu, Barsoum, and Kalidindi, 2006), $LiNbO_3$ (Basu, Zhou, and Barsoum, 2008), and others (Barsoum and Basu, 2010) – this response is characteristic of KNE solids.

9.7.4 Other Hardness Related Complications

Another wrinkle in what constitutes hardness in the MAX phases has recently been exposed by Bei *et al.* (2012). These authors, working with Ti_3AlC_2 and $Ti_3Al_{0.8}Sn_{0.2}C_2$, showed that when the locations of the nanoindenter were restricted to a single grain – that is, kept away from grain boundaries – an indentation size effect, wherein the hardness decreased with increasing indentation depth, was observed (squares in Figure 9.17). They also showed that the hardness reached limiting values of 11.4 ± 0.7 and 10.2 ± 0.6 GPa for Ti_3AlC_2 and $Ti_3Al_{0.8}Sn_{0.2}C_2$, respectively. The authors labeled this limiting value the *intrinsic* hardness. When the hardness of the same surface was measured over a larger area, there was a distinct drop in hardness (denoted by triangles in Figure 9.17). This drop was ascribed to the presence of grain boundaries.

When the results of Kooi *et al.* (2003) for Ti_3SiC_2 grains loaded along [0001] are plotted on the same plot (solid circles in Figure 9.17) as those for Ti_3AlC_2 it is reasonable to conclude that the intrinsic hardness for the latter is lower than that of Ti_3AlC_2. Why that should be the case is not clear, but it is quite likely that

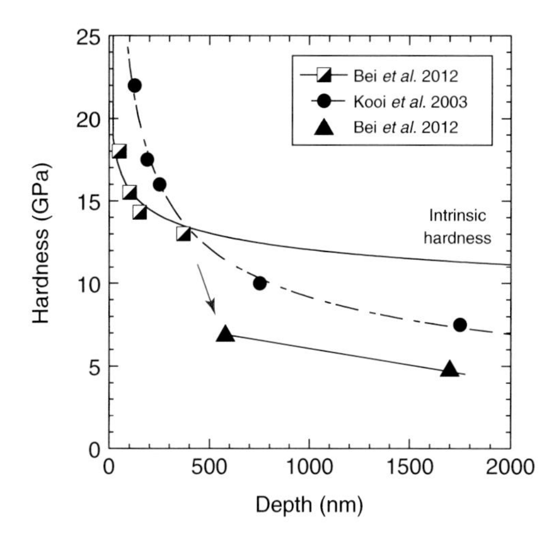

Figure 9.17 Replot of the hardness results of Bei *et al.* (2012) and Kooi *et al.* (2003) for Ti_3AlC_2 and Ti_3SiC_2, respectively. The drop in hardness shown by arrow was attributed to the presence of grain boundaries.

had Bei *et al.* (2012) reported their hardness values of single grains at penetration depths of the order of 1700 nm, their intrinsic hardness values would have been lower. Clearly more work is needed to better understand how and what influences the intrinsic hardness of the MAX phases. This is especially true given that at some level the idea of intrinsic hardness contradicts the fact that FG MAX phases are harder than their CG counterparts (e.g., Figure 9.11). This apparent paradox vanishes when one stipulates that the intrinsic hardness is only applicable for a single crystal or single grain. Said otherwise, when the indentation is confined to a single grain the hardness will be higher than when the indentation spreads across several grains. Another obvious complication is the fact that the intrinsic hardness will be a function of basal plane orientation.

Note that the hardness values measured on MAX thin films are typically much higher than those measured on bulk, polycrystalline solids of the same composition. The reason why is not entirely clear could be related to a Hall–Petch type effect since, typically, the grains in thin films are at the nanoscale (Eklund *et al.*, 2010).

On the basis of the totality of the hardness results and microstructural observations made to date the following scenario for what happens when a MAX phase is indented can be reconstructed. When a polycrystalline sample is loaded, at low loads, the indenter can penetrate into the surface – beyond the elastic limit – and leave no trace when removed (Murugaiah *et al.*, 2004). In that case, the deformation occurs by the formation of fully reversible IKBs under the indenter. At intermediate loads, the indented grain's orientation plays a role. When the deformation occurs parallel to thc basal planes, no pileup of material is observed; instead the basal planes cleave (see top left inset in Figure 9.13a). If, on the other hand, the loading direction is along [0001], that is, parallel to the basal planes, pop-ins are generally observed. For grains with orientations in between these two extremes, pile-ups – that are typically not symmetric – and sometimes cracks are observed. For a given grain size, grain boundaries result in softening since they can – like a free surface – allow kinking to occur more readily.

Another way to look at the problem – and to the mechanical response of the MAX phases in general – is from a confinement point of view. If delaminations and kinking can be suppressed by confinement – whether by the matrix around an indenter or in a triaxial experiment, for example – the MAX phases are quite resistant to deformation. The effect of confinement on the compressive behavior of the MAX phases was discussed above.

When that confinement is lost, these solids will fail at much lower stresses. This conclusion has far-reaching practical implications. It explains, for example, why the MAX phases are significantly weaker in tension than in compression. It also explains why they are quite brittle when formed into thin shapes.

Fracture toughness, damage tolerance, and fatigue resistance are considered next. In each case the properties are presented. The micromechanical reasons the properties are what they are, are discussed separately in Section 9.11.

9.8 Fracture Toughness and R-Curve Behavior

For any material, the fracture toughness, K_{1c}, is given by:

$$K_{1c} = \sqrt{EG_c} \tag{9.8}$$

where E is Young's modulus and G_c is the toughness of the material. The latter is a measure of the energy needed to extend a crack. For purely brittle solids, G_c is twice the surface energy, γ. Assuming K_{1c} is 5 MPa m$^{1/2}$ and E is 300 GPa, results in a γ of about 40 J m^{-2}. As this value is at least an order of magnitude higher than the surface energies of the MAX phases (Chapter 2), it is reasonable to conclude that the fracture is not purely brittle, which is not surprising since dislocations are known to multiply and glide at room temperature.

Fracture will occur when:

$$K_1 = \sigma\sqrt{\pi a} > K_{1c} \tag{9.9}$$

where K_1 is the stress intensity at the crack tip and a is the crack length.

Table 9.4 summarizes the room temperature K_{1c} values, measured using various techniques. The latter range from about 5 MPa m$^{1/2}$ to almost 20 MPa m$^{1/2}$ (Table 9.4). In general, the MAX phases' fracture toughness values – herein the focus is on K_{Ic} exclusively – are quite respectable when compared with other structural ceramics in general and layered ceramics in particular.

El-Raghy *et al.* (1997) were the first to report on K_{1c} for a MAX phase. Using a single edge notched beam (SENB) technique they reported a K_{1c} for Ti_3SiC_2 of 6 MPa m$^{1/2}$. This value was significantly higher than other structural ceramics such as

Table 9.4 Summary of room temperature K_{1c} values for select MAX phases measured using the single-edged notched beam, SENB, chevron notched beam (CNB), or compact tension, CT specimens. R-curve refers to R-curve behavior (see text).

Sample	Test	Grain size (μm)	K_{1c} (MPa m$^{1/2}$)	References
		413 phases		
Nb_4AlC_3	SENB[a]	50*D*; 17*W*	7.1	Hu *et al.* (2008c)
Nb_4AlC_3	SENB[a]	—	12 ± 2 to 18 ± 5	Hu *et al.* (2011a)
β-Ta_4AlC_3	SENB[a]	10*D*; 3*W*	7.7	Hu *et al.* (2007)
		312 phases		
Ti_3SiC_2	SENB	—	6.0	El-Raghy *et al.* (1997)
Ti_3SiC_2	CT	CG	16 (R-curve)	Gilbert *et al.* (2000)
(2 vol% SiC)		FG	9.5 (R-curve)	
Ti_3SiC_2 (<3 vol% TiC)	SENB[b]	FG	4.5 ± 0.1	Li *et al.* (2001)
Ti_3SiC_2	SENB	10	6.2	Wang, Jin, and Miyamoto (2002)
Ti_3SiC_2	SEPB[a]	NA	6.2 ± 3.0	Bao and Zhou (2005)
(7 wt% TiC)	CNB[b]		6.6 ± 0.1	
	SENB[a]		6.6 ± 0.1	
Ti_3SiC_2 (TiC)	SENB[a]		7.2	Li *et al.* (2002)
$Ti_3Si(Al)C_3$	SENB[b]	16*D*; 4*W*	6.2	Zhou *et al.* (2006c)
$Ti_3Si(Al)C_3$	CNB[b]	15*D*; 4*W*	6.4	Wan *et al.* (2008)
		50*D*; 16*W*	6.8	
Ti_3AlC_2	SENB[a]	—	7.2	Wang and Zhou (2002a)
Ti_3AlC_2	CNB[b]	28*D*; 6*W*	7.8 ± 0.1	Wan *et al.* (2008) (Figure 10.1)
		80*D*; 20*W*	9.5 ± 0.1	
Ti_3AlC_2 (TiC)	SENB[a]	—	4.6 ± 0.3	Peng (2007)
		211 phases		
Ti_2AlC	SENB[a]	CG	6.5	Wang and Zhou (2002b)
V_2AlC	SENB[a]	(≈100*D*; 40*W*)	5.7	Hu *et al.* (2008a)
Cr_2AlC	SENB[a]	10	5.8	Zhou, Mei, and Zhu (2009)
Cr_2AlC	SENB[a]	10–40	6.2 ± 0.3	Ying *et al.* (2011b)
Cr_2AlC	SENB[a]	(CG; 35 μm)	6.2	Li *et al.* (2011)
		(FG; 2 μm)	4.7	
Cr_2AlC	SENB[a]	30	6.2	Yu, Li, and Sloof (2010)
$Cr_2(Al_{0.96},Si_{0.13})C$	SENB[a]	50	6.6	Yu, Li, and Sloof (2010)
Nb_2AlC	SENB[a]	17	5.9	Zhang *et al.* (2009)
Ta_2AlC	SENB[a]	15*D*; 3*W*	7.7	Hu *et al.* (2008b)

[a]Three-point bend bars.
[b]Four-point bend bars.

Al_2O_3, Si_3N_4, and SiC, at comparable stages of development. A year later, Zhou *et al.* (1998) reported a K_{1c} value of 7.9 MPa $m^{1/2}$, confirming the relatively high value.

9.8.1
Effect of Grain Size

A perusal of the results listed in Table 9.4, indicate that, in general, like other structural ceramics, the larger the grain size the higher the K_{1c} values. Gilbert *et al.* (2000) were the first to show this effect in Ti_3SiC_2. At 8 MPa $m^{1/2}$, K_{Ic} of the FG microstructure (Figure 9.18a) was lower than the 8.5–11 MPa $m^{1/2}$ of its CG counterpart. More recently, Li *et al.* (2011) showed that K_{1c} of Cr_2AlC increased from 4.7 to 6.2 MPa $m^{1/2}$ as the grain size increased from 2 to 35 μm (Figure 9.18b).

Hu *et al.* (2008a) measured K_{1c} of V_2AlC over a wide range of grain sizes. When three of their four data points are plotted as a function of grain diameter (Figure 9.18b) it is clear that the K_{1c} increases with grain diameter up to about 100 μm. K_{1c} for a sample with a grain diameter of 400 μm (denoted by arrow in Figure 9.18b) was 3.7 MPa $m^{1/2}$. The reason for the drop is not clear at this time, but can be due to several reasons, including the formation of microcracks because of thermal expansion anisotropies and/or incipient thermal dissociation as a result of the long annealing required to grow the grains. These comments notwithstanding, more work is needed, to understand the reasons for this decrease in K_{1c}.

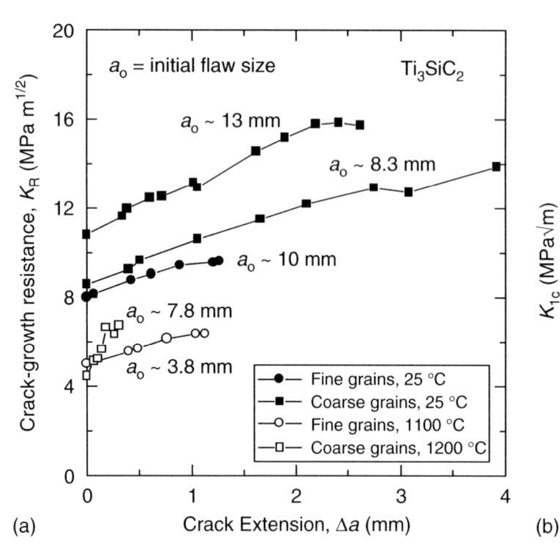

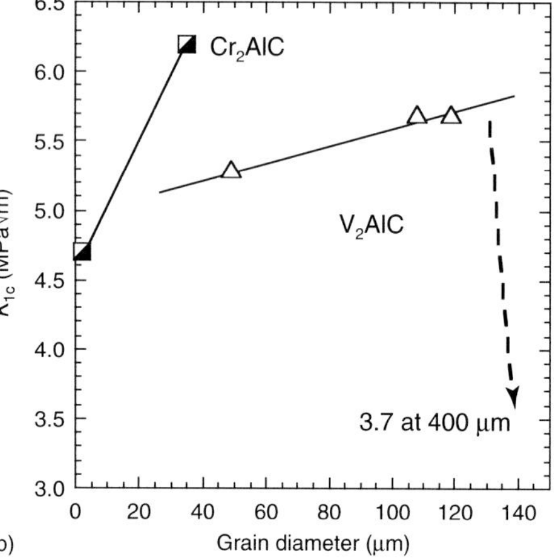

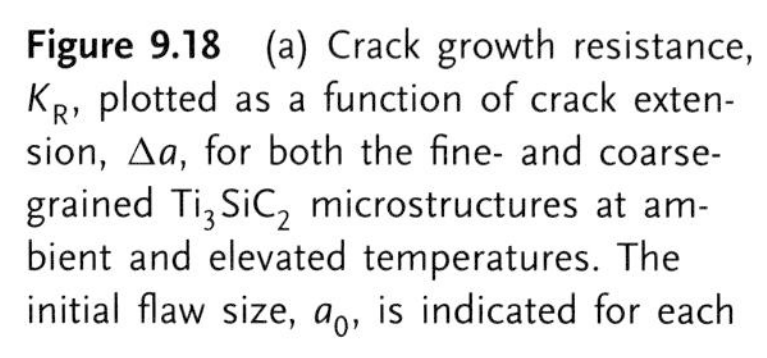

Figure 9.18 (a) Crack growth resistance, K_R, plotted as a function of crack extension, Δa, for both the fine- and coarse-grained Ti_3SiC_2 microstructures at ambient and elevated temperatures. The initial flaw size, a_0, is indicated for each measured R-curve. Note K_R *drops* above the brittle to plastic transition temperature (Chen *et al.*, 2001). (b) Effect of grain diameter on K_{1c} values for Cr_2AlC (Li *et al.*, 2011) and V_2AlC (Hu *et al.*, 2008a).

9.8.2
Anisotropic and Other Effects

As noted above, Hu *et al.* (2011a) were able to fabricate highly oriented Nb_4AlC_3 samples (see microstructure in Figure 9.10) and reported three-point flexural strengths of the order of 1.2 GPa that were *not* a strong function of orientation. Interestingly, the K_{1c} values were. When the load was applied along [0001], K_{1c} was 18 ± 5 MPa $m^{1/2}$; when the basal planes were loaded edge-on, however, K_{1c} dropped to ≈11.5 ± 2 MPa $m^{1/2}$ (Hu *et al.*, 2011a). And while this is certainly a fruitful avenue of research it is important to try and understand what endows these textured solids with their record properties. The simple explanation – that the properties are a result of orienting the grains – is too facile, for the simple reason that, as discussed above, deflections of the propagating cracks occur at the millimeter scale (Figure 9.10a) despite the fact that the grains are ≈2 μm in diameter (Fig. 9.10b).

It is instructive to place the results of Hu *et al.* (2011a) in perspective. When the combination of flexural strengths and K_{1c} values are compared with other structural ceramics (Figure 9.19a) they stand out. Along the same lines, after decades of research, the Si_3N_4 community was able to fabricate oriented polycrystalline Si_3N_4 samples with fracture stresses of the order of with ≈2 GPa and K_{1c} values of ≈10 MPa $m^{1/2}$ (see dashed line labeled seeded/oriented Si_3N_4 in Figure 9.19b). It follows that one of the first attempts to orient MAX phase grains resulted in a combination of a fracture stresses of ≈1.2 GPa and K_{Ic}values of ≈18 MPa $m^{1/2}$ (Figure 9.19b) is remarkable and suggests that this approach is a very fruitful one indeed. Figure 9.19b also plots the aforementioned results for V_2AlC (Hu *et al.*, 2008a) and self-reinforced Si_3N_4.

We note in passing that Chen *et al.* (2001), used compact tension (CT) specimens, to show that the K_{1c}'s of both FG and CG Ti_3SiC_2 *decreased* with increasing temperature (Figure 9.18a). This was an important result because it established that what is occurring at higher temperature is *not* a brittle-to-ductile transition, wherein supplementary slip systems are activated, but rather a *brittle-to-plastic* transition, the nature of which is discussed in more detail in Chapter 10. This conclusion was confirmed by Wan *et al.* (2008) who measured K_{Ic} of Ti_3AlC_2 as a function of temperature. Their results (Figure 10.1) clearly show that K_{Ic} indeed decreased at temperatures above 1000 °C. It is also clear that the K_{Ic} values of the CG samples were higher. Similar results (not shown) were reported for $Ti_3Si(Al)C_3$ samples. The K_{1c} of the latter, however, were lower (Table 9.4).

9.8.3
R-Curve Behavior

Solids for which K_{1c} increases with increasing crack length are said to exhibit R-curve behavior. That this is occurring in the MAX phases is clear from the results shown in Figure 9.18a (Gilbert *et al.*, 2000). The FG samples initiated at a K_{1c} of ≈8 MPa $m^{1/2}$ and then rose to ≈9.5 MPa $m^{1/2}$ after 1.5 mm of crack extension. The

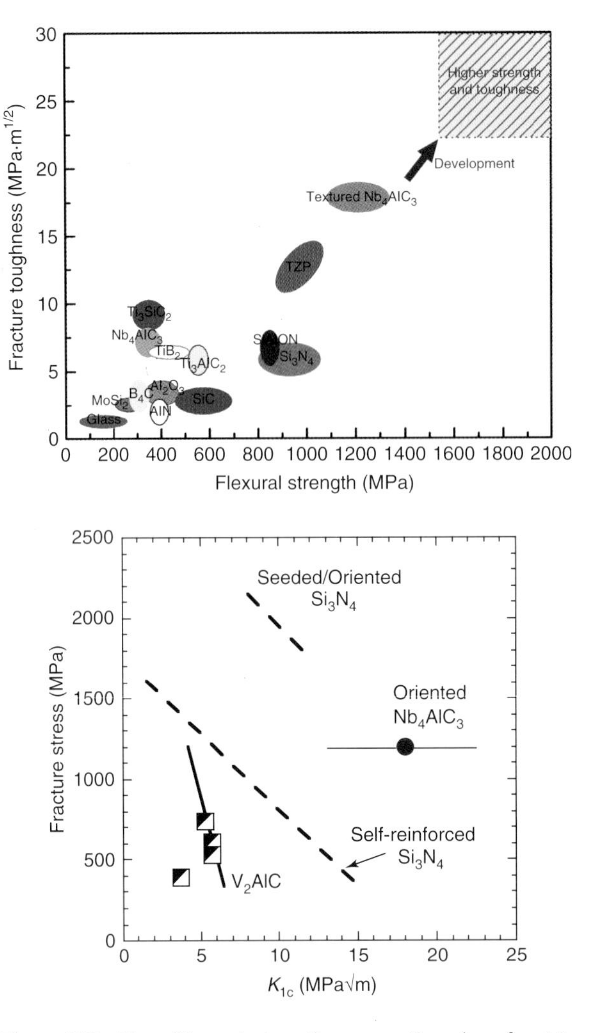

Figure 9.19 Plot of flexural strengths versus K_{1c} values for, (a) select structural ceramics including highly oriented Nb_4AlC_3 (Hu *et al.*, 2011a) and (b) V_2AlC, Nb_4AlC_3, and Si_3N_4. The results for Si_3N_4 were taken from Ohji (2010).

CG samples exhibited substantially stronger R-curve behavior, with crack growth initiating between 8.5 and 11 MPa $m^{1/2}$ and peaking at 14–16 MPa $m^{1/2}$, after a 2.7–4 mm crack extension (Figure 9.18a). At the time, the latter values were believed to the highest K_{Ic} values reported for a monolithic, single-phase nontransforming ceramic. In 2007 Sarkar *et al.* (2007) confirmed the R-curve behavior of Ti_3SiC_2. Working with a Ti_3SiC_2 sample with plate-like elongated grains, 50–200 μm in diameter, with an aspect ratio of ≈8 they showed that the K_{Ic} steadily increased from a threshold, K_{th}, of ≈5.3 MPa $m^{1/2}$, before reaching a steady-state crack growth resistance value of ≈9 MPa $m^{1/2}$, when the precrack length was more than 1.25 mm.

9.9 Fatigue Resistance

When dealing with fatigue, there are generally two domains that are important, viz. the short and long crack domains. Each is discussed separately below.

9.9.1 Long Crack Fatigue Resistance

Damage associated with cyclic loading in many ceramics is generally attributed to cycle-dependent frictional wear at grain bridging sites. As such, ceramic microstructures designed for high damage tolerance – like fiber-reinforced ceramics – are generally more prone to cyclic fatigue degradation, a fact that has been well documented for a number of monolithic and composite ceramics (Gilbert and Ritchie, 1998; Ritchie, Gilbert, and McNaney, 2000).

Fatigue, because of long cracks in solids is typically characterized by plotting crack growth rates, da/dN, where N is number of cycles, versus ΔK_1 (Fig. 9.20a) where ΔK_1 is the stress intensity factor range given by:

$$\Delta K_1 = \zeta \left(\sigma_{max} - \sigma_{min}\right) \sqrt{\pi a} \tag{9.10}$$

and where σ_{max} and σ_{min} are the maximum and minimum stresses applied. Note that Eq. (9.10) implies that

$$K_{max} = \frac{\Delta K}{1 - R} \tag{9.11}$$

where K_{max} is the maximum stress intensity, that is, at σ_{max} and R is the $\sigma_{min}/\sigma_{max}$ ratio.

For ceramic materials, in general, and the MAX phases in particular, the following relationship has been shown to apply (Zhang *et al.*, 2003b):

$$\frac{da}{dN} = \nu_o K_{max}^p (\Delta K)^q \tag{9.12}$$

where ν_o, p, and q are constants. For a fixed K_{max}, Eq. (9.11) reverts to the well-known Paris power law expression, viz.:

$$\frac{da}{dN} = D_s (\Delta K)^q \tag{9.13}$$

where $D_s = \nu_o\ K_{max}^p$. For ceramic materials, q is typically quite high (Fig. 9.20b).

There have been relatively few – we are only aware of two room temperature (Gilbert *et al.*, 2000; Zhang *et al.*, 2003b) and one elevated temperature (Chen *et al.*, 2001) – studies that have explored the fatigue resistance of the MAX phases.

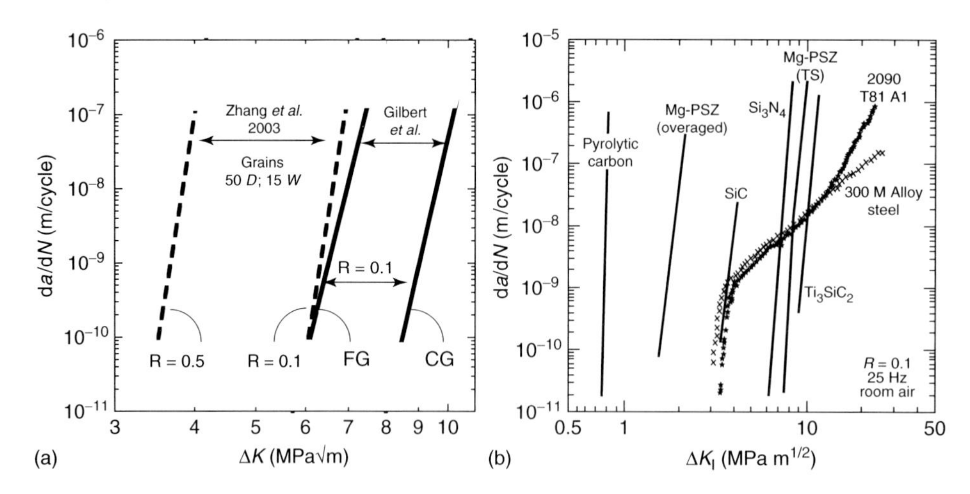

Figure 9.20 Functional dependence of cyclic crack growth rate, d*a*/d*N*, on stress intensity range, ΔK, (a) as a function of grain size and *R* (data taken from Gilbert *et al.* 2000 and Zhang *et al.* 2003b) and (b) of CG Ti_3SiC_2 as compared to other materials tested under identical conditions (Gilbert *et al.*, 2000).

The first cyclic-fatigue study of CG and FG Ti_3SiC_2 samples (Gilbert *et al.*, 2000) showed that, like other structural ceramics (Gilbert and Ritchie, 1998; Ritchie, Gilbert, and McNaney, 2000), d*a*/d*N*, was a strong function ΔK (Figure 9.20b). However, the thresholds, ΔK_{th}, were comparatively higher than those for typical ceramics, and some metals, tested under the same conditions (Figure 9.20b). The fatigue thresholds, ΔK_{th}, of the CG samples (9 $\text{MPa}\sqrt{\text{m}}$) were some of the highest values ever-observed in monolithic, non-transforming ceramics. For the FG micro-structure, ΔK_{th} was $\approx$6.5 MPa $m^{1/2}$ (Figure 9.20a).

In 2003, Zhang *et al.* (2003a) investigated the fatigue of Ti_3SiC_2 polycrystalline samples with $D \approx 50\,\mu m$ and $W = 15\,\mu m$. They showed that while indeed, Eq. (9.12) was applicable, d*a*/d*N* depended more strongly on *R* and K_{max} than on ΔK. By varying *R*, Zhang *et al.* showed that while *q* in Eq. (9.12) was 3.4, *p* was 22.6. Said otherwise, da/dN is much more sensitive to k_{max} than to ΔK. The effect of *R* is clearly shown in Figure 9.20a, where d*a*/*dN* for $R = 0.1$ and 0.5, with the latter having significantly lower ΔK_{th} values.

9.9.2
Short Crack Fatigue Resistance

Zhang *et al.* (2003b) were also the first to investigate the short crack fatigue behavior of polycrystalline Ti_3SiC_2 samples with $D \approx 50\,\mu m$ and $W{=}15\,\mu m$. They reported that for relatively short crack lengths, d*a*/d*N* often *decreased* with the number of cycles, reflecting a faster increase in crack growth resistance than the applied crack driving force (Figure 9.21a). Such behavior has been observed during the fatigue of short cracks in some metals.

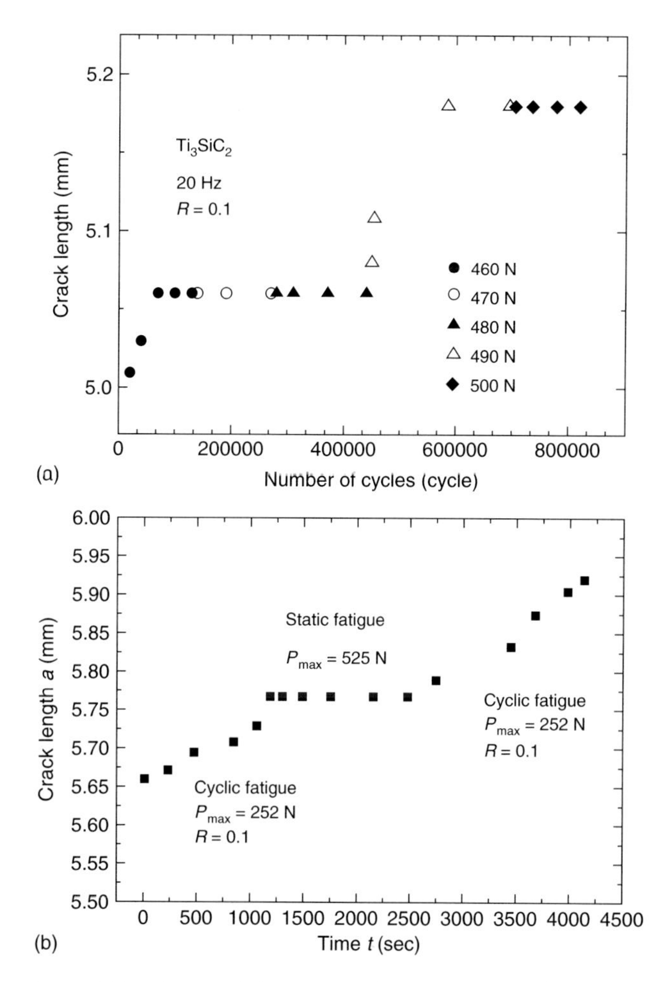

Figure 9.21 Functional dependence of cyclic fatigue crack length, a, on, (a) number of cycles, *N*, at a load ratio *R* of 0.1 in Ti_3SiC_2 for relatively shorter crack lengths. The maximum applied loads are indicated in figure and (b) time, *t*, showing the contributions of cyclic and static fatigue to crack growth (Zhang *et al.*, 2003b).

Zhang *et al.* (2003b) also studied the static fatigue of cracks in Ti_3SiC_2. The results shown in Figure 9.21b, clearly show that the crack growth rates under cyclic fatigue were significantly faster than those under static fatigue. It follows that, at least for Ti_3SiC_2 at room temperature, static fatigue can be ignored compared to cyclic fatigue.

In summary, the fatigue resistance of polycrystalline Ti_3SiC_2 – and likely most other MAX phases – is characterized by high Paris exponents. Further, the crack growth rates are a stronger function of K_{max} than ΔK. However, since these conclusions are based on only a few studies, more work is indicated, especially if the MAX phases are to be used in applications where fatigue is a concern. In the latter case, the design criterion should be to ensure that the maximum values of K_1 a component would ever be exposed to in service never exceed K_{th}.

9.10 Damage Tolerance

Another hallmark of the MAX phases is their damage tolerance. The latter is usually determined by preindenting the tensile surfaces of bend bars with Vickers indentations made at increasing indentation loads. The damage tolerance of Ti_3SiC_2, Ti_3GeC_2, and $Ti_3Si_{0.5}Ge_{0.5}C_2$ solid solutions is best exemplified in Figure 9.22a,

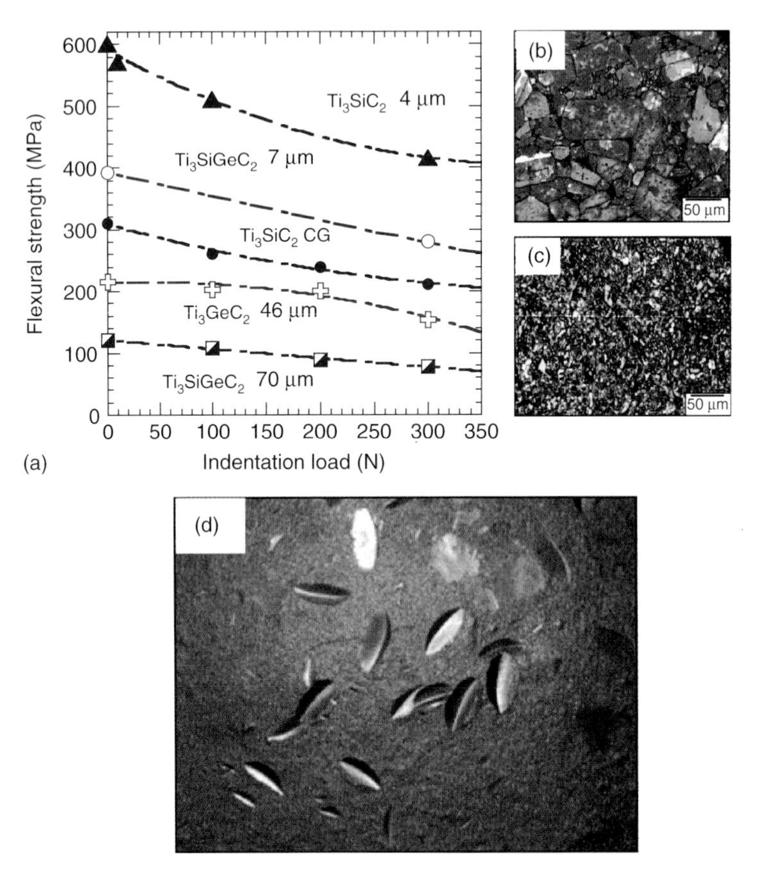

Figure 9.22 (a) Functional dependence of four-point flexural strengths on Vickers indentation loads for Ti_3SiC_2 and Ti_3GeC_2 and $Ti_3Si_{0.5}Ge_{0.5}C_2$ as a function of the average grain diameter indicated (Ganguly *et al.*, 2004). (b) Microstructure of 46 μm Ti_3GeC_2 sample. (c) microstructure of 7 μm $Ti_3Si_{0.5}Ge_{0.5}C_2$ sample. (d) Picture of Ti_2AlC block repeatedly hit with a steel hammer. The only effect of such hitting was the formation of surface dents not unlike dents one would observe in a metal plate. The block shown was roughly the size of a small laptop.

in which the functional dependencies of the post-indentation four-point, flexural strengths are plotted versus the Vickers indentation loads and grain sizes (El-Raghy *et al.*, 1999, 1997; Ganguly, Zhen, and Barsoum, 2004). Figure 9.22b,c show two of the microstructures that were tested; two of the others are shown in Figure 9.6a,b. From these, and many others results since (Amini, Barsoum, and El-Raghy, 2007; Amini *et al.*, 2008; Barsoum, Ali, and El-Raghy, 2000; Ganguly, Zhen, and Barsoum, 2004; Hu *et al.*, 2008b, 2007; Procopio, Barsoum, and El-Raghy, 2000; Tzenov and Barsoum, 2000; Wang and Zhou, 2002a; Zhang *et al.*, 2009) the following conclusions can be reached concerning the damage tolerance of the MAX phases:

1. The post-indentation flexural strengths are considerably less dependent on the indentation loads than typical structural ceramics (El-Raghy *et al.*, 1999).
2. Similar to other structural ceramics, the damage tolerance of CG microstructures is superior to that of their finer-grained counterparts.

This damage tolerance is intimately related to the ability of the MAX phases to contain and confine the extent of damage to small areas around the indentations by plastic deformation. In other words, it is indirectly related to the fact that no cracks emanate from the corners of Vickers indents (e.g., Figures 9.12 and 9.13). A practical example of such damage tolerance is shown in Figure 9.22d that shows a picture of Ti_2AlC block repeatedly hit with a steel hammer. The only effect of such hitting was the formation of surface dents, not unlike ones one would observe in a metal.

The importance of this attribute cannot be overemphasized since it implies that the MAX phases are much more tolerant to processing and service flaws, that are typically quite detrimental to the mechanical properties of brittle solids. This, in turn, should also greatly increase manufacturing yields, since the need for full density is somewhat relaxed.

9.10.1 Weibull Statistics

Enhanced damage tolerance should translate to high Weibull moduli. Bao *et al.* showed that indeed to be the case, at least for Ti_3SiC_2 loaded in flexure (Bao, Zhou, and Zhang, 2007). At $\approx$28, the Weibull modulus measured were quite respectable. It should be noted that the authors polished the tensile surfaces of their electro-discharged machined, three-point bend bars to a 1 μm finish. And while it would have been instructive to determine the Weibull moduli without such polishing, this result is still important and valid. We note in passing that we seldom polish samples after machining and before mechanically testing them, even in tension.

9.11 Micromechanisms Responsible for High K_{1c}, R-Curve Behavior, and Fatigue Response

The K_{1c} values reported above are about an order of magnitude higher than those of other layered solids, such as mica or graphite. This useful attribute stems from, (i) their layered nature, (ii) the fact that basal dislocations are mobile and multiply at room temperatures, (iii) the metallic nature of the bonding that allows for very sharp radii of curvature associated with the KBs that form that, in turn, form tenacious bridging ligaments, (iv) the crack-arresting properties of kink boundaries, and (v) the formation of IKBs. Each of these attributes is discussed separately below.

9.11.1 Layered Nature

It is not the layered nature of the MAX phases per se that endows them with their attractive mechanical properties. After all, graphite, mica, and high temperature superconductors, to name a few, are also layered, but their K_{1c} and ΔK_{th} values are quite low indeed. For example, ΔK_{th} of pyrolytic graphite is <1 MPa $m^{1/2}$ (Figure 9.20b). Clearly, layering, although important, by itself is not sufficient.

9.11.2 Dislocations and the Metallic Nature of the Bonding

Indubitably, one of the major contributors to the high K_{1c} values and excellent damage tolerance of the MAX phases has to be the fact that dislocations nucleate and are mobile at ambient temperatures. In principle, and at least over a limited volume especially during *confined deformation*, dislocations, like in metals, must absorb considerable energy by their multiplication and motion. However, dislocations and their mobility are, again, insufficient to explain the mechanical properties. For example, basal dislocations multiply and are quite mobile in graphite and mica and yet polycrystalline samples of these solids are quite brittle. The comparison with graphite is illuminating for another reason. The bonding in graphite, while not metallic, certainly does have metallic characteristics.

This thus begs the question: why do the MAX phases behave the way they do? The answer is – pun intended – multilayered.

9.11.3 Crack Deflection

It is experimentally well established that K_{1c} of a polycrystalline ceramic is appreciably higher than that of a single crystal of the same composition. One of the reasons invoked to explain this effect is crack deflection at the grain boundaries. In a polycrystalline material, if the crack is deflected along a weak grain boundary, the average stress intensity at its tip, K_{tip}, is reduced, because the stress is no longer necessarily normal to the crack plane. In general, it can be shown that K_{tip} is related to the applied stress intensity K_{app} and the angle of deflection, θ by Barsoum (2003):

$$K_{tip} = K_{app}\left(\cos^3\frac{\theta}{2}\right) \tag{9.14}$$

On the basis of this equation, if one assumes an average θ value of, say, 45°, the increase in K_{1c} expected should be ≈1.25 above the single-crystal value. It follows that crack deflection can account for some of the fracture toughness enhancements, but probably only a small fraction. In polycrystalline materials, crack bifurcation around individual or grain clusters (e.g., Figures 9.23 and 9.24), can lead to a much more potent toughening mechanism, namely, crack bridging.

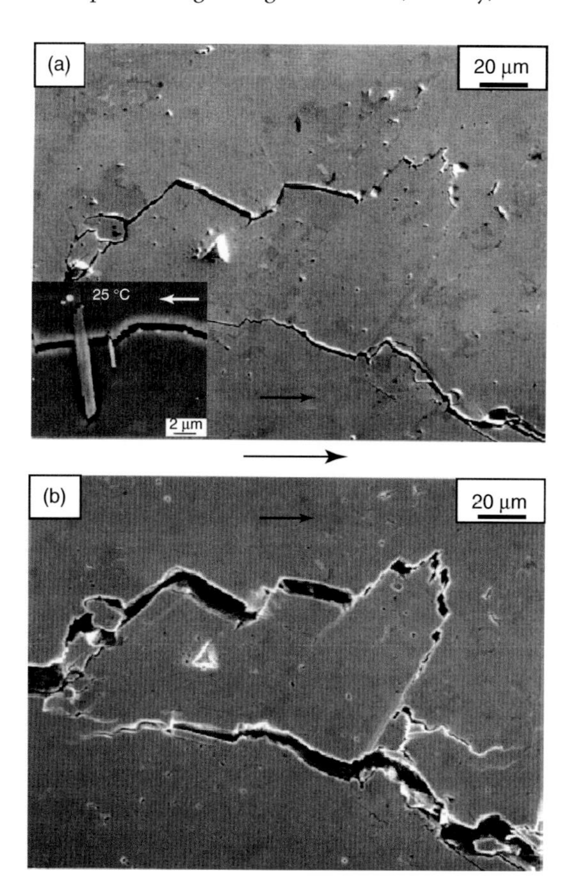

Figure 9.23 SEM micrographs of crack-wake bridging in a CG Ti_3SiC_2 sample. The same region is depicted in, (a) at 1.7 mm behind the crack tip and in (b) after the crack advanced under monotonic loading such that it is now at 5.4 mm behind the crack tip. Arrows indicate the direction of crack propagation (Gilbert *et al.*, 2000). Inset in (a) shows crack wake shielding from a grain in which the basal were normal to the direction of propagation of the crack (Chen *et al.*, 2001).

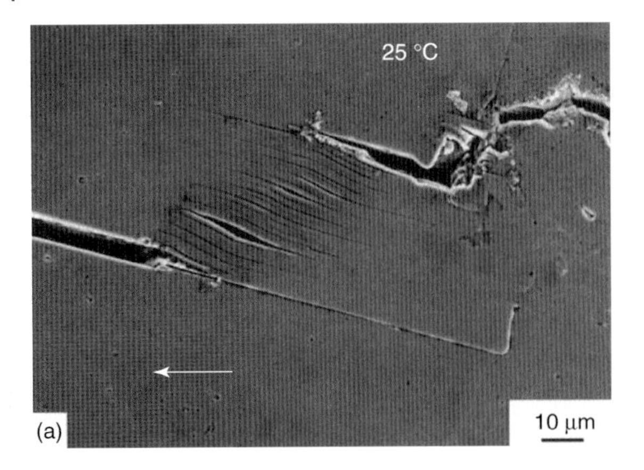

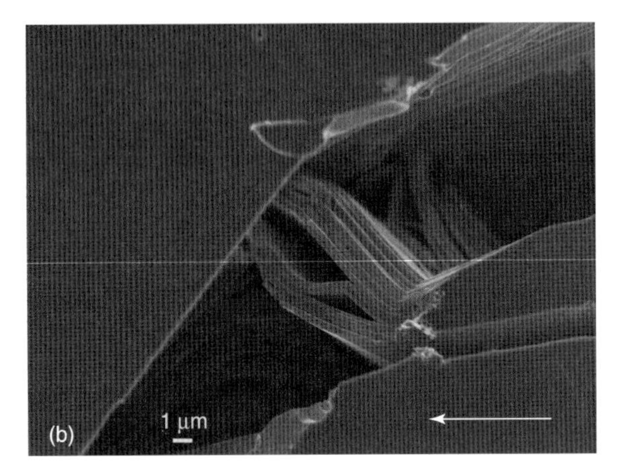

Figure 9.24 SEM images of profiles of fatigue cracks, growing from right to left (as denoted by arrows), in a coarse-grained Ti_3SiC_2 sample tested in compact tension, (a) just after the central crack branched across a grain and (b) later in the process, where now the crack opening is larger and the torque on the bridging grain resulted in its delamination and the creation of tenacious TiC-based ligaments (Chen *et al.*, 2001; Gilbert *et al.*, 2000).

9.11.4 Delaminations and Crack Bridging

In this mechanism, the toughening results from bridging of the crack surfaces behind the crack tip by strong bridging ligaments. Typical crack bridges in Ti_3SiC_2 can be seen in Figure 9.23. In these SEM micrographs the same region in the crack wake is depicted twice: First at 1.7 mm behind the crack tip (Figure 9.23a); the second time after the crack advanced, under monotonic loading, such that it is now at 5.4 mm behind the crack tip (Figure 9.23b). As the crack advances, delaminations around the grain boundaries and their bifurcation result in the bridge observed. Another example of a more classic crack bridge is shown in the inset in Figure 9.23a. In all figures the arrows indicate the direction of crack propagation.

Crack bridges result in closure forces on the crack face that reduce K_{tip}. In other words, by providing some partial support of the applied load, the bridging constituents reduce K_{tip}. A useful way to think of the problem is to imagine the unbroken ligaments or bridges in the crack wake as tiny springs that have to be stretched – and hence consume energy and shield the crack tip – as the crack front advances.

In general, for structural ceramics the energy consumed arises from elastically deforming the ligaments and the need to overcome the friction between them and the matrix as they pullout. A good example of such a bridging ligament is shown in inset in Figure 9.23a. Such ligaments have been observed in well-studied systems such as Al_2O_3, Si_3N_4, and SiC (Gilbert and Ritchie, 1998; Ritchie, Gilbert, and McNaney, 2000). What is unique to the MAX phases, however, are heavily plastically deformed lamellae bridging the crack, together with significant amounts of delamination and bending (Figure 9.24).

9.11.5
Plastically Deformable Bridging Ligaments

As noted above most bridging ligaments in structural ceramics and ceramic/ceramic composites are brittle and hence their energy absorbing mechanisms are limited to elastic deformation and friction. In sharp contradistinction the ligaments in the MAX phases can plastically deform as clearly seen in Figure 9.24. The significant amounts of delamination and bending observed and the presence of strong and tenacious M–X bonds has to play an important role in shielding the crack tip. As these processes typically only occur at the individual grain level, the K_{1c} plateau for CG microstructures is typically higher than their finer grained counterparts (Figure 9.18a).

Because of the nonmetallic character of the bonding and the lack of mobile dislocations, this mechanism is not observed in typical ceramics. This fact renders the deformation needed to form these ligaments – including the very sharp radii of curvature needed for bridging–impossible.

In general if the grains are too small, the extent of bridging is reduced and intergranular failure becomes the default mechanism. The notable exception to this general conclusion are the results of Hu *et al.* (2011b) shown in Figure 9.10a. Clearly in this case, by orienting the grains, the deflections of the main cracks occur at a length scale that is almost two orders of magnitude higher than that of the grain size. Why this is the case is not understood at this time. Interestingly, at higher magnifications (Figure 9.10b) evidence for both inter- and intragranular failure can be seen.

9.11.6
Crack Arresting Properties of Kink Boundaries

Another crucial mechanism that allows the aforementioned crack bridges to form is the fact that when kink boundaries form they become very potent crack arrestors. The best example of this mechanism and its vital importance is shown Figure 9.2b. As discussed above, upon loading of this highly oriented sample, a KB is formed at the lower left hand corner. However, in sharp contradistinction to most all other layered solids, including laminated composites or single crystals, the delamination did not extend past $\approx 200\,\mu m$. Instead the deformation was deflected toward the center of the sample and resulted in the formation of the shear band shown in the micrograph. From these results it is reasonable to conclude that kink boundaries are potent suppressors of damage.

Another excellent example of this type of damage containment can be seen in Figure 8.6c. In this TEM micrograph of a KB, it is apparent that the area between the two kink boundaries is significantly more fractured/delaminated than the areas beyond them. At some level the kink boundaries can be thought of as barriers to damage propagation.

To understand these intriguing results one needs to further understand what happens to a low angle kink boundary or what we call MDWs when it is sheared. Stroh (1958) pointed out that if a dislocation wall terminated within a crystal, the stress concentration at its end could be large enough to cause a crack normal to that wall. In the same paper, Stroh considered the initiation of a crack at the end of a MDW of length 2α (Figure 9.25a), in which the top part (AB in Figure 9.25a), of the wall is held against an obstacle and the lower part (CE in Figure 9.25a) continues to shear to the left. Stroh showed that under such conditions, cleavage along the basal planes, viz. along CF in Figure 9.25a, would occur when:

$$\sigma_n \tau_s \geq \frac{k_S G \gamma_S}{2\pi\alpha} \tag{9.15}$$

where σ_n is remotely applied normal tensile stress, τ_s the resolved shear stress on the basal planes, and γ_S the surface energy of the cleavage planes. k_s is a numerical constant of the order of unity that depends on the elastic anisotropy of the crystal.

The Stroh crack is not symmetrical but extends much further in one direction than the other. Typical examples of such highly asymmetric cracks in Ti_3SiC_2 are shown in Figure 9.26a and b. Such cracks have also been observed in room temperature deformed Zn single crystals (Gilman, 1954), SiC single crystals deformed at elevated temperatures (Suematsu *et al.*, 1991), h-BN (Turan and Knowles, 1995), and many other layered solids. The reason the cracks are asymmetrical is simple: When a dislocation wall is terminated inside a crystal, one side of the wall (region around B in Figure 9.25a) will be in tension, the other (region around C in Figure 9.25a), however, will be in compression. In other words, it is energetically quite expensive to move section CE towards the left if section AB of the wall is immobilized.

The fact that it is difficult to move part of a kink boundary in one direction but not the other is fundamental and ultimately responsible for the crack bridges that

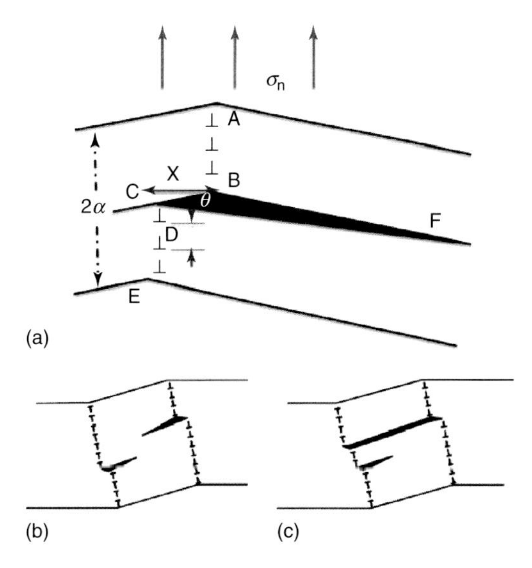

Figure 9.25 (a) Schematic of problem tackled by Stroh in which the top part, AB, of a MDW is held up at an obstacle and the bottom part, CE, continues to shear to the left (Stroh, 1958). Under certain conditions, given by Eq. (9.15), a highly asymmetric delamination or cleavage crack, CF, will nucleate. (b) Schematic of how the volume enclosed by a kink boundary – comprised of two MDWs or kink boundaries – will delaminate by the formation of asymmetrical cracks. (c) Same as (b), assuming two cracks link together (Barsoum *et al.*, 1999b).

form (e.g., Figure 9.24). Note that had the energy needed for delaminating the basal planes been low, such as in graphite, mica, or say polymer composites, splitting the MDW would have resulted in the total delamination of the grain into two halves, rather than the formation of a bridge as observed.

The ligaments shown in Figure 9.24 are examples of the extent by which individual grains of Ti_3SiC_2 can deform (Barsoum and El-Raghy, 1999). This flexibility, typically absent in other layered compounds such as mica and graphite, can be traced to the metallic nature of the bonding and the absence of strong in-plane Si–Si bonds (Barsoum *et al.*, 1999a). Such processes are highly unusual in ceramic systems and must, at least partially, account for the high steady-state K_{1c} values by promoting very sizable bridging zone lengths (well over 5 mm in the CG microstructure).

SEM images of fatigue damage behind the crack tip in a CG specimen revealed profuse amounts of wear debris along contacting surfaces (Figure 9.27). Significant bending is apparent, as well as delamination along basal planes (Figure 9.27a). Furthermore, in Figure 9.27b, sliding along the contacting surface of a bridge generated shear-faults along the basal planes of the grain on the right. Such shear-faulting should reduce the severity of frictional damage in these microstructures and help to account for the high fatigue thresholds observed (Figure 9.20b). Said otherwise, the high fatigue resistance can be traced to the laminated nature of

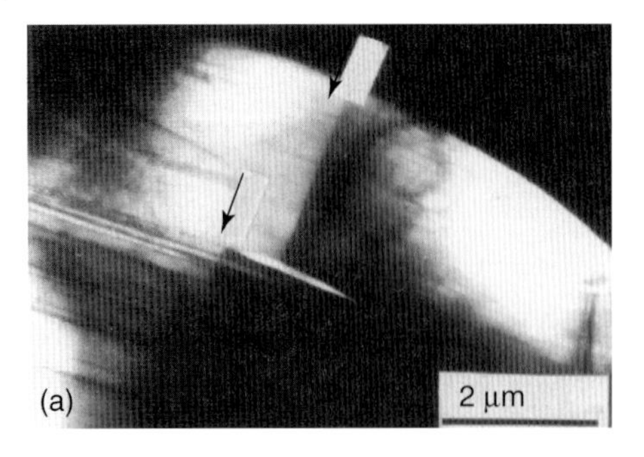

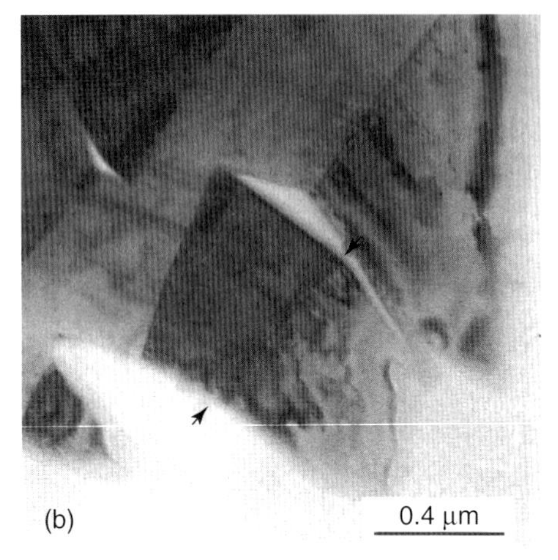

Figure 9.26 Direct TEM evidence for the formation of a delamination crack because of the partial movement of a kink boundary considered by Stroh (1958). In, (a) Ti_3SiC_2 deformed at room temperature and (b) same as (a), but in a different area showing multiple asymmetric cracks and delaminations (Barsoum *et al.*, 1999b).

Ti_3SiC_2 that, in turn, leads to extensive energy dissipation through microfaults, crack deflection, branching, and intact grain bridging. It is also more likely than not, that IKBs also play an important role in this enhanced fatigue resistance.

9.11.7 Incipient Kink Bands

At this time, the role of IKBs on the mechanical properties, except for damping discussed in Chapter 8, is less well documented. It can be shown that at small strains, the nonlinear (i.e., non-Hookean) energy, U_{NL}, stored in a solid because of DPs and IKBs is given by Barsoum *et al.* (2005):

$$U_{NL} = \frac{\pi G b}{2\lambda}\varepsilon_{DP} + \sqrt{\frac{G^2 \gamma_c}{2\pi(1-\nu)^2 N_k \alpha^3}}\, \varepsilon_{IKB}^{1.5} \tag{9.16}$$

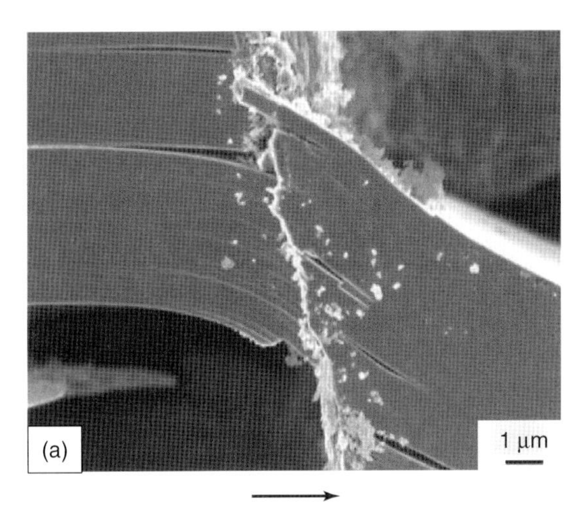

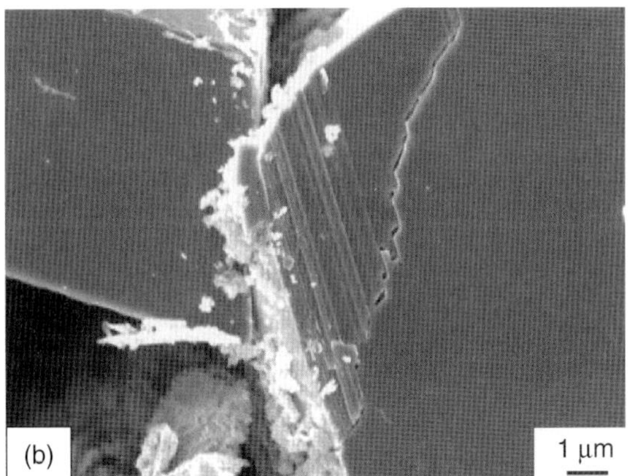

Figure 9.27 SEM images in, (a,b) of fatigue damage behind a crack tip in a CG Ti_3SiC_2 sample. Note the appearance of wear debris along contacting surfaces. In (a) significant bending is observed, as well as delamination along basal planes; in (b), sliding along the contacting surfaces generated shear-faults along the basal planes of the grain on the right. The arrow indicates the direction of crack propagation (Gilbert *et al.* 2000).

where λ is the grain diameter (i.e., the dimension along [100]). The other terms were defined in Chapter 8 (Eqs. (8.1–8.5)). From the results shown in Table 8.1, it is obvious that $N_k\alpha^3$ is of the order of unity. If one conservatively assumes $N_k\alpha^3$ to be 4, $\gamma_c = 0.01$, $G \approx 140$ GPa, and $\lambda \approx 8\,\mu$m, the two terms comprising Eq. (9.15) can be plotted as a function of nonlinear strain. When such a plot is generated (Figure 9.28) an important conclusion becomes patently clear: For a given strain, IKBs can store/absorb roughly 30 times more mechanical energy than DPs. Note that if $N_k\alpha^3$ is assumed to be unity and the grain diameter is assumed to be of the order of 100 μm, the ratio is of the order of 10^4! For a given strain, IKBs can roughly double the elastic energy stored in a solid (Figure 9.28). IKBs thus absorb

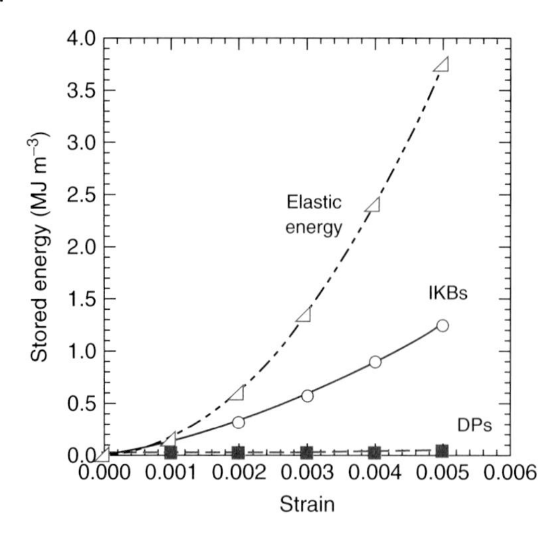

Figure 9.28 Plot of Eq. (9.15) for Ti_3SiC_2. The energy stored per unit of strain by IKBs (blue open circles) solid is significantly higher than the energy stored by dislocation pile-ups (solid red squares) and comparable to the elastic energy stored (open red triangles) (Barsoum *et al.* 2005).

significant amounts of energy at low strains; DPs result in large strains, but little stored energy. It is crucial to note here that U_{NL} is *not* simply the area under the stress–strain curves, but much more importantly, the energy *stored* in the solid that, upon unloading, is capable of doing work. Note that U_{NL} is the area OAC in Fig. 8.8c and thus neither includes the linear elastic energy or half W_d.

9.12 Thermal Sock Resistance

Another beneficial attribute of the MAX phases is their thermal shock resistance. One of the first experiments we carried out on Ti_3SiC_2 was to document its thermal shock resistance (Barsoum and El-Raghy, 1996). The response of Ti_3SiC_2 to thermal shock depends on grain size (El-Raghy *et al.*, 1999); the post-quench flexural strengths of CG samples are not a function of quench temperature and actually get slightly stronger when quenched from 1400 °C (Figure 9.29a). The response of FG Ti_3SiC_2 samples is different; instead of exhibiting a critical quenching temperature above which the strength is greatly reduced as typical for ceramics, their post-quenching strengths gradually decrease over a 500 °C range (Figure 9.29a). The same nonsusceptibility to thermal shock is also exhibited by $Ti_3(Si_{0.5}Ge_{0.5})C_2$ solid solutions and most other MAX phases that have been tested to date (Figure 9.29b). In one case, the post-quench flexural strengths of CG $Ti_3(Si_{0.5},Ge_{0.5})C_2$ samples were actually ≈20% higher than the as-received CG samples (Figure 9.29a).

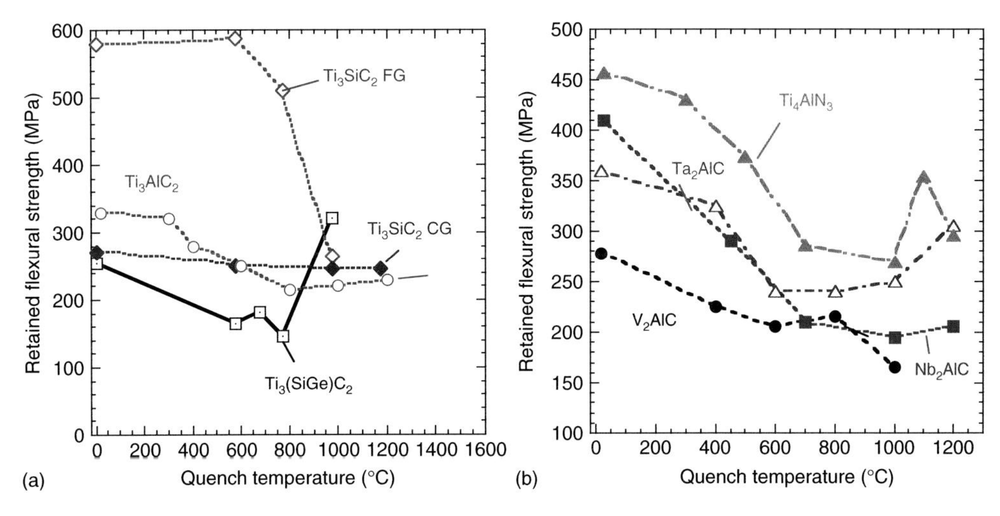

Figure 9.29 Post quench four-point flexural strength vs. quench temperature of (a) CG Ti_3GeC_2 samples and FG and CG $Ti_3Si_{0.5}Ge_{0.5}C_2$ and Ti_3SiC_2 samples. The quench was into ambient temperature water. Each point is the average of at least three separate tests. Note that the flexural strength of the 40 μm $Ti_3Si_{0.5}Ge_{0.5}C_2$ sample is *higher* than the as-fabricated, unquenched sample (Ganguly, Zhen, and Barsoum, 2004). (b) Ti_4AlN_3 (Procopio, Barsoum, and El-Raghy, 2000), Nb_2AlC (Salama, El-Raghy, and Barsoum, 2002), Ta_2AlC (Hu *et al.*, 2008b), and V_2AlC (Hu *et al.*, 2008a).

The reasons for this quench hardening is not entirely clear at this time, but are most probably related to the formation of smaller domains as a result of thermal residual stresses.

9.13 Strain Rate Effects

The importance and effect of strain rates on the mechanical properties on the MAX phases at high temperatures is reasonably well understood (Chapter 10). Somewhat surprisingly, the situation at ambient temperatures is much less clear. The few existing results are scattered and not well understood. For example, the effects of strain rates on the compressive stress–strain curves of Nb_2AlC and TiNbAlC, the latter for two different grain sizes, are shown in Figure 9.30. In the case of Nb_2AlC, the results (Figure 9.30a), while somewhat noisy, are typical of those obtained for the MAX phases at higher temperatures (Chapter 10) in that decreasing the strain rate increases the ductility at the expense of the ultimate compressive strengths. The stress–strain curves obtained for the TiNbAlC solid solution are quite intriguing. For both grain sizes, *decreasing* the strain rates resulted in an *increase* in the ultimate compressive strengths and a slight increase in ductility. These results were so surprising that they were confirmed by repeating the tests twice, with the same result each time. At this time, these results are not understood.

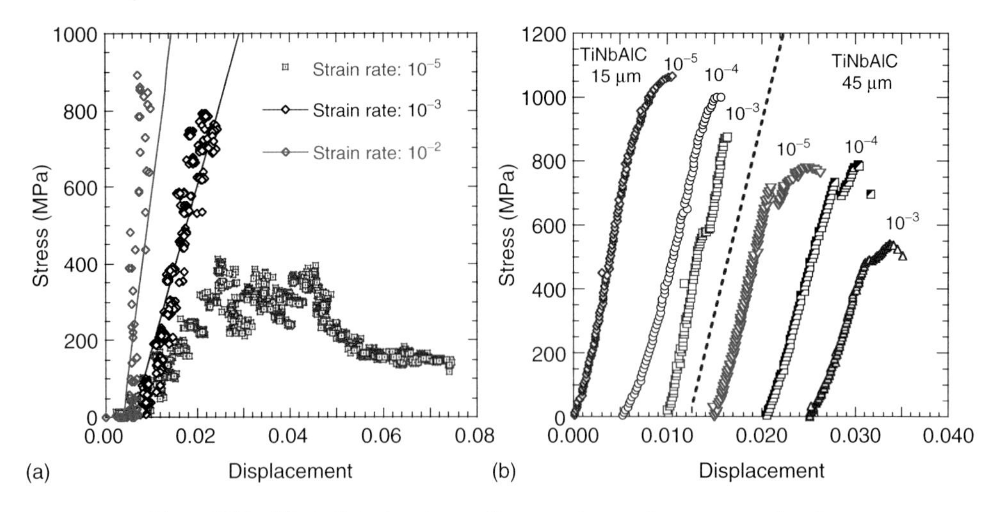

Figure 9.30 Effect of strain rates on the room temperature compressive stress–displacement curves of (a) Nb_2AlC and (b) TiNbAlC with two different grain sizes (Salama, El-Raghy, and Barsoum, 2002). The curves in (b) are shifted to the right for clarity.

9.14
Solid Solution Hardening and Softening

As noted above, since substitutions can be made on either of the three sub-lattices, the number of solid solutions possible is quite high. It is thus not unreasonable to assume, that once the effects of such substitutions have on the mechanical properties are understood, major improvements in the latter should accrue.

The relatively little work that exists on MAX phase solid solutions is intriguing. On the basis of a limited data set, it appears that the only substitutions that lead to dramatic enhancements in flexural strength are those on the X-sites. In an early paper on $Ti_2AlC_xN_{1-x}$ solid solutions (Barsoum, Ali, and El-Raghy, 2000a) it was shown that at 10 N, H_v of $Ti_2AlC_{0.5}N_{0.5}$ at 5.5 ± 0.5 GPa was higher than the 4.5 ± 0.3 ≈ GPa of the end member, Ti_2AlC, with comparable grain size. More importantly, the flexural strength increased from 540 ± 20 for Ti_2AlC to 810 ± 80 for $Ti_2AlC_{0.5}N_{0.5}$ (Figure 9.7a). To date, none of the substitutions on either the M or A sublattices have resulted in equivalent enhancements in flexural strengths.

The hardening, or lack thereof, appears to be a function of the MAX phases explored and solute atom chosen. Substitutions on the M-sites (e.g., $(Ti_{0.5},Nb_{0.5})_2AlC$, (Salama, El-Raghy, and Barsoum, 2002)) or the A-sites (e.g., $Ti_3(Si_{0.5},Ge_{0.5})C_2$ (Ganguly, Zhen, and Barsoum, 2004)) do not appear to lead to solid solution hardening. In contradistinction, Meng, Zhou, and Wang (2005) showed that H_v, at 10 N, of Ti_2AlC doped with 15 at% V increased from 3.5 to 4.5 GPa, as a result. The flexural and compressive strengths also increased. The same group showed that substituting 25 at% of the Al in Ti_3AlC_2 by Si, increased H_v, at 10 N, from ≈3 to 4 GPa (Zhou *et al.*, 2006b). Here again the flexural and compressive strengths increased by 12 and 29%, respectively, over those of the end member, Ti_3AlC_2. Wan *et al.* (2010) have shown that by substituting 7 at% of the Ti atoms in essentially Ti_3SiC_2 by Zr to form $(Ti_{0.93},Zr_{0.07})_3(SiAl)C_2$ solid solutions, H_v increased from about 4 to 5.5 GPa. The Young's moduli changes were more subtle. These atomic substitutions, on the other hand, had little effect on K_{1c} or the flexural strengths.

Another example of an increase in H_v with chemistry can be found in the $(Cr_{1-x},V_x)_2AlC$ system, where H_v increases – more or less linearly – from about 3.5 ± 0.2 GPa for $x = 0$, to 4.3 ± 0.1 GPa at $x = 0.5$ (Tian, Sun, and Hashimoto, 2009).

The effect of solid solutions on the mechanical properties of the MAX phases is a wide open field of study with potentially large dividends. When carrying out such studies it is crucial to compare compositions with identical grain sizes. Otherwise, the results can be masked or distorted by grain size effects.

9.15 Machinability

Lastly one would be remiss not to mention probably the most characteristic trait of the MAX phases and one that truly sets them apart from other structural ceramics or high temperature alloys: ease of machinability (Barsoum, Brodkin, and El-Raghy, 1997; Barsoum and El-Raghy, 1996, 1997; Barsoum, El-Raghy, and Radovic, 2000b). The MAX phases are quite readily machinable with regular high-speed tool steels or even manual hacksaws (e.g., Figure 1.3). It is important to note that the machining does not occur by plastic deformation, as in the case of metals, but rather by the breaking off of tiny microscopic flakes. In that respect, they are not unlike other machinable ceramics such as Maycor™. The analogy with ice is also apt (Barsoum *et al.*, 2001); the MAX compounds do not machine as one scoops ice cream (as in metals), but rather as in shaving ice (Barsoum *et al.*, 2001). With the exception of Be and Be-alloys, as shown in Figure 3.1, the MAX phases have some of the highest specific stiffness values for readily machinable solids.

There have been few studies on the machinability of the MAX phases. In 2012 Hwang *et al.* compared the cutting resistance of Ti_3SiC_2 to that of a middle-carbon steel, SM45C (Hwang *et al.*, 2012). The values of the principal forces measured during the machining of Ti_3SiC_2 were lower than those of SM45C. After machining, the surface roughness of the Ti_3SiC_2 was lower than that of SM45C. However, the damage to the tool bits used for machining the latter was less than the damage to those used for the machining of Ti_3SiC_2.

9.16 Summary and Conclusions

With only two independent slip systems operative at any time, the response of the MAX phases to stress falls somewhat between metals – with 5+ slip systems – and ceramics, with no slip systems at room temperature. If the grains are highly oriented and/or the deformation is confined, the MAX phases can exhibit significant ductility, even at room temperature, mostly because of the formation of shear bands.

With Vickers hardness values between 2 and 8 GPa, the MAX phases are relatively soft compared to other structural ceramics in general, and the MX binaries, in particular. In general the Vickers hardness is a function of applied load and only saturates at higher loads. Kinking and the formation of kink boundaries play a large role in confining the indentation damage to the near vicinity of the indentation mark. This in turn endows the MAX phases with a high degree of damage tolerance and relatively high K_{1c} values that range from 5 MPa to almost 20 MPa $m^{1/2}$. Highly textured Nb_4AlC_3 samples exhibit excellent combinations of flexural strengths and K_{1c} values.

Under cyclic loading, for the long crack regime, both the Paris exponents and threshold stress intensity factors are quite high. The latter, however are a strong function of K_{max}. In the short crack regime, and based on only one study on Ti_3SiC_2, the fatigue resistance appears to be good.

For reasons that are not entirely clear at this time, the MAX phases appear to be quite resistant to thermal shock, with coarser-grained microstructures being less susceptible than their finer-grained counterparts.

Many of unusual and/or advantageous mechanical properties of the MAX phases can be traced back to their layered and metallic nature and concomitant dislocation-based deformation mechanisms. The latter can result in tenacious bridges across crack interfaces. IKBs probably also play a role.

Lastly, if there is one property that epitomizes the MAX phases it must be the ease by which they can be machined.

References

Amini, S., Barsoum, M.W., and El-Raghy, T. (2007) Synthesis and mechanical properties of fully dense Ti_2SC. *J. Am. Ceram. Soc.*, **90**, 3953–3958.

Amini, S., Zhou, A., Gupta, S., DeVillier, A., Finkel, P., and Barsoum, M.W. (2008) Synthesis and elastic and mechanical properties of Cr_2GeC. *J. Mater. Res.*, **23**, 2157–2165.

Anasori, B., Sickafus, K.E., Usov, I.O., and Barsoum, M.W. (2011) Spherical nanoindentation study of effect of Ion irradiation on deformation micromechanism of $LiTaO_3$ single crystals. *J. Appl. Phys.*, **110**, 023516.

Bai, Y., He, X., Li, Y., Zhu, C., and Zhang, S. (2009) Rapid synthesis of bulk Ti_2AlC by self-propagating high temperature combustion synthesis with a pseudo–hot isostatic pressing process. *J. Mater. Res.*, **24**, 2528.

Bao, Y.W., Chen, J.X., Wang, X.H., and Zhou, Y.C. (2004) Shear strength and shear failure of layered machinable Ti_3AlC_2 ceramics. *J. Eur. Ceram. Soc.*, **24**, 855–860.

Bao, Y.W., Zhang, H.B., and Zhou, Y.C. (2004) Punch-shear tests and size effects for evaluating the shear strength of machinable ceramics. *Z. Metalkd.*, **95**, 372–376.

Bao, Y.W. and Zhou, Y.C. (2005) Effect of sample size and testing temperature on the fracture toughness of Ti_3SiC_2. *Mater. Res. Innovations*, **9**, 41–42.

Bao, Y.W., Zhou, Y.C., and Zhang, H.B. (2007) Investigation on reliability of

nanolayer-grained Ti_3SiC_2 via Weibull statistics. *J. Mater. Sci.*, **42**, 4470.
Barsoum, M.W. (2000) The $M_{n+1}AX_n$ phases: a new class of solids; thermodynamically stable nanolaminates. *Prog. Solid State Chem.*, **28**, 201–281.
Barsoum, M.W. (2003) *Fundamentals of Ceramics*, Institute of Physics, Bristol.
Barsoum, M.W., Ali, M., and El-Raghy, T. (2000a) Processing and characterization of Ti_2AlC, Ti_2AlCN and $Ti_2AlC_{0.5}N_{0.5}$. *Metall. Mater. Trans.*, **31A**, 1857–1865.
Barsoum, M.W., El-Raghy, T., and Radovic, M. (2000b) Ti_3SiC_2: a layered machinable ductile ceramic. *Interceram*, **49**, 226–233.
Barsoum, M.W. and Basu, S. (2010) in *Encyclopedia of Materials Science and Technology* (eds R.W.C.K.H.J. Buschow, M.C. Flemings, B. Ilschner, E.J. Kramer, S. Mahajan, and P. Veyssiere), Elsevier, Oxford, pp. 1–23.
Barsoum, M.W., Brodkin, D., and El-Raghy, T. (1997) Layered machinable ceramics for high temperature applications. *Scr. Metall. Mater.*, **36**, 535–541.
Barsoum, M.W. and El-Raghy, T. (1996) Synthesis and characterization of a remarkable ceramic: Ti_3SiC_2. *J. Am. Ceram. Soc.*, **79**, 1953–1956.
Barsoum, M.W. and El-Raghy, T. (1997) A progress report on Ti_3SiC_2, Ti_3GeC_2 and the H-phases, M_2BX. *J. Mater. Synth. Process.*, **5**, 197–216.
Barsoum, M.W. and El-Raghy, T. (1999) Room temperature ductile carbides. *Metall. Mater. Trans.*, **30A**, 363–369.
Barsoum, M.W., El-Raghy, T., Rawn, C.J., Porter, W.D., Wang, H., Payzant, A., and Hubbard, C. (1999a) Thermal properties of Ti_3SiC_2. *J. Phys. Chem. Solids*, **60**, 429.
Barsoum, M.W., Farber, L., El-Raghy, T., and Levin, I. (1999b) Dislocations, kink bands and room temperature plasticity of Ti_3SiC_2. *Metall. Mater. Trans.*, **30A**, 1727–1738.
Barsoum, M.W., Murugaiah, A., Kalidindi, S.R., and Gogotsi, Y. (2004a) Kink bands, nonlinear elasticity and nanoindentations in graphite. *Carbon*, **42**, 1435–1445.
Barsoum, M.W., Murugaiah, A., Kalidindi, S.R., and Zhen, T. (2004b) Kinking nonlinear elastic solids, nanoindentations and geology. *Phys. Rev. Lett.*, **92**, 255508–255501.
Barsoum, M.W., Radovic, M., Finkel, P., and El-Raghy, T. (2001) Ti_3SiC_2 and ice. *Appl. Phys. Lett.*, **79**, 479–481.
Barsoum, M.W., Zhen, T., Zhou, A., Basu, S., and Kalidindi, S.R. (2005) Microscale modeling of kinking nonlinear elastic solids. *Phys. Rev. B*, **71**, 134101.
Basu, S. and Barsoum, M.W. (2007) Deformation micromechanisms of ZnO single crystals as determined from spherical nanoindentation stress–strain curves. *J. Mater. Res.*, **22**, 2470–2477.
Basu, S. and Barsoum, M.W. (2009) On spherical nanoindentations, kinking nonlinear elasticity of mica single crystals and their geological implications. *J. Struct. Geol.*, **31**, 791–801.
Basu, S., Barsoum, M.W., and Kalidindi, S.R. (2006a) Sapphire: a kinking nonlinear elastic solid. *J. Appl. Phys.*, **99**, 063501.
Basu, S., Moseson, A., and Barsoum, M.W. (2006b) On the determination of indentation stress–strain curves using spherical indenters. *J. Mater. Res.*, **21**, 2628–2637.
Basu, S., Zhou, A.G., and Barsoum, M.W. (2008) Reversible dislocation motion under contact loading in $LiNbO_3$ single crystal. *J. Mater. Res.*, **23**, 1334–1338.
Bei, G.-P. (2011) PhD thesis. University of Poitiers, Poitiers.
Bei, G.P., Gauthier-Brunet, V., Tromas, C., and Dubois, S. (2012) Synthesis, characterization, and intrinsic hardness of layered nanolaminate Ti_3AlC_2 and $Ti_3Al_{0.8}Sn_{0.2}C_2$ solid solution. *J. Am. Ceram. Soc.*, **95**, 102–107.
Bei, G.-P., Laplanche, G., Gauthier-Brunet, V., Bonneville, J., and Dubois, S. (2013) The compressive behavior of Ti_3AlC_2 and $Ti_3Al_{0.8}Sn_{0.2}C_2$ MAX phases at room temperature. *J. Am. Ceram. Soc.*, **96**, 567–576.
Bradby, J.E., Kucheyev, S.O., Williams, J.S., Jagadish, C., Swain, M.V., Munroe, P., and Phillips, M.R. (2002) Contact-induced defect propagation in ZnO. *Appl. Phys. Lett.*, **80**, 4537–4539.
Chen, D., Shirato, K., Barsoum, M.W., El-Raghy, T., and Ritchie, R.O. (2001) Cyclic fatigue-crack growth and fracture properties in Ti_3SiC_2 ceramics at elevated temperatures. *J. Am. Ceram. Soc.*, **84**, 2914.

Eklund, P., Beckers, M., Jansson, U., Högberg, H., and Hultman, L. (2010) The $M_{n+1}AX_n$ phases: materials science and thin-film processing. *Thin Solid Films*, **518**, 1851–1878.

El-Raghy, T., Barsoum, M.W., Zavaliangos, A., and Kalidindi, S.R. (1999) Processing and mechanical properties of Ti_3SiC_2: II, effect of grain size and deformation temperature. *J. Am. Ceram. Soc.*, **82**, 2855–2860.

El-Raghy, T., Chakraborty, S., and Barsoum, M.W. (2000) Synthesis and characterization of Hf_2PbC, Zr_2PbC and M_2SnC (M=Ti, Hf, Nb or Zr). *J. Eur. Ceram. Soc.*, **20**, 2619–2625.

El-Raghy, T., Zavaliangos, A., Barsoum, M.W., and Kalidindi, S.R. (1997) Damage mechanisms around hardness indentations in Ti_3SiC_2. *J. Am. Ceram. Soc.*, **80**, 513–516.

Ganguly, A., Zhen, T., and Barsoum, M.W. (2004) Synthesis and mechanical properties of Ti_3GeC_2 and $Ti_3(Si_xGe_{1-x})C_2$ (x = 0.5, 0.75) solid solutions. *J. Alloys Compd.*, **376**, 287–295.

Gao, N.F., Miyamoto, Y., and Zhang, D. (1999) Dense Ti_3SiC_2 prepared by reactive HIP. *J. Mater. Sci.*, **34**, 4385–4392.

Gilbert, C.J., Bloyer, D.R., Barsoum, M.W., El-Raghy, T., Tomasia, A.P., and Ritchie, R.O. (2000) Fatigue-crack growth and fracture properties of coarse and fine-grained Ti_3SiC_2. *Scr. Mater.*, **42**, 761–767.

Gilbert, C.J. and Ritchie, R.O. (1998) On the quantification of bridging tractions during subcritical crack growth under monotonic and cyclic fatigue loading in a grain-bridging silicon carbide ceramic. *Acta Mater.*, **46**, 609–616.

Gilman, J.J. (1954) Mechanism of ortho kink-band formation in compressed zinc monocrystals. *Trans. AIME*, **200**, 621–629.

Goto, T. and Hirai, T. (1987) Chemically vapor deposited Ti_3SiC_2. *Mater. Res. Bull.*, **22**, 2292–2295.

Hu, C., Lin, Z., He, L., Bao, Y., Wang, J., Li, M., and Zhou, Y.C (2007) Physical and Mechanical Properties of Bulk Ta_4AlC_3 Ceramic Prepared by an In Situ Reaction Synthesis/Hot-Pressing Method. *J. Am. Ceram. Soc.*, **90**, 2542–2548.

Hu, C., He, L., Liu, M., Wang, X., Wang, J., Li, M., Bao, Y., and Zhou, Y. (2008a) In situ reaction synthesis and mechanical properties of V_2AlC. *J. Am. Ceram. Soc.*, **91**, 4029–4035.

Hu, C., He, L., Zhang, J., Bao, Y., Wang, J., Li, M., and Zhou, Y. (2008b) Microstructure and properties of bulk Ta_2AlC ceramic synthesized by an in situ reaction/hot pressing method. *J. Eur. Ceram. Soc.*, **28**, 1679–1685.

Hu, C., Li, F., He, L., Liu, M., Zhang, J., Wang, J., Bao, Y., Wang, J., and Zhou, Y. (2008c) In situ reaction synthesis, electrical and thermal, and mechanical properties of Nb_4AlC_3. *J. Am. Ceram. Soc.*, **91**, 2258–2263.

Hu, C., Lin, Z., He, L., Bao, Y., Wang, J., Li, M., and Zhou, Y.C. (2007) Physical and mechanical properties of bulk Ta_4AlC_3 ceramic prepared by an in situ reaction synthesis/hot-pressing method. *J. Am. Ceram. Soc.*, **90**, 2542–2548.

Hu, C., Sakka, Y., Grasso, S., Nishimura, T., Guo, S., and Tanaka, H. (2011a) Shell-like nanolayered Nb_4AlC_3 ceramic with high strength and toughness. *Scr. Mater.*, **64**, 765–768.

Hu, C., Sakka, Y., Nishimura, T., Guo, S., Grasso, S., and Tanaka, H. (2011b) Physical and mechanical properties of highly textured polycrystalline Nb_4AlC_3 ceramic. *Sci. Technol. Adv. Mater.*, **12**, 044603.

Hu, C., Sakka, Y., Tanaka, H., Nishimura, T., and Grasso, S. (2009) Low temperature thermal expansion, high temperature electrical conductivity, and mechanical properties of Nb_4AlC_3 ceramic synthesized by spark plasma sintering. *J. Alloys Compd.*, **487**, 675–681.

Hwang, S.-S., Lee, S.C., Han, J.-H., Lee, D., and Park, S.-W. (2012) Machinability of Ti_3SiC_2 with layered structure synthesized by hot pressing mixture of TiC_x and Si powder. *J. Eur. Ceram. Soc.*, **32**, 3493–3500.

Kooi, B.J., Poppen, R.J., Carvalho, N.J.M., De Hosson, J.T.M., and Barsoum, M.W. (2003) Ti_3SiC_2: a damage tolerant ceramic studied with nanoindentations and transmission electron microscopy. *Acta Mater.*, **51**, 2859–2872.

Kuroda, Y., Low, I.M., Barsoum, M.W., and El-Raghy, T. (2001) Indentation responses and damage characteristics of Hot isostatically pressed Ti_3SiC_2. *J. Aust. Ceram. Soc.*, **37**, 95–102.

Lane, N.J., Simak, S.I., Mikhaylushkin, A., Abrikosovv, I., Hultman, L., and Barsoum, M.W. (2011) A first principles study of dislocations in HCP metals through the investigation of the $(11\bar{2}1)$ twin boundary. *Phys. Rev. B*, **84**, 184101.

Li, J.F., Pan, W., Sato, F., and Watanabe, R. (2001) Mechanical properties of polycrystalline Ti_3SiC_2 at ambient and elevated temperatures. *Acta Mater.*, **49**, 937–945.

Li, S.B., Xie, J.X., Zhao, J.Q., and Zhang, L.T. (2002) Mechanical properties and mechanism of damage tolerance for Ti_3SiC_2. *Mater. Lett.*, **57**, 119–123.

Li, S.B., Yu, W.B., Zhai, H.X., Song, G.M., Sloof, W.G., and van der Zwaag, S. (2011) Mechanical properties of low temperature synthesized dense and fine-grained Cr_2AlC ceramics. *J. Eur. Ceram. Soc.*, **31**, 217–224.

Lin, Z.J., Zhou, Y.C., Li, M.S., and Wang, J.Y. (2005) In-situ hot pressing/solid–liquid reaction synthesis of bulk Cr_2AlC. *Z. Metallkd.*, **96**, 291–296.

Lis, J., Pampuch, R., Piekarczyk, J., and Stobierski, L. (1993) New ceramics based on Ti_3SiC_2. *Ceram. Int.*, **19**, 91–96.

Low, I.M. (1998) Vickers contact damage of micro-layered Ti_3SiC_2. *J. Eur. Ceram. Soc.*, **18**, 709–713.

Low, I.M., Lee, S.K., Lawn, B., and Barsoum, M.W. (1998) Contact damage accumulation in Ti_3SiC_2. *J. Am. Ceram. Soc.*, **81**, 225–228.

Meng, F.L., Zhou, Y., and Wang, J.Y. (2005) Strengthening of Ti_2AlC by substituting Ti with V. *Scr. Mater.*, **53**, 1369–1372.

Molina-Aldareguia, J.M., Emmerlich, J., Palmquist, J., Jansson, U., and Hultman, L. (2003) Kink formation around indents in laminated Ti_3SiC_2 thin-films studied in the nano scale. *Scr. Mater.*, **49**, 155–160.

Moseson, A.J., Basu, S., and Barsoum, M.W. (2008) Determination of the effective zero point of contact for spherical nanoindentation. *J. Mater. Res.*, **23**, 204–209.

Murugaiah, A., Barsoum, M.W., Kalidindi, S.R., and Zhen, T. (2004) Spherical nanoindentations in Ti_3SiC_2. *J. Mater. Res.*, **19**, 1139–1148.

Nickl, J.J., Schweitzer, K.K., and Luxenberg, P. (1972) Gasphasenabscheidung im Systeme Ti-C-Si. *J. Less-Common Met.*, **26**, 335–353.

Ohji, T. (2010) Structure–property relations, in *Ceramics Science and Technology*, Properties, Vol. **2** (eds R.R. Riedel and I.-W. Chen), Wiley-VCH Verlag GmbH.

Opeka, M., Zaykoski, J., Talmy, I., and Causey, S. (2011) Synthesis and characterization of Zr_2SC ceramics. *Mater. Sci. Eng., B*, **A528**, 1994–2001.

Orowan, E. (1942) A type of plastic deformation new in metals. *Nature*, **149**, 463–464.

Pampuch, R. and Lis, J. (1995) Ti_3SiC_2 – A Plastic Ceramic Material, Vol. 3B, Techna Srl, Faenza.

Pampuch, R., Lis, J., Stobierski, L., and Tymkiewicz, M. (1989) Solid combustion synthesis of Ti_3SiC_2. *J. Eur. Ceram. Soc.*, **5**, 283.

Peng, L.M. (2007) Fabrication and properties of Ti_3AlC_2 particulates reinforced copper composites. *Scr. Mater.*, **56**, 729–732.

Procopio, A., Barsoum, M.W., and El-Raghy, T. (2000) Characterization of Ti_4AlN_3. *Metall. Mater. Trans.*, **31A**, 333–337.

Radovic, M., Barsoum, M.W., El-Raghy, T., Seidensticker, J., and Wiederhorn, S.M. (2000) Tensile properties of Ti_3SiC_2 in the 25-1300C temperature range. *Acta Mater.*, **48**, 453.

Radovic, M., Barsoum, M.W., El-Raghy, T., Wiederhorn, S.M., and Luecke, W.E. (2002) Effect of temperature, strain rate and grain size on the mechanical response of Ti_3SiC_2 in tension. *Acts Mater.*, **50**, 1297–1306.

Ritchie, R.O., Gilbert, C.J., and McNaney, C.J. (2000) Mechanics and mechanisms of fatigue damage and crack growth in advanced materials. *Int. J. Solids Struct.*, **37**, 311–329.

Salama, I., El-Raghy, T., and Barsoum, M.W. (2002) Synthesis and mechanical properties of Nb_2AlC and $(Ti,Nb)_2AlC$. *J. Alloys Compd.*, **347**, 271–278.

Sarkar, D., Basu, B., Chu, M.C., and Cho, S.J. (2007) R-curve behavior of Ti_3SiC_2. *Ceram. Int.*, **33**, 789–793.

Stroh, A.N. (1958) The cleavage of metal single crystals. *Philos. Mag.*, **3**, 597.

Suematsu, H., Suzuki, T., Iseki, T., and Mori, T. (1991) Kinking and cracking caused by slip in single crystals of silicon carbide. *J. Am. Ceram. Soc.*, **74**, 173–178.

Sun, Z.-M. (2011) Progress in research and development on MAX phases - a family of metallic ceramics. *Int. Mater. Rev.*, **56**, 143–166.

Tian, W., Sun, Z.-M., and Hashimoto, H. (2009a) Synthesis, microstructure and properties of $(Cr_{1-x}V_x)_2AlC$ solid solutions. *J. Alloys Compd.*, **484**, 130–133.

Tian, W., Sun, Z.-M., Hashimoto, H., and Du, Y. (2009b) Compressive deformation behavior of ternary compound Cr_2AlC. *J. Mater. Sci.*, **44**, 102–107.

Tian, W., Wang, P., Zhang, G., Kan, Y., and Li, Y. (2007) Mechanical properties of Cr_2AlC ceramics. *J. Am. Ceram. Soc.*, **90**, 1663–1666.

Tian, W., Wang, P., Zhang, G., Kan, Y., Li, Y., and Yan, D. (2006) Synthesis and thermal and electrical properties of bulk Cr_2AlC. *Scr. Mater.*, **54**, 841–846.

Turan, S. and Knowles, K.M. (1995) High resolution transmission electron microscopy of the planar defect structure of hexagonal boron nitride. *Phys. Status Solidi A*, **150**, 227–237.

Tzenov, N. and Barsoum, M.W. (2000) Synthesis and characterization of $Ti_3AlC_{1.8}$. *J. Am. Ceram. Soc.*, **83**, 825–832.

Wan, D.-T., He, L.-F., Zheng, L.-L., Zhang, J., Bao, Y.-W., and Zhou, Y.-C. (2010) A new method to improve the high-temperature mechanical properties of Ti_3SiC_2 by substituting Ti with Zr, Hf, or Nb. *J. Am. Ceram. Soc.*, **93**, 1749–1753.

Wan, D.T., Meng, F.L., Zhou, Y.C., Bao, Y.W., and Chen, J.X. (2008) Effect of grain size, notch width, and testing temperature on the fracture toughness of $Ti_3Si(Al)C_2$ and Ti_3AlC_2 using the chevron-notched beam (CNB) method. *J. Eur. Ceram. Soc.*, **28**, 663–669.

Wang, H.J., Jin, Z.H., and Miyamoto, Y. (2002) Effect of Al_2O_3 on mechanical properties of Ti_3SiC_2/Al_2O_3 composite. *Ceram. Int.*, **28**, 931–934.

Wang, X.H. and Zhou, Y.C. (2002a) Microstructure and properties of Ti_3AlC_2 prepared by the solid liquid reaction synthesis and simultaneous in-situ hot pressing process. *Acta Mater.*, **50**, 3141–3149.

Wang, X.H. and Zhou, Y.C. (2002b) Solid–liquid reaction synthesis and simultaneous densification of polycrystalline Ti_2AlC. *Z. Metallkd.*, **93**, 66–71.

Ying, G., He, X., Li, M., Du, S., Han, W., and He, F. (2011a) Effect of Cr_7C_3 on the mechanical, thermal, and electrical properties of Cr_2AlC. *J. Alloys Compd.*, **509**, 8022–8027.

Ying, G., He, X., Li, M., Han, W., He, F., and Du, S. (2011b) Synthesis and mechanical properties of high-purity Cr_2AlC ceramic. *Mater. Sci. Eng., A*, **528**, 2635–2640.

Yu, W., Li, S., and Sloof, W.G. (2010) Microstructure and mechanical properties of a $Cr_2Al(Si)C$ solid solution. *Mater. Sci. Eng., A*, **527**, 5997–6001.

Zhang, Z.F., Eckert, J., and Schultz, L. (2003a) Difference in compressive and tensile fracture mechanisms of $Zr_{59}Cu_{20}Al_{10}Ni_8Ti_3$ bulk metallic glass. *Acta Mater.*, **51**, 1167–1179.

Zhang, H., Wang, Z.G., Zang, Q.S., Zhang, Z.F., and Sun, Z.M. (2003b) Cyclic fatigue crack propagation behavior of Ti_3SiC_2 synthesized by pulse discharge sintering (PDS) technique. *Scr. Mater.*, **49**, 87–92.

Zhang, Z.F. and Sun, Z.-M. (2005) Shear fracture behavior of Ti_3SiC_2 induced by compression at temperatures below 1000 °C. *Mater. Sci. Eng.*, **408**, 64–71.

Zhang, W., Travitzky, N., Hu, C., Zhou, Y., and Greil, P. (2009) Reactive Hot pressing and properties of Nb_2AlC. *J. Am. Ceram. Soc.*, **92**, 2396–2399.

Zhou, A.G., Barsoum, M.W., Basu, S., Kalidindi, S.R., and El-Raghy, T. (2006a) Incipient and regular kink bands in dense and 10 vol.% porous Ti_2AlC. *Acta Mater.*, **54**, 1631.

Zhou, Y.C., Chen, J.X., and Wang, J.Y. (2006b) Strengthening of Ti_3AlC_2 by incorporation of Si to form $Ti_3Al_{1-x}Si_xC_2$ solid solutions. *Acta Mater.*, **54**, 1317–1322.

Zhou, Y.C., Wan, D.T., Bao, Y.W., and Wang, J.Y. (2006c) In situ processing and high-temperature properties $Ti_3Si(Al)C_2/SiC$ composites. *Int. J. Appl. Ceram. Soc.*, **3**, 47–54.

Zhou, W.B., Mei, B.C., and Zhu, J.Q. (2009) On the synthesis and properties of bulk ternary Cr_2AlC ceramics. *Mater. Sci. Pol.*, **27**, 973–980.

Zhou, Y.C., Sun, Z.M., Chen, S.Q., and Zhang, Y. (1998) In situ hot pressing/solid–liquid reaction synthesis of dense titanium silicon carbide bulk ceramics. *Mater. Res. Innovations*, **2**, 142–146.

10
Mechanical Properties: High Temperatures

10.1
Introduction

As discussed in the previous chapter, the MAX phases possess an attractive combination of mechanical properties at room temperature. This chapter deals with their high temperature (HT) properties that may be even more attractive. At sufficiently HTs these phases go through a brittle-to-plastic transition temperature (BPTT). The key to understanding the HT mechanical properties lies in understanding the nature of that transition.

The BPTT varies from phase to phase, but for many Al-containing phases and Ti_3SiC_2, it is between 1000 and 1100 °C (Barsoum *et al.*, 2000; El-Raghy *et al.*, 1999; Tian *et al.*, 2009b; Wan *et al.*, 2008). Chen *et al.* (2001) were the first to document the drop in K_{1c} above the BPTT (Figure 9.18a), a phenomenon that was unusual for crystalline solids at that time. Typically, K_{1c} of most solids, that go through, what is more commonly labeled, a ductile-to-brittle transition (DBT), increases dramatically above the transition temperature because of the activation of slip systems that are inactive below that temperature. It follows that the fact that K_{1c} *drops* above the BPTT (Figure 10.1) *categorically rules out* the activation of additional slip systems. It is for this reason that it is more accurate to label the transition as a BPT, rather than a DBT – as some still do – as the latter implies the activation of more independent slip systems than at temperatures below the transition temperature.

Most of our understanding on the deformation of the MAX phases at HTs is based on early work carried out on Ti_3SiC_2, which is the most studied and best understood MAX phase to date (Barsoum *et al.*, 2001; Chen *et al.*, 2001; Radovic *et al.*, 2000, 2001, 2002). However, there is little doubt, and more recent work confirms, that what applies to Ti_3SiC_2 applies to other MAX phases as well. This chapter is structured as follows. First evidence for the presence of large internal stresses during both tensile and compressive loading of polycrystalline samples is presented. This is followed by a summary of the creep properties, again, both tensile and compressive. The last section compiles and summarizes the mechanical properties at HT.

MAX Phases: Properties of Machinable Ternary Carbides and Nitrides, First Edition. Michel W. Barsoum.

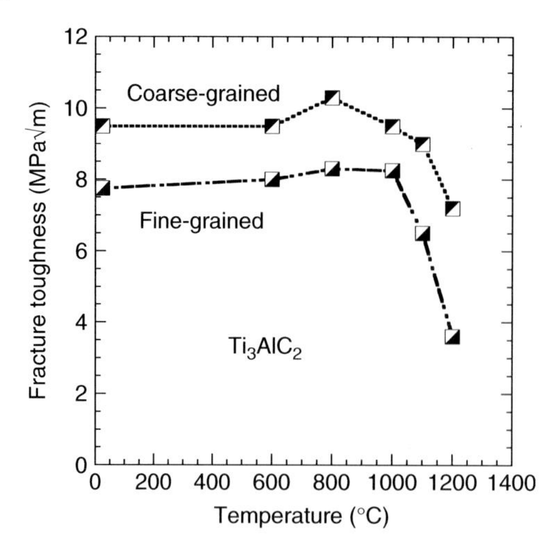

Figure 10.1 Effect of temperature on K_{1c} of both FG and CG Ti_3AlC_2 samples. The grains in the CG sample were 20 ± 8 μm thick plates, ≈75 ± 25 μm in diameter; those in the FG samples were 6.5 ± 2 μm thick plates and 28 ± 10 μm in diameter (Wan *et al.*, 2008).

10.2 Plastic Anisotropy, Internal Stresses, and Deformation Mechanisms

The MAX phases, ice (Duval *et al.*, 1983), graphite (Kelly, 1981) and layered minerals, such as mica (Bell and Wilson, 1981; Christoffersen and Kronenberg, 1993; Kronenberg *et al.*, 1990), are all quite plastically anisotropic in that the dislocations are confined to two-dimensions. This plastic anisotropy, combined with the fact that these solids lack the five independent slip systems needed for ductility, quickly leads to an uneven state of stress when polycrystalline samples are loaded (Barcelo *et al.*, 2010; Duval *et al.*, 1983; Radovic *et al.*, 2000). Glide of basal plane dislocations takes place only in favorably oriented, or soft grains, which rapidly transfer the load to hard grains, viz. those not favorably oriented to the applied stress, σ. Needless to say, this quickly leads to the generation of high internal stresses.

Understanding the nature of these internal stresses, why they accrue and how they dissipate, is the *key* to understanding the HT mechanical response of the MAX phases, as well as any other solids in which dislocations are confined to two dimensions. To put it most succinctly: the response of the MAX phases to stresses at elevated temperatures is related to the rate at which these internal stresses accrue vs. the rate at which they are relaxed. If the former is faster than the latter, brittle failure ensues. If the opposite is true, significant plasticity is possible and observed. It follows that at higher σs or strain rates, $\dot{\varepsilon}$, and/or lower Ts the failure is brittle. Conversely, higher Ts, together with lower σs and lower $\dot{\varepsilon}$s can result in significant plasticity. This is most dramatically seen in Figure 10.2 in which the response of a Ti_3SiC_2 sample loaded in tension at 1200 °C in air is plotted versus

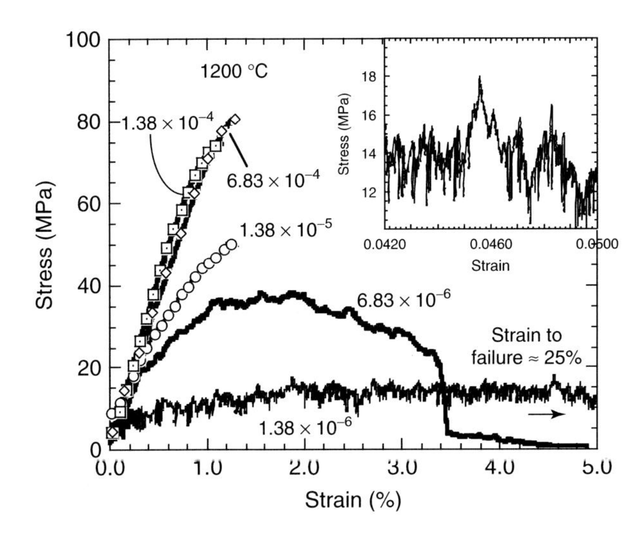

Figure 10.2 Stress–strain curves for coarse-grained Ti_3SiC_2 samples loaded in tension as a function of strain rate at 1200 °C in air. Reducing the strain rate results in a more plastic response and softer samples because of damage accumulation. Inset is a magnification of serrations observed in the sample loaded at the lowest strain rate. This is not noise (Barsoum *et al.*, 2001).

$\dot{\varepsilon}$ (Barsoum *et al.*, 2001). When pulled rapidly, the response is brittle-like. When pulled slowly, however, plastic strains of over 25% were observed.

In Chapter 8, the case was made that the fully reversible, strain rate independent stress–strain loops are due to incipient kink bands (IKBs). At HTs, the response becomes time dependent. It follows that new micro-mechanisms must be invoked. As discussed in this chapter the latter are mobile dislocation walls (MDWs) that lead to kink boundaries as well as dislocation pileups (DPs) (Barcelo *et al.*, 2010; Radovic *et al.*, 2002, 2003; Zhen, Barsoum, and Kalidindi, 2005a).

10.2.1 Origin of Internal Stresses

Before presenting evidence for the presence of large internal stresses, it is important to appreciate their origins. As noted above, there are at least two sources: IKBs and MDWs, on the one hand, and DPs, on the other. Each is discussed below.

10.2.2 Dislocation Pileups

It has long been appreciated that the stress concentration due to a DP is given by:

$$\sigma_x = C_o \left(\frac{\lambda}{x} \right)^{1/2} \tag{10.1}$$

where C_0 is a constant and x is the distance from the tip of a DP. The pileup length is typically assumed to scale with grain size, λ. The longer the DP, the larger the stress concentration at its tip. It is for this reason that the yield points of fine-grained (FG) solids are larger than their coarse-grained (CG) counterparts; a strengthening effect that is referred to as Hall–Petch. Note that for hexagonal solids, such as the MAX phases, in which the grains tend to grow into thin disks, λ is the diameter of the disk (i.e., the dimension along [100]).

When DPs occur on a single slip plane the resulting deformation shears the crystal on one side of the slip plane relative to that on the other side. A good illustration of the effects of DPs on the deformation of a Ti_3SiC_2 grain is shown in Figure 10.3a. It is reasonable to assume from this OM image that the steps at the grain boundary (denoted by thick black arrow) were caused by slip along the basal planes. Another example, at the submicron level, is shown in Figure 10.3b. Here again, steps at the grain boundaries, (GBs) are taken as evidence for basal slip.

10.2.3
Mobile Dislocation Walls

As discussed in Section 9.1.1.6, when part of an MDW is arrested but the remaining part keeps moving, large tensile and compressive stresses accrue near the tip of the part that is not arrested (Figure 9.25). If high enough, these Stroh cracks can lead to delaminations. Examples of such cracks in the MAX phases were shown in Figure 9.26. As discussed in Chapter 9, such cracks have been observed in many other plastically anisotropic solids such as Zn, ice, graphite, and SiC single crystals among many others.

It is crucial to note here that MDWs of opposite polarity will attract each other. The time constant of this attraction, however, is longer than that of the IKBs. As discussed in Chapter 8, the response of the latter is time independent at least at the rates they were tested.

According to Eq. (9.15), all else being equal, thicker grains (i.e., those with larger 2αs) should delaminate first. And as delamination is the critical event for the transition of IKBs to MDWs and kink bands (KBs), this transition should occur at lower stresses, and by extension lower temperatures, in CG samples than in FG ones.

Before presenting the evidence for this conjecture it is important to review what happens to the fully reversible cyclic stress–strain loops described in Chapter 8, with increasing temperature. Figure 10.4a,b show typical compression results for CG and FG Ti_3SiC_2 samples, respectively. From these results the following observations can be made as the temperature is increased:

1) The loops become significantly larger because they are open and no longer fully reversible.
2) The response is no longer strain rate independent.
3) After complete unloading, the strain continues to recover. This is better seen in Figure 10.5, where stress and strain are plotted as a function of time; the latter keeps recovering long after the load is removed. The recovered strain is not small – it is roughly two orders of magnitude higher than the strains at

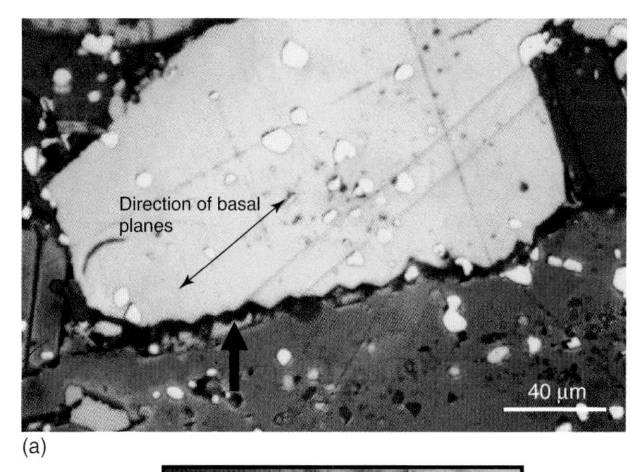

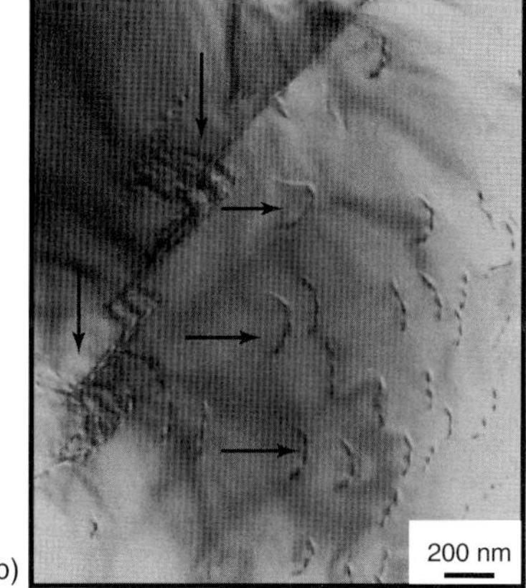

Figure 10.3 (a) OM image of polished and etched gage surface that was parallel to the vertically applied load. This sample, with a duplex microstructure, was loaded in tension at 1000 °C and 40 MPa. Decohesion of grain boundary is denoted by vertical arrow. Steps at grain boundary were presumably formed by slip along the basal planes whose direction is indicated on figure (Radovic *et al.*, 2003) and (b) TEM image of a coarse-grained Ti_3SiC_2 sample, again showing grain boundary steps (vertical arrows) and dislocation arrays (horizontal arrows) after creep testing (Barcelo *et al.*, 2010).

room temperature. Note that the recoverable strain in the CG sample is higher than that for the FG sample.

4) Both microstructures become less stiff, with the effect being more pronounced for the CG microstructure. The inset in Figure 10.4b compares the stress–strain curves of the FG and CG microstructures at 1200 °C, from which it is reasonable to conclude that for a given stress, the plastic deformation is significantly larger in the CG than in the FG samples.
5) For the FG samples, and up to ≈70 MPa, the stress–strain curves at room temperature, 900 and 1100 °C, are, within the resolution of the experimental results, identical (Figure 10.4b). Furthermore, up to 160 MPa, there is little difference between the RT and 900 °C curves. Furthermore, up to 900 °C, the

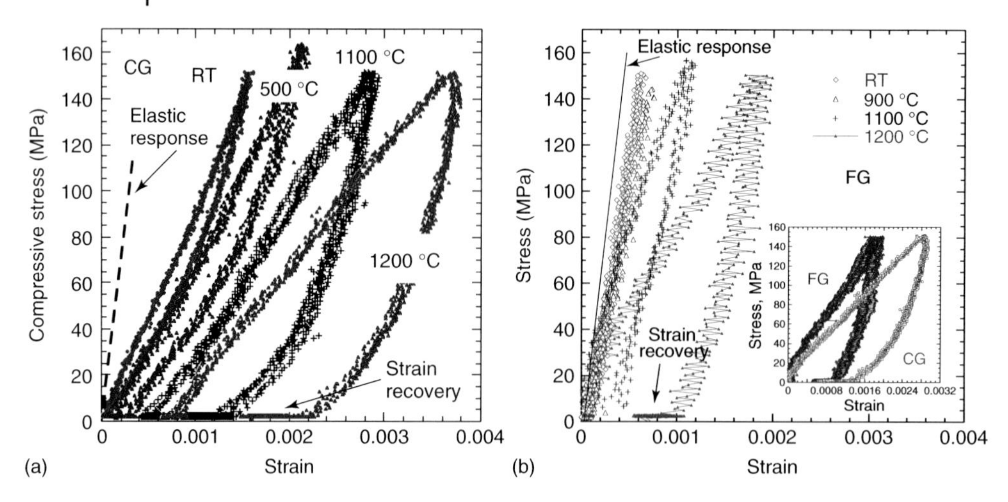

Figure 10.4 Temperature dependence of cyclic compressive stress–strain curves for, (a) CG and (b) FG Ti_3SiC_2 samples. Inset in (b) compares the first cycles for both microstructures at 1200 °C. Cycles in (a) are shifted to the right for clarity (Zhen, Barsoum, and Kalidindi, 2005a).

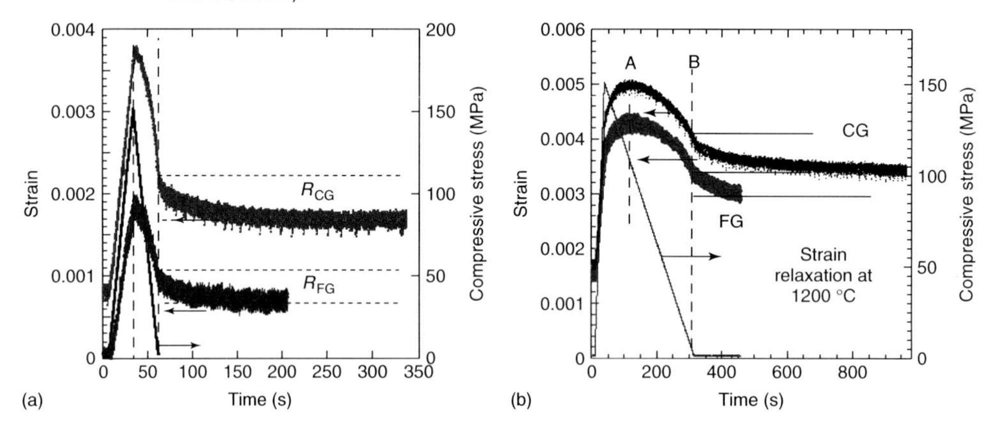

Figure 10.5 Time dependence of stress (right axis) and strain (left axis) for FG and CG samples, (a) same loading and unloading rates and (b) same as (a), but with 10 times slower unloading rates. Note time delay in the maximum positions of the stresses and strains in (b). Also note relaxation after load is fully removed (Zhen, Barsoum, and Kalidindi, 2005a).

initial slopes of the stress–strain cycles are close to the true elastic moduli (as determined from ultrasound measurements).

6) The response of the CG samples at 500 °C (Figure 10.4a) is comparable to that of the FG samples at 1100 °C (Figure 10.4b). At 900 °C, and up to 140 MPa, the response of the latter is almost linear elastic (Figure 10.4b).
7) Both the relaxation times and total recovered strain are higher in the CG microstructure than its FG counterpart (Figure 10.5). The differences in the responses of the FG and CG microstructures are more pronounced if the

unloading rates are fast (Figure 10.5a), than if they are slow (Figure 10.5b). The slower unloading rates result in distinct bow outs in the corresponding stress/strain curves (e.g., see Figure 3a in Barsoum *et al.*, 2003). These bow outs reflect the fact that the strain and stress maxima do not coincide (Figure 10.5b). In other words, the samples keep shrinking for a relatively long time after the applied load is totally removed.

The full implications of these results are discussed below.

10.2.4 Evidence for Internal Stresses

The most direct evidence for the presence of large internal stresses at high temperatures that relax with time are the results shown in Figures 10.4 and 10.5. Even after total unloading, the compression samples continue to expand; an expansion that cannot occur without the presence of residual internal stresses. Note that the presence of the latter is not confined to samples loaded in compression, but are also observed during tensile loading at higher temperatures (Radovic *et al.*, 2002, 2003). Additionally, the presence of large internal stresses is not restricted to the MAX phases, but have been observed in ice (Duval *et al.*, 1983), and other plastically anisotropic solids. These internal stresses are most probably stored in the hard grains, that is, those that are elastically deformed and or at any MDWs that are being partially sheared.

10.2.5 Evidence for Uneven Deformation

The fact that the number of independent slip systems is <5 results in uneven deformation. Barcelo *et al.* (2010) used electron backscattered diffraction (EBSD) maps of post-crept samples to show that when CG Ti_3SiC_2 samples (Figure 10.6a,b) were deformed in creep at elevated temperatures, the resulting microstructure (Figure 10.6c) was comprised of grains that were heavily deformed, adjacent to ones that appeared to be undeformed or intact. This is best seen in Figure 10.6c where individual misorientation maps within each grain are plotted. Blue grains denote ones in which little evidence for plastic deformation was observed. The rainbow colored grains, on the other hand, indicate ones in which plastic deformation occurred.

The work of Barcelo *et al.* (2010) further showed that the deformed grains came in two configurations: heavily kinked and banana-shaped shown in Figure 10.7a,b, respectively. In the former, a number of boundaries of opposite signs – viz. KBs depicted by the sketch in the left middle inset in Figure 10.7a – are observed. In the latter, the boundaries are all of the same sign resulting in a gradual bending of the grain into a banana-like shape (top inset in Figure 10.7b). Such grains were not present in the undeformed gauge section.

The most dramatic illustration of the complex relationship between kinking and the state of nonuniform stresses is shown in Figure 10.7a. When the misorientations within grain II, shown in Figure 10.6b, were mapped (Figure 10.7a),

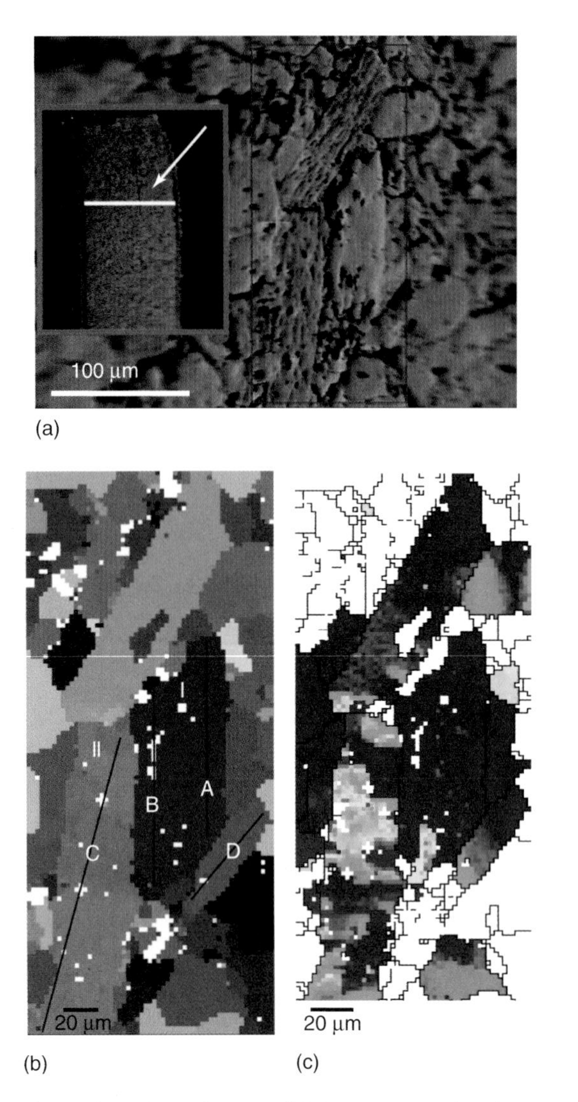

Figure 10.6 (a) Backscattered SEM image of gauge area of a CG Ti_3SiC_2 sample deformed at 1000 °C under a load corresponding to a stress of 80 MPa. The strain to failure was 2.5%. Inset shows a low magnification SEM image of entire sample that was loaded vertically in tension. The orientation image microscopy (OIM), information was taken at the location, denoted by the white horizontal line, that was ≈1.5 mm below the fracture plane, which is at the top of the figure. Width of sample is 2 mm; (b) OIM of area bounded by red rectangle in (a); and (c) Individual misorientation (MO), angle map within each grain, with blue = 0° and red = 10°. Note that in grain II, in which line labeled C appears, the basal planes are edge-on (Barcelo *et al.*, 2010).

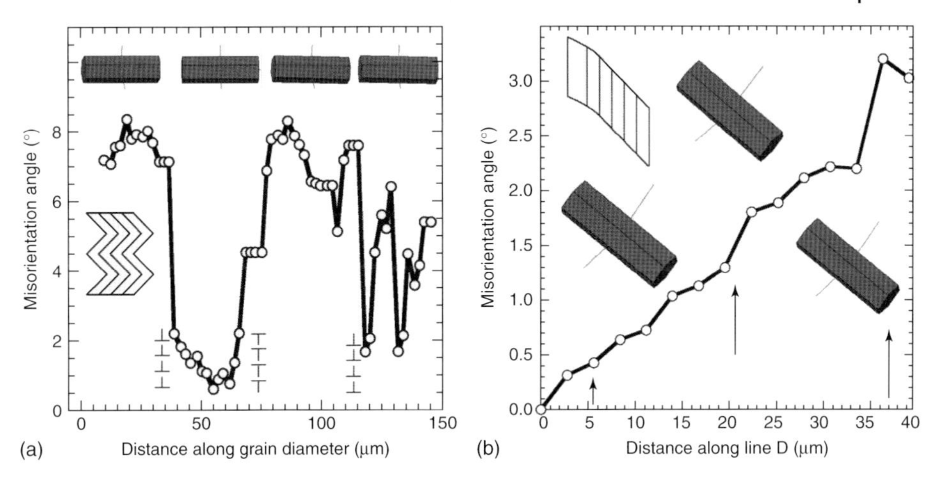

Figure 10.7 Traces of misorientation angle (MO) in deformed, (a) grain II along line C in Figure 10.6b and (b) grain III along line D, shown in Figure 10.6b. Unit cells in (a) and (b), denote their orientation with respect to the direction of the vertically applied load. Left middle inset in (a) is a schematic of the grain shape after deformation. Left top inset in (b) is a schematic of the grain shape after deformation. (Barcelo *et al.*, 2010).

it was concluded that the grain had kinked like an accordion along the axial direction, despite the fact that the sample was loaded in tension in the same direction! This is a *remarkable result* that is not easily reconciled with strain energy considerations or common sense. After all, when one pulls a deck of cards, the last response one would expect is a buckling of the cards and the formation of kink boundaries and yet that is what appears to have occurred to grain II in Figure 10.6b. The simplest explanation for this unusual observation has to be that DPs in adjacent soft grains caused grain C shown in Figure 10.7b, to kink the way it did. This deformation mode was found to be quite common.

Interestingly, decades ago Hauser *et al.* (1955) made a somewhat similar observation in Mg. In a paper in which they carefully studied the deformation of CG Mg loaded in tension they showed that in one specimen 10 twins caused extension and 23 caused contraction in the stress direction. Hauser *et al.* observed, "... the overall strain due to twinning in this specimen is actually *negative*, a fact not easily reconciled with strain energy considerations." The similarities between Ti_3SiC_2 and Mg run deeper than this commonality. We have recently shown that several hexagonal metals, including Mg, Co, Zn, and Ti, are like the MAX phases KNE solids (Zhou and Barsoum, 2009; Zhou *et al.*, 2010; Zhou, Basu, and Barsoum, 2008). It is thus not surprising that they behave similarly. Along those lines it is important to note that in an EBSD study of post-creep CG Mg loaded in tension – quite reminiscent of Barcelo's *et al.* paper – the central and key role of kink boundaries in the deformation process was clearly and unambiguously shown (Yang, Miura, and Sakai, 2003).

The aforementioned results are not the only evidence available for the deformation and breakup of individual grains by the formation of kink boundaries.

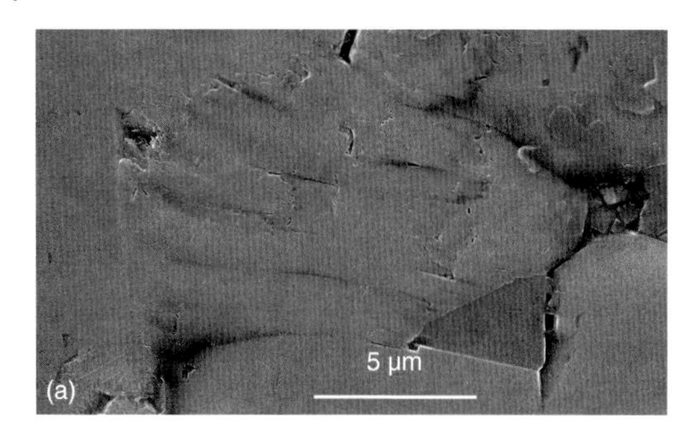

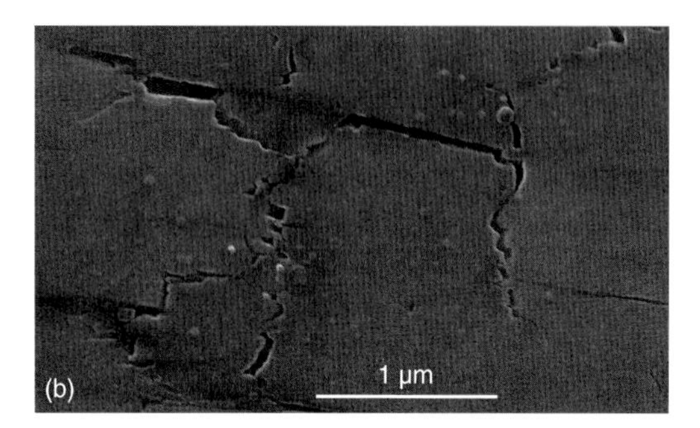

Figure 10.8 SEM images of coarse-grained Ti_3SiC_2 sample deformed 4% in compression under a load corresponding to a stress of 25 MPa at 1300 °C showing, (a) breakup of a large grain into a number of smaller rectangular-like grains and (b) same as (a), but at higher magnification (Zhen *et al.* 2005b).

Figure 10.8a shows a SEM image of a CG Ti_3SiC_2 sample deformed 4% in compression under a load corresponding to a stress of 25 MPa at 1300 °C. Figure 10.8b shows a grain in which a number of parallel kink boundaries were observed. Clearly, this modest deformation resulted in the breakup of a large grain into a number of smaller, rectangular-like grains. On the basis of what is known about how the MAX phases deform, the morphology of these subgrains and their relationship to the basal planes, it is reasonable to assume that the subgrains are separated from each other by kink boundaries, concomitant with basal plane delaminations. It is important to point out at this juncture that the delaminations most probably occur along DPs for the simple reason that such delaminations rid the crystal of the pileups' strain energy.

The TEM images of a CG Ti_3SiC_2 sample loaded in tension at 1200 °C shown in Figure 10.9 also confirm the presence of a large number of dislocations – in the form of DPs as a result of the HT deformation. MDWs and low angle grain boundaries (not shown) were also observed.

In the CG sample, deformation is more complex, wherein basal slip leads to the formation of KBs and parallel dislocation walls (not shown). The deformation is also quite nonhomogeneous; the small grains appeared undeformed; only the larger grains were deformed, and not all of them. In Figure 10.6b, it is apparent that many

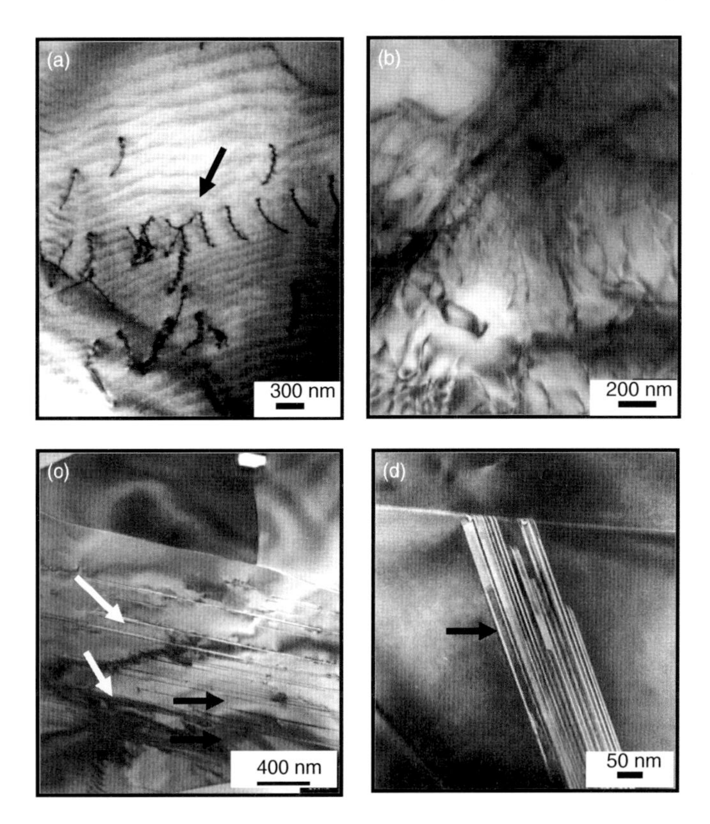

Figure 10.9 TEM images of a CG Ti_3SiC_2 sample crept in tension at 1200 °C showing, (a) basal dislocation arrays or pileups (denoted by arrow) in un-deformed grip area, (b) large number of dislocations in a highly deformed grain in gauge area, (c) shear band in deformed region, and (d) same as (c), but at higher magnification (Barcelo *et al.*, 2010).

of the grains were undeformed, with grain II taking up most of the deformation. In some cases, the majority of a large grain remained undeformed, with the deformation limited to a single kink boundary (e.g., grain I in Figure 10.6b).

Although, a relatively small number of the grains were studied by TEM, the deformation in the FG structure seems to be more homogeneous than in the CG structure. In the FG samples, deformation seemed to occur predominately by basal slip leading to GB steps and cavitations at the edges of these steps as well as bending of the basal planes as a result of MDWs or low angle grain boundaries. As importantly, the dislocation density in the gauge area of FG samples, after deformation, appeared to be *lower*, than in the grip section, suggesting the presence of an effective dislocation annihilation mechanism, the most likely being the grain boundaries. Another possibility is the nucleation of cavities (Figure 10.10) that cause stress relaxation and the concomitant annihilation of dislocations. It is important to note that these conclusions are tentative at this time and more

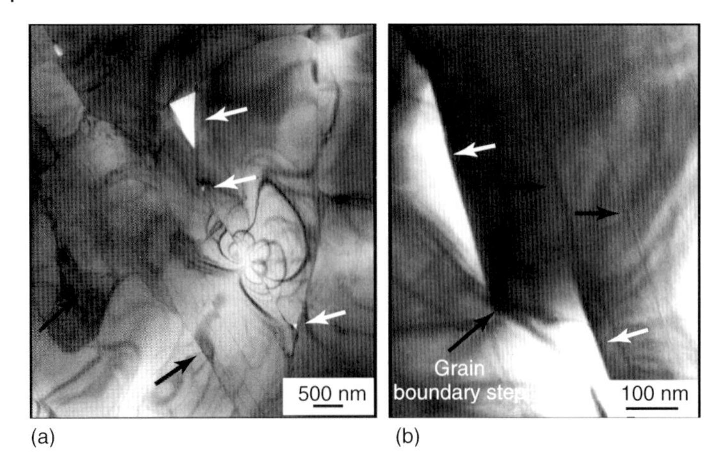

Figure 10.10 Post-creep TEM image of FG Ti_3SiC_2 sample showing, (a) triangular cavities (white arrows), and stacking faults or dislocation arrays (black arrow) and (b) triangular cavities (white arrows), grain boundary steps and dislocation arrays, and/or stacking faults (black arrows) (Barcelo *et al.*, 2010).

comprehensive TEM work on other samples deformed under different conditions need to be examined.

10.2.6
Softening of Grain Boundaries and Interfaces between Layers

According to the aforementioned discussion, it would appear that, paradoxically, the internal stresses at elevated temperatures are higher than those at room temperature. This, however, would be an incorrect conclusion. For a given microstructure, the internal stresses should be a weak function of temperature. What changes with temperature, however, are the grain boundary decohesion strengths and/or the strengths of the various interfaces between the nanolayers. In other words, what most probably changes is γ_S in Eq. (9.15).

Said otherwise, at higher temperatures, it is reasonable to assume that the GBs are softer, and consequently, the IKBs can now detach from each other more easily and create MDWs, which in turn results in plastic deformation, and ultimately KBs, such as those shown in Figure 10.8. This breakup of grains does not occur at lower temperatures.

Such softening would also allow the DPs to run into the GBs, causing the latter and/or triple points to slide and form triple point and other cavities. Direct microstructural evidence for the latter mechanism at two very different lengths scales is shown in Figures 10.3 and 10.10a, b.

Probably the strongest, and most direct, evidence for this softening can be found in the temperature dependencies of the slopes of the *initial* portions of tensile stress–strain curves, shown in Figure 10.11a. When the latter are compared to the values of E that one would expect based on ultrasound (shown as solid line

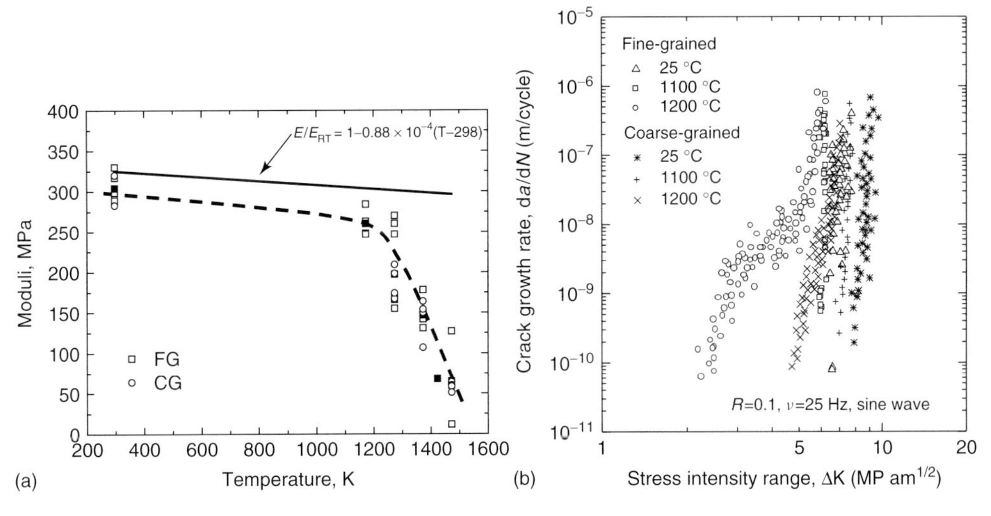

Figure 10.11 (a) Temperature dependence of Young's moduli obtained from the slopes of the initial parts of the tensile stress–strain curves. Solid line represents expected changes in E with temperature from sound velocity measurement. Dashed line is guide to the eyes (Radovic *et al.*, 2002). (b) Cyclic fatigue-crack growth rates, da/dN, at $R = 0.1$ in Ti_3SiC_2 with FG and CG microstructures, as a function of the applied stress-intensity range, ΔK, at 25, 1100, and 1200 °C. Note the large reduction in K_{th} at 1200 °C for the FG microstructure (Chen *et al.*, 2001).

in Figure 10.11a) there is little doubt that the microstructure softens considerably around the BPTT.

Another manifestation of the same phenomenon is the aforementioned decrease in K_{1c} observed at higher Ts (Figures 9.18a and 10.1). At 1200 °C, which is above the BPTT, there is a striking change in behavior, principally in the form of significantly reduced ΔK_{th} thresholds (Figure 10.11b) that can be attributed to the onset of significant deformation and associated microstructural damage.

At temperatures around the BPTT, macroscopic deformation was also observed in the form of significant widening of the notch. This is shown in Figure 10.12a,b, which compares the microstructure, near the crack tip/wake at 1100 and 1200 °C for a CG sample, respectively. The crack paths involved significantly more transgranular and/or translamellar cracking, especially in the FG microstructure (not shown) compared with that seen at lower temperatures, which severely diminished the propensity for grain bridging in the crack wake. More importantly, extensive microcracking and, to a lesser extent, cavitation, were apparent throughout both microstructures, although the intensity of damage was most severe in the vicinity of the crack tip (Figure 10.12b).

The decrease in K_{1c} above the BPTT not only rules out the activation of non-basal slip systems, but also indicates that the back-pressure from dislocation arrays, kink boundaries and MDWs, and so on at the crack tip is reduced. The most likely mechanism is delaminations and/or grain boundary decohesion. *To summarize, one – possibly the only – assumption needed to understand the HT response of the*

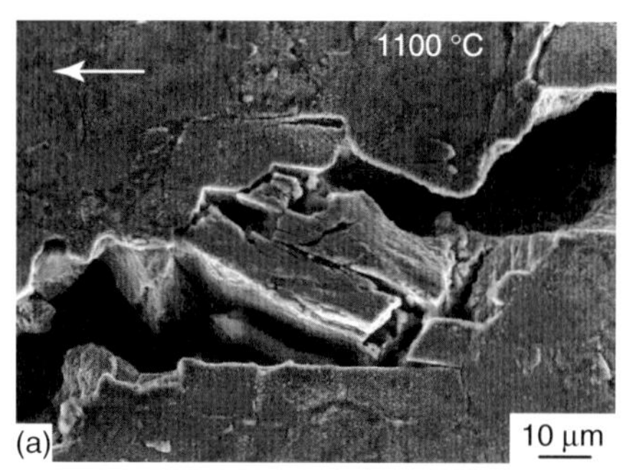

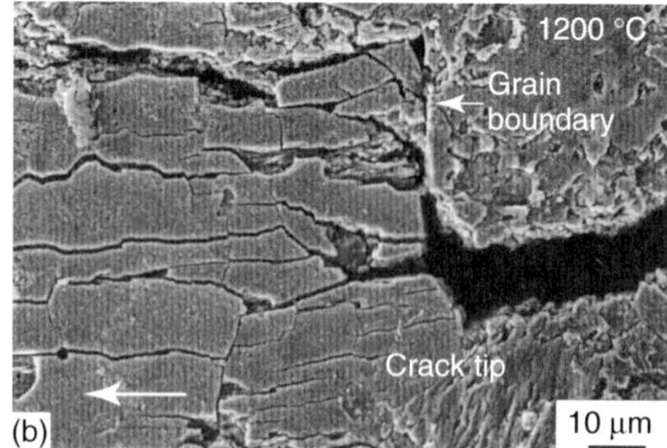

Figure 10.12 SEM images of the profiles of fatigue cracks propagating in CG Ti_3SiC_2 samples in the crack tip region at, (a) 1100 °C and (b) 1200 °C. The latter is above the BPTT of Ti_3SiC_2. Note the profuse density of microcracks and cavities near the main crack, especially trans-granular cracks. Arrows indicate direction of crack growth (Chen *et al.*, 2001).

MAX phases is to invoke a temperature dependent grain boundary decohesion and/or interplanar delamination stress.

Before presenting the evidence for this conjecture, it is important to point out that paradoxically, under *cyclic* loading, these HT delaminations and grain boundary decohesions do *not* lead to cyclic softening, but cyclic hardening instead!

10.2.7 Cyclic Hardening

Clear evidence for cyclic *hardening* at temperatures as high as 1200 °C has been observed for both FG and CG Ti_3SiC_2 samples (Figure 10.13a); the effect on the latter, however, is more dramatic (Barsoum *et al.*, 2003). The hardening is manifested two ways: first the areas enclosed by the loops become smaller (Figure 10.13a). Second, the *initial* slopes of the stress–strain curves approach the linear elastic limit – as measured by ultrasound – with increased cycling (Figure 10.13b). The latter clearly implies that cycling results in micro-domains that are more difficult to kink than the initial grains (i.e., a reduction of 2α in Eq. (8.2) or 2β in Eq. (8.4)).

Figure 10.14 compares the compressive stress–strain curves of a CG Ti_3SiC_2 sample tested at *room temperature before* and *after* a 2% deformation at 1300 °C. The latter was loaded twice; the first loading resulted in a slightly open loop (red open circles in Figure 10.14), the second (blue open squares in Figure 10.14) and subsequent cycles, resulted in only closed loops (Barsoum *et al.*, 2003; Zhen, Barsoum, and Kalidindi, 2005a). The simplest interpretation for this observation is that during the first cycle, the MDWs generated during the HT deformation stage are swept into the kink boundaries immobilizing them. On subsequent loading, only IKBs are activated and thus the loops are once again fully reversible and reproducible.

As important, when the loops before and after the 2% deformation at 1300 °C are compared to each other – and to the ones obtained on FG and CG Ti_3SiC_2 samples at room temperature (see e.g., Figure 8.7) – the inescapable conclusion is that a

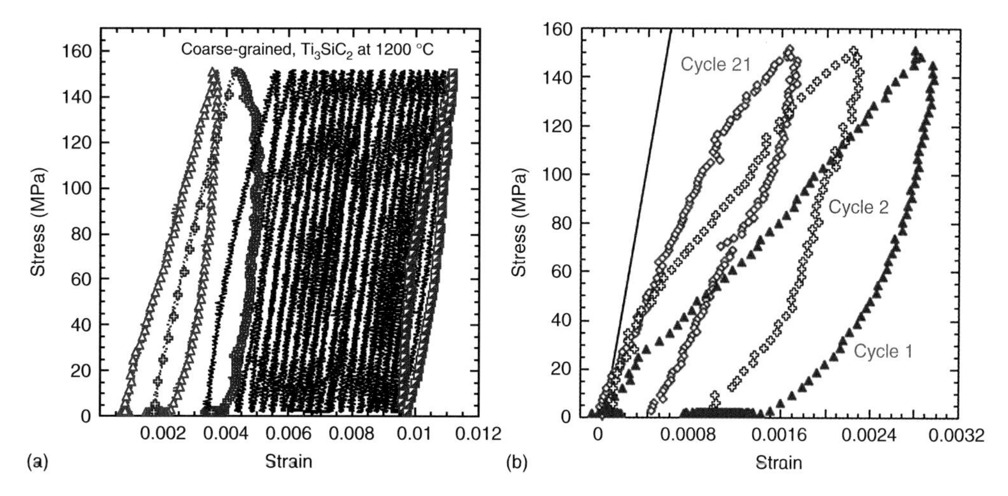

Figure 10.13 Stress–strain curves recorded when a CG Ti_3SiC_2 sample was loaded cyclically 21 times at 1200 °C in air; (a) totality of cycles and (b) comparison of cycles 1, 2, and 21, in (a). In (b) all cycles start at the origin. Cyclic hardening is unambiguous. Solid inclined line in (b) represents a Young's modulus of $E = 270$ GPa, which is the expected linear elastic response at that temperature based on ultrasound measurements (Zhen, Barsoum, and Kalidindi, 2005a).

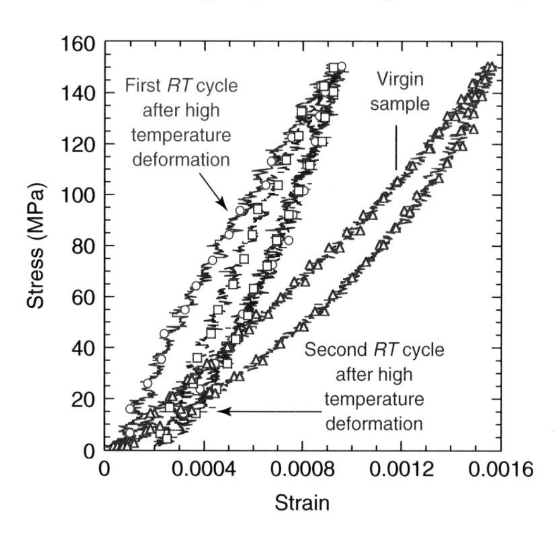

Figure 10.14 Comparison of compressive stress–strain curves of a coarse-grained Ti_3SiC_2 sample tested at room temperature before and after a 2% deformation at 1300 °C. The latter was loaded twice; the first loading resulted in a slightly open loop (red, open circles), the second and subsequent, only closed loops (blue, open squares) (Zhen, Barsoum, and Kalidindi, 2005a).

modest HT deformation effectively *reduces* the grain size of the CG sample in such a way that its room temperature response after the deformation becomes more reminiscent of a much finer grained microstructure (compare stress–strain loops

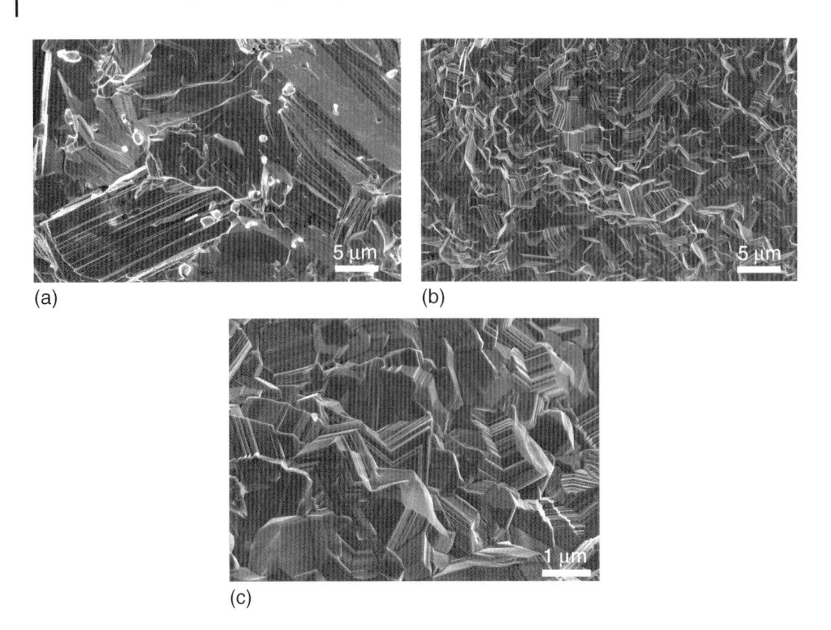

Figure 10.15 SEM images of a fractured Ti_2AlC surface in, (a) grip and, (b) gauge region; (c) is same as (b) but at a higher magnification. The sample was loaded in tension at 1100 °C, under a load corresponding to a stress of 30 MPa. The sample failed after 3 h; the total strain to failure was 16% (Tallman, Naguib, and Barsoum, 2012).

of the virgin CG sample with the ones after deformation in Figure 10.14). The clear stiffening after the HT deformation again strongly implies a reduction in the sizes of the micro-domains (i.e., a reduction of 2α in Eq. (8.2) or 2β in Eq. (8.4)) rendering them more difficult to kink than the virgin grains. The hardening observed also implies that most delaminations/microcracks formed during HT deformation are somewhat healed when the samples are cooled back to ambient temperatures.

Direct evidence for such grain refinement and massive formation of numerous KB can be seen when the fractured surfaces of the grip and gauge areas of a Ti_2AlC sample are compared in Figure 10.15a,b, respectively. Figure 10.15c is a higher magnification of Figure 10.15b, showing clear evidence for the formation of numerous KBs. This sample – loaded in tension at 1100 °C, under a load corresponding to a stress of 30 MPa – failed after 3 h; the total strain to failure was 16% (Tallman, Naguib, and Barsoum, 2012).

10.3
Creep

In the 1000–1300 °C temperature range, Ti_3SiC_2 samples loaded in tension (Figure 10.16a,b) or compression (Figure 10.16c) exhibit primary, secondary, and tertiary creep (Radovic *et al.*, 2001, 2003; Tallman, Naguib, and Barsoum,

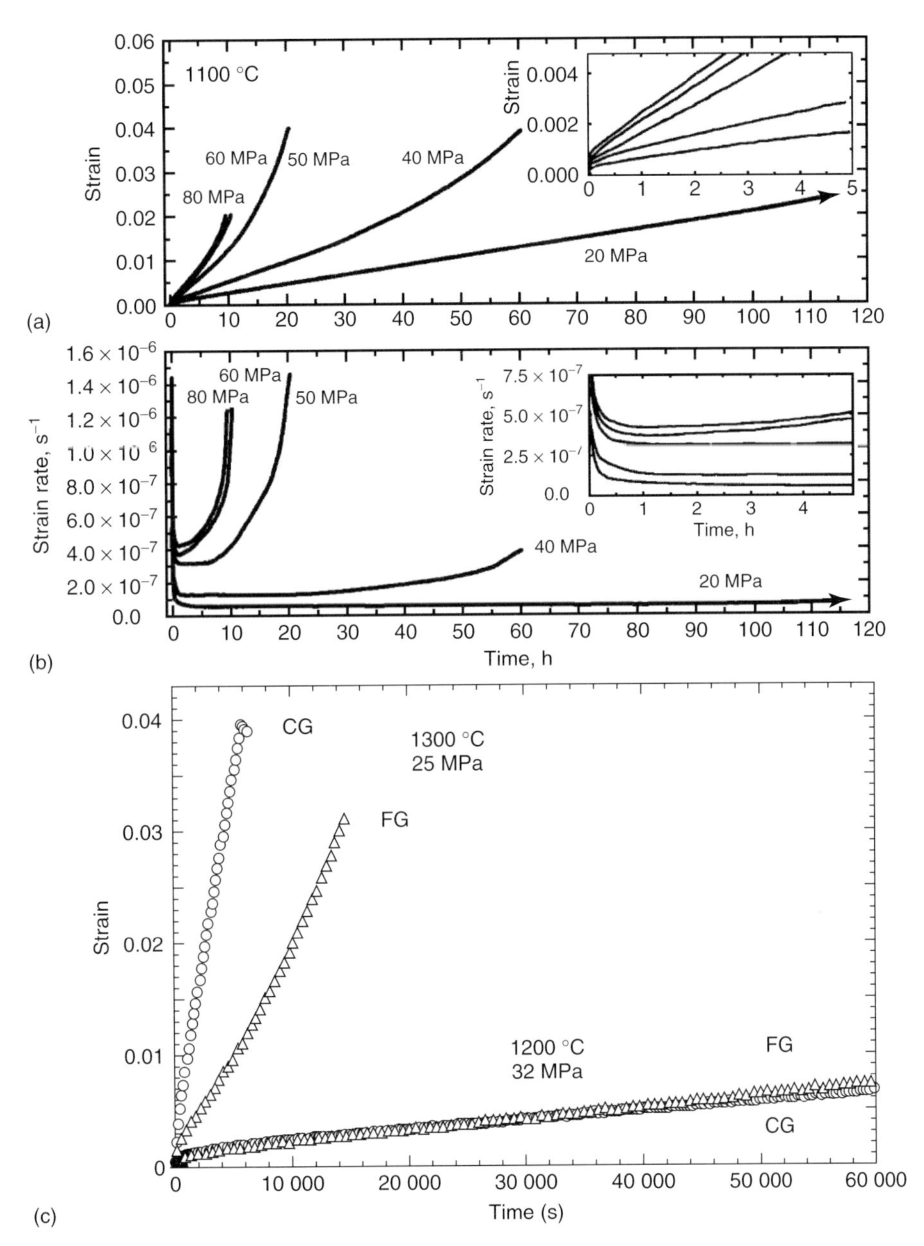

Figure 10.16 Typical time and stress dependencies of, (a) tensile strain and (b) strain rate of CG Ti_3SiC_2 samples tested at 1100 °C. Plot (b) is obtained by differentiation of curves shown in (a). Insets are enlargements of the initial parts of the plots shown in (a) and (b) (Radovic *et al.*, 2003), and (c) compressive strains for FG and CG Ti_3SiC_2 samples tested at 1200 and 1300 °C at stresses shown on figure (Zhen *et al.*, 2005b).

2012; Zhen *et al.*, 2005b). A short primary creep stage, where the deformation rate decreases rapidly, is followed by a secondary creep regime during which the creep rate, although not truly reaching a constant level, does not change significantly over a significant time period (Figure 10.16).

The intermediate or secondary stage can be characterized by a minimum creep rate, $\dot{\varepsilon}_{\min}$. As typically done with creep results, log $\dot{\varepsilon}_{\min}$ versus log σ plots as a function of temperature were generated (Figure 10.17). In tension, a series of almost parallel lines were obtained for both FG and CG microstructures shown, respectively, in Figure 10.17a,b. In compression, on the other hand, the situation is more complicated as two regimes can be discerned. One regime occurs at

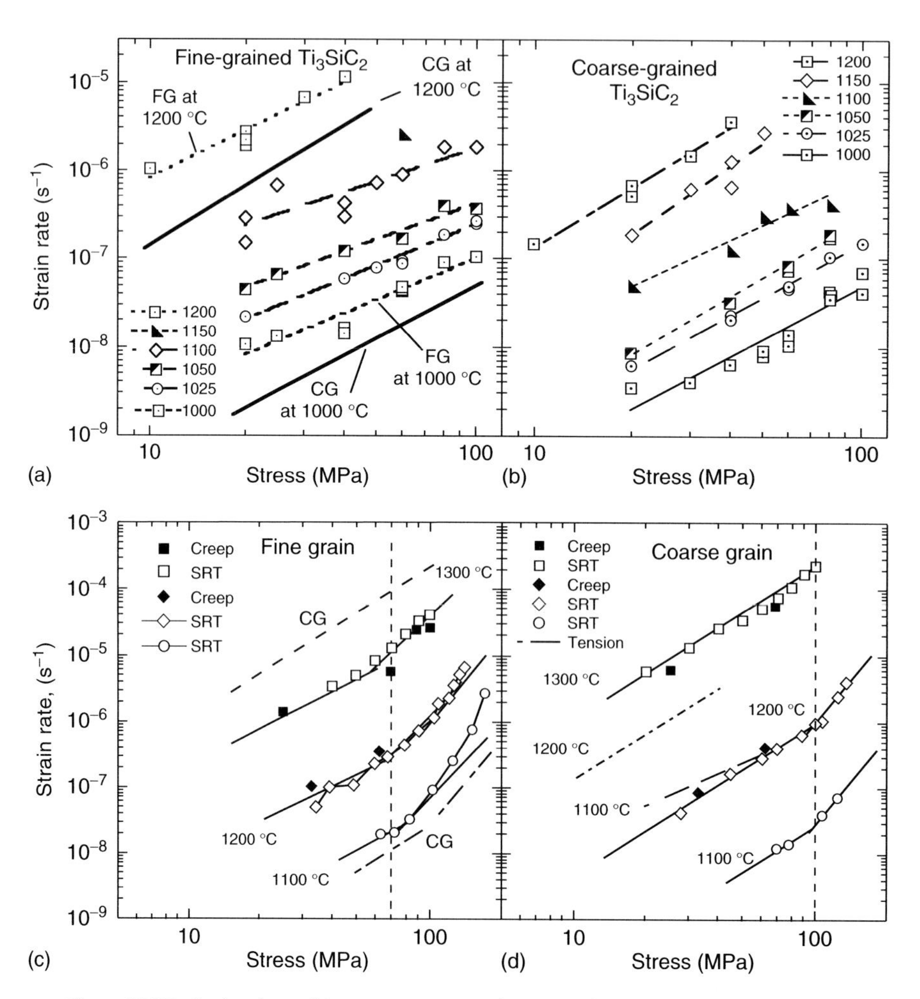

Figure 10.17 ln–ln plots of $\dot{\varepsilon}_{\min}$ versus σ as a function of T and testing technique for Ti_3SiC: (a) FG in tension, (b) CG in tension; (c) FG in compression, and (d) CG in compression. Results in (a) and (b) are taken from Radovic *et al.* (2003); in (c) and (d) from Zhen *et al.* (2005b). SRT refers to stress relaxation technique.

σs < 70 MPa for the FG microstructure (Figure 10.17c) and σ < 100 MPa for the CG microstructure (Figure 10.17d). In the second regime, the creep rate increases. Note that at 1100 °C, the creep rate of the FG microstructure is higher than its CG counterpart; at 1300 °C, the situation is reversed. The full implications of these results are discussed below.

The solid black lines shown in Figure 10.17a, plot the creep rates of the CG Ti_3SiC_2 samples at 1000 and 1200 °C. Comparing these lines to those of the FG samples at the same temperature, it is clear that the creep rates of the latter at low stresses are roughly an order of magnitude higher than the CG samples. At higher stresses, the differences decrease.

When log $\dot{\varepsilon}_{min}$*exp Q/RT is plotted versus log σ, all the data points appear to collapse on a single line (Figure 10.18). Similar observations were obtained when Ti_3SiC_2 was crept under compressive loads (Zhen *et al.*, 2005b). More recent results on the tensile creep of Ti_2AlC also resulted in similar observations (Tallman, Naguib, and Barsoum, 2012). These results suggest that during both tension and compression, the creep rates, at relatively low stresses, can be described by the following relationship (Radovic *et al.*, 2001, 2003; Zhen *et al.*, 2005b):

$$\dot{\varepsilon}_{min} = \dot{\varepsilon}_o A \left(\frac{\sigma}{\sigma_o}\right)^n \exp\left(-Q/_{RT}\right) \tag{10.2}$$

where A, n, and Q are, respectively, a stress-independent constant, stress exponent, and activation energy for creep; R and T have their usual meanings and $\dot{\varepsilon}_o = 1\ s^{-1}$ and σ_o = 1 MPa. The various creep parameters obtained for the two Ti_3SiC_2 microstructures and Ti_2AlC – with an initial microstructure shown in Figure 10.15a – are summarized in Table 10.1.

From these, and other results, the following points – that depend on the level and direction of the stresses applied – are salient (Figure 10.18).

Table 10.1 Summary of creep parameters given in Eq. (10.2) for Ti_3SiC_2 and Ti_2AlC loaded in compression (comp.) or tension (tens.) in air as a function of grain size. These parameters are only valid in compression for σ < 70–100 MPa and T < 1300 °C.

Material	Load	GS	ln A	*n*	Q (kJ mol^{-1})	$\dot{\varepsilon}_{min}$ (s)$^{-1}$ at 1200 °C and 50 MPa	References
Ti_3SiC_2	Comp.	CG	44	2.1	835 ± 40	1.2×10^{-7}	Zhen *et al.* (2005b)
Ti_3SiC_2	Comp.	FG	20	1.9	537 ± 30	7.4×10^{-8}	Zhen *et al.* (2005b)
Ti_3SiC_2	Tens.	CG	17	2	458 ± 12	3.5×10^{-6}	Radovic *et al.* (2003)
Ti_3SiC_2	Tens.	FG	19	1.5	445 ± 10	1.0×10^{-5}	Radovic *et al.* (2001)
Ti_2AlC	Tens.	Intermediate[a]	12.5	2.5	362 ± 88	6.8×10^{-4}	Tallman, Naguib, and Barsoum (2012)

[a] 14 ± 8 μm (Figure 10.15a).

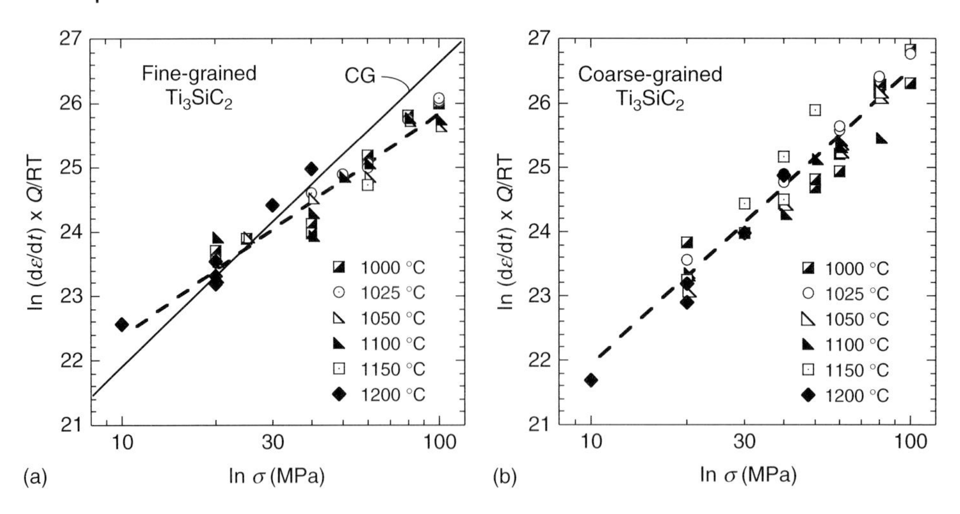

Figure 10.18 (a) ln–ln plot of $\dot{\varepsilon}_{min}$*exp$Q$/RT versus σ for, (a) FG and (b) CG Ti_3SiC_2 samples (Radovic *et al.* 2001 and 2003). Dashed lines are plots of Eq. (10.2) using the parameters listed in Table 10.1. In (a), the solid inclined line represents the results for the CG microstructure (i.e., dashed line in (b)).

10.3.1
Low Stress Regime

1) Based on the similarities in the values of n, especially at low stresses (see Table 10.1 or Figure 10.17), it is fair to conclude that the same creep mechanism is operative for all microstructures. This claim is made despite the differences in activation energies. Why the activation energies are a function of the type of loading and/or chemistry and their origin, is unclear at this time. More work is needed.
2) For Ti_3SiC_2, at any given stress, $\dot{\varepsilon}_{min}$ is at least an order of magnitude lower in compression than in tension (Table 10.1). The same is probably true of Ti_2AlC.
3) For Ti_3SiC_2 in tension, $\dot{\varepsilon}_{min}$ is a weak function of grain size, with grain size exponents <1 (Radovic *et al.*, 2003). The same is true in compression: when the ε versus t curves for FG and CG samples are compared (Figure 10.16c) they are almost indistinguishable. Note that this is only true at σ < 70 MPa. This weak dependence on grain size, in turn, implies that the dominant creep mechanism is most likely dislocation creep, despite the fact that n is less than the exponents (3–7) typically associated with dislocation creep (Barsoum, 2003).
4) All else being equal, $\dot{\varepsilon}_{min}$ for samples fabricated with commercially available powders (such as the Ti_2AlC samples tested by Tallman *et al.*) is about an order of magnitude higher, that is, less creep resistant, than those fabricated by reactive hot pressing (See Figure 11.2b). This is an important observation as it implies that the GB chemistries – that are presumably less clean in the case of samples fabricated with commercially available powders – also contribute

to creep. It also implies that if the MAX phases are to be used for structural applications at HTs, the commercial powders would have to be significantly cleaner and/or reactive hot pressing will be required to make parts (see Chapter 11).

10.3.2 Higher Stress Regime

In compression, especially at $\sigma > 70\,\text{MPa}$ and/or higher temperatures, $\dot{\varepsilon}_{\text{min}}$ can actually be *lower* for the FG than the CG microstructures (e.g., compare results labeled FG and CG at 1300 °C in Figure 10.16c) (Zhen *et al.*, 2005b). In that regime, the stress exponents also increase (Table 10.2) as compared to those at lower stresses.

The fact that at 1300 °C the creep of FG Ti_3SiC_2 samples is better than its CG counterpart is *unusual*. In the metals literature, for example, grain boundaries are the weakest link. The exact reason for this state of affairs is not totally understood at this time, but post-testing microstructural examinations of samples in that regime suggests a change in mechanism from dislocation creep to subcritical crack growth. The latter regime is more difficult to document in tension because the stress state is such as to rapidly cause fast fracture. These comments notwithstanding, more work is needed to better understand what is occurring.

10.3.3 Times to Failure

When log–log plots of σ versus time to failure, t_f, for both the FG and CG Ti_3SiC_2 microstructures are plotted (Figure 10.19a) a series of near parallel lines are obtained for each microstructure. These results suggest that the results can be fitted to a Monkman-Grant expression, viz.:

$$t_f\,(s) = t_o K_{MG} \left(\frac{\dot{\varepsilon}_{\text{min}}}{\dot{\varepsilon}_o} \right)^{-m} \qquad (10.3)$$

where K_{MG} and m, are a dimensionless constant and the Monkman–Grant exponent, respectively; $t_o = 1\,\text{s}$. A least squares fit of the results, shown in Figure 10.19b, obtained on CG Ti_3SiC_2 subjected to tensile creep, yields a m of 1.06 ± 0.06 and $K_{MG} = \exp(-2 \pm 0.3)$. Also shown in Figure 10.19b as a dotted line, are the results for FG samples for which $m = 0.9 \pm 0.1$ and $K_{MG} = \exp(-2 \pm 1)$ (Radovic *et al.*, 2001).

Table 10.2 Summary of creep parameters in Eq. (10.2) for Ti_3SiC_2 loaded in compression in air at $\sigma > 70$–$100\,\text{MPa}$ (Zhen *et al.*, 2005b).

Grain size	ln A	*n*	Q (kJ mol^{-1})
CG	42	2.5	812 ± 50
FG	12	3.7	518 ± 50

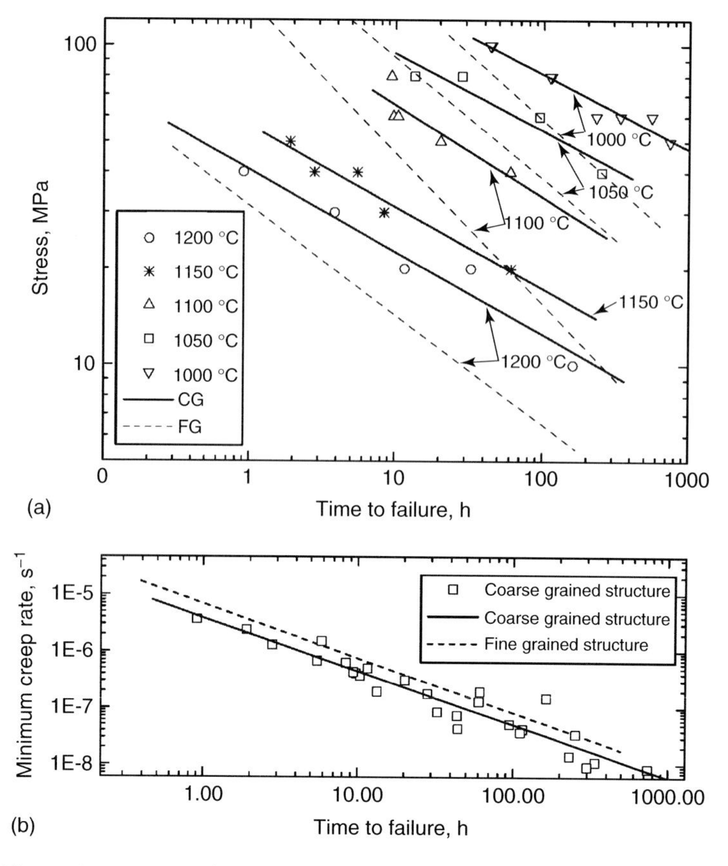

Figure 10.19 (a) Log–log plot of σ versus time to failure, t_f, for different temperatures and (b) log–log plot of t_f versus t (Monkman–Grant plot). The solid and dashed lines were obtained by least squares fit of the results for the CG and FG microstructures, respectively (Radovic *et al.*, 2001, 2003).

Despite the closeness of the parameters for the two microstructures, based on Figure 10.19a, it is obvious that t_f for the CG samples, especially at lower stresses, are longer than those for the FG microstructures. This observation can in turn, be attributed to the higher damage tolerance of the CG microstructure, that is, the ability of the CG microstructure to sustain higher bridging stresses than the finer grained ones.

10.3.4 Damage Tolerance

In Chapter 9, ample and compelling evidence was presented highlighting the damage tolerance of the MAX phases at room temperature. That the MAX phases are also quite damage tolerant at higher temperatures can be seen in the series of micrographs shown in Figure 10.20. From these micrographs – taken of samples

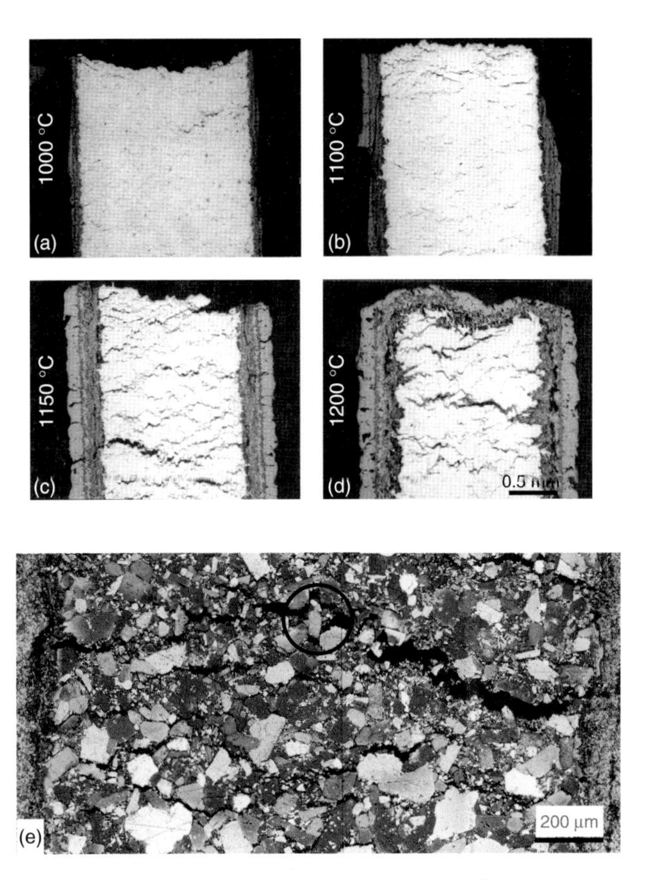

Figure 10.20 OM images of Ti_3SiC_2 surfaces that were parallel to vertically applied load. Fracture surfaces are located at the top of micrographs. Samples tested at, (a) 1000 °C, 80 MPa; t_f = 114 h, ε_f = 2.3%; (b) 1100 °C, 40 MPa; t_f = 60 h, ε_f = 4.7%; (c) 1150 °C, 20 MPa; t_f = 61 h, ε_f = 4.8%; (d) 1200 °C, 20 MPa; t_f = 11.6 h, ε_f = 8.8%; and (e) 1150 °C, 40 MPa; t_f = 61 h, ε_f = 8%. Encircled detail shows crack bridging. This micrograph was etched (Radovic *et al.*, 2003).

loaded to progressively higher temperatures – it is apparent that the lower the temperature, the more localized the damage zone is to the near vicinity of the fracture surface. Conversely, with increasing temperatures, the damage becomes much more distributed as a comparison of Figures 10.20a and d makes it amply clear.

The OM images shown in Figure 10.20 are of great importance because they clearly illustrate the remarkable ability of the MAX phases to tolerate damage even at temperatures as high as 1200 °C. That the cracks observed in Figure 10.20c–e remained stable and do not lead to catastrophic failure is unusual, to say the least. What these post-deformation micrographs clearly and unambiguously show is that the initiation of damage in a given region does *not* lead to a weakening of that region, but rather to its *hardening*. It is only by making this postulation that such distributed damage, especially in tension, can evolve. Said otherwise, had the initiation of damage, on a given plane, resulted in more damage and local softening, failure would have been very localized and shear bands, such as those shown in Figures 9.1 and 9.2, would have resulted. Along those lines it is important to note that the failure mode in compression, at HTs, indeed occurs by the formation of shear bands (Zhen *et al.* 2005a). Ultimate failure in tension, on the other hand, most probably occurs by the linking of smaller microcracks that run normal to the direction of the applied load as shown in Figure 10.20e. It is vital to note here that regardless of the strains to failure, no necking has ever been observed in the MAX phases. Instead past roughly 4% strain (see below) the majority of the strain is due to damage accumulation in the form of cavitation, pores, microcracks.

Lastly, in this section, the fact that the failure modes in tension and compression are so different would appear to contradict the conclusion reached above that dislocation creep is the dominant operative mechanism. This apparent paradox vanishes however, when it is appreciated that the combination of basal slip, IKBs, MDWs, and KB formation is sufficient to result in roughly 4% plastic deformation with little or no cavitation. This is best shown in Figure 10.21 where careful polishing of the gauge and grip areas of a Ti_3SiC_2 sample loaded at 1050 °C and 60 MPa are compared. The strain to failure for this sample was ≈4% and up to that strain there is little difference between the grip and gauge areas in terms of microcracks and cavitation, and so on. Above strains of ≈4% cavitation, microcracks and pore formation account for the majority of the strain. (Radovic *et al.*, 2000, 2001, 2002, 2003). This is an important result since it suggests that materials with

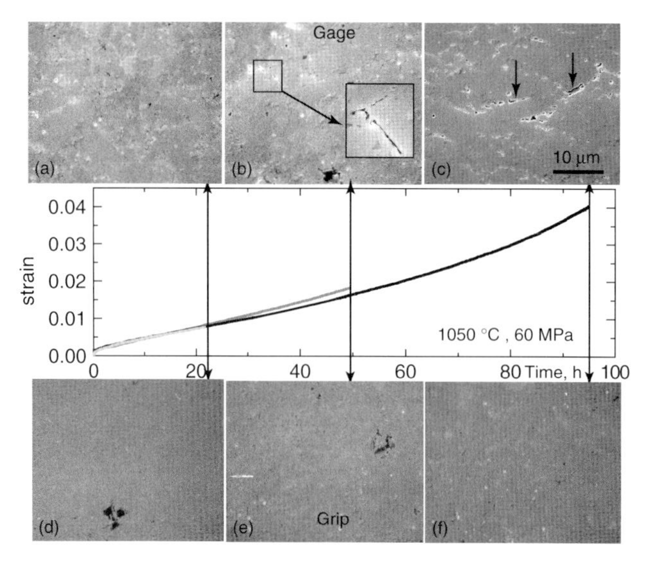

Figure 10.21 Sequence of damage formation for Ti_3SiC_2 specimens tested at 60 MPa and 1050 °C. Creep results are shown in the middle panel. SEM images of specimens gage are shown in the upper row; corresponding micrographs of grip area are shown in bottom row. For (a) and (d) test was stopped after 21 h; for (b) and (e) test was stopped after ≈50 h; and for (c) and (f) specimen failed after ≈95 h (Radovic *et al.*, 2003).

only basal slip can result in a ≈5% plastic deformation without apparent damage or cavitation.

The mechanisms that lead to local hardening and/or slow down crack growth are several and and are discussed in the following section.

10.3.5
Crack Arresting Properties of Kink Boundaries

In Chapter 9, evidence was presented for the crack arresting properties of kink boundaries. That the same occurs at high temperatures is probably one of the more remarkable properties of the MAX phases. To make the case, refer to the OM and SEM image shown in Figures 10.22a to d and 10.23a to c. Figures 10.22c and 10.23a provide examples of grains at, or near, the fracture plane that had only *partially* delaminated and thus acted as potent bridges. Furthermore, the importance of plastic deformation of individual grains (Figure 10.22b, d) cannot be overemphasized at this point. Such deformations must also act as potent strengthening mechanisms.

Like at room temperature, the microstructural evidence suggests the presence of at least two different crack bridging mechanisms. The first – in which grains

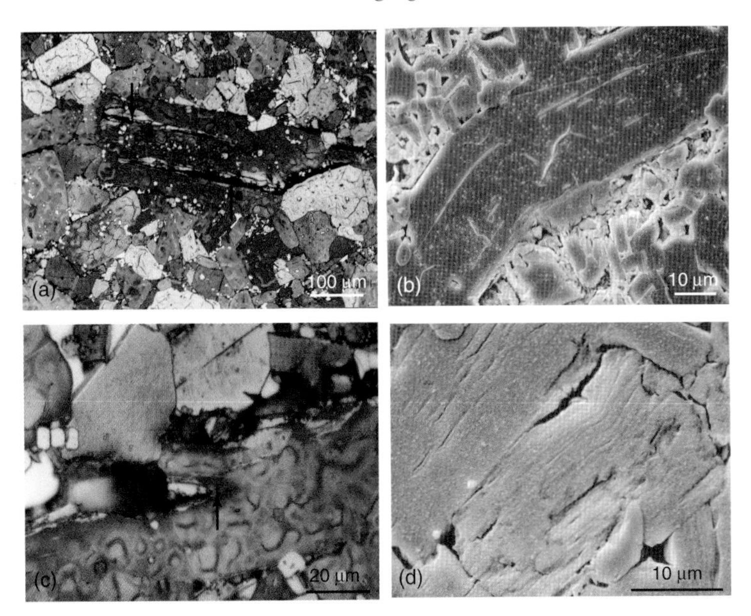

Figure 10.22 Micrographs of gage surfaces that were parallel to vertically applied load (a) etched OM of sample tested at 1000 °C and 80 MPa showing delamination denoted by arrows. Part of the grain between delamination cracks is slightly bent; (b) SEM of sample tested at 1150 °C and 20 MPa showing bent grain; (c) etched OM of sample tested at 1100 °C and 40 MPa. Bent lamella, denoted by arrow, serves as a crack bridge; and (d) SEM of sample tested at 1050 °C and 40 MPa, showing simultaneous delamination and kinking of a grain (Radovic *et al.*, 2003).

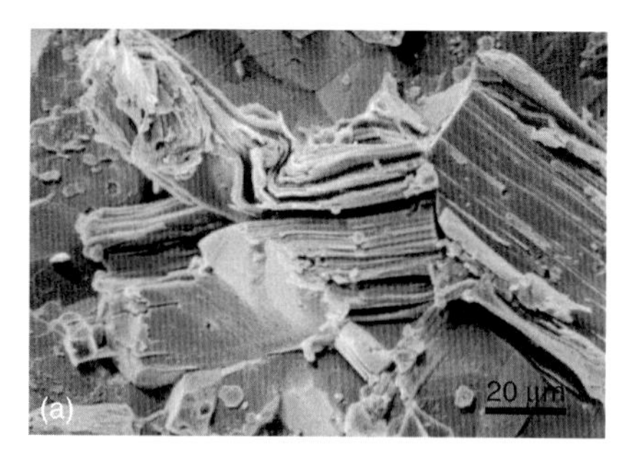

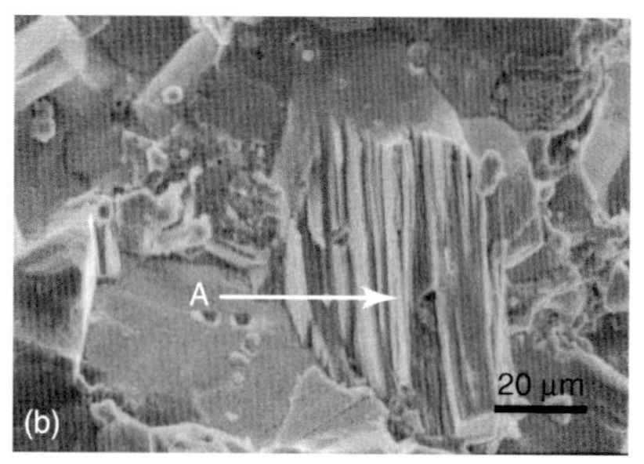

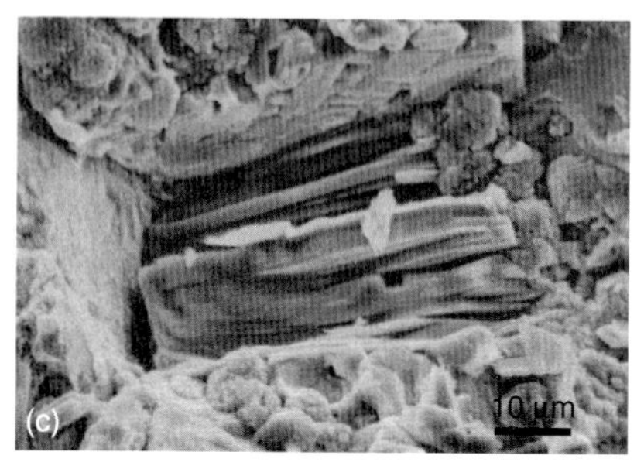

Figure 10.23 SEM of fracture surfaces of a CG specimen tested at 1050 °C and 80 MPa showing: (a) delamination of a single grain with part of it sticking out of, and part in the fracture plane; (b) same sample, ruptured grain sticking out from fracture plane; and (c) cavity remaining in fracture plane due to the pull-out of grain shown in (b) (Radovic *et al.*, 2003).

whose basal planes are parallel to the direction of applied load serve as classic crack bridges – can be seen in the encircled area in Figure 10.20e and in Figure 10.23b,c. These grains can absorb energy either by plastic deformation and/or by frictional pull out. Figure 10.23b,c show two mating fractured surfaces. As a portion of grain labeled A in Figure 10.23b is visible in the mating surface (Figure 10.23c) it is clear that this grain ruptured normal to the basal planes rather than pulled out. Given the strengths of the TiC bonds, the bridging stresses at one point must have been substantial.

The second mechanism begins with delaminations that are initiated on different basal planes at opposite ends of a single grain (e.g., Figure 10.22a), that with further deformation separate the lamellae between the two delamination cracks. This mechanism ultimately leads to crack bridges such as those shown in Figure 10.12a.

10.4 Response to Other Stress States

10.4.1 Compressive Stresses

As discussed in Chapter 9, when Ti_3SiC_2 is compressed at room temperature, the failure is either purely brittle or occurs at roughly 45° to the loading axis along a

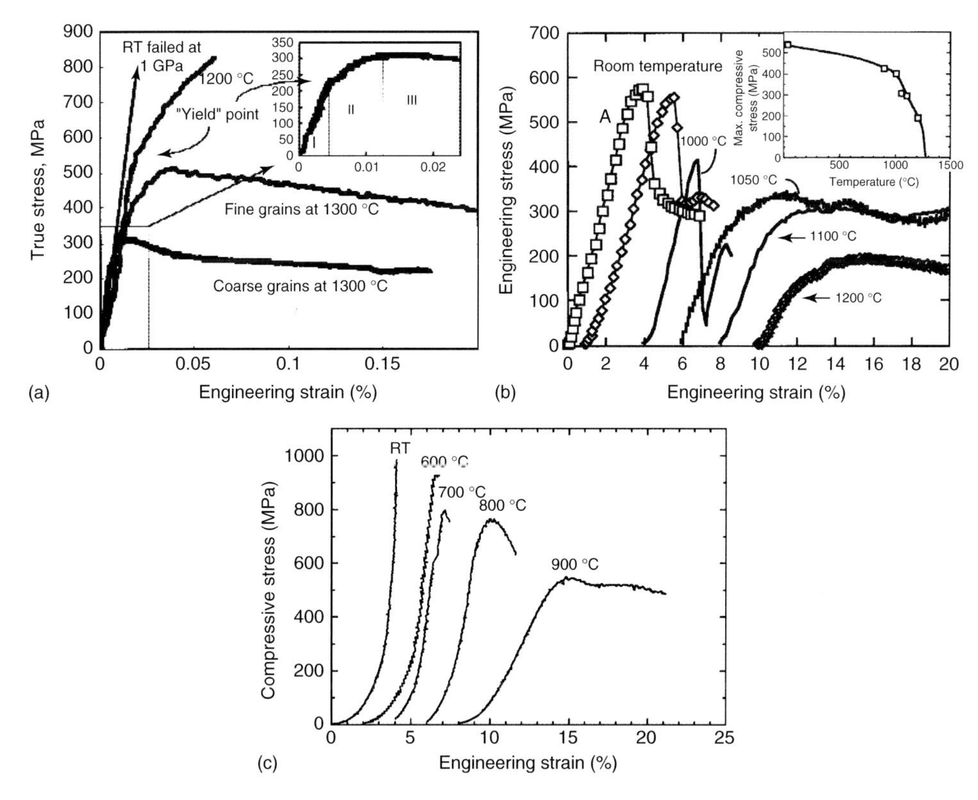

Figure 10.24 Dependence of compression stress–strain curves on temperature for, (a) Ti_3SiC_2 as a function of grain size. Below 1200 °C, the failure is brittle; above 1200 °C, the failure is plastic, with significant plasticity (El-Raghy *et al.*, 1999), (b) Ti_3AlC_2 loaded at an initial strain rate of 10^{-4} s^{-1}. Inset shows the effect of temperature on the ultimate compressive stresses. In (b), the curves are shifted to the right for clarity (Tzenov and Barsoum, 2000), and (c) Cr_2AlC (Tian *et al.*, 2009b).

shear band. When compressed at higher temperatures, however, the deformation is much more plastic. The strains to failure at 1300 °C can exceed 15% (Figure 10.24a). At 1200 °C, the stress–strain curves are characterized by a "yield point," followed by a regime of work hardening, before failure. At 1300 °C, the stress–strain curves also exhibit a yield point followed by a region of apparent work hardening, before softening is observed (El-Raghy *et al.*, 1999). Above the BPTT, the shape of the curves and the ultimate compressive strengths (UCSs), is a strong function of strain rate.

Similar results were obtained for Ti_3AlC_2 (Figure 10.24b) (Tzenov and Barsoum, 2000), and Cr_2AlC (Figure 10.24c) (Tian *et al.*, 2009b) and other MAX phases, such as Ti_2AlC, Ti_2AlN, Ti_2GeC, and Ti_3GeC_2 (Barsoum, Brodkin, and El-Raghy, 1997). Figure 10.25 summarizes some of the results.

As discussed in Chapter 9, at room temperature failure often, but not always, occurred by the formation of shear bands that typically resulted in graceful failure (Figure 10.24b). When deformation occurred at higher temperatures, however, the damage was more distributed. A good example is shown in Figure 10.26, which

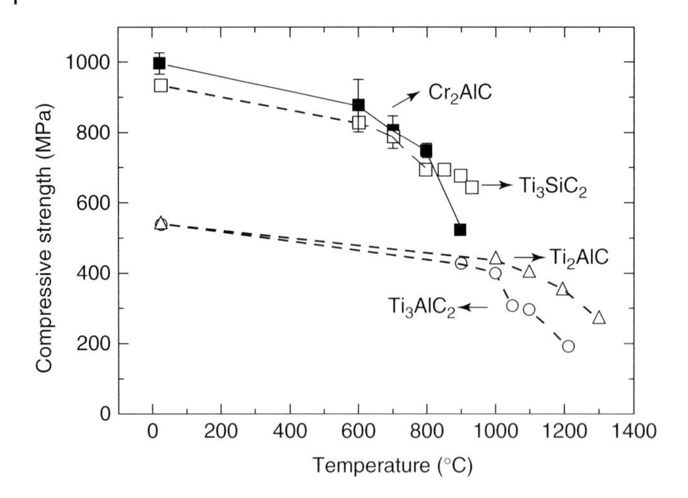

Figure 10.25 Effect of temperature on the compressive strengths of Cr_2AlC (Tian *et al.*, 2009b), Ti_3SiC_2, Ti_3AlC_2 (Tzenov and Barsoum, 2000), and Ti_2AlC.

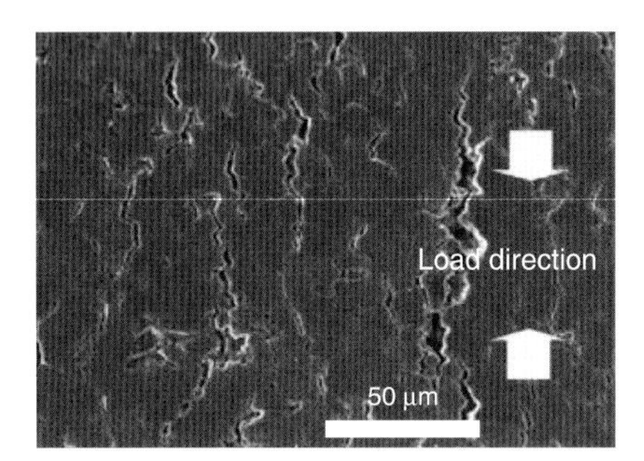

Figure 10.26 SEM image of Cr_2AlC sample loaded at 800 °C. Note extensive network of grain boundary cracks that formed parallel to the loading direction (Tian *et al.*, 2009b).

shows the microstructure of a Cr_2AlC sample after loading at 800 °C. Note the extensive network of intergranular cracks formed parallel to the loading direction, most probably as a result of the Poisson expansion normal to the direction of applied load. In this case, what appears to determine the BPTT is grain boundary softening rather than interlaminar strength.

10.4.2
Tensile Stresses

As noted above, the response of Ti_3SiC_2 to tensile stresses is a strong function of temperature and strain rate (Figure 10.2) (Barsoum *et al.*, 2001). At ≈0.5, the strain rate sensitivity for both CG and FG microstructures is quite high (Radovic *et al.*, 2000, 2002) and more characteristic of super-plastic solids, than typical metals or ceramics. This does not imply that the deformation mechanisms of the MAX phases are in any way comparable to those of super plastic solids. The latter have

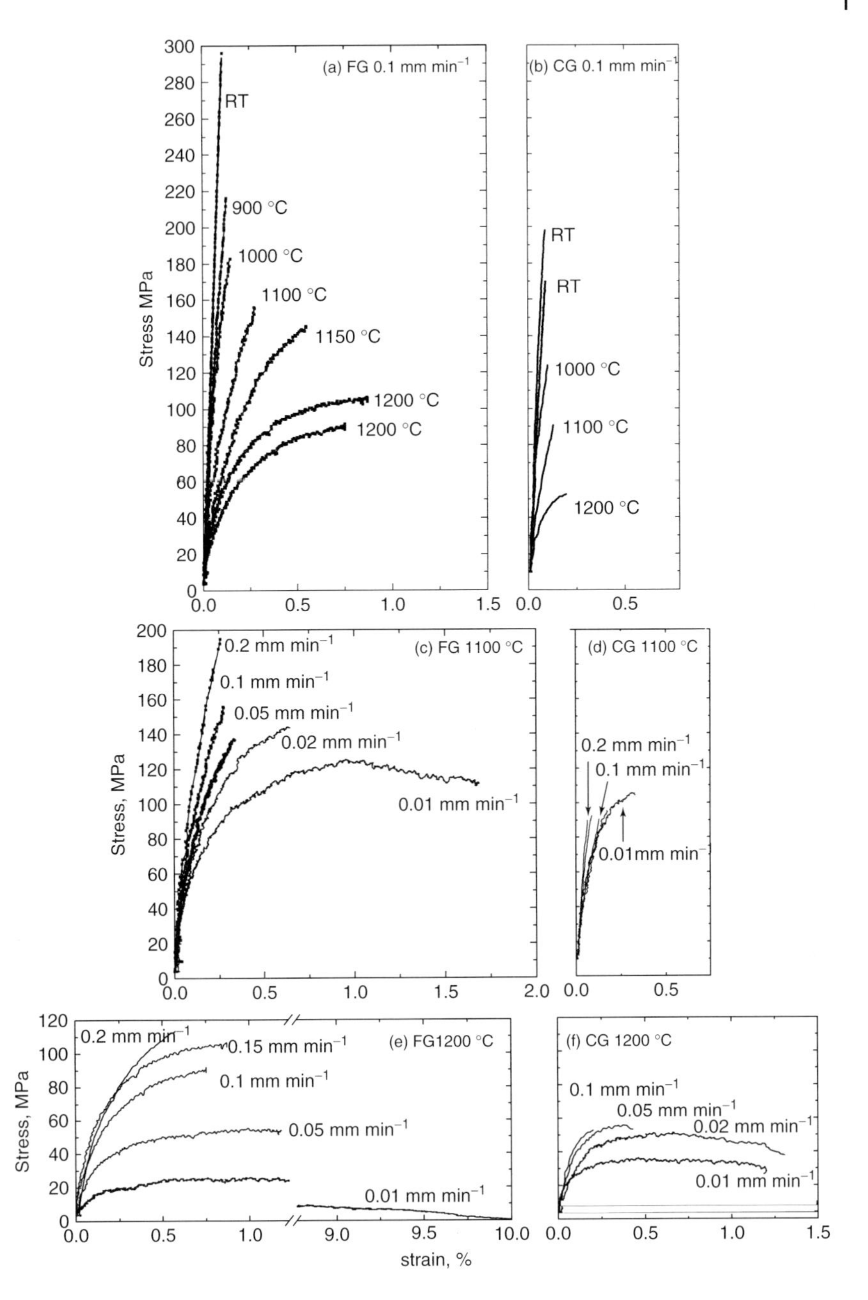

Figure 10.27 Engineering stress versus strain plots obtained from tensile tests for, (a) FG microstructure as a function of T at a cross-head displacement (CHD) of 0.1 mm min^{-1}; (b) CG microstructure as a function of T at a CHD of 0.1 mm min^{-1}; (c) FG at 1100 °C as a function of CHD rate; (d) CG at 1100 °C as a function of CHD rate; (e) FG at 1200 °C as a function of CHD rate, and (f) CG at 1200 °C as a function of CHD rate (Radovic *et al.*, 2002).

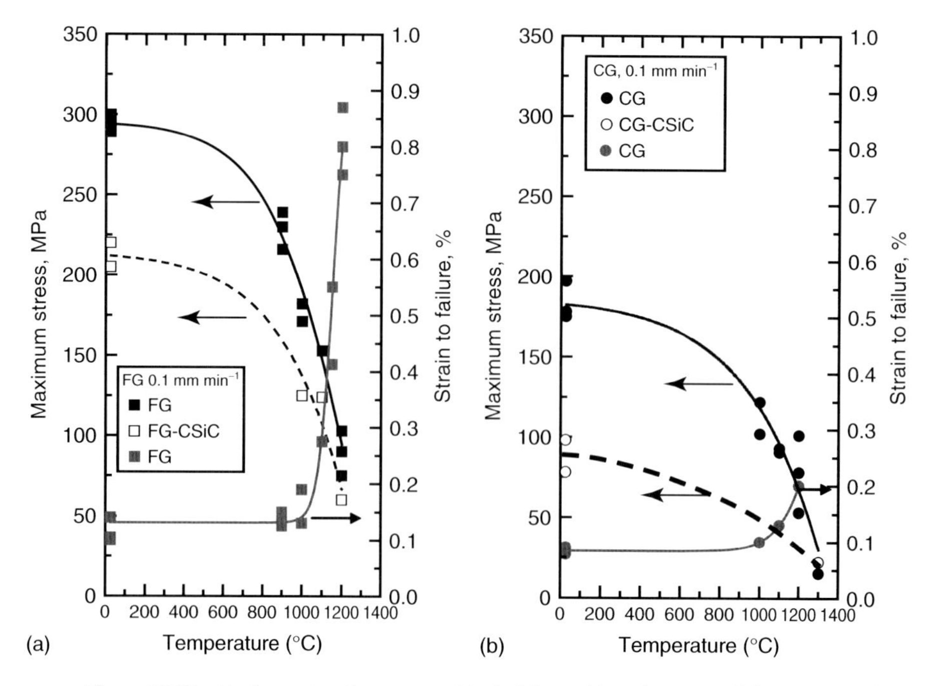

Figure 10.28 Maximum tensile stresses (black, left y-axis) and strain-to-failure (gray, right y-axis) as a function of temperature and microstructure. Results were obtained at a CHD rate of 0.1 mm min^{-1} (Radovic *et al.*, 2002).

comparable strain rate sensitivities, but only at grain sizes that are at least two orders of magnitude smaller than those of the CG Ti_3SiC_2 samples tested.

Figure 10.27a,c,e compare the response of FG Ti_3SiC_2 to its CG counterpart (Figure 10.27b,d,f). Figure 10.28 plots the maximum tensile stresses (black) and strains-to-failure (gray) as a function of temperature and microstructure. From these two figures, it is obvious that all else being equal, both the strains to failure, ε_f, and the ultimate tensile stresses (UTSs), of the FG microstructures are higher than those of CG ones. Note that this response is opposite to that encountered during creep and consistent with Hall–Petch like considerations. There are two possible explanations to these observations, both alluded to above: (i) the FG microstructure is able to dissipate the internal stresses generated more readily by grain boundary sliding and decohesion and/or (ii) the internal stresses are higher for the CG microstructure. The answer most probably is a combination of both.

10.4.3
Flexural Stresses

The response of the MAX phases to flexural stresses is intermediate to that for compression and tension. Typical results for Cr_2AlC are shown in Figure 10.29a. Below the BPTT, failure occurs with little deformation; at higher temperatures

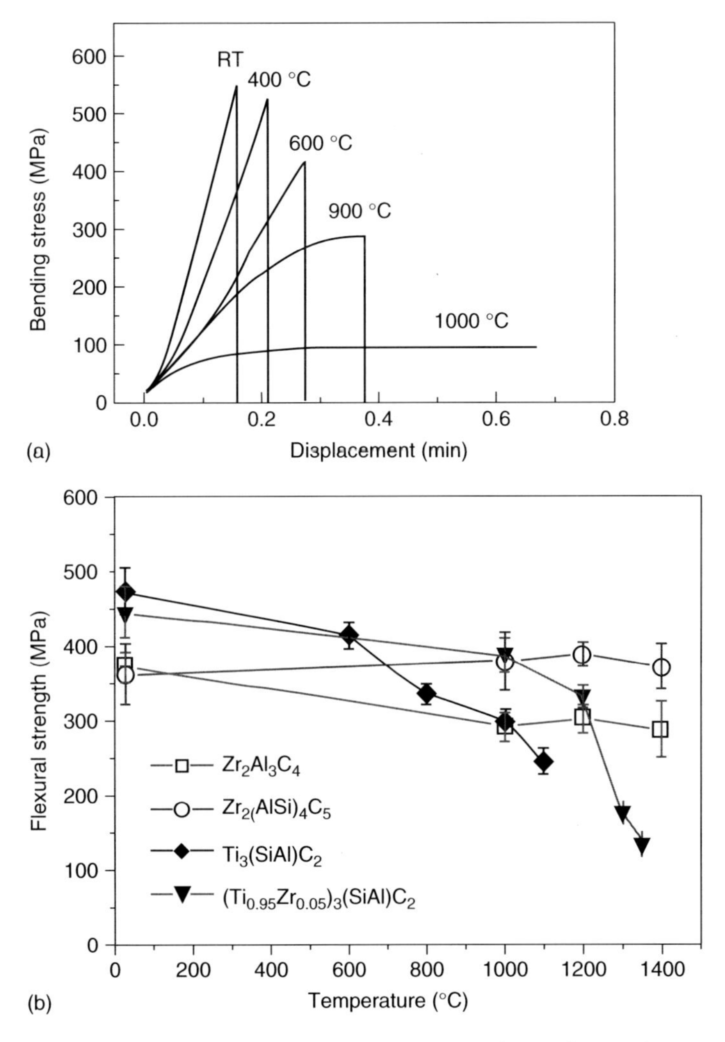

Figure 10.29 (a) Typical stress-displacement curves for Cr_2AlC tested at various temperatures (Tian *et al.*, 2009a). (b) Dependence of flexural strengths on test temperature for $Zr_2Al_3C_4$, $Zr_2(AlSi)_4C_5$, $Ti_3(Si,Al)C_2$, and $(Ti_{0.95}Zr_{0.05})_3(Si,Al)C_2$ (Wan *et al.*, 2010).

significant plastic deformation concomitant with a decrease in ultimate strengths is observed.

10.4.4
R-Curve Behavior and Fatigue

With increasing test temperature, the fatigue thresholds for both the FG and CG Ti_3SiC_2 microstructures decrease slightly up to $\approx 1100\,^\circ C$ (Figure 10.11b). At 1200 °C, which is above the BPTT, the da/dN versus ΔK curves show three distinctive regions that are more pronounced for the FG samples than for the CG samples (Figure 10.11b) (Chen *et al.*, 2001).

Referring to Figure 9.18a, at 1200 °C, the CG microstructure still displayed a rising *R*-curve, although the extent of stable crack growth was much reduced from 3 to 4 mm at ambient temperatures to 0.5 mm at 1200 °C. This is presumably caused by the onset of macro (creep) deformation of the specimen and extensive microcracking at the crack tips (Figure 10.12b) at these elevated temperatures. Chen *et al.* (2001) also concluded that the primary damage (intergranular and/or interlamellar cracking and the frictional wear degradation of the bridging zone in the crack wake and crack-tip shielding (grain and lamellae bridging) mechanisms governing high-temperature fatigue-crack growth behavior in Ti_3SiC_2 up to ≈1100 °C, are essentially unchanged from those at ambient temperature.

However, at 1200 °C, viz. above the BPTT, there was a striking change in behavior, principally in the form of significantly reduced ΔK_{th} thresholds that can be attributed to the onset of significant high-temperature deformation and associated microstructural damage (Figure 10.12b). At this temperature, macroscopic deformation of the test specimen was observed in the form of significant widening of the notch. More importantly, extensive microcracking and, to a lesser extent, cavitation, was apparent throughout the sample in both microstructures, although the intensity of damage was most severe in the vicinity of the crack tip (Figure 10.12b). Moreover, the crack paths involved significantly more transgranular and/or translamellar cracking, especially in the FG microstructure (not shown) as compared with that seen at lower temperatures, which severely diminished the propensity for grain bridging in the crack wake.

10.5 Summary and Conclusions

Like in ice (Barsoum *et al.*, 2001) and other plastically anisotropic solids – and for the same reason viz. the paucity of operative slip systems – glide of basal plane dislocations takes place only in favorably oriented grains or soft grains. During the short primary creep regime, basal plane dislocations glide and form pile-ups, MDWs, and kink boundaries, resulting in large internal stresses.

The BPTT can be explained by simply invoking that the grain boundary and/or interlaminar strengths are temperature dependent.

Despite the plethora of observations and results presented above, some of which may seem at odds with each other, the response of the MAX phases to deformation at both ambient and elevated temperatures is straightforward: The response is directly related to the rate at which internal stresses are generated versus the rate at which they are relaxed. This is best illustrated in the Ashby type map shown in Figure 10.30 (Radovic *et al.*, 2002), where the tensile response can be divided into two regimes: brittle and plastic. In the former, shown on the top right in Figure 10.30, the failure stresses are high and weakly dependent on strain rate. Furthermore, the FG samples' failure stresses are greater than those of the CG

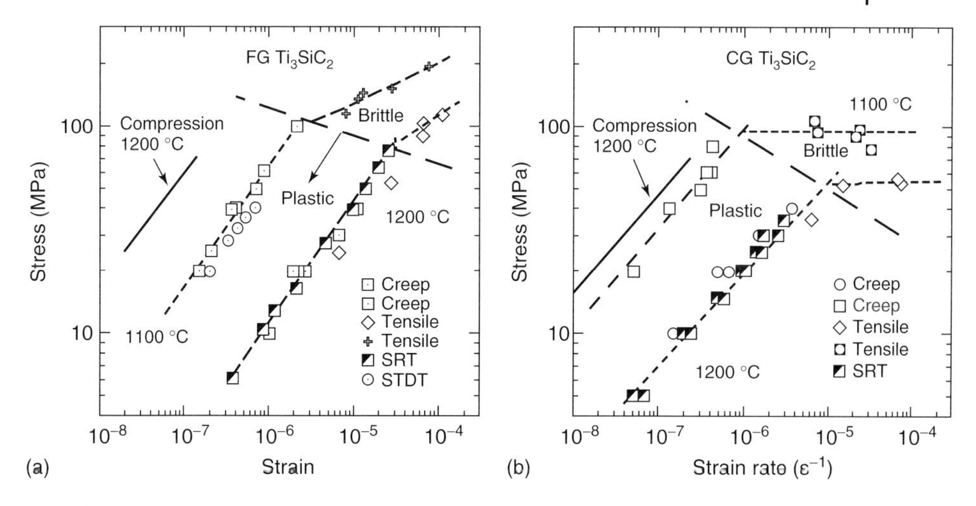

Figure 10.30 Summary log–log plots of stress versus strain rate results obtained from tensile tests at constant cross-head displacement rates, stress relaxation tests (SRT), creep tests, and strain transient dip tests (STDTs). Plotted lines are only guides for the eyes, for, (a) FG microstructure and (b) CG microstructure (Radovic *et al.*, 2002). Also plotted are the compression creep results at 1200 °C as solid lines on the left hand side of each plot.

samples signifying higher internal stresses in the latter and the presence of more efficient stress relaxation mechanisms in the FG samples.

At lower strain rates, shown in bottom left in Figure 10.30, the deformation is plastic. In this regime, the maximum stresses are a strong function of strain rate, with the CG microstructures outperforming their FG counterparts most probably because of the formation of tenacious bridges that tend to slow down creep.

Also plotted in Figure 10.30 (as solid black lines labeled compression) are the compression creep results for $\sigma < 70$ MPa. All else being equal, the MAX phases perform better in compression than in tension. This latter statement implies that while the creep of the MAX phases is dislocation mediated, it is also sensitive to the state of stress. Tensile stresses that tend to open microcracks, rather than close them, are more deleterious to deformation by creep.

If the rate at which the internal stresses accrue is significantly higher than the rate at which they can relax, failure, as shown in Figure 10.30, is brittle, and insensitive to strain rate or grain size (Radovic *et al.*, 2002). At the lowest strain rates and/or highest temperatures, the internal stresses can be accommodated by interlaminar decohesion, grain bending or kinking, and/or grain boundary decohesion and sliding. As the number of cavities and microcracks increase and coalesce, this naturally leads to an increase in $\dot{\varepsilon}_{\text{min}}$ and a transition from secondary to tertiary creep. In other words, dislocation creep gives way to subcritical crack growth rate (Radovic *et al.*, 2003; Zhen *et al.*, 2005b).

In general, and for the reasons outlined above, the CG samples are more resistant to deformation than the FG samples, but only up to ≈1200 °C and/or at lower strain rates; at 1300 °C, the opposite is true (e.g., Figure 10.16c and 10.17c). This is an important observation that implies that if the interfacial strengths of the basal planes become too weak with increasing temperatures, the large grains become a liability rather than an asset leading to faster, and larger deformations.

Lastly we note that despite the potential of using the MAX phases as structural materials at HTs, and the crucial importance of tensile creep in such applications it is somewhat surprising that there have been so few HT studies, in general, and so few creep studies in particular. These comments notwithstanding, it is important to place the creep response of Ti_3SiC_2 in context. When the creep results obtained for Ti_3SiC_2 are compared to other HT alloys (Figure 11.1a) it is reasonable to conclude that it is a very promising material indeed, especially given its early stage of development as compared to the others.

The preceding discussion was based on the MAX phases because it is in those materials that are best understood. The concepts and ideas presented, however, should be applicable to other KNE solids as well. For example, KBs have been observed in SiC (Suematsu *et al.*, 1991) and sapphire single crystals deformed at HTs. It is reasonable to assume that the ideas presented herein are relevant to these solids as well.

References

Barcelo, F., Doriot, S., Cozzika, T., Le Flem, M., Béchade, J.-L., Radovic, M., and Barsoum, M.W. (2010) Electron-backscattered diffraction and transmission electron microscopy microstructural study of post-creep Ti_3SiC_2. *J. Alloys Compd.*, **488**, 181–189.

Barsoum, M.W. (2003) *Fundamentals of Ceramics*, Taylor and Francis, London.

Barsoum, M.W., Ali, M., and El-Raghy, T. (2000) Processing and characterization of Ti_2AlC, Ti_2AlCN and $Ti_2AlC_{0.5}N_{0.5}$. *Metall. Mater. Trans.*, **31A**, 1857–1865.

Barsoum, M.W., Brodkin, D., and El-Raghy, T. (1997) Layered machinable ceramics for high temperature applications. *Scr. Metall. Mater.*, **36**, 535–541.

Barsoum, M.W., Radovic, M., Finkel, P., and El-Raghy, T. (2001) Ti_3SiC_2 and ice. *Appl. Phys. Lett.*, **79**, 479–481.

Barsoum, M.W., Zhen, T., Kalidindi, S.R., Radovic, M., and Murugahiah, A. (2003) Fully reversible, dislocation-based compressive deformation of Ti_3SiC_2 to 1 GPa. *Nat. Mater.*, **2**, 107–111.

Bell, I.A. and Wilson, C.J. (1981) Deformation of biotite and muscovite – TEM microstructure and deformation model. *Tectophysics*, **78**, 201–228.

Chen, D., Shirato, K., Barsoum, M.W., El-Raghy, T., and Ritchie, R.O. (2001) Cyclic fatigue-crack growth and fracture properties in Ti_3SiC_2 ceramics at elevated temperatures. *J. Am. Ceram. Soc.*, **84**, 2914.

Christoffersen, R. and Kronenberg, A.K. (1993) Dislocation interactions in experimentally deformed biotite. *J. Struct. Geol.*, **15**, 1077.

Duval, P., Ashby, M.F., and Andermant, I. (1983) Rate-controlling processes in the creep of polycrystalline ice. *J. Phys. Chem.*, **87**, 4066–4074.

El-Raghy, T., Barsoum, M.W., Zavaliangos, A., and Kalidindi, S.R. (1999) Processing and mechanical properties of Ti_3SiC_2: II, effect of grain size and deformation temperature. *J. Am. Ceram. Soc.*, **82**, 2855–2860.

Hauser, F.E., Starr, C.D., Tietz, L., and Dorn, J.E. (1955) Deformation mechanisms in polycrystalline aggregates of magnesium. *Trans. Am. Soc. Met.*, **47**, 102–134.

Kelly, B.T. (1981) *Physics of Graphite*, Applied Science Publishers, London.

Kronenberg, A., Kirby, S., and Pinkston, J. (1990) Basal slip and mechanical anisotropy in biotite. *J. Geophys. Res.*, **95**, 19257.

Radovic, M., Barsoum, M.W., El-Raghy, T., Seidensticker, J., and Wiederhorn, S.M. (2000) Tensile properties of Ti_3SiC_2 in the 25-1300 °C temperature range. *Acta Mater.*, **48**, 453.

Radovic, M., Barsoum, M.W., El-Raghy, T., and Wiederhorn, S.M. (2001) Tensile creep of fine-grained (3–5 μm) Ti_3SiC_2 in the 1000–1200 °C temperature range. *Acta Mater.*, **49**, 4103–4112.

Radovic, M., Barsoum, M.W., El-Raghy, T., and Wiederhorn, S.M. (2003) Tensile creep of coarse-grained (100–300 μm) Ti_3SiC_2 in the 1000–1200 °C temperature range. *J. Alloys Compd.*, **361**, 299–312.

Radovic, M., Barsoum, M.W., El-Raghy, T., Wiederhorn, S.M., and Luecke, W.E. (2002) Effect of temperature, strain rate and grain size on the mechanical response of Ti_3SiC_2 in tension. *Acta Mater.*, **50**, 1297–1306.

Suematsu, H., Suzuki, T., Iseki, T., and Mori, T. (1991) Kinking and cracking caused by slip in single crystals of silicon carbide. *J. Am. Ceram. Soc.*, **74**, 173–178.

Tallman, D., Naguib, M., and Barsoum, M.W. (2012) Tensile creep of Ti_2AlC in in Air in the 1000–1150 °C temperature range. *Scr. Mater.*, **66**, 805–808.

Tian, W., Sun, Z.-M., Du, Y., and Hashimoto, H. (2009a) Mechanical properties of pulse discharge sintered Cr_2AlC at 25–1000 °C. *Mater. Lett.*, **63**, 670–672.

Tian, W., Sun, Z.-M., Hashimoto, H., and Du, Y. (2009b) Compressive deformation behavior of ternary compound Cr_2AlC. *J. Mater. Sci.*, **44**, 102–107.

Tzenov, N. and Barsoum, M.W. (2000) Synthesis and characterization of $Ti_3AlC_{1.8}$. *J. Am. Ceram. Soc.*, **83**, 825–832.

Wan, D.-T., He, L.-F., Zheng, L.-L., Zhang, J., Bao, Y.-W., and Zhou, Y.-C. (2010) A new method to improve the high-temperature mechanical properties of Ti_3SiC_2 by substituting Ti with Zr, Hf, or Nb. *J. Am. Ceram. Soc.*, **93**, 1749–1753.

Wan, D.T., Meng, F.L., Zhou, Y.C., Bao, Y.W., and Chen, J.X. (2008) Effect of grain size, notch width, and testing temperature on the fracture toughness of $Ti_3Si(Al)C_2$ and Ti_3AlC_2 using the chevron-notched beam (CNB) method. *J. Eur. Ceram. Soc.*, **28**, 663–669.

Yang, X.Y., Miura, H.M., and Sakai, T. (2003) Dynamic evolution of new grains in magnesium alloy AZ31 during hot deformation. *Mater. Trans.*, **44**, 197–203.

Zhen, T., Barsoum, M.W., and Kalidindi, S.R. (2005a) Effects of temperature, strain rate and grain size on the compressive properties of Ti_3SiC_2. *Acta Mater.*, **53**, 4163–4171.

Zhen, T., Barsoum, M.W., Kalidindi, S.R., Radovic, M., Sun, Z.M., and El-Raghy, T. (2005b) Compressive creep of fine and coarse-grained T_3SiC_2 in Air in the 1100 to 1300 °C temperature range. *Acta Mater.*, **53**, 4963–4973.

Zhou, A. and Barsoum, M.W. (2009) Kinking nonlinear elasticity and the deformation of Mg. *Metall. Mater. Trans.*, **40A**, 1741–1756.

Zhou, A.G., Basu, B., and Barsoum, M.W. (2008) Kinking nonlinear elasticity, damping and microyielding of hexagonal close-packed metals. *Acta Mater.*, **56**, 60–67.

Zhou, A., Brown, D., Vogel, S., Yeheskel, O., and Barsoum, M.W. (2010) On the kinking nonlinear elastic deformation of polycrystalline cobalt. *Mater. Sci. Eng., A*, **527**, 4664–4673.

11
Epilogue

11.1
Outstanding Scientific Questions

As outlined in this book, our understanding of the properties of the MAX phases and what influences them – microstructurally and chemically – has come a long way in roughly fifteen years. That does not mean, however, that there are no outstanding scientific questions that need to be answered and technological hurdles that need to be overcome.

For the most part, and for many MAX phases, whenever compared, *ab initio* calculations and experiment are in decent agreement. This is especially true for the elastic constants, Raman spectroscopy peaks, and presumably, other spectroscopic techniques such as infrared spectroscopy.

The fact that the ratio of $N(E_F)$ determined from density functional theory (DFT) calculations and those measured for the majority of MAX phases is more or less a constant (Figure 4.18) is comforting and suggests that the calculated values are valid. Along the same lines, when $N(E_F)$ determined from DFT is plotted versus n_{val}, an R^2 value of 0.85 is obtained if Ti_2SC and Nb_2AsC are not included, and a value of 0.60 is obtained if they are (Figure 2.14). The agreement between the measured heat capacities and those calculated from DFT is also quite good (Figure 4.20).

Unfortunately, there is less DFT work in the literature on solid solutions. This, in part, reflects the fact that to carry out meaningful calculations on nonordered- or random-solid solutions is computationally quite intensive. This is slowly changing, however. In Sweden, there is at least one group trying to design magnetic MAX phases by doping Cr_2AlC with Mn, for example (Dahlqvist *et al.*, 2011). More such work is indicated.

On the basis of the work done to date, some of which is reviewed in this book, it is fair to conclude that some MAX phases – most notably Ti_3SiC_2 – are line compounds with very well-defined chemistries. Others, such as Ti_2SnC, Ti_2InC, and Nb_2AlC, exist over a range of stoichiometries. Still others, such as Ti_2SC, appear to be line compounds, wherein the elemental ratio is not 2 : 1 : 1. There has been very little systematic work on the effect of stoichiometry, or lack thereof on properties. There is also little DFT work in the literature on the subject. It follows

MAX Phases: Properties of Machinable Ternary Carbides and Nitrides, First Edition. Michel W. Barsoum.

that both theoretical and experiment results are currently lacking. This research direction is quite ripe for major inroads because nonstoichiometry can occur on the *X*, *A*, and possibly, the *M* sublattices. Point defects on multiple sublattices cannot be ruled out at this time. Anti-site disorder, where, for example, an M atom sits on an A site or vice versa, should also in principle be possible, especially at high temperatures. Whether such defects exist and, if they do, whether they affect properties is a wide open question.

In the realm of thermal properties, qualitatively we understand what affects the thermal conductivity: rattling of the A-atoms and the presence of lattice defects. And while theoretical progress is being made on the rattling aspect (N. Lane, PhD, Drexel University, 2013), less work exists on the latter. Thus, the nature of the lattice defects that affect thermal conductivity – presumed to be vacancies – has not been experimentally unequivocally ascertained. More work in which the vacancy concentration is systematically changed and its effect on thermal conductivity, by electrons and phonons, is examined is needed.

What determines the thermal expansion anisotropy is another area that is ripe for modeling. Many of the MAX phases are thermally, more or less, isotropic. However, some (such as V_2AsC) are not. A related question has to do with what determines the absolute average value of the thermal expansion coefficients. In Chapter 4 (Figure 4.21a), it was argued that as the thermal coefficients of expansion (TCE), were a weak function of n_{val} up to about 3.5×10^{29} m^{-3}. Beyond that threshold value, either the average TCE increases or the TCE anisotropy increase significantly. The latter include several Cr-containing phases. Why that is occurring is not clear.

One of the most fundamental and unanswered questions concerning the transport properties of at least some of the MAX phases is why $n \approx p$ and $\mu_e \approx \mu_p$. There is no fundamental reason for that to be the case. And while it is not too difficult to rationalize why $n = p$ – the crystal has to remain neutral after all – if indeed the electrons and holes are being transported in different bands, there is no *a priori* reason why their mobilities, in their respective bands, should be equal. This is certainly not true of most semiconductors or insulators but seems to be a pervasive property of many MAX phases.

Currently, the circumstantial evidence and some of our preliminary *in situ* diffraction studies under load all strongly suggest the existence of incipient kink bands (IKBs). Direct evidence, however, is still lacking. On a related note, we showed that the damping of ultrasound in the MAX phases can be enhanced by mild deformation, which presumably increases the dislocation density (Figure 8.17a). A more systematic study is indicated. Such a study would explore the limits of this approach to sound damping, and greatly enhance our understanding of the interaction of sound with dislocations.

In terms of mechanical properties, there are a number of outstanding questions. The first is why the MAX phases are as thermal shock resistant as they are. The second has to do with what affects the critical resolved shear stress (CRSS) of basal plane dislocations at room and elevated temperatures. In principle, the CRSSs in defect-free hexagonal solids should be close to zero (Lane *et al.*, 2011).

What determines the actual CRSSs is, thus most probably, the presence of other dislocations. It follows that careful deformation studies in which the CRSSs are carefully measured combined with post-deformation TEM studies of dislocation densities are needed. And while measuring the CRSSs of a solid requires large single crystals, the technique outlined in Chapter 8 – based on measuring the energy dissipated per cycle per unit volume – in which polycrystalline solids are used instead could prove very useful indeed.

A related question has to do with the Hall–Petch relationship. The CRSS of basal dislocations in the MAX phases seem to follow the Hall–Petch relationship (Figure 8.10). If that is the case, and on the basis of the aforementioned discussion, then there must be a relationship between grain size and dislocation density. Why that should be the case is unclear at this time, but careful TEM studies together with measuring the CRSSs would start answering the fundamental physics embedded – but currently lacking – in the Hall–Petch relationship.

Another question for which we currently have no real answer but one of fundamental importance is: Can solid solutions be used to strengthen the MAX phases? At this time, the few systematic results available suggest that in many cases, solid solutions do indeed result in hardening. In a few studies, however, either a solid solution softening effect or no effect is observed. Another important question is: Can second phases be used to dispersion harden the MAX phases? The few results available suggest that while the dispersion of a second hard phase increases hardness, and most probably wear resistance, it typically ends up reducing the strengths of the composites, especially in tension. For example, when we dispersed SiC in Ti_3SiC_2, the ultimate tensile strengths (UTSs) were significantly reduced (compare results labeled FGCSiC and CGCSiC to those labeled fine-grained (FG) and coarse-grained (CG) in Figure 10.28). This comment notwithstanding, too little work exists to date to reach any general conclusions.

On the basis of some recent studies by Hu *et al.* (2011a,c), probably one of the more fruitful approaches to enhance the mechanical properties of the MAX phases is to orient their grains while keeping them small. As shown in Chapter 9, the combinations of fracture toughness and bend strengths (Figure 9.19) for some of these oriented microstructures are quite impressive. What is less clear, however, is why the properties are as good as they are. In their work on Nb_4AlC_3, the authors used strong magnetic fields to orient the grains. Whether this technique can be used on other MAX phases, in general, and Ti_2AlC, in particular, and whether orienting the grains would lead to enhanced properties in other MAX phases remain outstanding questions. It is important to note that enhancing the properties in only one dimension in general does not lead to applications if the reduction in off axis mechanical properties is too high. The current results suggest that this can be tuned.

Fatigue is one of the least understood of the mechanical properties. As far as I am aware, there have been two studies on the long crack response of Ti_3SiC_2 (Chen *et al.*, 2001; Gilbert *et al.*, 2000) and one study on the long and short crack response of the same compound (Zhang *et al.*, 2003). What effects chemistry and structure, including microstructure, have on fatigue resistance are for the most part unknown. At this time, it has not even been established whether the MAX phases have a fatigue limit.

Figure 11.1 compares the fatigue resistance of Ti_3SiC_2 with a number of other ceramics, aluminum, and steel. On the basis of these results, it is reasonable to conclude that Ti_3SiC_2 has excellent long crack fatigue resistance. However, given that the results obtained are a strong function of R (see Figure 9.20a), more work is needed to understand the fatigue properties, let alone engineer fatigue resistant compounds or microstructures. Note that there are many reasons to believe that the fatigue properties of the MAX phases will be exceptional, not least of which is their propensity to form IKBs.

Indubitably, the use of the MAX phases for high-temperature structural applications will be creep limited. It is thus instructive to compare the creep properties of Ti_3SiC_2 to other high-temperature materials. Figure 11.2a compares the creep resistance of Ti_3SiC_2, in both tension (dashed line in bottom right in Figure 11.2a) and compression (solid line in bottom right in Figure 11.2a) to other high-temperature alloys. From this figure, it is clear that Ti_3SiC_2 falls in the high-temperature and low-stress – especially in tension – quadrant. This comment notwithstanding, these results are quite encouraging for many reasons, major among them:

1) The results shown in this figure compare mature alloys with a first generation MAX phase.
2) The results for Ti_3SiC_2 are for a single phase, pure solid; most commercial high-temperature alloys contain multiple elements and phases that have been added over the years, as better understanding was established. Said otherwise,

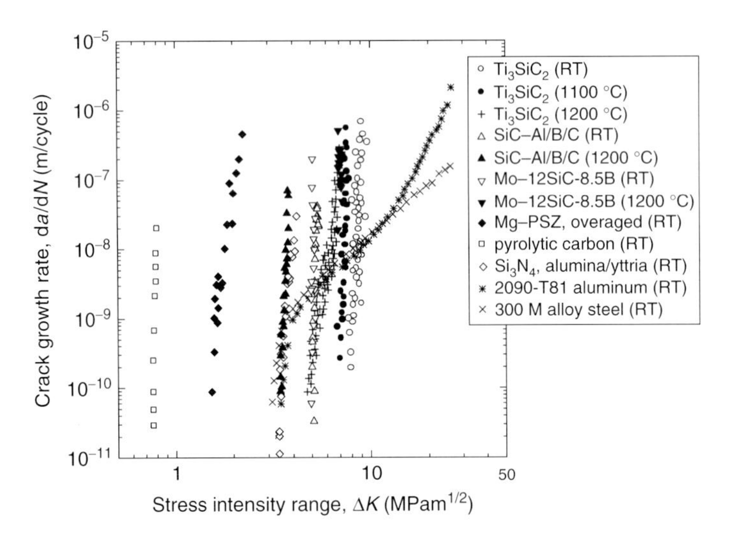

Figure 11.1 Dependence of cyclic-fatigue crack growth rates, da/dN, on applied stress-intensity range, ΔK. When the fatigue properties of coarse-grained Ti_3SiC_2 up to 1100 °C are compared to a number of ceramics and metals measured under identical conditions, it is clear that Ti_3SiC_2 has one of the highest fatigue thresholds (Gilbert *et al.*, 2000).

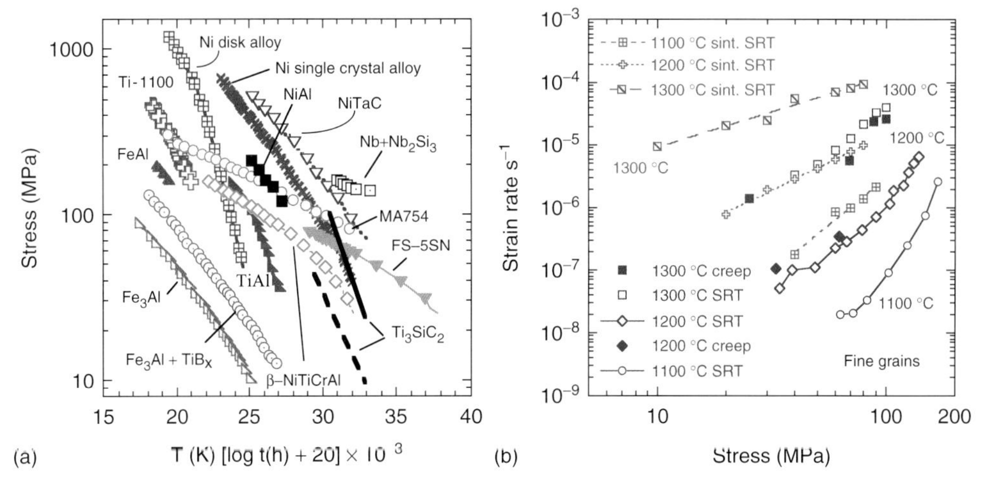

Figure 11.2 (a) Creep properties of select metallic, intermetallic alloys (Dimiduk, Mendiratta, and Subramanian, 1993), and Ti_3SiC_2 plotted as stress-to-rupture versus the Larson–Miller parameter (T is the creep temperature and t the time-to-rupture). The dashed line summarizes the tensile creep results; the solid line the compression creep results. (b) Comparison of creep rates of reactively hot-pressed Ti_3SiC_2 samples (blue lines and data points) with those fabricated by pressureless sintering of commercial powders (dashed red lines) (Zhen *et al.*, 2005).

the first generation Ni-based super alloys were not as creep resistant as the ones shown in Figure 11.2a.

3) This figure does not account for density. At 4.5 Mg m^{-3}, the density of Ti_3SiC_2 is roughly half that of many of the alloys shown in the figure.
4) This figure also does not take oxidation resistance into account. As discussed in Chapter 6, the most likely MAX phase to be used at high temperatures, because of its superb oxidation and crack healing capabilities, is Ti_2AlC. Few of the alloys listed in Figure 11.2a are alumina formers.

11.2 MAX Phase Potential Applications

The MAX phases have been proposed for numerous applications, some of which are discussed briefly below. This is by no means an exhaustive list and it is fair to assume that a thorough search of the patent literature would unearth many more. The purpose of this section is mostly to show the rich potential for applications and the range of industries that the MAX phases could in principle impact.

11.2.1 Replacement for Graphite at High Temperatures

Graphite is an important high-temperature material used extensively in many industries. For example, graphite is used as connectors, heaters, furnace linings, shields, rigid insulation, curved heating elements, and fasteners in vacuum furnaces. Graphite is also the material of choice for the hot pressing of diamond-cutting tools and other materials, for the construction industry. In general, some of the MAX phases such as Ti_2AlC have several advantages over graphite such as better wear and oxidation resistance. The ease by which Ti_2AlC can be machined to high tolerances is also an important consideration. The high strengths, moduli, and thermal conductivities of the MAX phases are also positive attributes. Figure 11.3 shows examples of MAX phase inserts that were actually tested in industrial dies and performed quite well.

11.2.2
Heating Elements

As discussed below, in the late 1990s, Kanthal Corp. licensed the MAX technology from Drexel University. Given that one of Kanthal's core businesses is heating elements, it was not surprising that MAX phase heating elements were one of the first applications targeted by the Swedish company. The heating element shown in Figure 11.4 was heated up to 1350 °C and cooled down to room temperature for ≈10 000 cycles (Sundberg *et al.*, 2004). The resistance of the element was found to be very stable and the protective oxide that formed was quite adherent and protective. This heating element is quite versatile and can be used up to 1400 °C in air, argon, hydrogen, or vacuum.

(a)

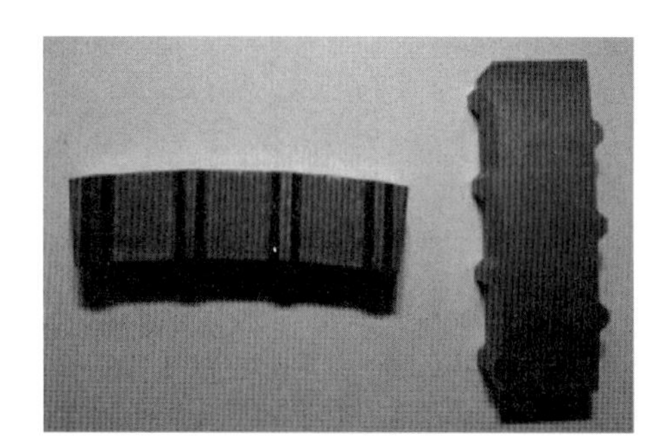

(b)

Figure 11.3 (a, b) MAX-based insets that were tested in industrial dies at elevated temperatures. (Courtesy of 3-ONE-2, LLC.)

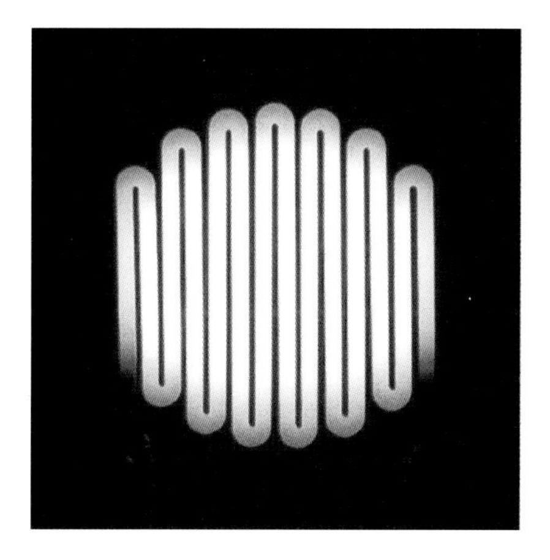

Figure 11.4 Example of Ti_2AlC-based heating element resistively heated to 1450 °C in air. (Courtesy of Kanthal.)

11.2.3 High-Temperature Foil Bearings and Other Tribological Applications

In our very first report on the properties of Ti_3SiC_2 (Barsoum and El-Raghy, 1996), we noted that the shavings felt lubricous and were reminiscent of graphite, which led us to speculate that the material is probably also self-lubricating. We also described it as the Si equivalent of graphite. A few years later, working with the samples we supplied them, Myhra, Summers, and Kisi (1999), using lateral force microscopy, measured the friction coefficients, μ, of the basal planes of Ti_3SiC_2. They showed that the basal planes had one of the lowest μ of any solid measured at that time. As important, the low μ values were maintained even after six months exposure to the atmosphere. Interestingly, these results have to date not been reproduced.

The low μ values, however, did not translate to low friction in polycrystalline samples (El-Raghy, Blau, and Barsoum, 2000). In pin-on-disk tests, irrespective of grain size, it was found that Ti_3SiC_2 undergoes an initial transition stage where μ increases linearly from 0.15–0.45. After this transition stage, μ rises to steady state values of about 0.8 for both microstructures. The transition from low to high μ was ascribed to the accumulation of debris entrapped between the disk and the pin, resulting in third-body abrasion. It is important to note that despite having low friction along the basal planes, MAX phase powders are quite abrasive.

In 2009, two patents were issued (Gupta *et al.*, 2009a; Palanisamy *et al.*, 2009) for the use of Ag-based MAX composites for foil bearing applications. In this application, the loads are low, but the parts are rotated at very high speeds. The bearings must also be able to function at both ambient temperatures and at temperatures around 500 °C. In a series of papers (Filimonov *et al.*, 2009; Gupta *et al.*, 2007, 2009b), it was shown that Ag/Ta_2AlC and Ag/Cr_2AlC composites worked well for this application. Further, these composites survived a rigorous

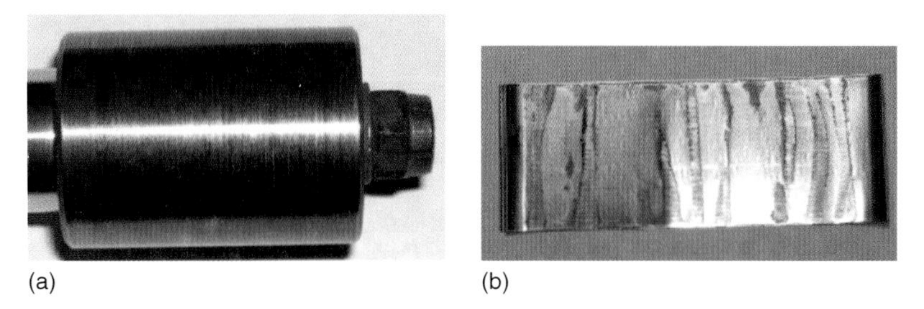

Figure 11.5 (a) Picture of Ag/Ta_2AlC composite cylinder mounted on a shaft after 10 000 stop–start cycles in a rig test and (b) picture of superalloy foil after rig testing (Gupta *et al.*, 2007).

rig test, where parts (Figure 11.5) were spun at 50 000 revolutions per minute and thermally cycled. After 10 000 stop–start cycles, the surface roughness of the Ag/Ta_2AlC composite increased to between 0.13 and 0.18 μm, from its initial value of 0.10 μm. It is important to note that one reason this application was even remotely possible is the very high tolerances to which these composites could be machined in the first place.

11.2.4 Gas Burner Nozzles

Owing to its excellent high-temperature properties and because it forms a protective alumina layer, Ti_2AlC-based MAX phases can be used in gas burning applications where traditional metallic alloys show limited service life (Figure 11.6). The same MAX compound can also replace metallic alloys to increase the burner process

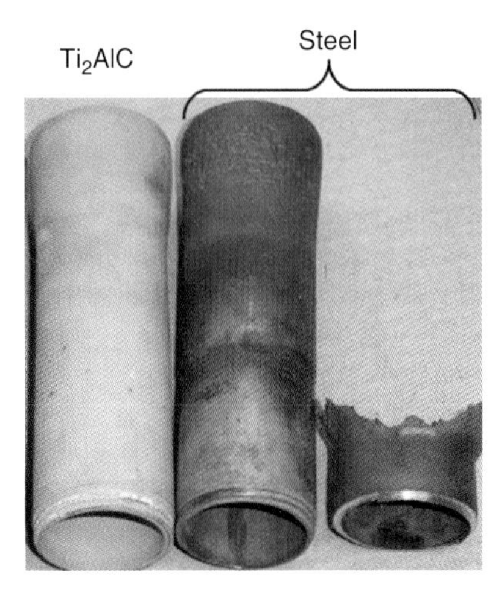

Figure 11.6 Pictures of Ti_2AlC and steel nozzles used in gas burners. The testing conditions that heavily corroded the steel nozzles did not appear to affect the MAX-based one. (Courtesy of Kanthal.)

temperatures up to 1400 °C. Note that, in contrast to traditional ceramics, joining problems to existing equipment are easily overcome as MAX nozzles can be readily threaded and thus can directly replace metallic nozzles.

11.2.5 Tooling for Dry Drilling of Concrete

In the early 2000s, 3-ONE-2 (a small company Dr. T. El-Raghy and I founded) and Hilti developed tooling for the dry drilling of concrete. The tooling was composed of industrial diamonds embedded in a Ti_3SiC_2 matrix hot pressed into small segments. The latter were then brazed to a steel shaft (Figure 11.7a). The performance of the MAX-phase-based segments was found to be superior to that of current diamond/Co segments (compare Figure 11.7a,b). There were two problems, however. The first was the smearing of the concrete powder due to the high temperatures generated at the tool/concrete interface (Figure 11.7c). The second was the inadequate toughness of the segments.

11.2.6 Glove and Condom Formers

Ansell Healthcare and Drexel University signed a research agreement for developing gloves and condom formers to make latex products. The work resulted in the development of the slip casting technology for manufacturing large, thin-walled complex parts (see Figure 11.8a–c) and a patent (Gromelski, Caciolo, and Cox, 2007).

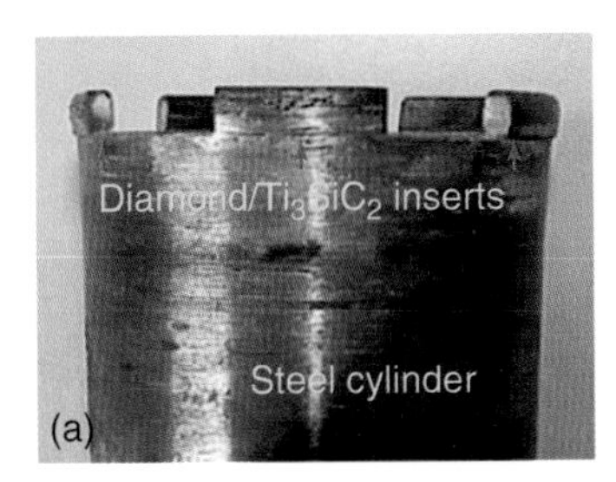

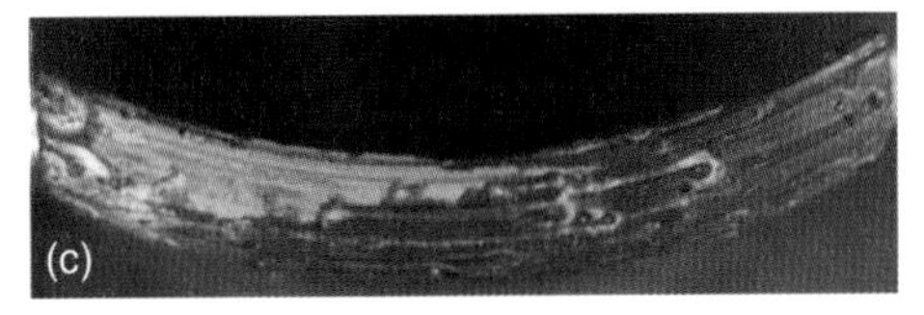

Figure 11.7 Pictures of, (a) steel hollow cylinder to which diamond/Ti_3SiC_2 inserts were brazed, after dry drilling of concrete, (b) same as (a), but where the diamonds were embedded in a Co matrix; and (c) higher magnification of insert after dry drilling. (Courtesy 3-ONE-2.)

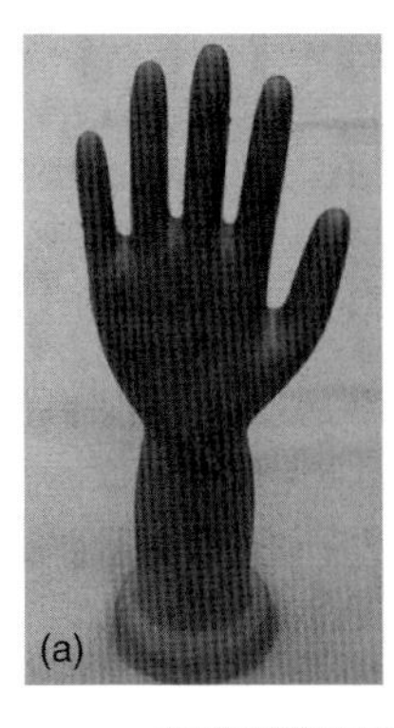

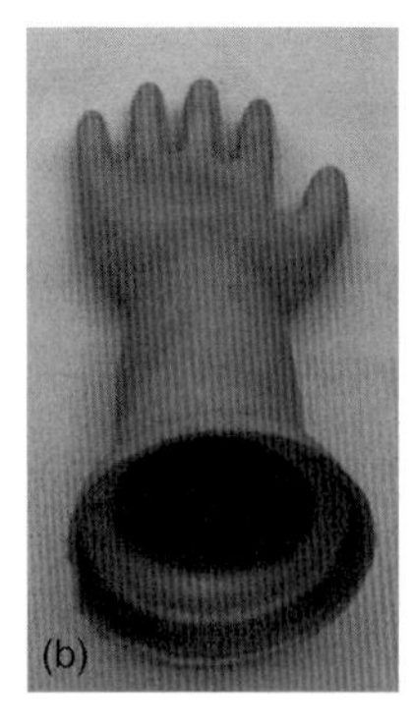

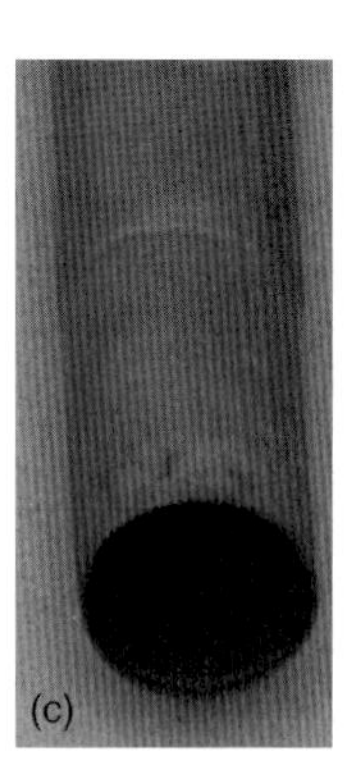

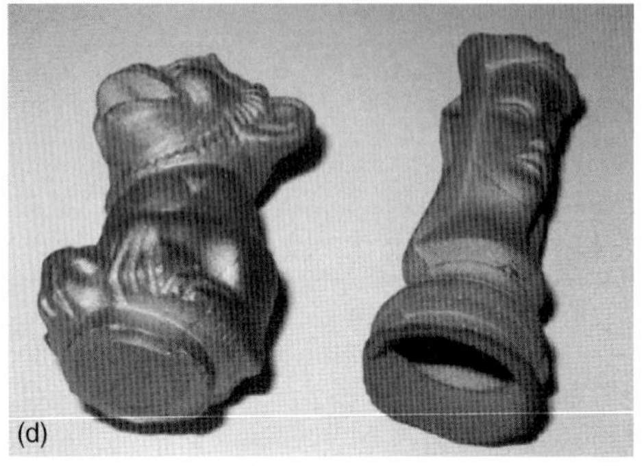

Figure 11.8 Picture of (a) large, hollow, slip cast Ti_3SiC_2 glove former; (b) and (c) same as (a), but showing thin walls possible; and (d) complex solid and hollow slip cast parts starting with Maxthal powders. (Courtesy 3-ONE-2.)

11.2.7
Nonstick Cookware

In 2007, a patent was issued (El-Raghy and Lyons, 2007) for the use of the MAX phases and their coatings as durable, stick and stain and thermal shock resistant, dishwasher safe, cookware, cutlery, and other cooking utensils.

11.2.8
Use in the Nuclear Industry

Although not discussed in this book, some of the MAX phases, most notably Ti_3AlC_2 and Ti_3SiC_2, are quite resistant to radiation damage (Le Flem *et al.*, 2010; Liu *et al.*, 2010; Nappé *et al.*, 2009; Whittle *et al.*, 2010). There is also a growing body of evidence that dynamic recovery may be occurring at temperatures of 700 °C or lower.

In a post-Fukushima world, it is imperative to build some accident tolerance to the Zircaloy tubes that hold the nuclear fuel. The simplest is to spray a thin coating of Ti_2AlC or Ti_3AlC_2 onto the Zircaloy tubes. If the coatings are thin enough and are designed in such a way that in the presence of oxygen they form a thin cohesive and adhesive alumina layer, then it is possible to protect the Zircaloy tubes in the case of accident due to loss of coolant . The challenge, however, is to thermally spray the Ti_2AlC such that it does not lose its ability to form thin alumina layers. The current literature on the subject shows that after high-velocity oxy-fuel, HVOF, spraying of Ti_2AlC, heating in air results in the formation of titania instead of alumina (Frodelius *et al.*, 2008; Sonestedt *et al.*, 2010a,b).

The fact that Ti_3SiC_2 does not react with molten Pb and or Pb–Bi alloys (Barnes, Dietz Rago, and Leibowitz, 2008; Heinzel, Müller, and Weisenburger, 2009; Utili *et al.*, 2011) renders it a good material to be used for containing molten Pb or Pb–Bi alloys in nuclear reactors. In a recent paper, Sienicki *et al.* (2011) suggested that Ti_3SiC_2 could be used in an improved natural circulation Pb-cooled small modular fast reactor.

11.2.9
Ignition Devices

In 2010, a patent (Walker, 2010) was issued for the use of the MAX phases as the conductive element in spark plugs and other such ignition devices.

11.2.10
Electrical Contacts

One of the first applications for Ti_3SiC_2 was by a small Swedish company, Impact Coatings, as sputtering targets for the deposition of electrical contacts. Although the company now uses other cheaper targets, this early application is significant. Currently, others are sputter depositing Cr_2AlC thin films on steels and turbine blades (e.g. Hajas *et al.*, 2011). In some cases, MAX phase targets are used. On the basis of the growing worldwide interest in the MAX phases, it is reasonable to assume that a market for sputter targets should emerge soon.

11.2.11
Electrical Contact for SiC-based Devices

Presently, SiC-based devices are under development for use in electronic devices that are subjected to high temperatures and/or corrosive environments. Silicon carbide's material properties (large band-gap, high-thermal conductivity, extremely high-melting and decomposition temperatures, excellent mechanical properties, and exceptional chemical stability) are far superior to those of Si and render it suitable for operation in hostile environments. Because its band gap is nearly three times larger than that of Si, it is possible to use SiC as a semiconductor, for example, to temperatures as high as 1000 °C because of its low intrinsic carrier concentrations and operation within the dopant-controlled saturation regime required for semiconducting devices. One of the challenges of this technology is to find a material to use as an electrical contact that does not react with SiC at elevated temperatures and results in low contact resistances. In 2003, a patent was issued (Tuller, Spears, and Mlcak, 2003) for the use of Ti_3SiC_2 as an electrical contact for SiC electronic components. The major advantage of Ti_3SiC_2 is that it is in thermodynamic equilibrium with SiC. This would in turn allow the SiC devices to be operated in high-temperature environments without the contact material reacting with the SiC, and deteriorating the performance of the device.

Some sensor applications are based on SiC field effect transistors that exploit the wide band gap of SiC and its chemical inertness. In such applications, there is also a need for a compatible inert electrode material such as Ti_3SiC_2.

11.2.12
Electrochemical Applications

In 2006, a patent was issued (Jovic and Barsoum, 2006) for the use of Ti_3SiC_2 and other MAX phases for chlorine production by the electrolysis of HCl. In 2010, a patent application was filed (US 2010/0236937 A1) for the use of the MAX phases as substrates for ruthenium oxide and sulfide-based catalysts.

11.3
Forming Processes and Sintering

One of the crucial advantages of working with the MAX phases is that they can be pressureless sintered to full density by heating green preforms in inert atmospheres such as argon.

The first report in the open literature for the pressureless sintering of a MAX phase to full density can be found in a patent that issued in 2002 (El-Raghy *et al.*, 2002). In 2004, the first paper (Murugaiah *et al.*, 2004) not only on the pressureless sintering of Ti_3SiC_2 but also on its tape casting, a process that, not surprisingly, led to orienting the flaky prereacted hexagonal grains was published. In the same paper, it was shown that the simplest method to producing highly oriented microstructures was to gently shake or tap the dies containing the prereacted Kanthal powders prior to sintering. In 2006, the first paper on the pressureless sintering of Ti_2AlC was published (Zhou *et al.*, 2006).

In general, the fact that the MAX phases can be sintered to full density, without the application of pressure is an important attribute that greatly enhances their chances for commercialization.

The methods for forming the green bodies are quite varied. They range from slip casting, to produce complex, thin wall shapes (e.g., Figure 11.8), extrusion to form tubes (Figure 11.6), cold pressing to form simple shapes (Figure 11.3) and cold isostatic pressing to metal injection molding. It is worth noting that the inherent ductility of the MAX phase powders results in green bodies that are quite robust vis-à-vis handling. Said otherwise, in certain green forming methods, no binders are needed.

Spark plasma sintering (SPS) is also emerging as a viable method to fabricate and densify the MAX phases (Cui *et al.*, 2012; Hu *et al.*, 2009, 2011b; Zhang *et al.*, 2003). Currently there is little work, however, comparing the high-temperature properties of samples made by SPS with those made by reactive hot pressing or pressureless sintered commercial powders.

11.4 Outstanding Technological Issues

In the late 1990s, Kanthal AB of Sweden licensed the MAX phase technology from Drexel University, and, together with 3-ONE-2, proceeded to scale up the production of Ti_3SiC_2 and Ti_2AlC powders. The cooperation resulted in two patents (El-Raghy *et al.*, 2002; Sundberg *et al.*, 2006).

Figure 11.9a summarizes the progress, which was swift as both the batch weight and weight of parts manufactured increasing roughly by an order of magnitude

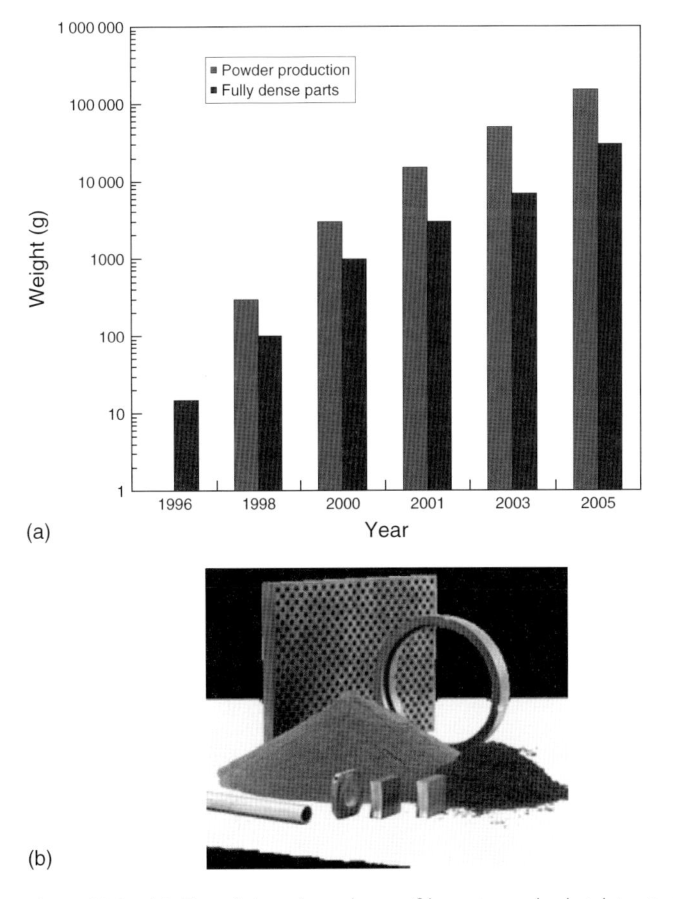

Figure 11.9 (a) Plot of time dependence of largest powder batch/parts in grams. Note semilog nature of the plot. (b) Picture of powders and parts made by Kanthal and sold under the trademark MAXTHAL. (Courtesy of Kanthal.)

every year. Currently, Ti_3SiC_2 and Ti_2AlC powders are commercially sold under the trade name of MAXTHAL. Figure 11.9b show examples of parts and powders commercialized by Kanthal.

11.4.1 Cost of Raw Materials

At this time, it is reasonable to assume that the most pressing technological hurdle that is slowing down MAX phases applications is the price of the powders. Currently the cost of Ti_3SiC_2 and Ti_2AlC powders is of the order of \$500 per kg. Among the reasons for this state of affairs, two stand out. The first is the classic conundrum of any new material: the cost is high because the demand is low and economies of scale do not kick in. The second is that one of the starting raw materials for these compounds is Ti powder, which relative to say steel, aluminum, or graphite is expensive. The MAX phase powders are also significantly more expensive than alumina, SiC, and Si_3N_3 powders that are used to make other structural ceramics. And while the relative ease by which the MAX phases can be sintered to full density, their damage tolerance and their machinability to very close tolerances – in their final fully dense state – should mitigate the cost of the final products made, the high cost of the raw material remains an obstacle. An obvious solution is to manufacture Ti-containing powders, starting with TiO_2 – a cheap abundant powder. Over the years, we have attempted to do so, but to date we have not been able to make predominantly pure MAX phase powders.

11.4.2 Creep Resistance of Parts Made with Prereacted Powders

Another problem of commercial powders is highlighted in Figure 11.2b. When commercial powders are used instead of reactively hot-pressed ones, the creep rates are roughly an order of magnitude higher (Figure 11.2b). It follows that one technological challenge at this time is to produce powders that, when sintered, yield solids that are as creep resistant as those made by reactive hot pressing. What is unknown at this time – and would be important to understand if better high-temperature solids are to be fabricated by pressureless sintering – is why the sintered samples have higher creep rates. The simplest conjecture – for which evidence is lacking – is to invoke the presence of a soft grain boundary phase in the sintered samples.

11.5 Some Final Comments

The MAX phases have attracted considerable attention and are garnering new devotees because they have a quite unusual combination of properties. The fact that their chemistries can be altered while keeping the structure the same allows for relatively rapid understanding. The emergence of possibly magnetic MAX phases is an exciting development that would greatly enhance their potential applications.

On the mechanical side, the fact that dislocations are confined to 2D is proving invaluable in elucidating the deformation behavior of layered solids in general, and ones that are also conductive in particular. At the end of 2012, a conference was held in Sapporo, Japan, on a new class of layered Mg alloys, also known as long-period stacked ordered phases, that exhibit some quite remarkable mechanical properties. Not surprisingly, much of what applies to the MAX phases also applies to these layered solids.

Lastly, the fact that it is now possible to produce 2D transition metal carbide and nitride phases by simply immersing select MAX phases in a simple acid such as HF (see Chapter 7) should not be underestimated.

References

Barnes, L.A., Dietz Rago, N.L., and Leibowitz, L. (2008) Corrosion of ternary carbides by molten lead. *J. Nucl. Mater.*, **373**, 424–428.

Barsoum, M.W. and El-Raghy, T. (1996) Synthesis and characterization of a remarkable ceramic: Ti_3SiC_2. *J. Am. Ceram. Soc.*, **79**, 1953–1956.

Barsoum, M.W. and El-Raghy, T. (1999) Process for making a dense ceramic workpiece. US Patent 6,231,969, U.P. Office, ed. (Drexel University).

Chen, D., Shirato, K., Barsoum, M.W., El-Raghy, T., and Ritchie, R.O. (2001) Cyclic fatigue-crack growth and fracture properties in Ti_3SiC_2 ceramics at elevated temperatures. *J. Am. Ceram. Soc.*, **84**, 2914.

Cui, B., Sa, R., Jayaseelan, D.D., Inam, F., Reece, R., and Lee, W.E. (2012) Microstructural evolution during high-temperature oxidation of spark plasma sintered Ti_2AlN ceramics. *Acta Mater.*, **60**, 1079–1092.

Dahlqvist, M., Alling, B., Abrikosov, I.A., and Rosén, J. (2011) Magnetic nanoscale laminates with tunable exchange coupling from first principles. *Phys. Rev. B.*, **84**, 220403(R).

Dimiduk, D.M., Mendiratta, M.G., and Subramanian, P.R. (1993) Development approaches for advanced intermetallic materials – historical perspective and selected successes. Paper presented at *Structural Intermetallics: Proceedings of the 1st International Symposium on Structural Intermetallics*, Champion, PA, The Minerals, Metals and Materials Society.

El-Raghy, T., Barsoum, M.W., Sundberg, M., and Pettersson, H. (2002) Process for forming 312 phase materials and process for sintering the same. US Patent 6,461,989, U.P. Office, ed. (US Kanthal and Drexel U).

El-Raghy, T., Blau, P., and Barsoum, M.W. (2000) Effect of grain size on friction and wear behavior of Ti_3SiC_2. *Wear*, **42**, 761.

El-Raghy, T. and Lyons, P. (2007) Stick resistant cooking utensils. US Patent 7,217,907, USPTO, ed. (USA).

Filimonov, D., Gupta, S., Palanisamy, T., and Barsoum, M.W. (2009) Effect of applied load and surface roughness on the tribological properties of Ni-based superalloys versus Ta_2AlC/Ag or Cr_2AlC/Ag composites. *Tribol. Lett.*, **33**, 9–20.

Frodelius, J., Sonestedt, M., Björklund, S., Palmquist, J.-P., Stiller, K., Högberg, H., and Hultman, L. (2008) Ti_2AlC coatings deposited by high velocity Oxy-fuel spraying. *Surf. Coat. Technol.*, **202**, 5976–5981.

Gilbert, C.J., Bloyer, D.R., Barsoum, M.W., El-Raghy, T., Tomasia, A.P., and Ritchie, R.O. (2000) Fatigue-crack growth and fracture properties of coarse and fine-grained Ti_3SiC_2. *Scr. Mater.*, **42**, 761–767.

Gromelski, S.J., Caciolo, P., and Cox, R.L. (2007) Carbide and nitride ternary ceramic glove and condom formers. US Patent 7,235,505, P. Office, ed. (US).

Gupta, S., Barsoum, M.W., Li, C.-W., and Palanisamy, T.G. (2009a) Ternary carbide and nitride materials having tribological applications and methods of making same (US). US Patent 7,553,564 B2.

Gupta, S., Filimonov, D., Zaitsev, V., Palanisamy, T., El-Raghy, T., and Barsoum, M.W. (2009b) Study of tribofilms formed during dry sliding of Ta_2AlC/Ag or Cr_2AlC/Ag composites against Ni-based superalloys and Al_2O_3. *Wear*, **267**, 1490–1500.

Gupta, S., Filimonov, D., Palanisamy, T., El-Raghy, T., and Barsoum, M.W. (2007) Ta_2AlC and Cr_2AlC Ag-based composites – new solid lubricant materials for use over a wide temperature range against Ni-based superalloys. *Wear*, **262**, 1479–1489.

Hajas, D.E., Baben, M.t., Hallstedt, B., Iskandar, R., Mayer, J., and Schneider, J.M., (2011) Oxidation of Cr_2AlC coatings in the temperature range of 1230 to 1410 °C. *Surface Coating Tech*, **206**, 591–598.

Heinzel, A., Müller, G., and Weisenburger, A. (2009) Compatibility of Ti_3SiC_2 with liquid Pb and PbBi containing oxygen. *J. Nucl. Mater.*, **392**, 255–258.

Hu, C., Sakka, Y., Grasso, S., Nishimura, T., Guo, S., and Tanaka, H. (2011a) Shell-like nanolayered Nb_4AlC_3 ceramic with high strength and toughness. *Scr. Mater.*, **64**, 765–768.

Hu, C., Sakka, Y., Grasso, S., Suzuki, T., and Tanaka, H. (2011b) Tailoring Ti_3SiC_2 ceramic via a strong magnetic field alignment method followed by spark plasma sintering. *J. Am. Ceram. Soc.*, **94**, 742–748.

Hu, C., Sakka, Y., Nishimura, T., Guo, S., Grasso, S., and Tanaka, H. (2011c) Physical and mechanical properties of highly textured polycrystalline Nb_4AlC_3 ceramic. *Sci. Technol. Adv. Mater.*, **12**, 044603.

Hu, C., Sakka, Y., Tanaka, H., Nishimura, T., and Grasso, S. (2009) Low temperature thermal expansion, high temperature electrical conductivity, and mechanical properties of Nb_4AlC_3 ceramic synthesized by spark plasma sintering. *J. Alloys Compd.*, **487**, 675–681.

Jovic, V.D. and Barsoum, M.W. (2006) Electrolytic cell and electrodes for use in electrochemical processes. US Patent 7,001,494, U.P. Office, ed. (USA).

Lane, N.J., Simak, S.I., Mikhaylushkin, A., Abrikosovv, I., Hultman, L., and Barsoum, M.W. (2011) A first principles study of dislocations in HCP metals through the investigation of the $(11\bar{2}1)$ twin boundary. *Phys. Rev. B*, **84**, 184101.

Le Flem, M., Liu, X., Doriot, S., and Cozzika, T. (2010) Irradiation damage in $Ti_3(Si,Al)C_2$: a TEM investigation. *Int. J. Appl. Ceram. Technol.*, **7**, 766–775.

Liu, X.M., Le Flem, M., Béchade, J.L., and Monnet, I. (2010) Nanoindentation investigation of heavy ion irradiated $Ti_3(Si,Al)C_2$. *J. Nucl. Mater.*, **401**, 149–153.

Murugaiah, A., Souchet, A., El-Raghy, T., Radovic, M., and Barsoum, M.W. (2004) Tape casting, pressureless sintering and grain growth in Ti_3SiC_2 compacts. *J. Am. Ceram. Soc.*, **87**, 550–556.

Myhra, S., Summers, J.W.B., and Kisi, E.H. (1999) Ti_3SiC_2 – A layered ceramic exhibiting ultra-low friction. *Mater. Lett.*, **39**, 6–11.

Nappé, J., Grosseau, P., Audubert, F., Guilhot, B., Beauvy, M., Benabdesselam, M., and Monnet, I. (2009) Damages induced by heavy ions in titanium silicon carbide: effects of nuclear and electronic interactions at room temperature. *J. Nucl. Mater.*, **385**, 304–307.

Palanisamy, T.G., Gupta, S., Barsoum, M.W., and Li, C.-W. (2009) Ternary carbide and nitride composites having tribological applications and methods of making same (US). US Patent 7,572,313 B2.

Sienicki, J.J., Moisseytsev, A., Bortot, S., Lu, Q., and Aliberti, G. (2011) SUPERSTAR: an improved natural circulation, lead-cooled, small modular fast reactor for international deployment. Paper presented at Proceedings of ICAPP 2011, Nice, France.

Sonestedt, M., Frodelius, J., Palmquist, J.-P., Hogberg, H., Hultman, L., and Stiller, K. (2010a) Microstructure of high velocity oxy-fuel sprayed Ti_2AlC coatings. *J. Mater. Sci.*, **45**, 2760–2769.

Sonestedt, M., Frodelius, J., Sundberg, M., Hultman, L., and Stiller, K. (2010b) Oxidation of Ti_2AlC bulk and spray deposited coatings. *Corros. Sci.*, **52**, 3955–3961.

Sundberg, M., Lindgren, K., El-Raghy, T., and Barsoum, M.W. (2006) Method of producing a metal-containing single-phase composition. US Patent 6,986,873, U.P. Office, ed.

Sundberg, M., Malmqvist, G., Magnusson, A., and El-Raghy, T. (2004) Alumina forming high temperature silicides and carbides. *Ceram. Int.*, **30**, 1899–1904.

Tuller, H.L., Spears, M.A., and Mlcak, R. (2003) Stable electrical contact for silicon carbide devices (US). US Patent 6,544,674.

Utili, M., Agostini, M., Coccoluto, G., and Lorenzini, E. (2011) Ti_3SiC_2 As a candidate material for lead cooled fast reactor. *Nucl. Eng. Des.*, **241**, 1295–1300.

Walker, W.J. (2010) Ceramic electrode and ignition device therewith. US Patent 7816845 B2, USPO, ed. (US, Federal Mogul Ignition Co.).

Whittle, K.R., Blackford, M.G., Aughterson, R.D., Moricca, S., Lumpkin, G.R., Riley, D.P., and Zaluzec, N.J. (2010) Radiation tolerance of $M_{n+1}AX_n$ phases, Ti_3AlC_2 and Ti_3SiC_2. *Acta Mater.*, **58**, 4362–4368.

Zhang, H., Wang, Z.G., Zang, Q.S., Zhang, Z.F., and Sun, Z.M. (2003) Cyclic fatigue crack propagation behavior of Ti_3SiC_2 synthesized by pulse discharge sintering (PDS) technique. *Scr. Mater.*, **49**, 87–92.

Zhen, T., Barsoum, M.W., Kalidindi, S.R., Radovic, M., Sun, Z.M., and El-Raghy, T. (2005) Compressive creep of fine and coarse-grained T_3SiC_2 in air in the 1100 to 1300 °C temperature range. *Acta Mater.*, **53**, 4963–4973.

Zhou, A.G., Barsoum, M.W., Basu, S., Kalidindi, S.R., and El-Raghy, T. (2006) Incipient and regular kink bands in dense and 10 vol.% porous Ti_2AlC. *Acta Mater.*, **54**, 1631.

Index

MAX Phases: Properties of Machinable Ternary Carbides and Nitrides, First Edition. Michel W. Barsoum.